AF333989

Geotechnical Modeling and Applications

Gulf Publishing Company
Book Division
Houston, London, Paris, Tokyo

Geotechnical

Modeling and

Applications

Sayed M. Sayed, Ph.D., P.E.
Editor

Geotechnical Modeling and Applications

Library of Congress Cataloging in Publication Data

Geotechnical modeling and applications.

Includes bibliographies and index.
1. Soil mechanics—Mathematical models.
2. Foundations—Mathematical models. 3. Structural dynamics—
Mathematical models. I. Sayed, S. M.
TA710.G446 1987   624.1′51   87-7453
ISBN 0-87201-787-7

Aleksandar S. Vesic
August 8, 1924–May 3, 1982

The geotechnical profession offers this special tribute to the memory of Professor Aleksandar Sedmak Vesic for his outstanding and lasting contributions in the fields of soil mechanics, rock mechanics and foundation engineering.

Professor Vesic was born in Yugoslavia. He graduated with a Bachelor of Civil Engineering with highest honors at the University of Belgrade in 1950. He immediately became an assistant of Professor Trojanovic, obtaining his doctorate at the University of Belgrade in 1956.

During the 1950–1956 period he was intensely involved in the design of bridges and dams, and assisted Professor Trojanovic in courses of stability of structures and soil mechanics. With the support of Professor Trojanovic, Vesic went to Belgium to enlarge his view in the field of soil mechanics and foundations. There he worked two years at the Belgian Geotechnical Institute as a research engineer. His theoretical and experimental research improved knowledge of the bearing capacity of piles and shallow foundations.

In 1956 Vesic was called to Atlanta by Professor Sowers to become assistant professor at the Georgia Institute of Technology. Great were the regrets of Professor Trojanovic and the University of Belgrade in losing one of their most eminent and promising young scientists. Because of the larger possibilities provided by the United States, Vesic's choice was beneficial for the international geotechnical community. In 1960 Vesic became associate professor at Georgia Tech.

In 1964 he became a professor at Duke University, where he organized and led research in soil mechanics. He was chairman of the Department of Civil Engineering for the period 1968–1974, held the J. A. Jones Distinguished Professorship from 1971–1982, and in 1974 became Dean of the School of Engineering.

By theory and experiment, Professor Vesic markedly increased our knowledge of failure phenomena of the soil underneath shallow and deep foundations. He demonstrated that the modes of failure on cohesionless soils depend not only on the relative density but also on the relative depth of the foundation. He clarified the role of general shear, local shear, and punching shear in failure.

Vesic considerably increased our knowledge of the behavior of piled foundations. Because of his very clear thinking on failure phenomena of soils, he became involved in the problems of cratering by underground nuclear explosions. Together with other scientists, he made theoretical deductions, and small-scale tests concerning the behavior of soils under very high confining pressures. Vesic was the first to take into account the compressibility of the soil on the failure phenomena, introducing the notion of the rigidity index. His papers on this subject are authoritative. The problem of the distribution of soil reactions underneath foundation rafts was greatly clarified by his papers on this topic.

For his contributions and achievements Vesic was granted the Highway Research Board Award of the U.S.A. National Research Council in 1969. In 1974 he was honored with the Thomas A. Middlebrooks Award of the American Society of Civil Engineers. In 1981 the honorary degree of Doctor of Applied Sciences of the University of Ghent (Belgium) was conferred upon him.

In 1975 Vesic organized a symposium on the bearing capacity and settlement of foundations, and edited the proceedings of this Duke symposium. He published a series of lectures he gave in New York on the behavior of pile foundations: one of the most complete and clearly written expositions on this subject.

His sudden death at the age of 57 was for all his friends and admirers a painful shock. For the geotechnical community it was a big loss. Undeniably he would still have brought many more advancements in several problem areas had he continued his research.

All who have had the privilege of working with Professor Vesic, of exchanging thoughts and ideas with him, and sharing his friendship wished that besides the Aleksandar Sedmak Vesic Memorial scholarship fund established at Duke University, a book on geotechnical engineering should be edited and dedicated to his memory. This book is the consequence of these feelings and is presented in courtesy to Milena Sedmak Vesic, the beloved wife of Professor Vesic, who encouraged him in all his activities and surrounded him with all her love.

E. E. De Beer
State University of Ghent
Ghent, Belgium

Contents

Dedication v

Acknowledgments ix

Contributors x

Foreword xi

Preface xii

Chapter 1
Geotechnical Engineering   Sayed M. Sayed 1

Chapter 2
Modeling the Behavior of Geomaterials   Jean H. Prevost 8
 Brief Review of Plasticity Theory. Integration Algorithms for Elastic-Plastic
 Constitutive Relations. A Simple Plasticity Model for Soils. References.

Chapter 3
Dynamics of Porous Media   Jean H. Prevost 76
 Basic Theory. Formulation. Examples. Concluding Remarks. References.

Chapter 4
**Static and Cyclic Response of Interfaces for Analysis and
Design of Soil-Structure Interaction Problems**   C. S. Desai 147
 Review of Developments. Proposed Models. Verification and Application.
 Laboratory Tests. Practical Problems. Simquake Test. Conclusions. References.

Chapter 5
**Stress, Degradation, and Shear Strength of Granular
Material**   Donald J. Murphy 181
 Purpose and Scope. Experimental Program. Test Results. Discussion.
 Conclusions. References.

Chapter 6
Analysis of Shallow Foundations   E. E. De Beer **212**

Ultimate Bearing Capacity. Problem of Deformations. Problem of the Distribution of the Soil Reactions Underneath a Foundation Raft. Numerical Methods. References.

Chapter 7
Static and Dynamic Analyses of Axially and Laterally Loaded Piles and Pile Groups   P. K. Banerjee, R. Sen, and T. G. Davies **322**

Statement of Problem. Methods of Analysis. Solution for Piles. Results of the Analysis for Static Loading. Results of Dynamic Analysis. Conclusions. Appendix—Lateral Vibration: Constants of Integration. References.

Chapter 8
Geotextiles and Related Products   J. P. Giroud **355**

Definitions and Types. Properties. Functions and Applications of Geotextiles. Design. References.

Chapter 9
Soft Soil Stabilization Using High-Strength Geotextiles and Strip Drains   Robert M. Koerner and Jack Fowler **391**

Overview of the Problem. Background. Current Polymeric Strip Drains. Stability and Reinforcement Concerns. Strip Drain Installation Considerations. Maryland Port Authority's "Seagirt" Project. Conclusions. Recommendations. References.

Conversion Factors **446**

Subject Index **447**

Author Index **453**

Acknowledgments

I want to express to the contributors my sincere appreciation for their loyal support and their genuine cooperation in these past years. They have made it possible for me to pass the milestones one by one and have contributed largely toward making this book possible.

I wish to thank Duke University for granting me financial support during the early stages of this project.

I wish to acknowledge the many valuable discussions and comments of Dr. Robert J. Melosh throughout the preparation of this book. His constant interest, encouragement, and availability are greatly appreciated. He has been an inspiration to me since my graduate days at Duke.

My sincere gratitude is extended to Mrs. Aleksandar S. Vesic for her courtesy and continuous support since I came to Duke in 1978.

Thanks to Melissa Lewis Beck and William J. Lowe of Gulf Publishing Company for their interest and patience during the preparation and production of the manuscript.

Finally I would like to thank my colleagues at the University of New Orleans for their friendship and support, especially Dr. R. M. A. Azzam.

Sayed M. Sayed
University of New Orleans
New Orleans, Louisiana

Contributors

P. K. Banerjee, professor of civil engineering, State University of New York at Buffalo, Buffalo, New York, USA.

T. G. Davies, lecturer in civil engineering, University of Glasgow, Glasgow, Scotland, UK.

E. E. De Beer, Professor Emeritus, State University of Ghent, Ghent, Belgium.

C. S. Desai, professor of civil engineering and engineering mechanics, University of Arizona, Tucson, Arizona, USA.

Jack Fowler, research civil engineer, U.S. Army Corps of Engineers, Waterways Experiment Station, Vicksburg, Mississippi, USA.

J. P. Giroud, principal, GeoServices, Inc., Consulting Engineers, Boynton Beach, Florida, USA.

Robert M. Koerner, professor of civil engineering, Drexel University, Philadelphia, Pennsylvania, USA.

Donald J. Murphy, vice-president, Langan Engineering Associates, Inc., Elmwood Park, New Jersey, USA.

Jean H. Prevost, professor of civil engineering, Princeton University, Princeton, New Jersey, USA.

Sayed M. Sayed, assistant professor of civil engineering, University of New Orleans, New Orleans, Louisiana, USA.

R. Sen, assistant professor of civil engineering and engineering mechanics, University of South Florida, Tampa, Florida, USA.

Foreword

This book honors Aleksandar Sedmak Vesic's contributions to his profession, geotechnical engineering; to researchers seeking to advance the technology; and to his students.

The idea of developing this book as a tribute to Professor Vesic germinated in 1982. Its fruition is largely a consequence of the diligence and persistence of Professor Sayed M. Sayed. Professor Sayed, like many of Professor Vesic's students, developed a deep respect and affection for Dr. Vesic because of Vesic's exceptional mastery of geotechnical engineering, his caring leadership in research, and his intellectual discipline.

The authors and the ideas they have contributed to this volume reflect Vesic's high standards of technical excellence. The content, like Professor Vesic's practice of engineering, represents a meaningful mix of science and engineering experiences.

R. J. Melosh
Department of Civil Engineering
Duke University

Preface

This book presents up-to-date, state-of-the-art soil models and their geotechnical applications. This text seeks to present the most relevant and recent information about soil modeling and problems of soil-structure interaction.

In the last twenty-five years, knowledge in the field of geotechnical engineering has increased immensely as a result of the development of large, high-speed computers. Numerical methods such as finite element and finite difference have provided solutions for several complex problems in geotechnical engineering. However, due to the wide range of geotechnical applications, numerous constitutive, deformation, and boundary models have been developed. Each model has been formulated to represent the soil behavior as accurately as possible.

Although much literature about these models has been written, researchers have provided little assessment of the relative value of the models. Thus young researchers and graduate students have a difficult time selecting the most appropriate model for their own studies. This book will sort and describe the potential and merit of the most relevant soil models and their applications.

Graduate students, research fellows, and practicing geotechnical engineers will find the book of interest. It can serve the practicing engineer as a reference for the methodology used in analysis and design, or be used as a textbook for a two-semester course in geotechnical engineering.

The chapters are written by recognized researchers in the field in which each has made practical applications of theory.

The first chapter is an introduction to the book. It describes the many aspects of geotechnical engineering and the various methods of analysis.

Chapter 2 covers the fundamentals of modeling the behavior of geomaterials. After a brief review of the plasticity theory, various integration algorithms for elastic-plastic constitutive relationships are discussed. Finally a simple plasticity model for frictionless soils is presented.

Chapter 3 discusses the mechanics of porous media. The general mixture equation is summarized and then applied to describe the behavior of saturated inelastic porous soil media. The solution techniques for hyperbolic problems are discussed.

Chapter 4 covers the behavior of the interfaces or junctions between the structure and foundation soils. In geotechnical engineering and soil-structure interaction, perhaps one of the most important topics is modeling the behavior of interfaces between structural and geologic materials including various modes of deformation: no-slip, slip, debonding, and rebonding. For realistic simulation of soil-structure interaction problems in the presence of static and dynamic loadings, the (nonlinear) behavior of interfaces must be characterized appropriately. Analytical models that allow for the relative motions at the interfaces are proposed for the shear and normal responses. The models are implemented in numerical (finite element) procedures, which can be used to predict behavior in various practical problems.

Chapter 5 covers the shear strength of granular material. The influence of a normal stress level on the shearing resistance of soil was recognized more than two-hundred years ago. Until the late 1960s, data used in applying this concept have been limited generally to those obtained from conventional triaxial compression tests conducted at confining pressures of 150 lb/in.^{2} (1034 kPa) or less. However, stress conditions in many geotechnical situations are outside this limitation. The solution to related engineering problems has required extensive extrapolation from low-pressure results. As the nonlinear nature of the relationship between normal stress and shearing resistance precluded rational extrapolation, prudence necessitated the employment of high "factors of uncertainty." Elimination of the need for such factors is possible only if strength envelope curvature and other pressure-dependent phenomena can be related to fundamental material characteristics. A substantial part of past research on the effects of high stress levels was directed toward this goal. This chapter presents the behavior of soils and rocks over wide ranges of stress levels and mineralogical composition.

Chapter 6 is devoted to shallow foundations. The contents discuss the computation of bearing capacity based on theoretical considerations and field tests, the effect of inhomogeneity of the soil, dynamic loading, and the choice of safety factors. Tolerable deformations and the methods of analysis are also described. Finally, the chapter treats the problem of distribution of the soil reactions underneath a foundation raft and the numerical methods.

Chapter 7 deals with boundary element formulation for the static and steady-state dynamic analyses of piles and pile groups embedded in an inhomogeneous soil stratum. The boundary element method offers considerable advantages over other numerical methods for this problem primarily because of its ability to take into account the three-dimensional effects of soil continuity and boundaries at infinity. In applications to static and dynamic analyses of pile groups, it has proved to be the most rigorous, versatile, and economical method of analysis. The analysis is applied to pile groups of complex geometry and loading.

Chapter 8 deals with geotextiles and other related products. In ten years, geotextiles have progressively pervaded all branches of geotechnical engineering, and this may be one of the most important developments to date in the history of geotechnical engineering. They are used as a practical means of solving construction problems and, at the same time, have opened up new opportunities for creativity to the geotechnical engineer. Geotextiles are used for drainage against basement walls, but also in large dams. They are used for temporary erosion control, but also in sophisticated reinforced structures with a 120-year design life. They are used not only in traditional geotechnical applications such as landslide rehabilitation, but also in the harsh chemical environment of hazardous-waste containment facilities.

Chapter 9 relates to a totally new concept of using geosynthetic materials with extremely weak subsoils. Soils with shear strengths of less than 50 lb/ft^2 (2.4 kPa) have effectively been handled by using high-strength geotextiles for reinforcement and vertical strip drains for drainage. When suitably surcharged, the weak soils consolidate into a significantly improved condition. The design of such systems using limit equilibrium concepts is illustrated along with proper strip drain design procedures. A specific case history of a recently completed project is used to numerically illustrate the procedures. Areas where additional research is required as well as the potential impact that the results might have on the area of stabilizing weak soils are discussed.

I am indebted to the contributors who prepared materials for this volume, because it is their contribution that makes this book an authoritative and valuable milestone in geotechnical engineering.

My close relationship with Dean Aleksandar Vesic during my graduate years at Duke University caused me, upon his death, to search for a way to express my feelings for his help. I am pleased that his colleagues in geotechnical engineering, whom Professor Vesic respected, have joined me in preparing this memorial volume.

Sayed M. Sayed
University of New Orleans
New Orleans, Louisiana

1

Geotechnical Engineering

Sayed M. Sayed
Department of Civil Engineering
University of New Orleans
New Orleans, Louisiana USA

Geotechnical engineering encompasses analysis, design, and construction involving earth materials. Its main branches, foundation engineering and earthwork engineering, deal with the interaction between engineering structures and their supporting soil subgrades and/or subbase materials. Engineers ascertain if the bearing capacity of soils is adequate for supporting structures without the soils settling excessively or failing. In earthwork engineering, for example, the soil must be properly selected, treated, and placed to fit the function of a certain earth structure. The methods of construction, in view of the wide range of environmental conditions, constitute a delicate balance between science and art. This, indeed, is the hallmark of geotechnical engineering.

Because geotechnical engineering applications vary significantly, including such projects as buildings, bridges, tunnels, earth dams, embankments, marine structures, space structures, highway pavements, and airport pavements, the geotechnical engineer must be adequately prepared and have a strong background in the necessary fundamentals.

The geotechnical engineer is mainly concerned with the design of foundations and with the stability of earth structures. Adequate knowledge of the subsurface conditions is needed to properly design an interactive soil-structure system. Whether the soil is used to support a foundation or as a construction material itself, the engineer must be able to assess its properties and behavior. He or she predicts the soil's behavior as a function of design parameters, and uses basic science, engineering science, soil mechanics, foundation engineering and experience to arrive at the solutions of soil engineering problems.

In solving these problems, the geotechnical engineer is confronted with formidable challenges. These result mainly from the random nature of the material itself, i.e., soil. Typically composed of solid, liquid, and gaseous phases, soil is more complex to deal with than ordinary structural materials such as steel, aluminum, concrete, and wood. To be sure, soil is not a quality-controlled product. It is neither homogeneous nor isotropic and extracting soil samples of a natural soil without altering (to some degree) its properties is very difficult, if not impossible. Furthermore, the analytical approach for most geotechnical problems is neither easy nor readily available. There simply is no means for fully testing the earth material.

Besides the difficulties in extracting representative soil samples from the ground, testing facilities invariably have their own limitations in simulating in-situ conditions. In addition, testing a limited number of soil samples does not tell much about the soil mass as a whole. These and other considerations make geotechnical problems especially difficult to adequately assess. Also the engineer often lacks complete information on the job he or she is dealing with. Therefore a good understanding of a broad spectrum of engineering problems as well as considerable intuition and common sense is required to produce an acceptable design.

In spite of the wide variety of geotechnical applications, the complex nature of soil behavior, and the usually incomplete information available, the geotechnical engineer is still able to arrive at acceptable solutions to the problem at hand. In fact, the geotechnical engineer often reaches several solutions to the same problem and accordingly has to choose the safest and most economic solution.

Since most soil engineering problems do not possess unique solutions, and mathematical models are generally abstractions of reality, engineering judgment is needed to separate optimal designs from ordinary or even inadequate ones. Both experience and intuition are major components of sound engineering judgment.

Terzaghi and Peck [1967] classified the causes of fatal misjudgment into three categories, namely:

1. Decisions based on test results marred by excessive sample disturbance or significant differences between test and field conditions.
2. Failure to recognize or judge correctly the most unfavorable subsoil conditions compatible with the field data.
3. Inadequate contact between the design and construction organizations, resulting in failure to detect significant departures of conditions or of construction procedures from those the designer anticipated or specified [Terzaghi, 1958a, 1963].

It should be emphasized that designing on the basis of the most unfavorable assumption is wasteful, and modifying the design as the construction progresses can help achieve maximum economy and assurance of safety. Basing the design on the most probable rather than the most unfavorable conditions is called the observational or "learn-as-you-go" method [Peck, 1969]. The geotechnical community owes the formulation, development, and application of this method mainly to Karl Terzaghi and Ralph Peck. In spite of its drawbacks, the observational method can provide the assurance of safe construction. However, the full merit of the method cannot be realized unless the engineer is well informed about the problem, makes continuous design and procedures modifications as the information becomes available, and has the courage to act quickly upon his decisions and conclusions. Several examples of the observational method can be found in References 2, 13, 14, 15, 18. The competence and success of a geotechnical engineer is measured by how comparable the predicted behavior of the soil-structure system is with the measured performance.

As in any other engineering field, the geotechnical engineer needs to simulate the behavior of the material involved. The application determines what analyses are to be made. If mathematical models are used to predict the soil's response, the soil parameters that best describe its properties (typically strength and deformation characteristics) must be available. Gathering such information usually requires experimental determination. The geotechnical engineer must use the appropriate tests for determining the soil parameters, along with the appropriate model to predict the performance of the soil-structure system. Occasionally, full-scale field tests are performed to guide and/or assess the methods used in the prediction.

As noted, the idealization of soil behavior and the response of a soil-structure system is difficult because of the complexity of the soil. This probably explains the reasons behind the development of so many models for various applications [Atkinson and Bransby, 1978; Desai and Christian, 1977; Schofield and Wroth, 1968; Scott, 1981; Tschebotarioff, 1973]. These models are intended to simulate the particular problem as closely as possible. The significance of numerical and physical modeling has been discussed in Desai and Christian [1977]. The choice of a certain model depends on the specific application for which the model is believed to provide the best possible prediction. However, there is a great need to assess the predictive capability and the reliability of the existing models. The methods of analysis and the associated models can be grouped, in general, into three main categories: elementary, advanced, and high technology analyses.

The elementary methods are mainly based on experience. Empirical and semi-empirical rules have been developed by correlating what is observed in the laboratory or the field with the real problem and used in analysis. Most of these analyses have little general validity and should be used cautiously. Certainly extrapolation to situations not encountered previously or upon which the methods are based is not warranted. These solutions are often favored because the analysis can be performed easily and in a relatively short time. Because experience preceded theoretical developments [Terzaghi and Peck, 1967], these methods were the only tools of analysis used in the early stages of geotechnical engineering.

An example of this class of solutions is the computation of the bearing capacity of piles using a classical dynamic formula. It has been observed [Mansour and Focht, 1953] that such a value of bearing capacity can be in error as much as 100 percent with field data. In spite of the convenience of using these methods, more elaborate methods that avoid such unreliable predictions are needed.

The advanced methods of analysis are based mainly on the linear theory of elasticity or plasticity. These methods adopt simple assumptions regarding the material and a geometric idealization of soil behavior. However, this class of models does not take into account many aspects of real soil behavior, such as the nonlinear nature of behavior, stress history, and time dependence. These methods are more favorable than the elementary methods because their mathematical formulation is more justified. An example of the advanced analyses is the computation of the soil deformation modulus using the experimental results of the pressuremeter test. This computation is commonly based on the adoption of the linear theory of elasticity. However, it has been shown [Sayed, 1982] that neglecting the nonlinear effects of the soil behavior, as is implied by the linear theory of elasticity, introduced as much as 100% overestimation in the magnitude of the soil deformation modulus.

As a result of the many shortcomings suffered by both the elementary and advanced methods, many investigators devote their efforts to developing more capable and reliable models. High technology models are in this category and are the most generalized methods of analysis. The majority of these models are so versatile and general that most of the shortcomings of the elementary and advanced models have been reduced or completely eliminated. Effects of stress path, stress history, time, and nonlinear effects are introduced to arrive at more realistic material descriptions. As may be expected, more soil parameters and more sophisticated procedures are needed to make use of these models.

The finite element method is one such procedure that has been used extensively to obtain elaborate, still approximate, solutions for a wide range of complex problems in geotechnical engineering. An example of this method is the analysis performed by Duncan and Clough [1971] of the Port Allen Lock, located on the right bank of the Mississippi River opposite Baton Rouge, Louisiana. The lock is a reinforced concrete U-frame structure with a navigation chamber 84 ft wide and 1,200 ft long, capable of a 50-ft navigation lift. The behavior of the Port Allen Lock was studied using incremental finite element techniques that closely sim-

ulated the construction history of the lock. The cross section of the lock
along with the finite element modeling of the river lock and its foundation
is shown in Figure 1-1. Comparisons between the results of these analyses and the observed behavior proved that the finite element method is of
great value in understanding the complex nature of soil-structure interaction problems.

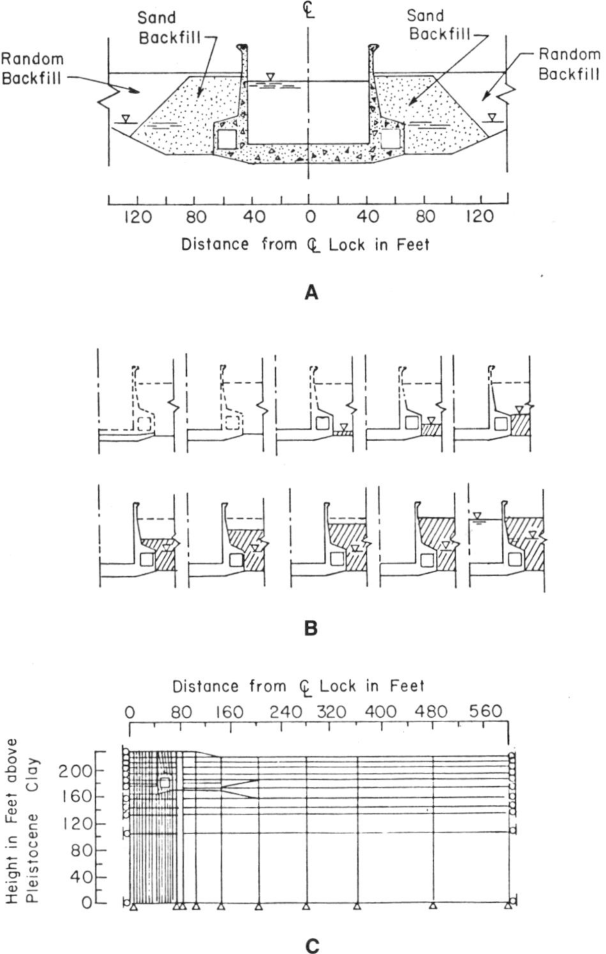

Figure 1-1. Finite element analysis of Port Allen Lock. (A) Aerial view and cross
section. (B) Schematic representation of finite element simulation of concrete
and fill placement. (C) Finite element mesh employed for incremental analyses.
[Duncan and Clough, 1971.]

Since there are limitations on the capabilities of testing soil specimens, these most often constrain the mathematical formulations of these models. These models are often the most realistic, reliable, and comprehensive ones that are available.

The successful use of high technology models depends more on understanding the physical behavior of the soils involved than on being exceptionally sophisticated in applying the techniques themselves. To accurately interpret the results obtained from these models, the sources of uncertainty involved in the analysis must be identified.

It should be reemphasized that the job of the geotechnical engineer is not easy and straightforward. Each problem faced is unique. Therefore one must be prepared to know what tools of analysis are to be employed in solving each problem. Which soil parameters are needed to utilize a particular model? What tests are to be used to determine these parameters? All these questions need sound judgment to arrive at the optimal solution for a specific problem.

REFERENCES

1. Atkinson, J. H., and Bransby, P. L., *The Mechanics of Soils—An Introduction to Critical State Soil Mechanics,* McGraw-Hill Book Company (UK) Ltd., 1978.
2. Casagrande, A., "Role of the Calculated Risk in Earthwork and Foundation Engineering," *Journal of the Soil Mechanics and Foundation Division,* ASCE Vol. 91, No. SM4, pp. 1–40, 1965.
3. Desai, C. S., and Christian, J. T., (Eds.), *Numerical Methods in Geotechnical Engineering,* McGraw-Hill Book Company, New York, 1977.
4. Duncan, J. M., and Clough, G. W., "Finite Element Analysis of Port Allen Lock," *Journal of the Soil Mechanics and Foundation Division,* ASCE, Vol. 97, No. SM8, pp. 1053–1068, 1971.
5. Mansour, C. I., and Focht, J. A., "Pile Loading Tests," Morganza Floodway Control Structure, Proc. ASCE, Vol. 79, No. 324, pp. 1–31, 1953.
6. Mitchell, J. K., *Fundamentals of Soil Behavior,* John Wiley & Sons, Inc., New York, 1976.
7. Peck, R. B., "Advantages and Limitations of the Observational Method in Applied Soil Mechanics," *Geotechnique* 19, No. 2, pp. 171–187, 1969.
8. Sayed, S. M., "Expansion of Long Cylindrical Cavities in Nonlinear Ditatant Media," Ph.D. Dissertation, Duke University, 1982.
9. Schofield, A. N., and Wroth, C. P., *Critical State Soil Mechanics,* McGraw-Hill Publishing Company, Ltd., London, 1968.

10. Scott, R. F., *Foundation Analysis,* Prentice-Hall, Inc., NJ, 1981.
11. Terzaghi, K., "Consultants, Clients, and Contractors," *J. Boston Soc. Civil Engrs.,* Vol. 45, No. 1, pp. 1–15, 1958a.
12. Terzaghi, K., "Discussion: Atomic Power, a Failure in Engineering Responsibility," *Trans. ASCE,* Vol. 128, Part 5, pp. 56–57, 1963.
13. Terzaghi, K., "Stabilization of Landslides," series of memoranda contained in *From Theory to Practice in Soil Mechanics,* John Wiley & Sons, New York, pp. 409–415, 1960d.
14. Terzaghi, K., and Leps, T. M., "Design and Performance of Vermilion Dam," *Trans. ASCE,* Vol. 125, pp. 63–100, 1960.
15. Terzaghi, K., and Lacroix, Y., "Mission Dam, an Earth and Rock-fill Dam on a Highly Compressible Foundation," *Geot.,* Vol. 14, pp. 14–50, 1964.
16. Terzaghi, K., and Peck, R. B., *Soil Mechanics in Engineering Practice,* 2nd ed. John Wiley & Sons, New York, 1967. (The first edition was published in 1948.)
17. Tschebotarioff, G. P., *Foundations, Retaining, and Earth Structures,* 2nd ed., McGraw-Hill Book Company, New York, 1973.
18. Zeevaert, L., "Foundation Design and Behavior of Tower Latino Americana in Mexico City," *Geot.,* Vol. 7, No. 3, pp. 115–133, 1957.

2

Modeling the Behavior of Geomaterials

Jean H. Prevost
School of Engineering/Applied Science
Department of Civil Engineering
Princeton University
Princeton, New Jersey USA

INTRODUCTION

The deformational behavior of all real materials, and in particular of geomaterials, under stress is highly complex and also very difficult to obtain by testing methods. It is sometimes possible to describe very accurately the measured behavior of a given material in a given test (e.g., in an axisymmetric or a pure shear experiment) by mathematical relationships. However, it is not always possible to extend and/or make use of these relations for the purpose of predicting the behavior of the same material under different loading conditions. A useful mathematical model of a material stress-strain-strength behavior is one that can be employed to *predict* satisfactorily the material performance in all the circumstances at hand. Such a mathematical model does not purport to describe the *real material* behavior exactly, but is said to represent an *ideal material*. It is important to realize the distinction between the ideal material model representation and the behavior of the real material which is being modeled.

Recent advances in digital computer technology and in numerical techniques such as the finite difference or the finite element methods have rendered possible, at least in principle, the solution of any properly posed boundary value problem in mechanics. Further progress in expanding analytical capabilities in geomechanics now depends upon consistent mathematical formulations of generally valid and realistic material constitutive relations. An increasing effort has thus been devoted in the past few years to a more comprehensive description of material behavior. Numerous formulations have been proposed in the soil mechanics literature. However, many of the constitutive laws presented seem unnecessarily arbitrary, and at this stage, it is important to emphasize that, to be satisfactory, a material model idealization should possess the following *necessary* properties:

1. The model should be *complete,* i.e., able to make statements about the material behavior for all stress and strain paths, and not merely restricted to a single class of paths (e.g., axial symmetry or pure shear).
2. It should be possible to identify the model parameters by means of a *small number* of standard, or simple material tests.
3. The model should be founded on some *physical interpretation* of the ways in which the material is responding to changes in applied stress or strain (e.g., the material should not be modeled as elastic if permanent deformations are observed upon unloading).

The first property is clearly essential if the model is to be of a practical application, and the second property is very desirable. The third property linked with the other two helps to ensure that time is spent developing a useful model and not merely on an elaborate curve-fitting exercise of limited application—of application in fact, no wider than the data that are being fitted. It is particularly with regard to these three basic properties that the approach adopted in this section differs from others which are basically curve-fitting and do not deal adequately with the nonlinear, anisotropic, hysteretic, and path-dependent stress-strain-strength properties of soil materials. Finally, it must be added that a material model may only be deemed to be satisfactory when with its aid, it is possible first to *determine* the stress-strain-strength behavior of the material at hand in one piece of testing equipment (e.g., in triaxial tests), and then to *predict* the observed behavior of the same material in some other type of testing equipment (e.g., in simple shear tests).

Although very complex when examined on the microscale and/or on the atomic scale, many materials (such as metals, polymers, concrete and geomaterials) may be idealized as behaving like continua with separable increments of recoverable (i.e., elastic) and irrecoverable (i.e., plastic) deformations. Therefore, while still recognizing the particulate nature of these materials, instead of studying them as discrete systems consisting of an assemblage of particles, it is more convenient to consider them as *continua,* and to use the methods and results of continuum mechanics for analyzing and modeling their behavior. The phenomenological behavior of the materials is therefore the standard of reference in this approach, and it is the one adopted here. The theory of plasticity allows for a differentiation between recoverable and irrecoverable deformations. It is therefore convenient to model geomaterials as elasto-plastic materials. However, as plasticity theory was originally developed for metals (for example, see [Hill, 1950]), modifications and extensions must be incorporated in order to apply it to model the behavior of a broader class of materials. In particular, when applied to model the behavior of soils, the theory must account for such experimental observations as:

1. That the yielding of soils is anisotropic and dependent upon the prevailing confining stress.
2. That there is a coupling between volumetric changes and changes in shear stress.
3. That dense soils expand in volume during pure shear whereas loose soils contract.

An attempt to develop such a general plasticity theory is to be presented in this section.

Plasticity theory requires consideration of three basic ingredients: the yield criterion, the flow rule, and the hardening rules. Therefore, for completeness, a brief description of these three basic concepts is given first. Integration of the rate elastic-plastic constitutive equations is thereafter considered, and the algorithmic implementation is detailed. Finally, a simple plasticity model applicable to both cohesive and cohesionless soils is presented.

BRIEF REVIEW OF PLASTICITY THEORY

Yield Criterion

A typical stress-strain-curve obtained in an axial compression test is shown in Figure 2-1. Three salient features are apparent:

- An initial rising *linear* range, followed by
- *Yield,* and then
- A *nonlinear* range in which the stress increases more slowly with strain than initially, and where a permanent (plastic) strain remains after unloading.

The yield stress is defined as the highest stress that can be reached before any permanent deformations occur upon unloading. This is often for real materials more a matter of definition than of reality since the yield stress is difficult to pinpoint precisely in experiments. Indeed, if observations were sufficiently precise, it would become questionable whether an exactly linear range or yield stress exists for real materials (in particular, for sands). However, the concept of yield plays a central role in the theory of plasticity, and it is convenient to idealize the material behavior by postulating the existence of a linear elastic range (which can possibly be shrunk to a point) outside of which yielding is to take place.

Although axial compression/extension tests serve to define some as-

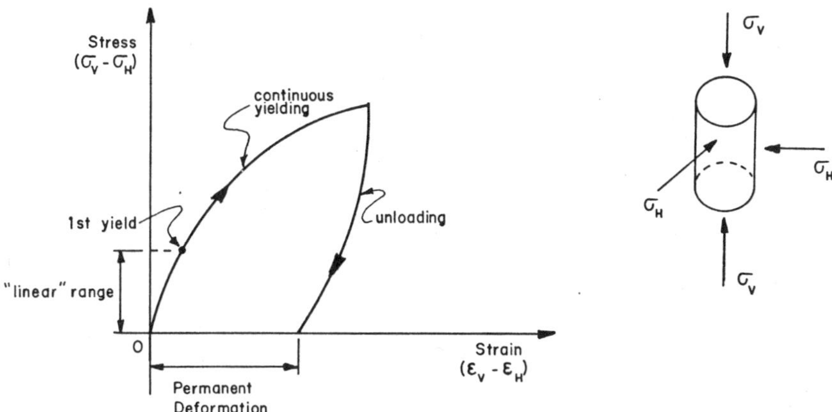

Figure 2-1. Stress-strain curve.

pects of the material behavior, they only give limited information applicable to the two- and three-dimensional situations encountered in practice. The following question must now be answered: "What happens if there are several stresses acting on a material element in different directions?" viz., What combination of these stresses will cause yielding? The criterion for deciding which combination of stresses will cause yielding of the material is called the *yield criterion.*

In the most general case, the yield criterion should depend on the applied state of stress and should reflect the effects of past loading events on the microstructure of the material element. In the macroscopic approach, dislocation density, preferred orientations, and permanent damages (such as grain slips) to the microstructure are to be taken into account indirectly by means of macroscopically defined structure (or "internal") variables which are collectively denoted by **S.** These may include, for example, scalars (to measure dislocation densities), vectors (to indicate some preferred orientation), and second-order tensors (to indicate permanent damages). Tensorial structure variables of higher order than two could also be considered (see e.g., Fardshished and Onat [1974]). The notion of vector directors connected to the microstructure was first introduced by E. and F. Cosserat [1909], and was later further expanded by Mandel [1971, 1973, 1974]. In the following, for simplicity, attention is restricted to small deformations, and direction vectors need not be considered. The macroscopic material state can then be described by the Cauchy stress σ (observed or "external" variable) and the collection of **S** structure variables consisting of second-order tensors α and scalars k.

The yield criterion is expressed mathematically by the following scalar yield function f:

$$f(\boldsymbol{\sigma}, \mathbf{S}) = 0 \tag{2-1}$$

where

$$\mathbf{S} = [\boldsymbol{\alpha}, k] \tag{2-2}$$

Objectivity Requirements. The functional form which the yield function can take (in terms of $\boldsymbol{\sigma}$ and $\mathbf{S}$) is restricted by the objectivity requirement, namely that f must be form invariant with respect to rigid body motions of the spatial frame of reference. Let $\mathbf{Q}$ be a proper orthogonal transformation, viz.,

$$\mathbf{Q}\mathbf{Q}^{\mathrm{T}} = \mathbf{Q}^{\mathrm{T}}\mathbf{Q} = \boldsymbol{\delta} \qquad \det \mathbf{Q} = +1 \tag{2-3}$$

where $\boldsymbol{\delta}$ = unit tensor. Then f being a scalar function, objectively requires that:

$$f(\boldsymbol{\sigma}, \mathbf{S}) = f(\mathbf{Q}\boldsymbol{\sigma}\mathbf{Q}^{\mathrm{T}}, \mathbf{Q}[\mathbf{S}]) = 0 \tag{2-4}$$

where the notation $\mathbf{Q}[\mathbf{S}]$ implies the corresponding transformation for the structure variables in $\mathbf{S}$, viz., $\mathbf{Q}[\boldsymbol{\alpha}] = \mathbf{Q}\boldsymbol{\alpha}\mathbf{Q}^{\mathrm{T}}$, $\boldsymbol{\alpha}$ being of second order. The consequences of this statement had already been employed by Boehler [1979] and have recently been restated as a general theorem by Liu [1982], which states that if f must satisfy Equation 2-4, then f must be an *isotropic function of the variables of both sets $\boldsymbol{\sigma}$ and $\mathbf{S}$*. The general functional form of f can thus be found by using the representation theorems for isotropic functions [Wang, 1970]. It is important to note that the fact that isotropic functions are used does *not* imply that the material is *isotropic*. This is further discussed later.

It is a classical result in the theory of tensor functions [Wang, 1970; Smith, 1971; Liu, 1982; Spencer, 1972] that for the scalar function f to be isotropic, it must depend on its second-order tensor arguments $\boldsymbol{\sigma}$ and $\boldsymbol{\alpha}$ only through their invariants and joint invariants. The complete and irreducible set of invariants for the two symmetric second-order tensors $\boldsymbol{\sigma}$ and $\boldsymbol{\alpha}$ are given by (see e.g., [Spencer, 1972]):

$$\overline{\mathrm{I}}_1 = \mathrm{tr}\,\boldsymbol{\sigma} \qquad \overline{\mathrm{II}}_1 = \mathrm{tr}\,\boldsymbol{\alpha} \tag{2-5}$$

$$\overline{\mathrm{I}}_2 = \mathrm{tr}\,\boldsymbol{\sigma}^2 \qquad \overline{\mathrm{II}}_2 = \mathrm{tr}\,\boldsymbol{\alpha}^2 \tag{2-6}$$

$$\overline{I}_3 = \mathrm{tr}\,\boldsymbol{\sigma}^3 \qquad \overline{II}_3 = \mathrm{tr}\,\boldsymbol{\alpha}^3 \tag{2-7}$$

$$\overline{K}_1 = \mathrm{tr}\,\boldsymbol{\sigma\alpha} \tag{2-8}$$

$$\overline{K}_2 = \mathrm{tr}\,\boldsymbol{\sigma\alpha}^2 \tag{2-9}$$

$$\overline{K}_3 = \mathrm{tr}\,\boldsymbol{\sigma}^2\boldsymbol{\alpha} \tag{2-10}$$

$$\overline{K}_4 = \mathrm{tr}\,\boldsymbol{\sigma}^2\boldsymbol{\alpha}^2 \tag{2-11}$$

$\overline{I}_i$ and $\overline{II}_i$, $i = 1,3$, are called direct invariants, and $\overline{K}_j$, $j = 1,4$ are called joint invariants. Other cases (vectors, etc.) are considered by Spender [1972]. Therefore, Equation 2-1 can now be rewritten as:

$$f(\boldsymbol{\sigma},\mathbf{S}) = f(\overline{I}_i,\overline{II}_i,\overline{K}_j,k)_{j=1,4}^{i=1,3} = 0 \tag{2-12}$$

The direct invariants $\overline{I}_i$ can be expressed in terms of the more familiar invariants I_i, which appear in the characteristic equation as shown in the appendix. Other useful results, such as derivatives of the invariants are also given in the appendix.

Material Symmetries. The type of symmetries (i.e., the level of anisotropy/isotropy) which the material exhibits is determined by the class of symmetry operations (rotations and reflections) which leaves the material response unchanged. Let $\mathbf{q}$ denote an orthogonal transformation, viz.,

$$\mathbf{q}\mathbf{q}^T = \mathbf{q}^T\mathbf{q} = \boldsymbol{\delta} \qquad \det\mathbf{q} = \pm 1 \tag{2-13}$$

The material is said to exhibit a symmetry with respect to the transformation $\mathbf{q}$ if:

$$f(\boldsymbol{\sigma},\mathbf{S}) = f(\boldsymbol{\sigma},\mathbf{q}[\mathbf{S}]) = 0 \tag{2-14}$$

where the notation $\mathbf{q}[\mathbf{S}]$ implies the corresponding transformation for the structure variables in $\mathbf{S}$, viz., $\mathbf{q}(\boldsymbol{\alpha}) = \mathbf{q}^T\boldsymbol{\alpha}\mathbf{q}$, $\boldsymbol{\alpha}$ being of second order. The transformation does not affect the scalars in $\mathbf{S}$, and therefore Equation 2-14 requires

$$\boldsymbol{\alpha} = \mathbf{q}^T\boldsymbol{\alpha}\mathbf{q} \tag{2-15}$$

$\boldsymbol{\alpha}$ being a symmetric second-order tensor, it possesses in general three distinct orthogonal principal directions. Therefore, in general, the only transformation which satisfies Equation 2-15 is a reflection through the planes formed by the principal axes of $\boldsymbol{\alpha}$. This type of symmetry is called *orthotropy* (three perpendicular axes of anisotropy). If $\boldsymbol{\alpha}$ has two equal principal values, the type of symmetry is called *cross-anisotropy* (or transverse isotropy) and the material then possesses a principal axis of anisotropy perpendicular to a plane of isotropy. Finally, if the three principal values of $\boldsymbol{\alpha}$ are all equal, the material is *isotropic*. Therefore, it is clear that it is the presence of the *joint invariants* in Equation 2-12 that renders the material *anisotropic* in general.

Remarks

An important case to be dealt with extensively hereafter is the case of *kinematic* plasticity. In that case $\boldsymbol{\alpha}$ is identified as the back-stress, and the functional dependence of the yield function upon the sets $\boldsymbol{\sigma}$ and $\mathbf{S}$ is taken as:

$$f(\boldsymbol{\sigma},\mathbf{S}) = f((\boldsymbol{\sigma} - \boldsymbol{\alpha}),\mathrm{k}) = 0 \tag{2-16}$$

Objectivity requires f to be an isotropic function of its arguments and therefore:

$$f((\boldsymbol{\sigma} - \boldsymbol{\alpha}),\mathrm{k}) = f(\mathrm{L}_1,\mathrm{L}_2,\mathrm{L}_3,\mathrm{k}) = 0 \tag{2-17}$$

where

$$\mathrm{L}_1 = \mathrm{tr}(\boldsymbol{\sigma} - \boldsymbol{\alpha}) \tag{2-18}$$

$$\mathrm{L}_2 = \mathrm{tr}(\boldsymbol{\sigma} - \boldsymbol{\alpha})^2 \tag{2-19}$$

$$\mathrm{L}_3 = \mathrm{tr}(\boldsymbol{\sigma} - \boldsymbol{\alpha})^3 \tag{2-20}$$

Simple algebra shows that:

$$\mathrm{L}_1 = \overline{\mathrm{I}}_1 - \overline{\overline{\mathrm{II}}}_1 \tag{2-21}$$

$$\mathrm{L}_2 = \overline{\mathrm{I}}_2 - 2\overline{\mathrm{K}}_1 + \overline{\overline{\mathrm{II}}}_2 \tag{2-22}$$

$$\mathrm{L}_3 = \overline{\mathrm{I}}_3 - 3\overline{\mathrm{K}}_2 - 3\overline{\mathrm{K}}^2{}_3 - \overline{\overline{\mathrm{II}}}_3 \tag{2-23}$$

therefore both L_2 and L_3 depend on the joint invariants $\overline{\mathrm{K}}_\mathrm{j}$, and the materials to be characterized by such yield functions are *anisotropic* (orthotropic) in general.

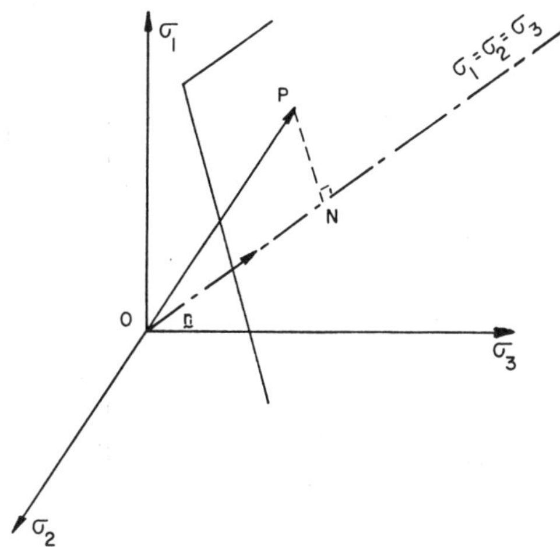

Figure 2-2. Principal stress space.

Principal Stress Space. It is customary and convenient to picture the yield criterion as a surface in stress space, and to refer to it as the yield surface. Since the stress tensor $\boldsymbol{\sigma}$ has six distinct components, this would in general require a six-dimensional space. A more convenient representation is achieved by using a cartesian reference coordinate system whose axes are the three principal stresses. This defines the so-called Haig Westergaard [1920] stress space, and is shown in Figure 2-2.

The hydrostatic axis (or space diagonal) is the line that makes equal angles with all three axes. The unit vector **n** along the hydrostatic axis has components

$$\mathbf{n} = \left\{ \frac{1}{\sqrt{3}}, \frac{1}{\sqrt{3}}, \frac{1}{\sqrt{3}} \right\}$$

Any plane perpendicular to the hydrostatic axis is called a *deviatoric* plane. The deviatoric plane which passes through the origin is called the *π-plane*.

Any given stress state plots as a point P with coordinates σ_1, σ_2 and σ_3; and the vector $\mathbf{OP} = \{\sigma_1, \sigma_2, \sigma_3\}$ is referred to as the *stress vector*. The stress vector may be decomposed into two components,

$$\mathbf{OP} = \mathbf{ON} + \mathbf{NP} \tag{2-24}$$

where the vector $\mathbf{ON}$ is carried by the space diagonal, and the vector $\mathbf{NP}$ is contained in the deviatoric plane passing through N and P. Then:

$$|\mathbf{ON}| = \mathbf{OP \cdot n} = \frac{1}{\sqrt{3}}(\sigma_1 + \sigma_2 + \sigma_3) = \sqrt{3}p \qquad (2\text{-}25)$$

where

$$p = \frac{1}{3}I_1 = \text{mean stress} \qquad (2\text{-}26)$$

The vector $\mathbf{ON}$ has components $\mathbf{ON} = \{p,p,p\}$ along the three reference axes. The components of $\mathbf{NP}$ are simply obtained from Equation 2-24 as:

$$\mathbf{NP} = \mathbf{OP} - \mathbf{ON} = \{s_1, s_2, s_3\} \qquad (2\text{-}27)$$

where $s_i = \sigma_i - p = $ principal values of the deviatoric stress tensor associated with $\boldsymbol{\sigma}$. The length of the vector $\mathbf{NP}$ is given by:

$$R = |\mathbf{NP}| = \{s_1^2 + s_2^2 + s_3^2\}^{1/2} = \{\text{tr}s^2\}^{1/2} = \sqrt{2J_2} \qquad (2\text{-}28)$$

Figure 2-3 shows a view at right angle to the deviatoric plane through N, so that the three principal stress axes appear at angles of $2\pi/3\,(=120°)$ to each other. The projections of P onto the axes are given by

$$PH_i = \sqrt{\frac{3}{2}}\,s_i \qquad (2\text{-}29)$$

Let θ denote the polar angle measured from the pure shear axis $\sigma_2 = (\sigma_1 + \sigma_3)/2$ as shown in Figure 2-3, then (from Equations 2-28 and 2-29).

$$\sin\theta = \frac{\sqrt{3}}{2}\frac{s_2}{\sqrt{J_2}} \qquad \frac{-\pi}{6} \le \theta \le +\frac{\pi}{6} \qquad (2\text{-}30)$$

Using the trigonometric identity: $\sin3\theta = 3\sin\theta - 4\sin^3\theta$ simple algebraic manipulations yield:

$$\sin3\theta = -\frac{3\sqrt{3}}{2}\frac{J_3}{J_2^{3/2}}; \quad J_3 = \det s = \text{tr}\,s^3/3 \qquad (2\text{-}31)$$

The three principal stresses are then given by:

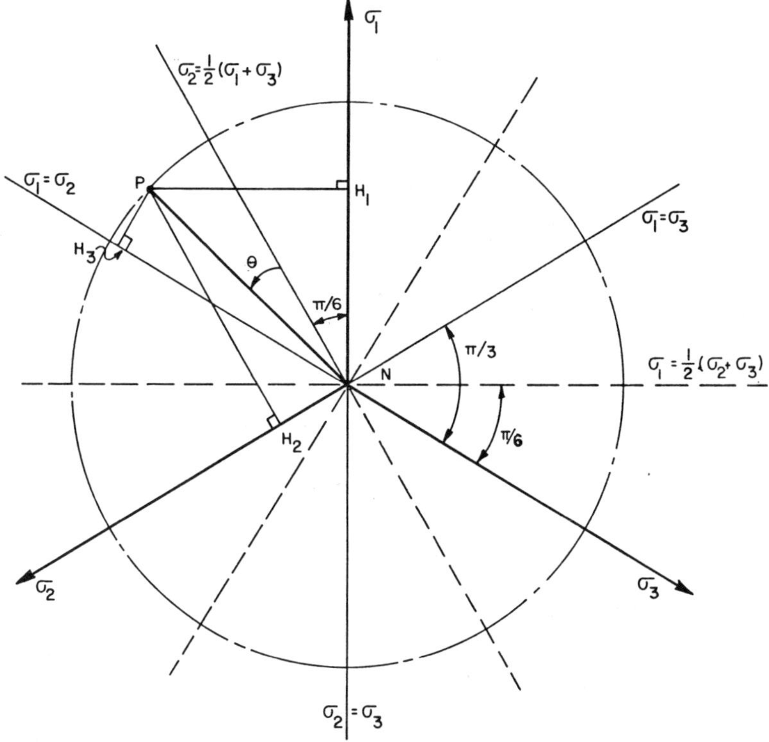

Figure 2-3. Deviatoric stress plane.

$$\sigma_I = p + \frac{2}{\sqrt{3}}\,\sqrt{J_2}\sin\left(\theta + \frac{2\pi}{3}\right)$$

$$\sigma_{II} = p + \frac{2}{\sqrt{3}}\,\sqrt{J_2}\sin\theta \qquad\qquad (2\text{-}32)$$

$$\sigma_{III} = p + \frac{2}{\sqrt{3}}\,\sqrt{J_2}\sin\left(\theta - \frac{2\pi}{3}\right)$$

where

$$\sigma_I \geqslant \sigma_{II} \geqslant \sigma_{III}$$

A path in stress space indicates successive states of stressing a material specimen homogeneously, and is called a *stress path*. If the path lies entirely within the plane passing through the σ_1-axis and making equal angles of $\pi/4$ (45°) with both the σ_2 and σ_3 axes, such a stress path is called

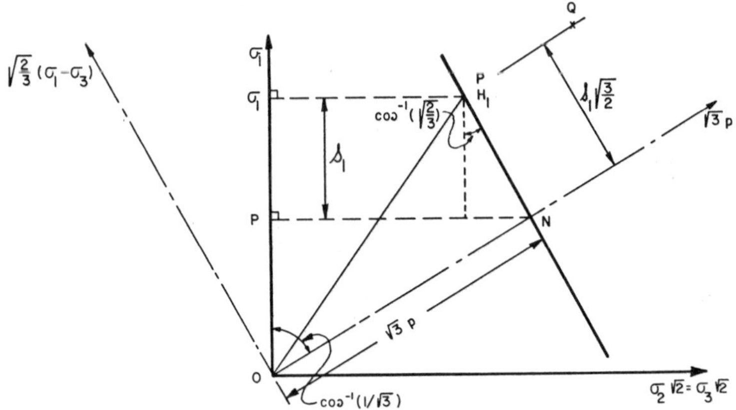

Figure 2-4. Axial stress plane.

an *axial stress path* (since $\sigma_2 = \sigma_3$). Figure 2-4 shows the axial stress plane, and Figure 2-5 identifies various commonly used stress paths in geomechanics material testing. Note that in Figure 2-5 following common usage in geomechanics, compressive stresses are counted as positive.

Examples of Yield Functions/Surfaces. Over the years, dating as far back as Coulomb in 1773, numerous criteria have been proposed for the

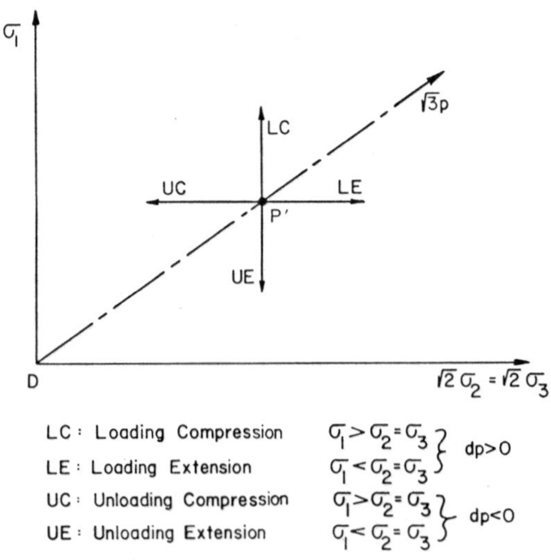

Figure 2-5. Axial stress paths (compression positive).

yielding of solids. A historical sketch on the development of plasticity theory is available in Scott [1985]. Many of the criteria had originally been suggested as failure criteria (i.e., for the ultimate state of yielding), and have later been adopted as yield criteria for ductile materials. The most commonly known yield criteria are described hereafter.

Tresca Criterion. The Tresca's yield criterion stipulates that yielding occurs when the maximum shear stress reaches the critical intensity k, i.e., when

$$f = (\sigma_I - \sigma_{III}) - k = 0 \qquad (2\text{-}33)$$

An invariant expression is obtained by substituting Equation 2-32 in Equation 2-33 as:

$$f(J_2, J_3, k) = 2\sqrt{J_2}\cos\theta - k = 0 \qquad (2\text{-}34)$$

Since the yield function is independent of I_1, the yielding of the material is assumed to be unaffected by the mean (confining) stress, and the material is said to be "pressure insensitive." Further, since f is not a function of the joint invariants, the material is assumed *isotropic*. For axial loading conditions ($\theta = -\pi/6$) Equation 2-34 simplifies to:

$$\sigma_2 = \sigma_3 \qquad |\sigma_1 - \sigma_3| = k \qquad (2\text{-}35)$$

and therefore k represents the yield stress measured in axial tests. The Tresca's yield surface plots in principal stress space as a regular hexagonal cylinder whose axis is the space diagonal as shown in Figure 2-6.

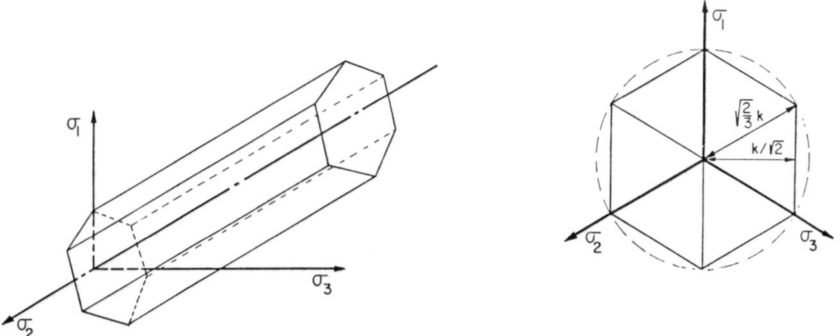

Figure 2-6. Tresca's yield surface.

Huber-von Mises Criterion. The Huber-von Mises-yield criterion is defined by the following equation:

$$f(J_2, k) = \sqrt{3J_2} - k = 0 \tag{2-36}$$

Again, the material is assumed "pressure insensitive," and isotropic. For axial loading condition,

$$\sqrt{3J_2} = |\sigma_1 - \sigma_3| \, ,$$

and therefore k is again the yield stress measured in axial experiments. For pure shear

$$(\sigma_1 = \sigma_2 = \sigma_3) \qquad \sqrt{3J_2} = \tau\sqrt{3}, \text{ and thus } k/\sqrt{3}$$

is the shear stress at yield in pure shear experiments. The Huber-von Mises-yield surface plots in principal stress space as a circular cylinder whose axis is the space diagonal as shown in Figure 2-7. Figure 2-7 also shows a comparison between the Tresca and Huber-von Mises-yield surfaces. The Tresca hexagon is inscribed in the von Mises circle. The maximum difference between the two criteria occurs in pure shear.

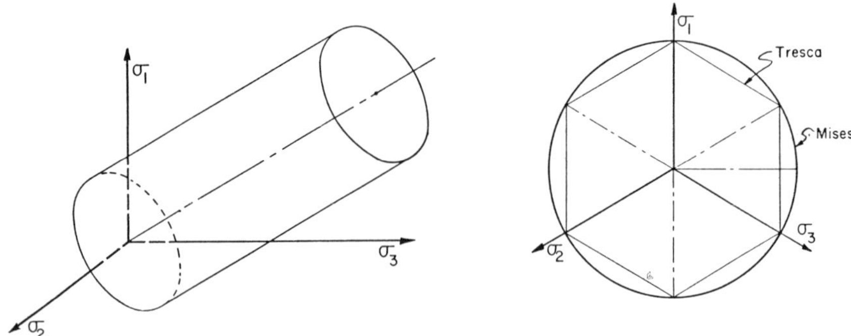

Figure 2-7. Huber-von Mises yield surface.

Mohr-Coulomb Criterion. The Mohr-Coulomb yield criterion is widely used in geomechanics. According to the Mohr-Coulomb criterion, yielding occurs when the Mohr circle becomes tangent to the line

$$\tau = c + \sigma\tan\phi \tag{2-37}$$

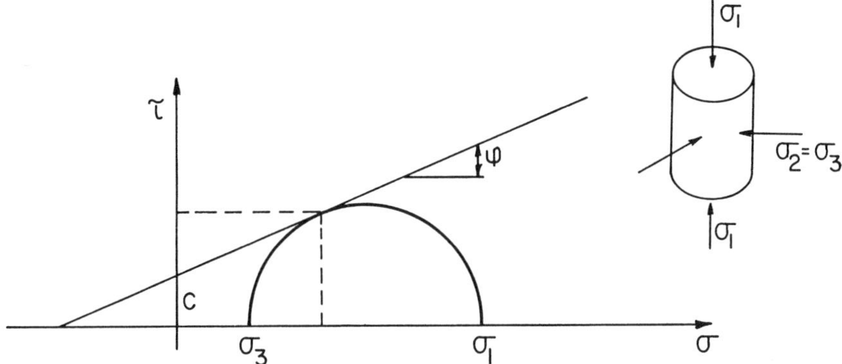

Figure 2-8. Mohr-Coulomb yield criterion (compression positive).

as shown in Figure 2-8 where c is the cohesion and ϕ is the material friction angle. Compressive stresses are taken as positive when writing Equation 2-37. An invariant expression for the Mohr-Coulomb yield function can be obtained as follows. Equation 2-37 is equivalently written as:

$$f = \frac{\sigma_1 - \sigma_3}{2} - \frac{\sigma_1 + \sigma_3}{2}\sin\phi - c\cos\phi = 0 \tag{2-38}$$

Now (from Equation 2-32)

$$\frac{\sigma_1 - \sigma_3}{2} = \sqrt{J_2}\cos\theta \tag{2-39}$$

$$\frac{\sigma_1 + \sigma_3}{2} = \frac{1}{3}I_1 - \frac{1}{\sqrt{3}}\sqrt{J_2}\sin\theta \tag{2-40}$$

substituting into Equation 2-38 yields:

$$f(I_1, J_2, J_3, c, \phi) = -\frac{I_1}{3}\sin\phi + \sqrt{J_2}\cos\theta$$
$$+ \frac{1}{\sqrt{3}}\sqrt{J_2}\sin\theta\,\sin\phi - c\cos\phi = 0 \tag{2-41}$$

The yielding of the material is now assumed to be affected by the mean stress ("pressure sensitive"), but is still assumed *isotropic*. The Mohr-Coulomb yield surface plots in stress space as an irregular hexagonal pyramid whose axis is the space diagonal as shown in Figure 2-9.

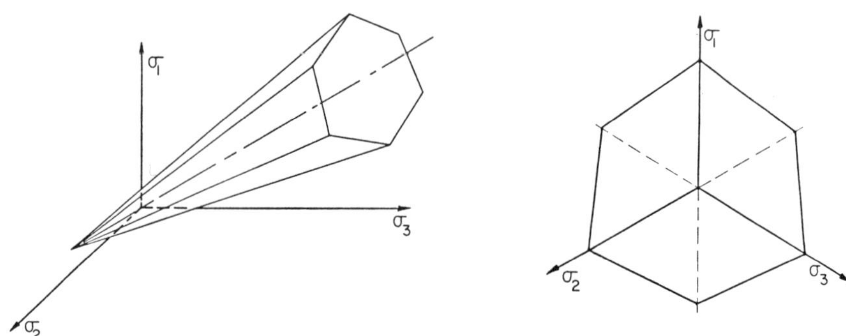

Figure 2-9. Mohr-Coulomb yield surface (compression positive).

The maximum and minimum radii of the circles passing through the corners of the trace of the pyramid in the π-plane ($I_1 = 0$) are given by:

$$\theta = -\frac{\pi}{6} \qquad R_{max} = \sqrt{2J_2} = \frac{2\sqrt{6}\ c\ \cos\phi}{3 - \sin\phi} \tag{2-42}$$

$$\theta = +\frac{\pi}{6} \qquad R_{min} = \sqrt{2J_2} = \frac{2\sqrt{6}\ c\ \cos\phi}{3 + \sin\phi} \tag{2-43}$$

Drucker-Prager Criterion. In 1952 Drucker and Prager proposed a simple generalization of the Mohr-Coulomb criterion as follows:

$$f(I_1, J_2, \alpha, k) = -\alpha I_1 + \sqrt{J_2} - k = 0 \tag{2-44}$$

where α and k are positive material parameters. The criterion plots in stress space as a circular cone whose axis is the space diagonal as shown in Figure 2-10. Figure 2-11 shows the trace of both the Mohr-Coulomb and the Drucker-Prager surfaces onto the π-plane. The values of α and k can be expressed in terms of the cohesion c and friction angle ϕ, by matching the predictions of both yield criteria in any given test condition. For instance, in axial compression ($\sigma_2 = \sigma_3$ and $\theta = -\pi/6$) one finds (from Equations 2-42 and 2-44)

$$\alpha = \frac{2\sin\phi}{\sqrt{3}(3 - \sin\phi)} \tag{2-45a}$$

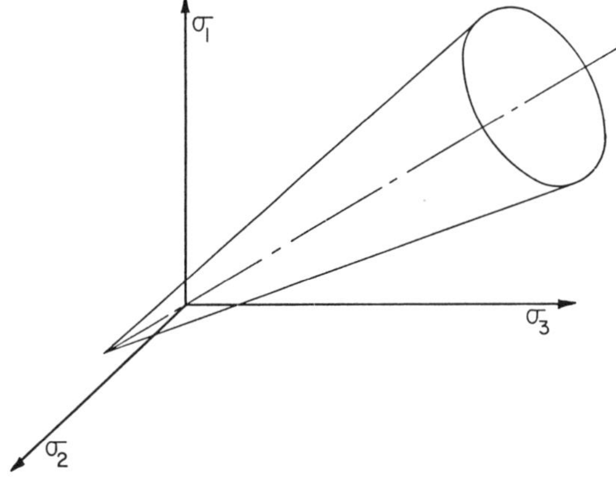

Figure 2-10. Drucker-Prager yield surface (compression positive).

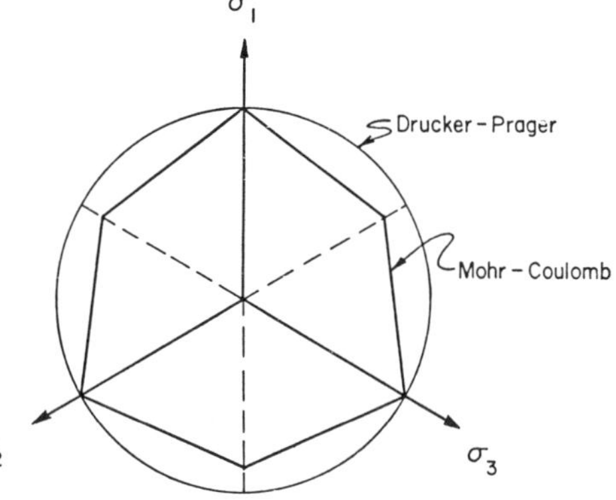

Figure 2-11. Traces of Mohr-Coulomb and Drucker-Prager yield surfaces onto the π-plane (compression positive).

$$k = \frac{6\,c\,\cos\phi}{\sqrt{3}(3 - \sin\phi)} = \frac{c}{\tan\phi}\,\alpha \qquad (2\text{-}45b)$$

Plastic Flow Rules

If yielding does occur, then

$$f(\boldsymbol{\sigma},\mathbf{S}) = 0$$

and it is assumed that the total strain rate $\overset{\circ}{\boldsymbol{\epsilon}}$ (viz., the symmetric part of the spatial velocity gradient) can be expressed as the sum of elastic (recoverable) $\overset{\circ}{\boldsymbol{\epsilon}}{}^{e}$, and plastic (irrecoverable) $\overset{\circ}{\boldsymbol{\epsilon}}{}^{p}$ strain rates as

$$\overset{\circ}{\boldsymbol{\epsilon}} = \overset{\circ}{\boldsymbol{\epsilon}}{}^{e} + \overset{\circ}{\boldsymbol{\epsilon}}{}^{p} \qquad (2\text{-}46)$$

The plastic flow rule gives the *direction* of the plastic part of the strain rate, and one writes:

$$\overset{\circ}{\boldsymbol{\epsilon}}{}^{\mathrm{p}} = \,< \lambda > \mathbf{P}(\boldsymbol{\sigma},\mathbf{S}) \tag{2-47}$$

where λ is a positive scalar quantity, called the plastic loading functions (also loading index), $\mathbf{P}$ is a symmetric second-order tensor, and the symbol $< >$ denotes the Macauley bracket, viz., $<\lambda> = \lambda$ if $\lambda \geqslant 0$ otherwise $<\lambda> = 0$. It is convenient to normalize $\mathbf{P}$ such that

$$\mathrm{tr}\mathbf{P}^2 = 1 \tag{2-48}$$

Let $\mathbf{Q}$ denote the normal to the yield surface, viz.

$$\mathbf{Q} = \frac{\partial f/\partial \boldsymbol{\sigma}}{\{\mathrm{tr}(\partial f/\partial \boldsymbol{\sigma})^2\}^{1/2}} \tag{2-49}$$

Then if $\mathbf{P}$ is selected such that: $\mathbf{P} = \mathbf{Q}$ the plastic flow rule is called *associative*, whereas if $\mathbf{P} \neq \mathbf{Q}$ the plastic flow rule is called *non-associative*.

Objectivity again requires that $\mathbf{P}$ be an isotropic function of the state variables $\boldsymbol{\sigma}$ and $\mathbf{S}$ in order to satisfy the proper invariance requirements under superimposed rigid body rotations.

Although not necessary by any means in general, it is sometimes convenient to write

$$\mathbf{P} = \frac{\partial g/\partial \boldsymbol{\sigma}}{\{\mathrm{tr}(\partial g/\partial \boldsymbol{\sigma})^2\}^{1/2}} \tag{2-50}$$

where g is a scalar function called the *plastic potential*. The same restrictions as discussed previously for the yield function apply on the functional form which the plastic potential can take, and in general

$$g = g(\overline{I}_i,\overline{II}_i,\overline{K}_j,k)_{j=1,4}^{i=1,3} = 0 \tag{2-51}$$

Remarks

1. Let g have the form given by Equation 2-51, then

$$\frac{\partial g}{\partial \boldsymbol{\sigma}} = \sum_i \frac{\partial g}{\partial \overline{I}_i}\frac{\partial \overline{I}_i}{\partial \boldsymbol{\sigma}} + \sum_j \frac{\partial g}{\partial \overline{K}_j}\frac{\partial \overline{K}_j}{\partial \boldsymbol{\sigma}}$$
$$= \mathbf{A} + \mathbf{B} + \mathbf{C} \tag{2-52}$$

where

$$\mathbf{A} = \frac{\partial g}{\partial \bar{I}_1}\delta + \frac{\partial g}{\partial \bar{I}_2}2\sigma + \frac{\partial g}{\partial \bar{I}_3}3\sigma^2$$

$$\mathbf{B} = \frac{\partial g}{\partial \bar{K}_1}\alpha + \frac{\partial g}{\partial \bar{K}_2}\alpha^2 \tag{2-53}$$

$$\mathbf{C} = \frac{\partial g}{\partial \bar{K}_3}(\sigma\alpha + \alpha\sigma) + \frac{\partial g}{\partial \bar{K}_4}(\sigma\alpha^2 + \alpha^2\sigma)$$

and therefore $\mathbf{A}$ is colinear to $\boldsymbol{\sigma}$ (i.e., has the same principal directions as $\boldsymbol{\sigma}$), $\mathbf{B}$ is colinear to $\boldsymbol{\alpha}$, whereas $\mathbf{C}$ is a complex combination of $\boldsymbol{\sigma}$ and $\boldsymbol{\alpha}$. Therefore, if the coaxiality of $\boldsymbol{\sigma}$ and $\overset{\circ}{\boldsymbol{\epsilon}}{}^p$ is to be imposed, then

$$g = g(\bar{I}_i, \bar{II}_i, k)^{i=1,3} \tag{2-54}$$

and the joint invariants cannot appear in the plastic potential as noted by Baker and Desai [1984].

2. If

$$\mathrm{tr}\mathbf{P} = 0 \tag{2-55}$$

then (from Equation 2-47)

$$\overset{\circ}{\epsilon}_v{}^p = \mathrm{tr}\overset{\circ}{\boldsymbol{\epsilon}}{}^p = \lambda\,\mathrm{tr}\mathbf{P} = 0 \tag{2-56}$$

and no plastic volumetric flow can take place. For instance, let

$$g = g(I_1, J_2, J_3, k) = 0 \tag{2-57}$$

then

$$\frac{\partial g}{\partial \sigma} = \left(\frac{\partial g}{\partial I_1} - \frac{2}{3}J_2\frac{\partial g}{\partial J_3}\right)\delta + \frac{\partial g}{\partial J_2}s + \frac{\partial g}{\partial J_3}s^2 \tag{2-58}$$

and

$$\mathrm{tr}\frac{\partial g}{\partial \sigma} = 3\frac{\partial g}{\partial I_1} + \frac{\partial g}{\partial J_3}[\mathrm{tr}s^2 - 2J_2] = 3\frac{\partial g}{\partial I_1} \tag{2-59}$$

Therefore, for plastic volumetric flow to take place, the plastic potential must be a function of $I_1(= 3p)$. It is therefore clear that if an associative flow rule is adopted for the Tresca or Huber-von Mises models, then $\overset{\circ}{\epsilon}_v{}^P = 0$ and the plastic flow is purely *deviatoric*. On the other hand, plastic *dilation* is to be predicted if an associative flow rule is used with either the Mohr-Coulomb or the Drucker-Prager models.

3. It is convenient to split $\mathbf{Q}$ and $\mathbf{P}$ into their deviatoric and volumetric components as follows

$$\mathbf{Q} = \mathbf{Q}' + Q''\boldsymbol{\delta} \qquad Q'' = \frac{1}{3}\,\mathrm{tr}\mathbf{Q} \tag{2-60}$$

$$\mathbf{P} = \mathbf{P}' + P''\boldsymbol{\delta} \qquad P'' = \frac{1}{3}\,\mathrm{tr}\mathbf{P} \tag{2-61}$$

Then most plasticity models used in geomechanics today assume that

$$\mathbf{P}' = \mathbf{Q}' \text{ and } P'' \neq Q'' \tag{2-62}$$

i.e., the deviatoric plastic flow is assumed associative, whereas the volumetric plastic flow is nonassociative. The necessity of having to resort to a nonassociative flow rule for the volumetric component of the deformations stems from the requirement to model behavior such as compaction followed by dilation upon loading in materials such as sands.

Plastic Hardening Rules

When yielding occurs

$$f(\boldsymbol{\sigma},\mathbf{S}) = 0 \text{ and } \overset{\circ}{\boldsymbol{\epsilon}}{}^P = \lambda\mathbf{P}$$

The macroscopic evolution rate equations for the structure variables $\mathbf{S}$ are to be used to describe macroscopically the resulting changes of the microstructure. The mathematical description is called the plastic hardening rule, and it may be viewed as causing changes in the yield surface. It is expressed collectively as:

$$\overset{\circ}{\mathbf{S}} = h(\lambda)\bar{\mathbf{S}}(\boldsymbol{\sigma},\mathbf{S}) \tag{2-63}$$

where $\bar{\mathbf{S}}$, depending upon the tensorial character of the variable of interest, must again satisfy objectivity requirements. In equation 2-63 $h(\lambda)$ denotes the Heaviside step function defined zero at $\lambda = 0$. Several cases can be considered:

1. *Isotropic hardening* (see e.g., Hill [1950]): In that case the yield surface is assumed to expand uniformly but not to change its orientation nor position in stress space. This is illustrated in Figure 2-12A where for convenience the Huber-von Mises yield surface is represented.
2. *Kinematic hardening* [Prager, 1959]: In that case the stress point drags along the yield surface as shown in Figure 2-12B. The yield surface is assumed not to change in size or orientation.
3. *Local effects:* Hardening could possibly also develop as a purely local effect. A corner (vertex) would then be formed on the yield surface by the stress point as shown schematically in Figure 2-12C.

On "unloading" (i.e., when the stress point moves back into the yield surface), the yield surface is to be left "frozen" in its deformed state since the material response to stress paths inside the new yield surface is considered to be elastic (i.e., no further damage can take place). The various hardening assumptions have important implications in the calculation of plastic deformations as will become apparent in the following.

Isotropic Hardening Rule. The simplest possible assumption is that the yield surface expands uniformly without changing its position in stress space. The yield function then depends on a single parameter k (the size of the yield surface), and usually takes the form

$$f = \hat{f}(\boldsymbol{\sigma} - \boldsymbol{\alpha}) - k^n(\boldsymbol{\epsilon}^P) = 0 \tag{2-64}$$

where $\hat{f}$ is a homogeneous function of its arguments, and n is the degree of $\hat{f}$. If the yield surface is to remain centered at the origin, then $\boldsymbol{\alpha} = \boldsymbol{0}$ in Equation 2-64. In Equation 2-64, $\boldsymbol{\epsilon}^P$ is used to indicate the dependence of the size parameter k on the plastic deformations. This is achieved by taking k function of invariant measures of the past plastic deformation history. These measures are called hardening parameters and are usually selected such that they increase monotonically during the deformation process. Two different measures have been used: the plastic work W^P, and the plastic volumetric and deviatoric strain invariants ϵ^P_v and $\bar{\epsilon}^P$,

$$W^P = \int \overset{\circ}{W} dt \qquad \epsilon_v{}^P = \int \overset{\circ}{\epsilon_v}{}^P dt \qquad \bar{\epsilon}^P = \int \overset{\circ}{\bar{\epsilon}}{}^P dt \tag{2-65}$$

where the integrations are carried along the strain path:

$$\overset{\circ}{W}{}^P = \text{tr}(\boldsymbol{\sigma}\overset{\circ}{\boldsymbol{\epsilon}}{}^P) = p\overset{\circ}{\epsilon}{}^P_v + \text{tr}(s\overset{\circ}{\boldsymbol{e}}{}^P) \tag{2-66}$$

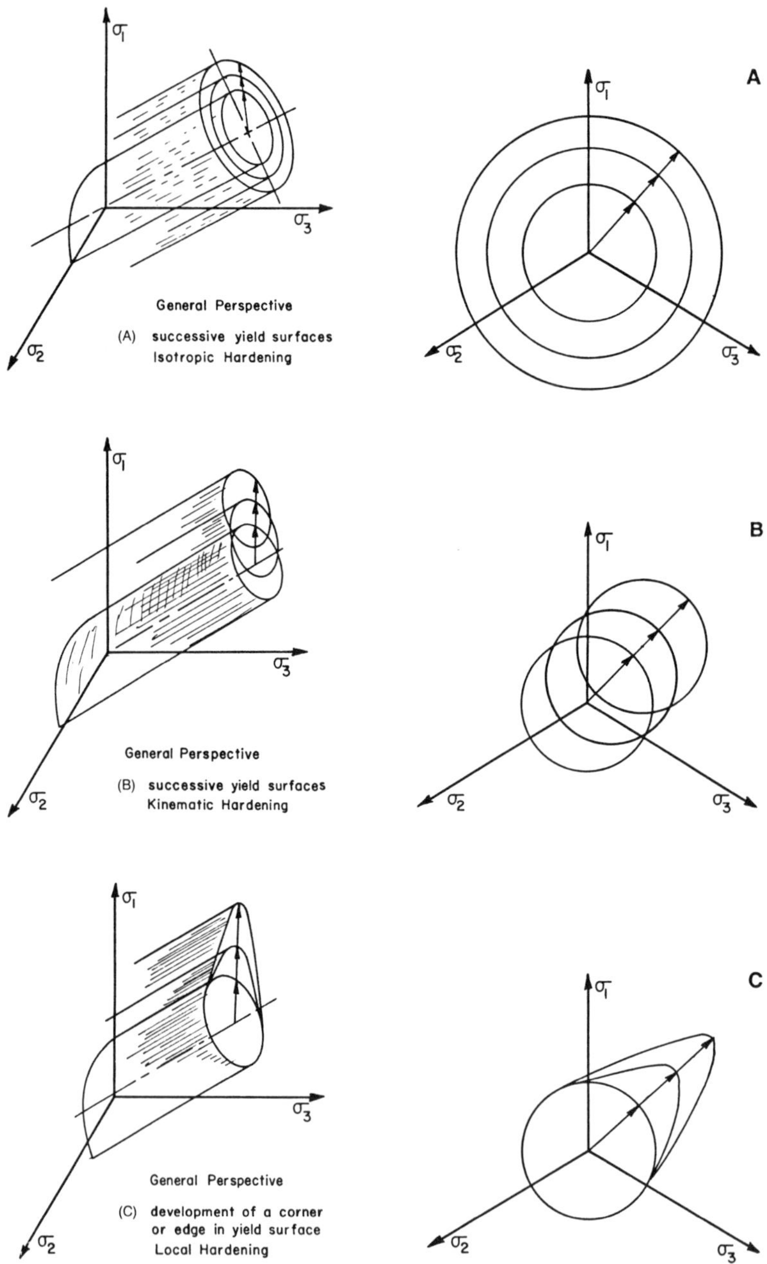

Figure 2-12. Hardening models: (A) isotropic; (B) kinematic; (C) local effect.

where

$$p = \frac{1}{3} \, \text{tr} \, \boldsymbol{\sigma} \qquad \overset{\circ}{\mathbf{e}}^P = \overset{\circ}{\boldsymbol{\epsilon}}^P - \frac{1}{3} \, \overset{\circ}{\epsilon}_v{}^P \boldsymbol{\delta} \tag{2-67}$$

and

$$\overset{\circ}{\epsilon}^P{}_v = \text{tr}\,\overset{\circ}{\boldsymbol{\epsilon}}^P \tag{2-68}$$

$$\overset{\circ}{\bar{\epsilon}}{}^P = \left\{ \frac{2}{3} \, \text{tr}(\overset{\circ}{\mathbf{e}}^P)^2 \right\}^{1/2} \tag{2-69}$$

is a measure of plastic shear distortion. The plastic work is now rarely used.

Remarks

For the Huber-von Mises plastic model with associative flow:

$$\overset{\circ}{\boldsymbol{\epsilon}}^P = \overset{\circ}{\mathbf{e}}^P = \lambda \, \sqrt{\frac{3}{2}} \, \frac{\mathbf{s}}{k} \tag{2-70}$$

then

$$\overset{\circ}{W}{}^P = \lambda \, \sqrt{\frac{3}{2}} \, \frac{\text{tr} \, \mathbf{s}^2}{k} = \lambda \, \sqrt{\frac{2}{3}} \, q \tag{2-71}$$

where

$$q = \sqrt{3J_2} = \text{equivalent stress} \tag{2-72}$$

and

$$\overset{\circ}{\bar{\epsilon}}{}^P = \lambda \, \frac{\{\text{tr} \, \mathbf{s}^2\}^{1/2}}{k} = \lambda \, \sqrt{\frac{2}{3}} \tag{2-73}$$

Therefore combining Equations 2-71 and 2-73,

$$\overset{\circ}{W}{}^P = q \overset{\circ}{\bar{\epsilon}}{}^P \tag{2-74}$$

Clearly, the assumption that one universal formula $k^n(W^P)$ or $k^n(\epsilon^P_v, \bar{\epsilon}^P)$ governs all possible stress loading conditions of a given material is a very strong idealization, especially when the loading deviates significantly from proportional loading and involves rotation of the principal stress directions. Further, the symmetries observed in the initial yield function cannot legitimately be imposed on the subsequent yield functions in general, as the very process of plastic flow itself is anisotropic in character, even for an initially isotropic material. Finally, isotropic hardening implies that the material is strengthened for all paths to yield by any path which raises the yield boundary. In particular upon unloading and reverse loading, the model predicts that the material is now stronger than it would have been if no loading had taken place in the first place. It has been argued that since isotropic hardening exhibits its greatest deficiency by providing erroneous results for problems involving unloading, its use in connection with situations where only loading occurs would not a priori be objectionable. However, it is unlikely that such restrictions can be met at every point in general boundary value problems. In order to describe more accurately a material behavior when subjected to complex loading paths, a nonisotropic hardening rule is needed. The yield surface must be allowed to change its size as well as its shape and position in stress space in order to account for anisotropic effects and permanent damages to the microstructure induced by the plastic flow. The simplest hardening rule incorporating such effects was proposed in 1955 by Prager and is called kinematic hardening.

Kinematic Hardening Rule. In order to account for the effects of loading/unloading/reloading events on the material behavior, the yield surface is allowed to be translated in stress space by the stress point and to change in size and shape simultaneously. The yield function is then represented by an equation of the form

$$f((\boldsymbol{\sigma} - \boldsymbol{\alpha}), k) = 0 \tag{2-75}$$

where the scalar structure variables k are function of ϵ^P_v and $\bar{\epsilon}^P$, (viz., $k = k(\epsilon^P_v, \bar{\epsilon}^P)$) and $\boldsymbol{\alpha}$ (the back-stress) is identified geometrically as the position vector of the center (or other geometric entity) of the yield surface. If the yield surfaces are not allowed to change in shape, Equation 2-75 simplifies to:

$$f = \hat{f}(\boldsymbol{\sigma} - \boldsymbol{\alpha}) - k^n(\epsilon^P) = 0 \tag{2-76}$$

If the material is initially isotropic, $\boldsymbol{\alpha} = \mathbf{0}$ initially in Equation 2-75. However, in general

$$\overset{\circ}{\boldsymbol{\alpha}} = h(\lambda)\boldsymbol{\mu}(\boldsymbol{\sigma},\mathbf{S}) \tag{2-77}$$

where $\boldsymbol{\mu}$ defines the translation direction (symmetric, second-order tensor). Initially, Prager [1955] proposed that the translation occurs in the direction of local normal to the yield surface at the stress point, i.e., $\boldsymbol{\mu} = \mathbf{Q}$. Other proposals have since been made and are discussed hereafter.

Elastic-Plastic Rate Equations

In order to complete the constitutive formulation it remains to specify the elastic rate relations for combination with the plastic rate relations. This would in general require specifying a complementary free energy density function of the state variables $\boldsymbol{\sigma}$ and $\mathbf{S}$ (see, for example, Nemat-Nasser [1983]). However, since attention herein is restricted to small deformations, it is assumed for simplicity that the elasticity of the material is *linear* and *isotropic*. Nonlinear and anisotropic effects are to be accounted for by the material plasticity. One then writes:

$$\overset{\circ}{\boldsymbol{\sigma}} = \mathbf{E} : (\overset{\circ}{\boldsymbol{\epsilon}} - \overset{\circ}{\boldsymbol{\epsilon}}^{\mathrm{P}}) \tag{2-78}$$

where $\mathbf{E}$ is the fourth-order isotropic elastic coefficient tensor, viz.,

$$E_{ijkl} = \left(B - \frac{2G}{3}\right) \delta_{ij}\delta_{kl} + G \, (\delta_{ik}\delta_{jl} + \delta_{il}\delta_{jk}) \tag{2-79}$$

where B = elastic bulk modulus
      G = elastic shear modulus

In recent years, stress-strain relations for elastic-plastic materials have been obtained by different authors in various ways and by using different assumptions. For example, Drucker [1959] advanced a fundamental postulate from which some basic ingredients for general elastic-plastic equations may be deduced. However, this approach resting entirely on Drucker's definition of a stable material is far too restrictive for a general treatment of elastic-plastic materials. In particular, frictional materials like soils "violate" Drucker's postulate, and this approach will not be considered here. Instead, a more general approach resting entirely on the consistency condition is adopted in the following. The yield surface is visualized as dividing the stress space into two regions. Inside the yield surface, the material behavior is elastic. Permanent plastic deformations can only occur when the stress point is on the yield surface. During the plastic flow, the stress point must remain on the yield surface. This is expressed mathematically by:

$$f(\boldsymbol{\sigma}, \mathbf{S}) = 0$$

and

$$\overset{\circ}{f}(\boldsymbol{\sigma}, \mathbf{S}) = \frac{\partial f}{\partial \boldsymbol{\sigma}} : \overset{\circ}{\boldsymbol{\sigma}} + \frac{\partial f}{\partial \boldsymbol{\alpha}} : \overset{\circ}{\boldsymbol{\alpha}} + \frac{\partial f}{\partial \mathbf{k}} \overset{\circ}{\mathbf{k}} = 0 \tag{2-80}$$

General Derivations. The key constitutive assumptions are summarized as follows:

1. Yield function:

$$f(\boldsymbol{\sigma}, \boldsymbol{\alpha}, \mathbf{k}) = 0 \tag{2-81}$$

2. Flow rule:

$$\overset{\circ}{\boldsymbol{\epsilon}}^P = \, <\lambda> \mathbf{P} \tag{2-82}$$

3. Hardening Rules:

$$\overset{\circ}{\boldsymbol{\alpha}} = (a \overset{\circ}{\epsilon}^P_v + b \overset{\circ}{\bar{\epsilon}}^P) \boldsymbol{\mu} \tag{2-83}$$

$$\overset{\circ}{\mathbf{k}} = c \overset{\circ}{\epsilon}^P_v + d \overset{\circ}{\bar{\epsilon}}^P \tag{2-84}$$

where

$$\overset{\circ}{\epsilon}^P_v = \, <\lambda> 3P'' \tag{2-85}$$

$$\overset{\circ}{\bar{\epsilon}}^P = \, <\lambda> \left\{ \frac{2}{3} \mathbf{P}' : \mathbf{P}' \right\}^{1/2} \tag{2-86}$$

and

$$\mathbf{P} = \mathbf{P}' + P'' \boldsymbol{\delta} \qquad\qquad \mathrm{tr}\, \mathbf{P}' = 0 \tag{2-87}$$

The consistency condition is then used to find λ, viz., substituting Equations 2-83–2-86 into Equation 2-80 yields:

$$NQ:\overset{\circ}{\boldsymbol{\sigma}} + \lambda \left\{ \left[a3P'' + b\left(\frac{2}{3}\,\mathbf{P}':\mathbf{P}'\right)^{1/2} \right] \frac{\partial f}{\partial \alpha} : \boldsymbol{\mu} \right.$$

$$\left. + \left[c3P'' + d\left(\frac{2}{3}\,\mathbf{P}':\mathbf{P}'\right)^{1/2} \right] \frac{\partial f}{\partial k} \right\} = 0 \tag{2-88}$$

where

$$N = \left\{ \mathrm{tr}\left(\frac{\partial f}{\partial \boldsymbol{\sigma}}\right)^2 \right\}^{1/2}$$

Further, from Equations 2-78 and 2-82

$$\mathbf{Q}:\overset{\circ}{\boldsymbol{\sigma}} = \mathbf{Q}:\mathbf{E}:\overset{\circ}{\boldsymbol{\epsilon}} - \mathbf{Q}:\mathbf{E}:\overset{\circ}{\boldsymbol{\epsilon}}^{\mathrm{P}}$$

$$= \mathbf{Q}:\mathbf{E}:\overset{\circ}{\boldsymbol{\epsilon}} - \lambda \mathbf{Q}:\mathbf{E}:\mathbf{P} \tag{2-89}$$

Let

$$H_o = \mathbf{Q}:\mathbf{E}:\mathbf{P} = B\,\mathrm{trP\,trQ} + 2G\mathbf{P}':\mathbf{Q}' \tag{2-90}$$

then combining Equations 2-88–2-90,

$$\lambda = \frac{\mathbf{Q}:\mathbf{E}:\overset{\circ}{\boldsymbol{\epsilon}}}{H_o - \dfrac{1}{N}\left[A\,\dfrac{\partial f}{\partial \alpha}:\boldsymbol{\mu} + B\,\dfrac{\partial f}{\partial k} \right]} \tag{2-91}$$

where

$$A = a3P'' + b\left\{ \frac{2}{3}\,\mathbf{P}':\mathbf{P}' \right\}^{1/2} \tag{2-92}$$

$$B = c3P'' + d\left\{ \frac{2}{3}\,\mathbf{P}':\mathbf{P}' \right\}^{1/2} \tag{2-93}$$

let

$$H' = -\frac{1}{N}\left[A\,\frac{\partial f}{\partial \alpha}:\boldsymbol{\mu} + B\,\frac{\partial f}{\partial k} \right] \tag{2-94}$$

where $H' =$ plastic modulus, then (from Equations 2-88 and 2-91–2-94)

$$\lambda = \frac{1}{H' + H_o} \mathbf{Q} : \mathbf{E} : \overset{\circ}{\boldsymbol{\epsilon}} = \frac{1}{H'} \mathbf{Q} : \overset{\circ}{\boldsymbol{\sigma}} \qquad (2\text{-}95)$$

which should be positive for plastic flow to occur. Combining Equations 2-78, 2-82, and 2-95, the rate stress-strain relation writes:

$$\overset{\circ}{\boldsymbol{\sigma}} = \mathbf{E} : \overset{\circ}{\boldsymbol{\epsilon}} - <\frac{\mathbf{Q} : \mathbf{E} : \overset{\circ}{\boldsymbol{\epsilon}}}{H' + H_o}> \mathbf{E} : \mathbf{P} \qquad (2\text{-}96)$$

The first contribution to the stress rate is often referred to as the *elastic (trial) stress,* and the second one as the *plastic stress relaxation.* Note that if a nonassociative plastic flow rule is used $\mathbf{P} \neq \mathbf{Q}$ and the resulting material stiffness matrix is nonsymmetric (from Equation 2-96).

Remarks

1. For the case of *isotropic hardening,* $a = b = 0$ in Equations 2-83, and Equation 2-94 simplifies to:

$$H' = -\frac{1}{N} B \frac{\partial f}{\partial k} \qquad (2\text{-}97)$$

Whereas for the case of purely *kinematic hardening,* $c = d = 0$ in Equation 2-84, and Equation 2-94 simplifies to:

$$H' = -\frac{1}{N} A \frac{\partial f}{\partial \alpha} : \mu \qquad (2\text{-}98)$$

It is therefore clear that Equation 2-94 places a constraint on H', A (a and b), and B (c and d). To allow for an arbitrary *linear* combination of isotropic and kinematic hardening, define

$$A = (\beta - 1) N \frac{H'}{\dfrac{\partial f}{\partial \alpha} : \mu} \qquad (2\text{-}99)$$

where β is a parameter. Equations 2-94 and 2-99 then serve to define B

$$B = -\beta N \frac{H'}{\partial f / \partial k} \qquad (2\text{-}100)$$

As may be seen from Equations 2-99 and 2-100, $\beta = 0$ corresponds to kinematic hardening, and $\beta = 1$ to isotropic hardening.

Substituting Equations 2-99 and 2-100 into Equations 2-83 and 2-84, respectively, leads to:

$$\overset{\circ}{\alpha} = \;<\lambda>(\beta - 1)\; N\; \frac{H'}{\dfrac{\partial f}{\partial \alpha} : \mu}\, \mu \tag{2-101}$$

$$\overset{\circ}{k} = \; -<\lambda>\beta N\; \frac{H'}{\partial f / \partial k} \tag{2-102}$$

2. As will become apparent later, it is convenient to be able to identify H' as a material parameter, and to specify directly the isotropic hardening rule. In that case, Equation 2-94 serves to compute the consistent motion of the yield surface as

$$\overset{\circ}{\alpha} = \;<\lambda>A\boldsymbol{\mu} \tag{2-103}$$

with

$$A = -\frac{NH' + B\partial f / \partial k}{\dfrac{\partial f}{\partial \alpha} : \boldsymbol{\mu}} \tag{2-104}$$

3. The preceding formulation has been described within the content of small-deformation plasticity theory. The generalization to the large deformation case (see e.g., Green and Naghdi, [1965]) may be facilitated by replacing the stress rate in Equation 2-78 by a corotational stress rate. However, as discussed by Dafalias [1984] and Loret [1983] constitutive relations must then also be provided for the plastic spin (defined as the antisymmetric part of the plastically induced velocity gradient). This is currently the subject of intensive research, and falls beyond the scope of this presentation.

Applications/special cases. Consider the simple case of the Huber-von Mises yield function with associative plastic flow, viz.,

$$f = \frac{3}{2}\, \text{tr}\; (s - \alpha)^2 - k^2 = 0 \tag{2-105}$$

$$\overset{\circ}{\boldsymbol{\epsilon}}{}^{\mathrm{P}} = \overset{\circ}{\mathbf{e}}{}^{\mathrm{P}} = \;<\lambda>\mathbf{Q} = \;<\lambda>\sqrt{\frac{3}{2}}\,\frac{(\mathbf{s}-\boldsymbol{\alpha})}{k} \tag{2-106}$$

$$\overset{\circ}{\boldsymbol{\alpha}} = \;<\lambda> \mathrm{A}\boldsymbol{\mu} \tag{2-107}$$

$$\overset{\circ}{k} = \mathrm{B}\,\overset{\circ}{\bar{\epsilon}}{}^{\mathrm{P}} = \;<\lambda>\sqrt{\frac{2}{3}}\,\mathrm{B} \tag{2-108}$$

Then consistency imposes that

$$\mathrm{H}' = \mathrm{A}\mathbf{Q}:\boldsymbol{\mu} + \frac{2}{3}\,\mathrm{B} \tag{2-109}$$

and

$$\lambda = \frac{1}{\mathrm{H}'}\,\sqrt{\frac{3}{2}}\,\frac{(\mathbf{s}-\boldsymbol{\alpha}):\overset{\circ}{\boldsymbol{\sigma}}}{k} = \frac{2\mathrm{G}}{\mathrm{H}'+2\mathrm{G}}\,\sqrt{\frac{3}{2}}\,\frac{(\mathbf{s}-\boldsymbol{\alpha}):\overset{\circ}{\boldsymbol{\epsilon}}}{k} \tag{2-110}$$

For pure *isotropic hardening,* $\mathrm{A} = 0$, and Equation 2-109 imposes

$$\mathrm{H}' = \frac{2}{3}\,\mathrm{B} = \frac{2}{3}\,\frac{\partial k}{\partial \bar{\epsilon}^{\mathrm{P}}} \tag{2-111}$$

whereas for pure kinematic hardening, $\mathrm{B} = 0$, and

$$\mathrm{A} = \frac{\mathrm{H}'}{\mathbf{Q}:\boldsymbol{\mu}} \tag{2-112}$$

The stress-strain relations write:

$$\overset{\circ}{\boldsymbol{\sigma}} = \left(\mathrm{B} - \frac{2\mathrm{G}}{3}\right)\overset{\circ}{\epsilon}_v\boldsymbol{\delta} + 2\mathrm{G}\overset{\circ}{\boldsymbol{\epsilon}}$$

$$\qquad - (2\mathrm{G}-\mathrm{H})\,\frac{3}{2}\,\frac{(\mathbf{s}-\boldsymbol{\alpha})}{k^2} < (\mathbf{s}-\boldsymbol{\alpha}):\overset{\circ}{\boldsymbol{\epsilon}}> \tag{2-113}$$

where

$$\mathrm{H} = \left(\frac{1}{2\mathrm{G}} + \frac{1}{\mathrm{H}'}\right)^{-1} \tag{2-114}$$

is the elasto-plastic shear modulus.

Attention is now turned to the axisymmetric testing environment ("triaxial") in which two lateral stresses are equal ($\sigma_2 = \sigma_3$), and a cylindrical specimen is subjected to axial stress changes (σ_1). In order for the material specimen to deform in an axisymmetric fashion ($\epsilon_2 = \epsilon_3$), the axes of loading must coincide with the principal axes of the anisotropic tensor $\boldsymbol{\alpha}$ and $\alpha_2 = \alpha_3$. Equation 2-105 then simplifies to:

$$f = (q - \alpha)^2 - k^2 = 0 \tag{2-115}$$

where

$$q = (\sigma_1 - \sigma_3) = \text{shear stress} \tag{2-116}$$

$$\alpha = (\alpha_1 - \alpha_3) = 3\alpha_1/2 \tag{2-117}$$

and Equation 2-113 simplifies to:

$$\overset{\circ}{q} = H\overset{\circ}{\bar{\epsilon}} \tag{2-118}$$

where

$$\overset{\circ}{\bar{\epsilon}} = (\epsilon_1 - \epsilon_3) = \text{shear strain} \tag{2-119}$$

The yield surface plots in principal stress space as an open-ended circular cylindrical surface of revolution whose axis is parallel to the space diagonal. The cylinder is perpendicular to the deviatoric planes and is thus completely described by the circle made by its intersection with such planes. This is shown in Figure 2-13 where the scaling factor $\sqrt{3/2}$ is used so that the lengths in Figures 2-13A and B are easily comparable. Note that because of the material initial cross-anisotropy ($\alpha_2 = \alpha_3$), the surface is centered along the σ_1-axis in Figure 2-13B. Compressive stresses and strains are counted as positive in the following. During both axial compression ($\sigma_1\uparrow$) and axial extension ($\sigma_1\downarrow$) tests, the stress point travels along the σ_1-axis. Upon loading in compression ($\overset{\circ}{q} > 0$), the stress point moves up the σ_1-axis, and when it reaches the yield surface at M the slope of the shear stress-strain curve is H (Equation 2-118). Similarly, upon loading in extension ($\overset{\circ}{q} < 0$) the stress point moves down the σ_1-axis, and when it reaches the yield surface at M$'$, the slope is again H. This provides a procedure for identifying the location and size of the yield surface, viz., (from Equation 2-115)

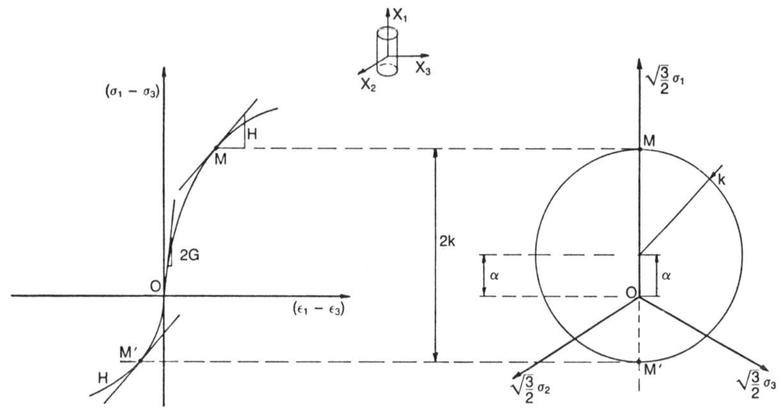

Figure 2-13. Interpretation of axial stress-strain tests: (A) Stress-strain curve. (B) Representation in stress space.

$$\alpha = \frac{1}{2}\,[q_C + q_E]$$

$$k = \frac{1}{2}\,[q_C - q_E] \tag{2-120}$$

where q_C and q_E are the measured shear stress in compression and extension for which the slope of the shear stress-strain curve is the same. Further, the plastic modulus H' associated with the yield surface can be determined (from Equation 2-114)

$$H' = \left(\frac{1}{H} - \frac{1}{2G}\right)^{-1} \tag{2-121}$$

once the elastic shear modulus is known. The elastic shear modulus is identified as the steepest initial slope (which may be measured either in compression or in extension) of the shear stress-strain plot.

From the foregoing, it is now clear that by comparing the experimental shear stress-strain curves obtained in both axial compression and extension tests, the position α, size k, and associated plastic shear modulus H' of the yield function f (Equations 2-105 and 2-115) can be determined. The subsequent yield function upon further loading in compression or extension is obtained by integrating Equations 2-107 and 2-108. One possibility, which shall in fact be rejected later, is then to select specific hardening rules for the translation and expansion/contraction of the yield surface by fitting an algebraic equation to the stress-strain curve obtained in axial compression and extension loading conditions, e.g, by selecting functions for both $H'(\bar{\epsilon}^P)$ (Equation 2-109) and $k(\bar{\epsilon}^P)$ (Equation 2-108) to fit the material behavior measured experimentally in axial tests, and thereafter to use these equations to model the material behavior for more complex loading conditions. This procedure has been widely used in metal and soil plasticity. However, it is not adequate in general because of its inherent limitations, namely that the selected functions are never general enough to describe the behavior of a large number of materials but are at best limited to a restricted class of materials. Therefore, in order to further expand the capabilities of the theory, and to allow for the adjustment of the plastic hardening rule to any experimental data, obtained for instance in axial compression/extension tests, a collection of nested yield surfaces [Mroz, 1967] $f^{(1)}, f^{(2)}, ..., f^{(p)}$ with respective sizes $k^{(1)}, k^{(2)}, ..., k^{(p)}$ is introduced. The yield functions, $f^{(m)}$, are then represented by equations of the form (from Equation 2-105)

$$f^{(m)} = \frac{3}{2} \, \mathrm{tr}(\mathbf{s} - \boldsymbol{\alpha}^{(m)})^2 - [k^{(m)}]^2 = 0 \qquad (2\text{-}122)$$

for all m = 1,2, ..., p, in which $\boldsymbol{\alpha}^{(m)}$ represents the coordinates of the center of the yield surface in the deviatoric stress space, and $k^{(m)}$ its size. To each yield function, $f^{(m)}$ is to be associated a constant plastic modulus $H'^{(m)}$; in other words, it is now proposed to identify a given yield function $f^{(m)}$ with a given slope $H^{(m)}$ of the stress-strain curve. Thus, the smooth experimental stress-strain curves are to be approximated by linear segments along which the tangent shear modulus is constant. This is shown in Figure 2-14. Note that the degree of accuracy achieved by this representation of the experimental curves is obviously directly dependent upon the number of linear segments used. Also, note that secant rather than tangent moduli could just as well have been used for this piecewise linear approximation of the stress-strain curves. The identification of the initial positions, sizes, and plastic shear moduli of each yield surfaces is done as previously, by identifying each yield surface with a given slope $H^{(m)}$ measured in both compression and extension loading conditions. This is illustrated in Figure 2-14. It is now assumed for simplicity that the yield surfaces do not change in size (viz., k = 0 in Equation 2-108). Then, upon loading in compression, the stress point moves from O along the σ_1-axis, reaches $f^{(1)}$ at A and translates the circle $f^{(1)}$ along this axis until it comes into contact with the circle $f^{(2)}$ at B. All the other circles remain fixed during this period. Between O and A, the material behavior is elastic, but between A and B, plastic flow occurs and the plastic modulus is $H'^{(1)}$. When the stress point moves from B to C, the circles $f^{(1)}$ and

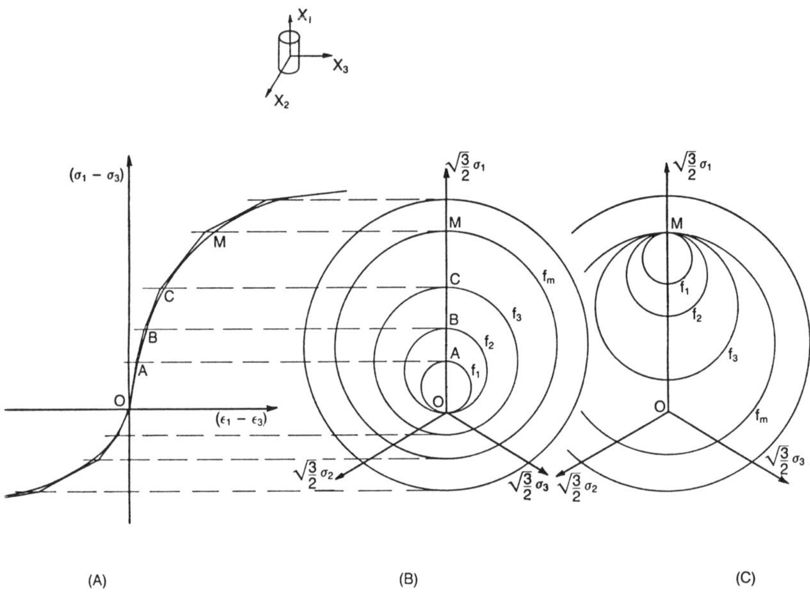

Figure 2-14. Approximation of axial compression and extension stress-strain curves. (A) stress-strain curves; (B) representation in stress space; (C) configuration of field of yield surfaces upon reaching point M in compression.

$f^{(2)}$ are translated together by the stress point until point C is reached where $f^{(1)}$ and $f^{(2)}$ touch $f^{(3)}$, which up to now was at rest, and so on as loading continues. For each increment of loading (from Equation 2-118)

$$\overset{\circ}{q} = H^{(m)} \overset{\circ}{\varepsilon} \qquad H^{(m)} = \left(\frac{1}{2G} + \frac{1}{H'^{(m)}} \right)^{-1} \tag{2-123}$$

where $H'^{(m)}$ is the plastic modulus of the outermost yield surface $f^{(m)}$ currently pushed by the stress point. Figure 2-15 illustrates the situation upon reaching the yield surface $f^{(m)}$ in compression. Similarly, upon loading in extension, the stress point moves along the σ_1-axis but in the opposite direction and translates the surfaces downwards.

Figure 2-15 presents the situation upon reaching point P in compression. Upon loading reversal, the stress point leaves point P, inverse plastic flow occurs, and the stress point translates the surfaces downwards. The reverse loading curve is shown in Figure 2-15.

The model can be further generalized by assuming that the yield surfaces are allowed to change in size as well as to be translated. The

$k^{(m)}(\bar{\epsilon}^P)$ functions are determined by comparing the stress differences corresponding to the segments PA'_1 and OA; $A'_1B'_1$, and AB etc. ... (Figure 2-15) since for the purely kinematic rule as described previously: $PA'_1 = 2k^{(1)}$, ... $PM'_1 = 2k^{(m)}$, etc, ... Experimental deviations from these equalities are attributed to variations in $k^{(m)}(\bar{\epsilon}^P)$. The functions $k^{(m)}(\bar{\epsilon}^P)$ can thus be determined by using cyclic axial test results by tracing the points on the successive hysteresis loops at which the plastic modulus $H'^{(m)}$ is activated.

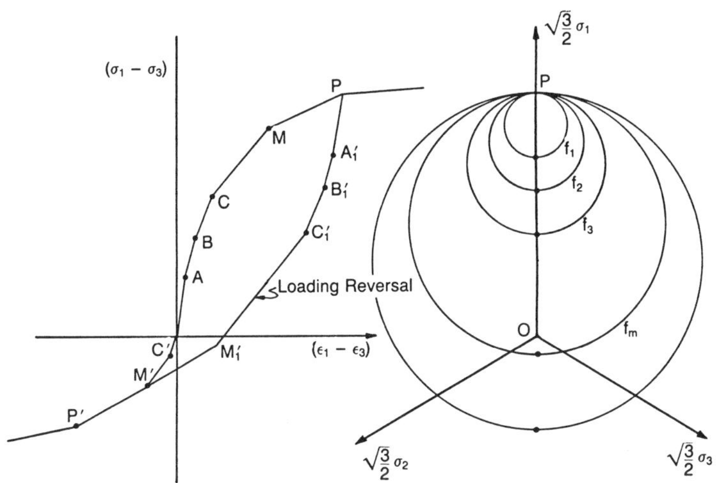

Figure 2-15. Approximation of axial stress-strain curves: loading reversal from P.

When the stress point is on $f^{(m)}$ and plastic loading occurs, the surfaces $f^{(1)}, f^{(2)}, ..., f^{(m)}$ are translated together by the stress point towards $f^{(m+1)}$. It is clear that the plastic stress-strain relations are uniquely defined if and only if the yield surfaces do not overlap (so that the plastic loading function is uniquely defined). Therefore, the surfaces $f^{(1)}, f^{(2)}, ..., f^{(m)}$ must remain tangent to each other at the stress point. Further, contact between $f^{(m)}$ and $f^{(m+1)}$ must only occur at points with the same direction of outwards normal, i.e., at conjugate points. Let R denote the point on $f^{(m+1)}$ at which the normal has the same direction as the M on $f^{(m)}$ as shown in Figure 2-16. Then, the instantaneous translation of $f^{(m)}$ towards $f^{(m+1)}$ must occur in the direction $\boldsymbol{\mu} = \mathbf{MR}$. For the case of von Mises surfaces,

$$\boldsymbol{\mu} = \frac{k^{(m+1)}}{k^{(m)}}\,(\mathbf{s} - \boldsymbol{\alpha}^{(m)}) - (\mathbf{s} - \boldsymbol{\alpha}^{(m+1)}) \qquad (2\text{-}124)$$

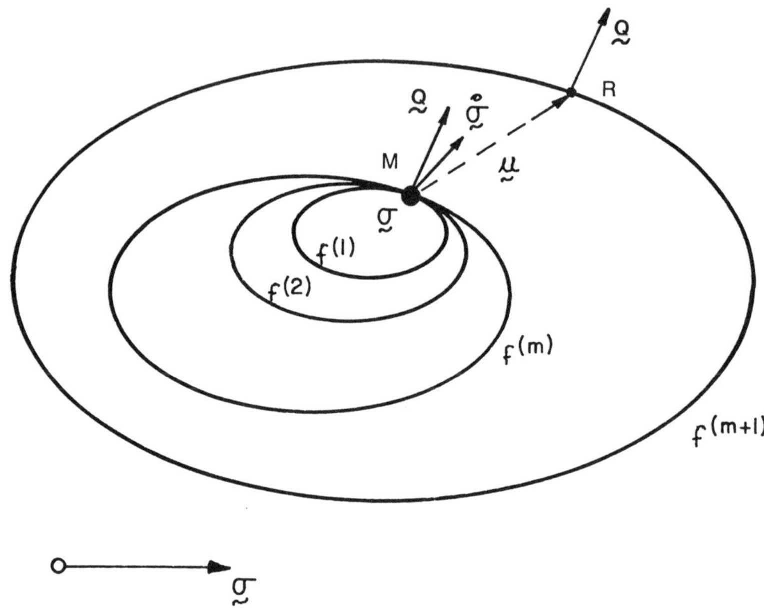

Figure 2-16. Definition of translation direction.

INTEGRATION ALGORITHMS FOR ELASTIC-PLASTIC CONSTITUTIVE RELATIONS

Plasticity theory has recently gained widespread acceptance in large-scale numerical simulations of practical geotechnical engineering problems, due to its extreme versatility and accuracy in modeling real engineering materials behavior. Building upon the pioneering works of Drucker and Prager [1952] on soil plasticity, the modern trend has been toward the development of more and more elaborate and complicated elastoplastic constitutive models which resemble the behavior of real engineering materials more closely.

In finite difference or finite element computer codes the elastoplastic constitutive equations are usually incorporated through a separate set of constitutive subroutines. The purpose of these subroutines is the integration of the elastic-plastic constitutive equations. That is, at every stress point, given a deformation history, the role of the constitutive-equation algorithm is to return the corresponding stress history. Evidently, the accuracy and stability of the global solution is to be strongly affected by the accuracy and stability of the stress-point algorithm. Also, the cost of the analysis is most strongly affected by the efficiency of the stress-point algorithm. The best algorithm, the one to be favored, is therefore the one which combines computational efficiency with accuracy.

Analytical solutions for the elastic-plastic evolution problem are not usually available, and all elastoplastic models are implemented in analysis programs with some error. However, despite the importance of understanding and controlling such errors, as of today the subject has received only little attention in geomechanics [Loret and Prevost, 1986; Ortiz and Simo, 1986]. On the other hand, considerable scrutiny [Krieg and Krieg, 1977; Schreyer et al., 1979; Yoder and Wirley, 1984; Ortiz and Popov, 1985] has recently been given in metal plasticity to the various algorithms commonly used to integrate the von Mises elastic-plastic equations, and for that purpose closed-form exact solutions [Krieg and Krieg, 1977; Yoder and Whirley, 1984] have been obtained for the evolution problem. However, such closed-form solutions are currently available only for the simplest of the traditional geomaterial models, i.e., the Drucker-Prager model [Loret and Prevost, 1986]. Approximate elaborate subincrementation strategies with successive radial stress corrections have been proposed (see e.g., [Nayak and Zienkiewicz, 1972]). Other somewhat arbitrary stress corrections have also been attempted (see e.g., [Chen, 1975; Vermeer, 1980]) to correct for the inherent stress drift away from the yield surface. However, all these procedures tend to be quite expensive and are *not error-free*. Without an exact solution available it is difficult to ascertain the relative merits and/or shortcomings of the various procedures.

In Loret and Prevost [1986], an *exact* solution for the Drucker-Prager elastic-plastic equations with an arbitrary degree of nonassociativity is presented. Although computationally too slow to be used routinely in actual calculations, the solution allows an assessment of the trade-offs between computational speed and accuracy for various methods. Further, for highly accurate solutions, being error-free, the solution would be advantageous to use against elaborate approximate subincrementation strategies.

In this section, stress-point numerical algorithms are developed as the basis for computer modules designed to interface with large-scale finite element/finite difference computer programs for solutions of boundary value problems. One yield surface and multiple-yield surface plasticity theories are considered.

Theory

The main equations of the theory are summarized as follows:

$$\overset{\circ}{\boldsymbol{\sigma}} = \mathbf{E} : (\overset{\circ}{\boldsymbol{\epsilon}} - \overset{\circ}{\boldsymbol{\epsilon}}^{\mathrm{p}}) \qquad \text{constitutive equation} \qquad (2\text{-}125)$$

$$f(\boldsymbol{\sigma},\boldsymbol{\alpha},k) = 0 \qquad \text{yield function} \tag{2-126}$$

$$\overset{\circ}{\boldsymbol{\epsilon}}{}^{p} = \,<\lambda>\mathbf{P} \qquad \text{flow rule} \tag{2-127}$$

$$\overset{\circ}{\boldsymbol{\alpha}} = \,<\lambda>(\beta-1)\,\mathrm{N}\,\frac{\mathrm{H}'}{\dfrac{\partial f}{\partial\boldsymbol{\alpha}}:\boldsymbol{\mu}}\,\boldsymbol{\mu}$$

$$\text{kinematic hardening rule} \tag{2-128}$$

$$\overset{\circ}{\mathrm{k}} = -<\lambda>\beta\,\mathrm{N}\,\frac{\mathrm{H}'}{\dfrac{\partial f}{\partial\mathrm{k}}}$$

$$\text{isotropic hardening rule} \tag{2-129}$$

and (from consistency)

$$\lambda = \frac{1}{\mathrm{H}'}\,\mathbf{Q}:\overset{\circ}{\boldsymbol{\sigma}} = \frac{1}{\mathrm{H}'+\mathrm{H_o}}\,\mathbf{Q}:\mathbf{E}:\overset{\circ}{\boldsymbol{\epsilon}} \tag{2-130}$$

where

$$\mathrm{N} = \left\{\mathrm{tr}\left(\frac{\partial f}{\partial\boldsymbol{\sigma}}\right)^{2}\right\}^{1/2} \tag{2-131}$$

$$\mathbf{Q} = \mathbf{Q}' + \mathrm{Q}''\boldsymbol{\delta} = \frac{1}{\mathrm{N}}\frac{\partial f}{\partial\boldsymbol{\sigma}} \tag{2-132}$$

$$\mathrm{H_o} = \mathbf{Q}:\mathbf{E}:\mathbf{P} = \mathrm{B(trP)(trQ)} + 2\mathrm{G}\mathbf{P}':\mathbf{Q}' \tag{2-133}$$

and, B, G = elastic bulk and shear moduli, respectively. For plastic flow to occur, the yield function Equation 2-126 must be satisfied and λ (Equation 2-130) must be positive.

Algorithms

Since the elastoplastic process is nonlinear, the loads on the elastoplastic solid are to be applied step-by-step. The problem to be addressed is therefore stated as follows: Given the known state variables: $(\boldsymbol{\sigma}_n, \boldsymbol{\alpha}_n, k_n)$

and (ϵ_n, ϵ_n^p) associated with a converged configuration at step n, and a new configuration at step (n + 1), via ϵ_{n+1}, find the new values of (σ_{n+1}, α_{n+1}, k_{n+1}) and ϵ_{n+1}^p at step (n + 1). In the process, the incremental strains $\Delta\epsilon = (\epsilon_{n+1} - \epsilon_n)$ defining the state update are assumed given. Further, it is assumed that the loading strain rate $\overset{\circ}{\epsilon}$ is constant over the time interval, viz.,

$$\Delta\epsilon = \Delta t\, \overset{\circ}{\epsilon} \tag{2-134}$$

where Δt = time step ($= t_{n+1} - t_n$).

As summarized by Ortiz and Popov [1985] an acceptable algorithm for the integration of Equations 2-125–2-128 should satisfy the following three basic requirements:

1. Consistency with the constitutive relations to be integrated (i.e., first-order accuracy).
2. Numerical stability.
3. Incremental plastic consistency.

Conditions 1 and 2 are necessary for attaining convergence of the numerical solution as the time step becomes vanishingly small. Condition 3 is the algorithmic counterpart of the plastic consistency condition which requires the yield function to be satisfied by the updated state variables.

Integration of Equations 2-125–2-129 is achieved by a stress relaxation procedure by which Equation 2-125 is first used to obtain an elastic predictor, hereafter referred to as the trial stress σ_{n+1}^{tr}, viz.,

$$\sigma_{n+1}^{tr} = \sigma_n + \mathbf{E} : \Delta\epsilon \tag{2-135}$$

which is then taken as an initial condition for the plastic relaxation:

$$\sigma_{n+1} = \sigma_{n+1}^{tr} - \Delta\sigma^p \tag{2-136}$$

where (from Equations 2-125 and 2-126)

$$\Delta\sigma^p = \int_0^{\Delta t} \mathbf{E} : \overset{\circ}{\epsilon}^p \, dt = \int_0^{\Delta t} \lambda \mathbf{E} : \mathbf{P} \, dt \tag{2-137}$$

The resulting procedure is shown schematically in Figure 2-17. In general, the return path defined by $\mathbf{P}$ in Equation 2-137 is not known in advance nor can it be determined analytically, and it is therefore necessary to integrate Equation 2-135 numerically. Several algorithms can be used as follows:

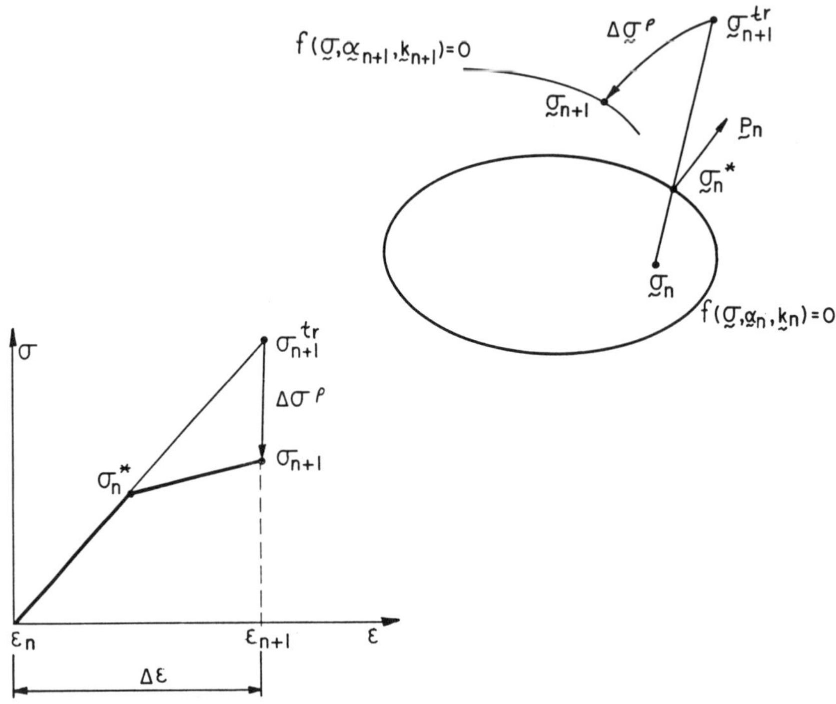

Figure 2-17. Schematic of elastic-plastic stress relaxation procedure.

Return Mapping Algorithm. An efficient procedure for performing the return mapping was proposed by Simo and Ortiz [1985] which can be defined as follows. The relaxation process for the stresses and the update of the plastic state variables are performed in an iterative fashion. At every iteration the yield function f is linearized around the current values of the state variables, $(\sigma_{n+1}^{(i)}, \alpha_{n+1}^{(i)}, k_{n+1}^{(i)})$ where i denotes the iteration counter, to obtain the following:

$$f(\sigma_{n+1},\alpha_{n+1},k_{n+1}) \approx f(\sigma_{n+1}^{(i)},\alpha_{n+1}^{(i)},k_{n+1}^{(i)}) +$$

$$\left.\frac{\partial f}{\partial \sigma}\right|_{n+1}^{(i)} : (\sigma_{n+1} - \sigma_{n+1}^{(i)}) +$$

$$\left.\frac{\partial f}{\partial \alpha}\right|_{n+1}^{(i)} : (\alpha_{n+1} - \alpha_{n+1}^{(i)}) +$$

$$\left.\frac{\partial f}{\partial k}\right|_{n+1}^{(i)} \cdot (k_{n+1} - k_{n+1}^{(i)}) = 0 \qquad (2\text{-}138)$$

where the derivatives are evaluated at $(\sigma_{n+1}^{(i)}, \alpha_{n+1}^{(i)}, k_{n+1}^{(i)})$. Furthermore, the plastic relaxation equations (Equations 2-125, 2-126–2-128) can be discretized as follows:

$$(\sigma_{n+1}^{(i+1)} - \sigma_{n+1}^{(i)}) = -\mathbf{E} : [(\epsilon^p)_{n+1}^{(i+1)} - (\epsilon^p)_{n+1}^{(i)}]$$

$$= -\tilde{\lambda}\,\mathbf{E} : \mathbf{P}_{n+1}^{(i)} \tag{2-139}$$

$$(\alpha_{n+1}^{(i+1)} - \alpha_{n+1}^{(i)}) = \tilde{\lambda}\,(\beta - 1)\,\mathbf{N}_{n+1}^{(i)}\;\frac{{H'}_{n+1}^{(i)}}{\left.\dfrac{\partial f}{\partial \alpha}\right|_{n+1}^{(i)} : \mu_{n+1}^{(i)}}\;\mu_{n+1}^{(i)} \tag{2-140}$$

$$k_{n+1}^{(i+1)} - k_{n+1}^{(i)} = -\tilde{\lambda}\,\beta\,\mathbf{N}_{n+1}^{(i)}\;\frac{{H'}_{n+1}^{(i)}}{\left.\dfrac{\partial f}{\partial k}\right|_{n+1}^{(i)}} \tag{2-141}$$

where $\tilde{\lambda}$ is the incremental plastic loading index. Combining Equations 2-138–2-141, the value of $\tilde{\lambda}$ can be obtained as follows:

$$\tilde{\lambda} = \frac{f_{n+1}^{(i)}}{\mathbf{N}_{n+1}^{(i)}[H_o + H']_{n+1}^{(i)}} \tag{2-142}$$

where

$$\left. H_o \right|_{n+1}^{(i)} = \mathbf{Q}_{n+1}^{(i)} : \mathbf{E} : \mathbf{P}_{n+1}^{(i)} \tag{2-143}$$

Combining Equations 2-139 and 2-142 finally yields the updated plastic state as

$$\sigma_{n+1}^{(i+1)} = \sigma_{n+1}^{(i)} - \tilde{\lambda}\,\mathbf{E} : \mathbf{P}_{n+1}^{(i)} \tag{2-144}$$

The resulting procedure is illustrated in Figure 2-18.

The iterations continue until plastic consistency is restored to within a prescribed tolerance, viz., $f_{n+1}^{(i+1)} \le \mathrm{TOL}\, f_{n+1}^{(0)}$ with $\mathrm{TOL} << 1$.

The algorithm is consistent, conditionally stable, and achieves a quadratic convergence rate for the update.

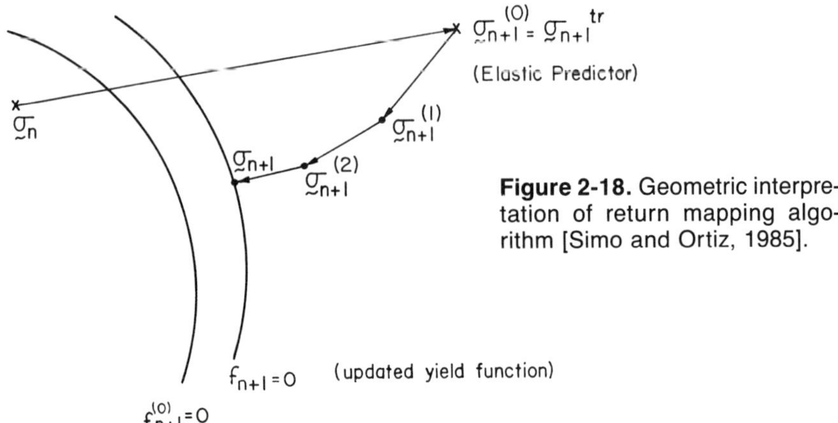

Figure 2-18. Geometric interpretation of return mapping algorithm [Simo and Ortiz, 1985].

Explicit (One-Step) Algorithm. A simpler (less expensive) algorithm can be obtained by performing an explicit forward (Euler) integration of the evolution equations [Nguyen, 1977], as follows:

$$\sigma_{n+1} = \sigma_{n+1}^{tr} - \Delta\sigma^p \tag{2-145}$$

with

$$\Delta\sigma^p = \mathbf{E} : \Delta\epsilon^p = \widetilde{\lambda}\, \mathbf{E} : \mathbf{P}_n \tag{2-146}$$

$$\alpha_{n+1} = \alpha_n + \widetilde{\lambda}\,(\beta - 1)H'_n \,\frac{N_n}{\left.\dfrac{\partial f}{\partial \alpha}\right|_n : \mu_n}\,\mu_n \tag{2-147}$$

$$k_{n+1} = k_n - \widetilde{\lambda}\,\beta\, H'_n \,\frac{N_n}{\left.\dfrac{\partial f}{\partial k}\right|_n} \tag{2-148}$$

where the trial stress is defined as previously (Equation 2-135). The incremental plastic loading index $\widetilde{\lambda}$ is to be obtained by requiring that Equation 2-126 holds.

A geometric interpretation of the algorithm is shown in Figure 2-19. The stress σ_{n+1}^{tr} is to be relaxed onto the updated yield surface along the initial plastic flow direction $\mathbf{P}_n$. If the stress point is initially inside the yield surface, viz., $f(\sigma_n, \alpha_n, k_n) < 0$, the elastic stress trajectory meets the yield surface at σ_n^c after which plastic flow takes place. Finding σ_n^c is not in general straightforward (except for simple yield functions), and the contact point at which the state variables are to be evaluated, is usually approximated by $\boldsymbol{\sigma}_n^*$ as shown in Figure 2-19. For instance, for the case of von Mises plastic models, $\boldsymbol{\sigma}_n^c$ is usually approximated by the intersection of the radial line $(\boldsymbol{\sigma}_{n+1}^{tr} - \boldsymbol{\alpha}_n)$ with the von Mises circle.

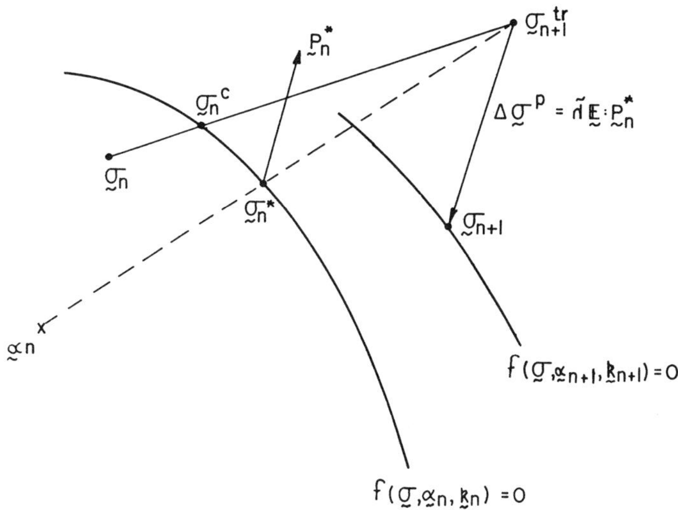

Figure 2-19. Geometric interpretation of explicit one-step algorithm.

The incremental plastic loading index $\tilde{\lambda}$ is obtained by linearizing the yield function as follows:

$$f(\sigma_{n+1},\alpha_{n+1},k_{n+1}) \approx f(\sigma_n^*,\alpha_n,k_n)$$

$$+ \left.\frac{\partial f}{\partial\sigma}\right|_n^* : (\sigma_{n+1} - \sigma_n^*)$$

$$+ \left.\frac{\partial f}{\partial\alpha}\right|_n^* : (\sigma_{n+1} - \alpha_n)$$

$$+ \left.\frac{\partial f}{\partial k}\right|_n^* : (k_{n+1} - k_n) \tag{2-149}$$

where the derivatives are to be calculated at the contact point (or its approximation). Now from Equations 2-145 and 2-186

$$(\sigma_{n+1} - \sigma_n^*) = (\sigma_{n+1}^{tr} - \sigma_n^*) - \tilde{\lambda}\,\mathbf{E}:\mathbf{P}_n \tag{2-150}$$

and combining Equations 2-147–2-150, the value of $\tilde{\lambda}$ can be obtained as follows:

$$\tilde{\lambda} = \frac{\mathbf{Q}_n^*:(\sigma_{n+1}^{\mathrm{tr}} - \sigma_n^*)}{[H_o + H']_n} \tag{2-151}$$

where

$$H_o\,|_n = \mathbf{Q}_n^*:\mathbf{E}:\mathbf{P}_n^* \tag{2-152}$$

The algorithm is consistent, first-order accurate and conceptually only conditionally stable. However, no stability difficulties have ever been encountered by the writer in practical situations.

Multi-Yield-Surface Plasticity Case. Within the context of multi-yield surface plasticity models, it may occur that after completion of the preceding calculations, it is found that the stress point lies outside the next larger yield surface. When this is the case, the value of $\sigma_{n+1}^{\mathrm{tr}}$ is reinitialized to σ_{n+1}. Calculations at this point begin again with respect to the next yield surface. However, some small, but crucial modifications must be made to take into account that $\sigma_{n+1}^{\mathrm{tr}}$ is no longer an elastic trial stress.

To see the origin of the necessary changes, assume that a stress relaxation was just performed with respect to the yield surface $f^{(m-1)}$, and it is now found that the resulting stress point is outside surface $f^{(m)}$. This is illustrated in Figure 2-20. In this case, the tensor $\mathbf{E}$ in Equation 2-146 must be replaced by the elastic-plastic tensor associated with surface $(m - 1)$, viz., by

$$\mathbf{E} - \frac{1}{H'^{(m-1)} + H_o^{(m-1)}} (\mathbf{E}:\mathbf{P}^{(m-1)})(\mathbf{Q}^{(m-1)}:\mathbf{E}) \tag{2-153}$$

where $\mathbf{P}^{(m-1)}$ and $\mathbf{Q}^{(m-1)}$ are to be evaluated at the contact point σ_n^* where the surface $f^{(m-1)}$ engaged the surface $f^{(m)}$. This point is usually approximated by, for instance, the intersection of the radial line $(\sigma_{n+1}^{\mathrm{tr}} - \alpha_n^{(m)})$ with the yield surface $f^{(m)}$ as shown in Figure 2-19. Note that $\mathbf{Q}^{(m-1)} = \mathbf{Q}^{(m)}$ at the contact point since $f^{(m-1)}$ and $f^{(m)}$ are to be tangent to each other upon contact.

If Equation 2-137 is used in place of $\mathbf{E}$, then for the explicit algorithm $\tilde{\lambda}$ becomes

$$\tilde{\lambda} = \frac{1}{H'^{(m)} + H_o^{(m)}} \frac{H'^{(m-1)} + H_o^{(m-1)}}{H'^{(m-1)}} \mathbf{Q}^{(m)}:(\sigma_{n+1}^{\mathrm{tr}} - \sigma_n^*) \tag{2-154}$$

and the plastic correction becomes

$$\Delta\sigma^P = \widetilde{\lambda}\,\mathbf{E}:\left(\mathbf{P}^{(m)} - \frac{H'^{(m)} + H_o^{(m)}}{H'^{(m-1)} + H_o^{(m-1)}}\,\mathbf{P}^{(m-1)}\right) \tag{2-155}$$

where the subscript n has been omitted to simplify the notation. It should be noted from the previous developments that calculations for subsequent yield surface within a step are essentially the same as for the surface first loaded within the step. This simplifies the implementation considerably.

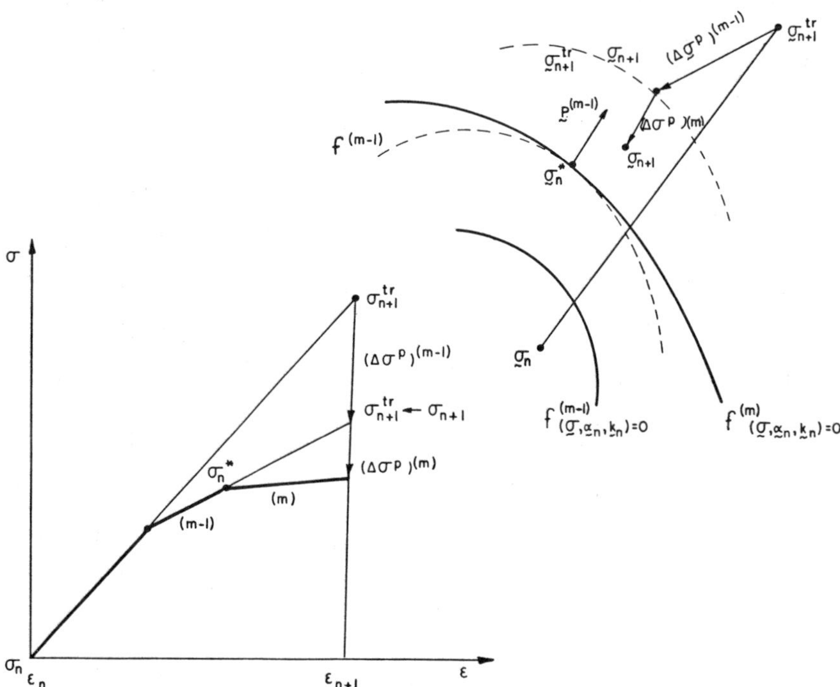

Figure 2-20. Multisurface plasticity case—schematic of correction.

Applications/Examples. Consider the simple case of yield functions of the type

$$f(\boldsymbol{\sigma},\boldsymbol{\alpha},k) = F(\boldsymbol{\sigma} - \boldsymbol{\alpha}) - k^2 = 0 \tag{2-156}$$

where F is a homogeneous function of its argument $(\boldsymbol{\sigma} - \boldsymbol{\alpha})$ and of degree two. Let NYS denote the number of yield functions used (NYS $\geq$ 1).

Then to avoid overlapping of the surfaces, the translation direction $\boldsymbol{\mu}$ is given by

$$\boldsymbol{\mu} = \frac{k^{(m+1)}}{k^{(m)}} (\boldsymbol{\sigma}^* - \boldsymbol{\alpha}^{(m)}) - (\boldsymbol{\sigma}^* - \boldsymbol{\alpha}^{(m+1)}) \tag{2-157}$$

when surface $f^{(m)}$ translates towards surface $f^{(m+1)}$. On the last surface $m = NYS$ and usually $H' = 0$, and no motion is to take place (the last surface then plays the role of a failure surface). Otherwise, $\boldsymbol{\mu} = \mathbf{Q}$ on the last surface. For an explicit one-step integration the flow of calculations proceeds as follows:

1. Initialize: $m = 0$
2. Elastic predictor:

$$\boldsymbol{\sigma}^{tr}_{n+1} = \boldsymbol{\sigma}_n + \mathbf{E} : \Delta\boldsymbol{\epsilon}$$

3. Check for yielding:

If $[m \geqslant NYS]$ GOTO 8

Otherwise: $\boldsymbol{\zeta} = (\boldsymbol{\sigma}^{tr}_{n+1} - \boldsymbol{\alpha}_n^{(m+1)})$ $k^2 = F(\boldsymbol{\zeta})$

If $[k \leqslant k_n^{(m+1)}]$ GOTO 8

Otherwise: $m \leftarrow (m+1)$

4. Compute contact point:

$$\boldsymbol{\sigma}_n^* = \frac{k_n^{(m)}}{k} \boldsymbol{\zeta} + \boldsymbol{\alpha}_n^{(m)}$$

5. Adjust $f^{(m-1)}$

If $[m = 1]$ GOTO 6

Otherwise: $\boldsymbol{\alpha}_n^{(m-1)} = \boldsymbol{\sigma}_n^* - \frac{k_n^{(m-1)}}{k_n^{(m)}} (\boldsymbol{\sigma}_n^* - \boldsymbol{\alpha}_n^{(m)})$

6. Compute stress correction

$$\widetilde{\lambda} = Q_n^{(m)}:(\sigma_{n+1}^{tr} - \sigma_n^*)/(H'^{(m)} + H_o^{(m)})$$

If $[m = 1]$: $\Delta\sigma^p = \widetilde{\lambda}\, E : P_n^{(m)}$

If $[m > 1]$: $\widetilde{\lambda} = \widetilde{\lambda}\,(H'^{(m-1)} + H_o^{(m-1)})/H'^{(m-1)}$

and: $\Delta\sigma^p = \widetilde{\lambda}\, E : \left[P_n^{(m)} - P_n^{(m-1)}\, \dfrac{H'^{(m)} + H_o^{(m)}}{H'^{(m-1)} + H_o^{(m-1)}} \right]$

7. Update:

$$\sigma_{n+1} = \sigma_{n+1}^{tr} - \Delta\sigma^p$$

$$k_{n+1}^{(i)} = k_n^{(i)} + \widetilde{\lambda}\,\beta H'^{(m)} N_n^{(m)}/2k_n^{(m)} \qquad\qquad (i = 1,\ldots,m)$$

$$\sigma_{n+1}^{tr} \leftarrow \sigma_{n+1}; \quad \text{GOTO 3}$$

8. Final updates:

$$\sigma_{n+1} \leftarrow \sigma_{n+1}^{tr}; \quad \text{If } [m = 0] \text{ Exit}$$

Otherwise compute translations:

a. If $[m = NYS]$:

$$\alpha_{n+1}^{(m)} = \alpha_n^{(m)} + \widetilde{\lambda}\,(1 - \beta)H'^{(m)}Q_n^{(m)}$$

b. If $[m < NYS]$:

$$\mu = \frac{k_n^{(m+1)}}{k_n^{(m)}}\,(\sigma_n^* - \alpha_n^{(m)}) - (\sigma_n^* - \alpha_n^{(m+1)})$$

$$\alpha_{n+1}^{(m)} = \alpha_n^{(m)} + \widetilde{\lambda}\,(1 - \beta)H'^{(m)}\mu/(Q_n^{(m)} : \mu)$$

c. Adjust inner surfaces:

If $[m > 1]$: $\alpha_{n+1}^{(i)} = \sigma_{n+1} - \dfrac{k_{n+1}^{(i)}}{k_{n+1}^{(m)}}\,(\sigma_{n+1} - \alpha_{n+1}^{(m)})$

$$(i = 1,\ldots,(m-1))$$

Exit

A SIMPLE PLASTICITY MODEL FOR SOILS

Considerable attention has been given in the past decade to the development of constitutive equations for soil media, but although many different models have been proposed, there is not yet firm agreement among researchers. Elastic (see, e.g., [Duncan and Chang, 1970; Coon and Evans; 1971], endochronic (see, e.g., [Valanis and Read, 1982]), and many elastic-plastic models with various degrees of sophistication and/or complexity have been proposed. Elastic-plastic models appear to be the most promising.

The most popular and most widely used soil models are Cap models [Roscoe and Burland, 1968; Schofield and Wroth, 1968; DiMaggio and Sandler, 1971; Baladi and Rohani, 1979] based on classical isotropic plasticity theory with associated flow, and are variations and refinements of the basic Cap model pioneered by Drucker, Gibson and Henkel [1955]. The most obvious limitations of these Cap models are:

1. They do not adequately model soil stress-induced anisotropy.
2. They are not applicable to cyclic loading conditions.

Similar limitations apply to the models presented in Lade and Duncan [1975] and Nemat-Nasser [1982]. It may be argued that plastic models based on isotropic plastic hardening rules are adequate for situations in which only loading (and moderate unloading) occurs, however, it is unlikely that such restrictions can be met at every point in general boundary value problems. In order to account for hysteretic effects, more elaborate plastic models based on a combination of isotropic and kinematic plastic hardening rules have recently been proposed. Some researchers prefer a two-yield surface plasticity (see e.g., [Ghaboussi and Momen, 1982; Mroz and Pietruszcak, 1983]), while others prefer a multi-yield surface plasticity (see e.g., [Prevost, 1977; 1978; 1985]). Both theories suffer inherent limitations namely: storage requirements for the multisurface theory, *a priori* selection of an evolution law for the two-surface theory. However, neither limitation is more degrading than the other. This is further discussed in Prevost [1982].

It is the purpose of this section to present a simple plasticity model for soils. The model is applicable to both cohesive and cohesionless soils. The model has been tailored:

1. To retain the extreme versatility and accuracy of the simple multisurface J_2-theory (see e.g., [Prevost, 1977; 1978]) in describing observed shear nonlinear hysteretic behavior, and shear stress-induced anisotropic effects.

2. To reflect the strong dependency of the shear dilatancy on the effective stress ratio in both cohesionless [Rowe, 1962, Luong, 1980; Luong

and Touati, 1983] and cohesive [Hicher, 1985] soils. Conical yield surfaces are used for that purpose.

The theory is applicable to general three-dimensional stress-strain conditions, but its parameters can be derived entirely from the results of conventional triaxial soil tests. The model versatility is illustrated by a number of examples.

In the following, all stresses are *effective stresses.*

Basic Theory

Yield Function. The yield function is selected from the following form:

$$f = \frac{3}{2}\,(s - \bar{p}\alpha):(s - \bar{p}\alpha) - M^2\,\bar{p}^2 = 0 \qquad (2\text{-}158)$$

where $s = \sigma - p\delta$ = deviatoric stress tensor
  $p = \frac{1}{3}\,\mathrm{tr}\,\sigma$ = effective mean normal stress
  $\bar{p} = p - a$ with a = attraction (= material parameter = $c/\tan\varphi$)
  c = cohesion
  φ = mobilized friction angle
  α = kinematic deviatoric tensor defining the coordinates of the yield surface center in deviatoric stress subspace
  M = material parameter.

The outer normal to the yield surface $\mathbf{Q}$ is normalized in the following:

$$\mathbf{Q} = \mathrm{grad}\,f/|\,\mathrm{grad}\,f| = \mathbf{Q}' + Q''\delta \qquad (2\text{-}159)$$

with (from Equation 2-158)

$$\mathrm{grad}\,f = \frac{\partial f}{\partial\sigma} = 3(s - \bar{p}\alpha) + \left[\bar{p}\left(\alpha:\alpha - \frac{2}{3}\,M^2\right) - s:\alpha\right]\delta \qquad (2\text{-}160)$$

so that

$$\mathbf{Q}:\mathbf{Q} = \mathbf{Q}':\mathbf{Q}' + \frac{1}{3}\,(3Q'')^2 = 1 \qquad (2\text{-}161)$$

The yield function plots as a conical yield surface in stress space with its apex located along the hydrostatic axis at the attraction. For cohesionless soils $a = 0$ and the apex of the cone is at the origin. Unless $\alpha = \mathbf{0}$, the axis of the cone does not coincide with the space diagonal. The cross section of the yield surface by any deviatoric plane ($\bar{p}$ = constant) is circular.

Its center does not generally coincide with the origin but is shifted by the amount $\bar{p}\alpha$. This is illustrated by Figure 2-21 in the principal stress space.

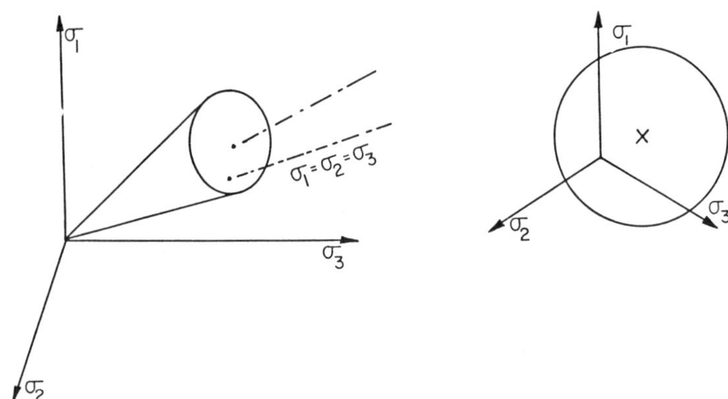

Figure 2-21. Yield surface in principal stress space (case a = 0).

Remark. If deemed necessary, the model may be easily modified to accommodate a non-circular cross section. For instance, a round cornered hexagonal cross section of the Mohr-Coulomb type can be obtained by using

$$\bar{f} = \frac{3}{2}\,(s - \bar{p}\alpha):(s - \bar{p}\alpha) - R^2(\theta)M^2\,\bar{p}^2 = 0 \tag{2-162}$$

in which (see, for example, [Argyris et al., 1973])

$$R(\theta) = \frac{2k}{(1+k)-(1-k)\sin 3\theta} \tag{2-163}$$

k = material parameter, and

$$\sin 3\theta = -\sqrt{6}\,\bar{J}_3/(\bar{J}_2)^{3/2} \tag{2-164}$$

$$\bar{J}_2 = \bar{s}:\bar{s} = \mathrm{tr}(\bar{s})^2 \tag{2-165}$$

$$\bar{J}_3 = 3\,\det(\bar{s}) = \mathrm{tr}(\bar{s})^3 \tag{2-166}$$

$$\bar{s} = 3(s - \bar{p}\,\alpha) \tag{2-167}$$

In the following $k = 1$.

Flow Rule. The plastic potential is selected such that the deviatoric plastic flow be associative. However, a non-associative flow rule is used for its dilatational component, and in the following:

$$\mathbf{P}' = \mathbf{Q}' \qquad 3P'' = \frac{(\eta/\bar{\eta})^2 - 1}{(\eta/\bar{\eta})^2 + 1} \qquad (2\text{-}168)$$

where

$$\eta = \left(\frac{3}{2}\,\mathbf{s}:\mathbf{s}\right)^{1/2}/\bar{p} = \text{mobilized stress ratio and } \bar{\eta} = \text{material parameter.}$$

When $\eta < \bar{\eta}$, $3P'' < 0$ and plastic compaction takes place, whereas when $\eta > \bar{\eta}$, $3P'' > 0$ and plastic dilation takes place. The case $\eta = \bar{\eta}$ corresponds to no plastic volumetric strains. This is illustrated in Figure 2-22. In the following, $\bar{\eta} = \eta_C$ when tr $\mathbf{s}^3 < 0$, and $\bar{\eta} = \bar{\eta}_E$ when tr $\mathbf{s}^3 > 0$.

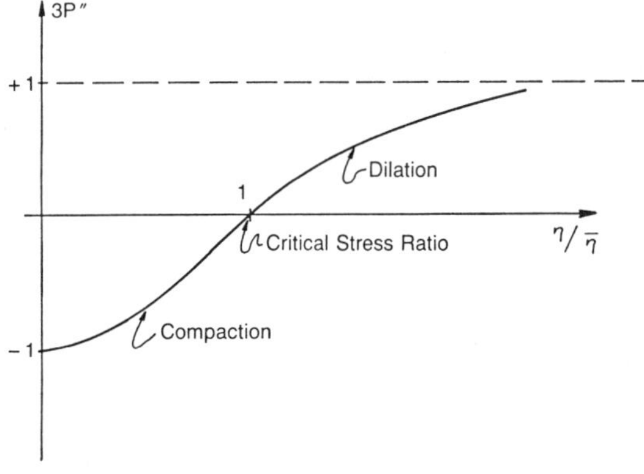

Figure 2-22. Dilatational plastic flow.

Hardening Rule: A purely deviatoric kinematic hardening rule is adopted and in the following,

$$\bar{p}\overset{\circ}{\boldsymbol{\alpha}} = A\boldsymbol{\mu} \qquad (2\text{-}169)$$

where $\boldsymbol{\mu}$ = (deviatoric) tensor defining the direction of translation; A = amount of translation determined through the consistency condition which emanates from time differentiation of Equation 2-158, viz.,

$$\overset{\circ}{f} = \mathbf{Q}:\overset{\circ}{\boldsymbol{\sigma}} - \bar{p}\mathbf{Q}':\overset{\circ}{\boldsymbol{\alpha}} = 0 \qquad (2\text{-}170)$$

from which

$$A = \frac{H'}{Q' : \mu} <\lambda> \qquad (2\text{-}171)$$

where λ is the plastic loading index, viz.

$$\lambda = \frac{1}{H'} \, Q : \overset{\circ}{\sigma} = \frac{1}{H' + H_o} \, Q : E : \overset{\circ}{\epsilon} \qquad (2\text{-}172)$$

with

$$H_o = Q : E : P = B(3P'')(3Q'') + 2GQ' : Q' \qquad (2\text{-}173)$$

and finally,

$$\overline{p}\overset{\circ}{\alpha} = \frac{H'}{Q' : \mu} <\lambda> \mu \qquad (2\text{-}174)$$

Note that the *direction* of translation remains arbitrary at this stage, and thus may be selected independently of any formal plasticity constraints.

In order to allow for the adjustment of the plastic hardening rule to any kind of experimental data, for example, data obtained from axial or simple shear soil tests, a collection of nested yield surfaces [Mroz, 1967] is used. The yield surfaces are all similar conical surfaces. Upon contact, the yield surfaces are to be translated by the stress point. In order to avoid overlappings of the surfaces (which would lead to a non-unique definition of the constitutive theory), the direction of translation μ of the active yield surface is selected such that

$$\mu = \frac{M'}{M} \, (s - \overline{p}\alpha) - (s - \overline{p}\alpha') \qquad (2\text{-}175)$$

where M' and α' are the plastic parameters associated with the next outer surface ($M' > M$). This is illustrated in Figure 2-23.

Remarks.

1. Under the assumptions just given, the elastic-plastic relations write in expanded form as

$$\overset{\circ}{\sigma} = 2G\overset{\circ}{\epsilon} + \left(B - \frac{2G}{3}\right) \overset{\circ}{\epsilon}_v \, \delta$$

$$-(2G \, Q' + B \, 3P''\delta) <\lambda> \qquad (2\text{-}176)$$

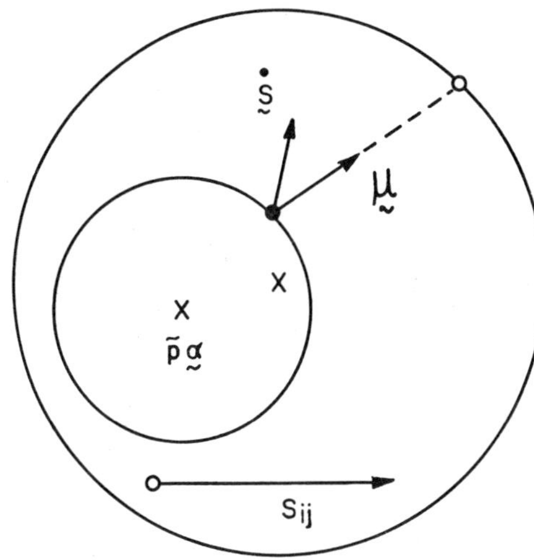

Figure 2-23. Yield surface translation by the stress point in deviatoric stress space.

with

$$\lambda = \frac{1}{H' + H_o}\,(2G\,\mathbf{Q}':\overset{\circ}{\boldsymbol{\epsilon}} + B\,3Q''\overset{\circ}{\epsilon_v}) \tag{2-177}$$

where $\overset{\circ}{\epsilon_v} = \mathrm{tr}\,\overset{\circ}{\boldsymbol{\epsilon}}$. Or equivalently, in terms of deviatoric and dilatational components

$$\overset{\circ}{s} = 2G\,\overset{\circ}{e} - 2G\,\mathbf{Q}'\,<\lambda> \tag{2-178a}$$

$$\overset{\circ}{p} = B\epsilon_v - B\,3P''\,<\lambda> \tag{2-178b}$$

where $\overset{\circ}{e} = \overset{\circ}{\boldsymbol{\epsilon}} - \dfrac{1}{3}\,\overset{\circ}{\epsilon_v}\,\boldsymbol{\delta} = $ deviatoric rate of deformation tensor.

2. It is assumed that no pure elastic domain exists. The first yield surface is thus chosen as a degenerate yield surface of size zero which coincides with the stress point. The normal associated with that yield surface is assumed to be purely dilatational (i.e., from Equation 2-161, $\mathbf{Q}' = \mathbf{0}$ and $3Q'' = -\sqrt{3}$). The plastic loading function associated with the stress point is defined through

$$\lambda = \frac{1}{H' + 3B}\,B\,3Q''\overset{\circ}{\epsilon_v} \tag{2-179}$$

3. The dependence of the model moduli upon the effective mean normal stress is assumed of the following form

$$G = G_1 \left(\frac{p}{p_1}\right)^n ; B = B_1 \left(\frac{p}{p_1}\right)^n ; H' = H_1' \left(\frac{p}{p_1}\right)^n \qquad (2\text{-}180)$$

respectively, where n = experimental parameter (n = 0.5 for most cohesionless soils, and n = 1 for cohesive soils [Richard et al., 1970]); p_1 = reference effective mean normal stress.

4. The dependence of the plastic modulus associated with any given yield surface upon the deviatoric stress is assumed of the following form

$$H' = \frac{H_C' - H_E'}{2}\,\theta + \frac{H_c' + H_E'}{2} \qquad (2\text{-}181)$$

where

$$\theta = \frac{1}{M\bar{p}}\frac{\text{tr }\bar{s}^3}{\text{tr }\bar{s}^2} \qquad \bar{s} = 3(s - \bar{p}\,\alpha) \qquad (2\text{-}182)$$

where H'_C and H'_E are material parameters. On the last outermost surface $H' = 0$, and the last surface therefore plays the role of a failure surface.

Application to the "Triaxial" Stress State/Parameters Identification

In this section, attention is restricted to the "triaxial" soil test for which the two effective (lateral) principal stresses are equal, $\sigma_2 = \sigma_3$. In order for the soil specimen to deform in an axisymmetric fashion ($\epsilon_2 = \epsilon_3$), the axes of loading must coincide with the principal axes of the anisotropic tensor α, and $\alpha_2 = \alpha_3$. In the following, in order to follow common usage in soil mechanics, compressive stresses and strains are counted as positive and the discussion is presented in terms of the following stress and strain variables;

$$q = (\sigma_1 - \sigma_3) \qquad\qquad p = (\sigma_1 + 2\sigma_3)/3 \qquad (2\text{-}183)$$

$$\bar{\epsilon} = (\epsilon_1 - \epsilon_3) \qquad \epsilon_v = \epsilon_1 + 2\epsilon_3$$

Equation 2-158 then simplifies to:

$$f = (q - \alpha\bar{p})^2 - M^2\,\bar{p}^2 = 0 \qquad (2\text{-}184)$$

where $\alpha = (\alpha_1 - \alpha_3) = 3\,\alpha_1/2$. The trace of the yield surface onto the triaxial (q,p) stress plane consists of two straight lines of slopes $(\alpha + M)$ and $(\alpha - M)$, respectively. This is illustrated in Figure 2-24.

When upon loading in compression, the stress point reaches the yield surface f,

$$\eta_C = \left(\frac{q}{p}\right)_C = \alpha + M \tag{2-185a}$$

whereas upon unloading in extension

$$\eta_E = \left(\frac{q}{p}\right)_E = \alpha - M \tag{2-185b}$$

and (from Equation 2-178)

$$\frac{\overset{\circ}{\epsilon}}{\overset{\circ}{q}} = \frac{1}{2G} + \frac{1}{H'}\frac{1 - \eta\overset{\circ}{p}/\overset{\circ}{q}}{1 + \dfrac{2}{9}\eta^2} \tag{2-186}$$

$$\frac{\overset{\circ}{\epsilon}_v}{\overset{\circ}{p}} = \frac{1}{B} \pm \frac{1}{H'}\frac{2}{\sqrt{9}}\frac{1 - (\eta/\bar{\eta})^2}{1 + (\eta/\bar{\eta})^2}\frac{\overset{\circ}{q}/\overset{\circ}{p} - \eta}{\left(1 + \dfrac{2}{\sqrt{9}}\eta^2\right)^{1/2}} \tag{2-187}$$

where $\eta = \eta_C$; $\bar{\eta} = \bar{\eta}_C$ and $H' = H'_C$ in compression; $\eta = \eta_E$, $\bar{\eta} = \bar{\eta}_E$ and $H' = H'_E$ in extension.

Given the experimental stress-strain curves obtained in a triaxial shear soil test, identification of the model parameters associated with any given yield surface f proceeds as follows. The smooth experimental *shear* stress-strain curves are approximated by linear segments along which the tangent (or secant) modulus is constant. (Evidently, the degree of accuracy achieved by such a representation of the experimental curves is directly dependent upon the number of linear segments used.) The yield surface f is identified by the condition that the slopes $\overset{\circ}{q}/\overset{\circ}{\epsilon}$ be the same in compression and extension, respectively, when the stress point has reached that yield surface. This is illustrated in Figure 2-24. The model parameters: α, M, H'_C, H'_E, $\bar{\eta}_C$ and $\bar{\eta}_E$ are then simply computed from Equations 2-185–2-187, where $\overset{\circ}{q}/\overset{\circ}{p}$ = slope of the effective stress path (given by the specific stress path followed in the test). Therefore, once the elastic shear G and bulk B moduli are known, the identification procedure is straightforward and can easily be automated.

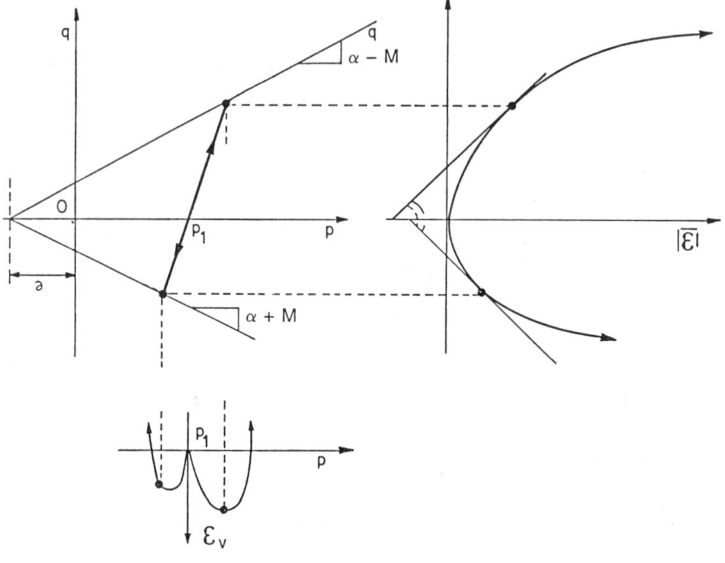

Figure 2-24. Model interpretation—triaxial soil test.

The elastic shear modulus (low strain modulus) is determined by the steepest slope measured at the origin of the shear stress-strain curve, or better, through seismic-type measurements. The elastic B, and plastic H′, bulk moduli associated with the stress point are determined by measuring the slopes $\overset{\circ}{p}/\overset{\circ}{\epsilon}_v$ of small hydrostatic load-unload cycles at selected hydrostatic pressures, viz.

$$\frac{\overset{\circ}{p}}{\overset{\circ}{\epsilon}_v} = B\,\frac{H'}{H' + 3B} \qquad \text{(load)} \tag{2-188}$$

$$\frac{\overset{\circ}{p}}{\overset{\circ}{\epsilon}_v} = B \qquad \text{(unload)}$$

Typically, $B = 2G/3$.

Remark: In general, the model allows different dilation parameters $\bar{\eta}_C$ and $\bar{\eta}_E$ to be associated with each yield surface. However, such a level of sophistication is usually unwarranted because of the rather inaccurate experimental measurements of the detailed volumetric strains observed in conventional triaxial soil tests (especially in extension). Therefore, in the following, one averaged value for $\bar{\eta}_c$ and $\bar{\eta}_E$ is used, and assumed to pertain to all yield surfaces.

Examples

Figure 2-25A shows typical stress-strain curves corresponding to drained axial compression/extension test. The curves shown in Figure 2-25 are synthesized data generated from real sand data. As shown in Figure 2-25B, the sand is assumed to exhibit first volumetric compaction then dilation in both compression and extension loading conditions. Table 1 shows the corresponding model parameter values generated using the procedure explained in the previous section. Ten yield surfaces were used in order to closely model the behavior depicted in Figures 2-25 and 2-26.

Figure 2-26 shows the model predictions for a uniaxial strain test ($\epsilon_2 = \epsilon_3 = 0$). Figure 2-26A shows the shear stress-strain response and Figure 2-26B shows the corresponding stress path. Note that the loading

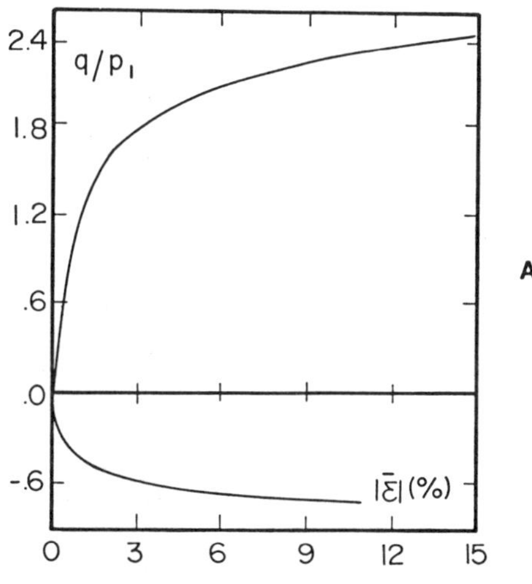

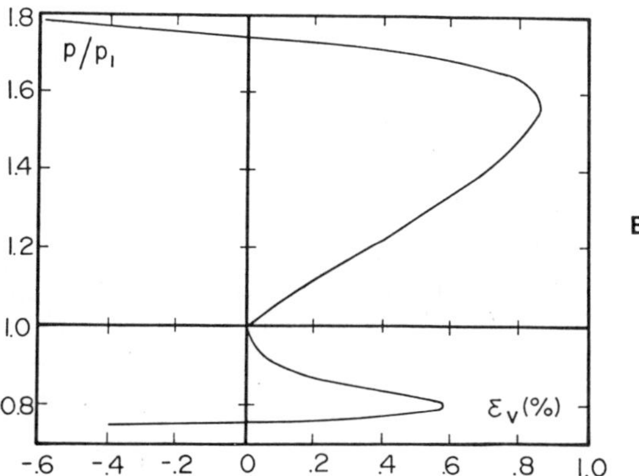

Figure 2-25. Sand behavior: tri-axial soil test. (A) shear stress-strain curve; (B) volumetric stress-strain curve.

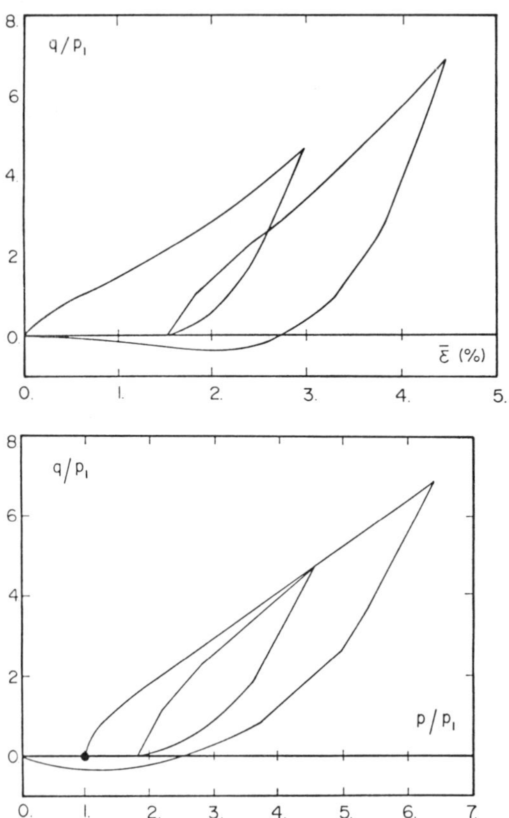

Figure 2-26. Sand behavior: model prediction uniaxial strain test (A) shear stress-strain curve; (B) stress path.

Table 2-1
Model Parameters

$G_1/p_1 = 126.70$	$B_1/p_1 = 84.45$	$(H_1')_o/p_1 = 337.80$
$n = 0.50$	$\bar{\eta}_C = 1.00$	$\bar{\eta}_E = 0.75$

YIELD SURFACE NUMBER	α	M	$(H_C')_1/p_1$	$(H_E')_1/p_1$
1	0.163	0.266	170.55	255.75
2	0.273	0.428	53.40	100.05
3	0.305	0.598	17.38	40.82
4	0.332	0.736	6.72	19.34
5	0.305	0.845	3.04	9.88
6	0.283	0.934	1.48	5.32
7	0.241	1.022	1.02	3.91
8	0.252	1.043	0.60	2.41
9	0.203	1.121	0.22	0.93
10	0.166	1.170	0.00	0.00

portion of the stress-strain curve is S-shaped. As the applied load increases, the material densifies and the curve exhibits an upward concavity typical of a locking tendency. Upon unloading-reloading, the stress-strain curve describes a hysteresis loop. The corresponding stress path (Figure 2-26B) indicates a tendency upon loading to reach a constant lateral stress ratio (q vs. p tends to become linear). Note that upon unloading to $\epsilon_1 = 0$:

- The corresponding stress path falls below the positive shear stress axis indicating shear reversal,
- The shear stress eventually vanishes when p = 0.

Such behavior has been observed in many experimental uniaxial strain test data.

The effects of the material anisotropy are illustrated in Figure 2-27 which shows the model prediction for a simple-shear strain test simulation ($\epsilon_1 = \epsilon_2 = \epsilon_3 = 0$). Figure 2-27 shows the shear stress-strain response τ_{12} vs. γ_{12}, and the lateral stress-difference response ($\sigma_{11} - \sigma_{22}$) vs. γ_{12}. Note that the model predicts that the lateral stress difference first increases, then decreases, upon shearing. Such behavior is typical of overconsolidated materials (see e.g., [Prevost, 1978]).

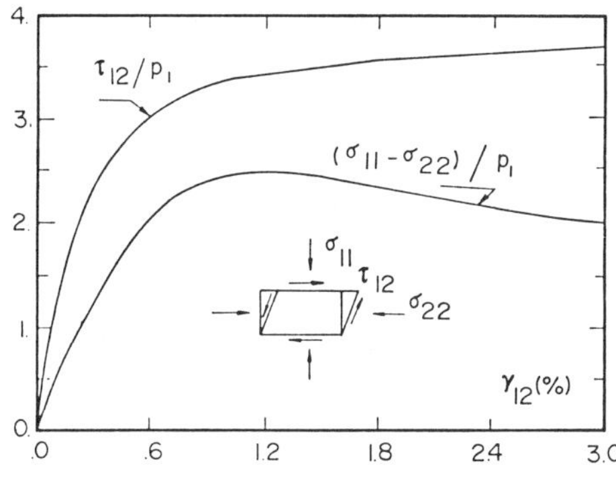

Figure 2-27. Sand behavior: model prediction monotonic simple shear strain test—shear stress-strain curves.

Figures 2-28 shows the model prediction of a cyclic simple shear strain test simulation. In that test, the shear strain amplitude $| \gamma_{12} | = 0.55\%$ is kept constant. Figure 2-28A shows the corresponding shear stress-strain hysteretic response of the material. As a result of cycling, the material softens and the shear stress amplitude decreases. Figure 2-28B shows the corresponding changes in lateral normal stresses and Figure 2-28C the associated stress path τ_{12} vs. p. Note that due to the compactive tendency of the material at low strain level, the effective mean stress decreases continuously as the load is cycled, and the material eventually "liquefies" (p = 0) after $4\frac{1}{2}$ cycles.

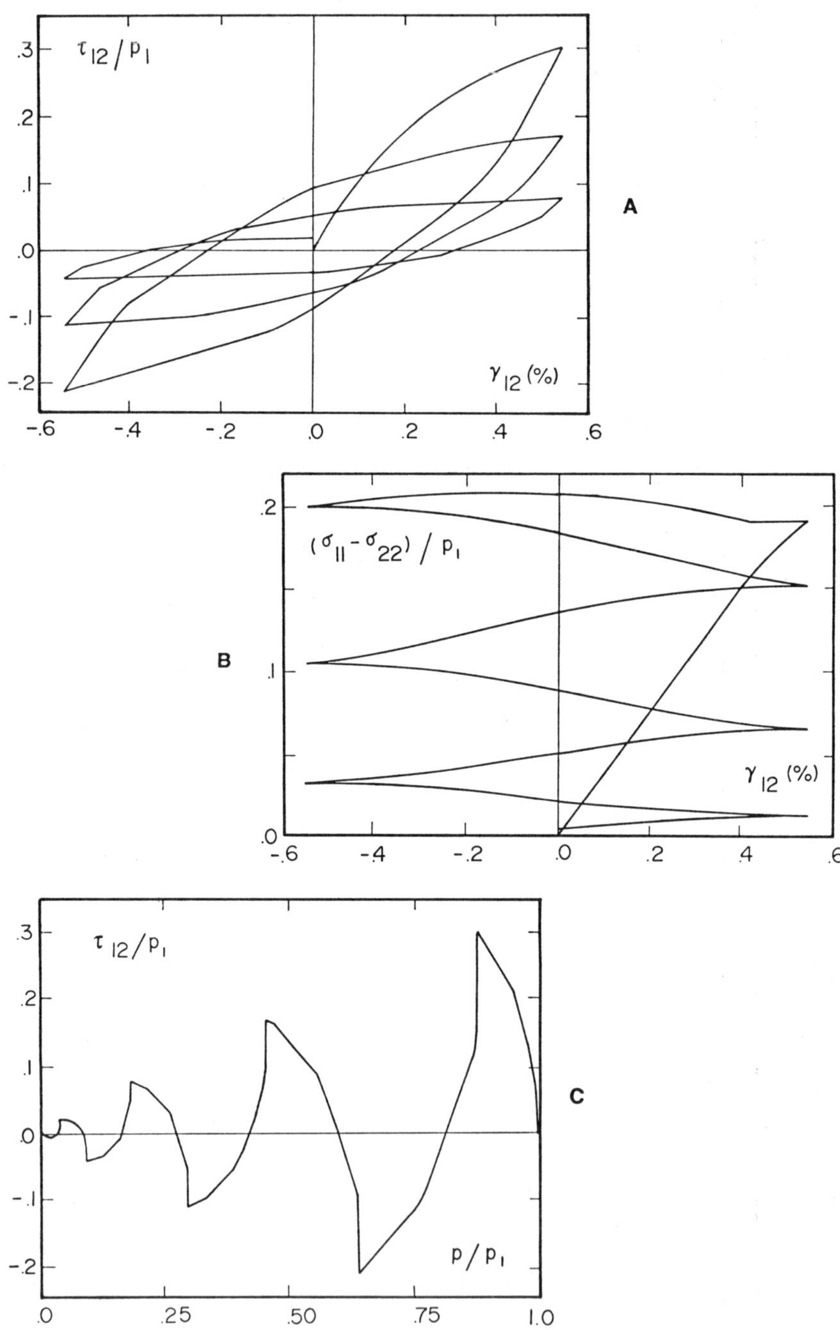

Figure 2-28. Sand behavior: model prediction cyclic simple shear strain test: (A) shear stress-strain curve; (B) changes in lateral normal stresses; (C) stress-path.

APPENDIX

Tensor Invariants

Let σ denote a symmetric second-order tensor. The direct invariants of σ are defined as follows:

$$\bar{I}_1 = \mathrm{tr}\,\sigma = \sigma : \delta = \sigma_{kk} \tag{2-189}$$

$$\bar{I}_2 = \mathrm{tr}\,\sigma^2 = \sigma : \sigma = \sigma_{ik}\sigma_{ki} \tag{2-190}$$

$$\bar{I}_3 = \mathrm{tr}\,\sigma^3 = \sigma_{ij}\sigma_{jk} = \sigma_{ki} \tag{2-191}$$

The usual invariants which appear in the characteristic equations, viz. from Cayley-Hamilton theorem:

$$\sigma^3 - I_1\sigma^2 + I_2\sigma - I_3\delta = 0 \tag{2-192}$$

are defined as follows:

$$I_1 = \mathrm{tr}\,\sigma = \bar{I}_1 \tag{2-193}$$

$$I_2 = (\det\,\sigma)(\mathrm{tr}\,\sigma^{-1}) = \frac{1}{2}\,(\bar{I}_1^2 - \bar{I}_2) \tag{2-194}$$

$$I_3 = \det\,\sigma = \frac{1}{3}\left(\frac{1}{2}\,\bar{I}_1^3 - \frac{3}{2}\,\bar{I}_1\,\bar{I}_2 + \bar{I}_3\right) \tag{2-195}$$

Conversely, one finds that

$$\bar{I}_1 = I_1 \tag{2-196}$$

$$\bar{I}_2 = I_1^2 - 2I_2 \tag{2-197}$$

$$\bar{I}_3 = I_1^3 - 3I_1\,I_2 + 3I_3 \tag{2-198}$$

Deviatoric Tensor Invariants:

Let s denote the deviatoric tensor associated with σ, viz.,

$$s = \sigma - \frac{1}{3}\,\mathrm{tr}\,\sigma\,\delta \tag{2-199}$$

The direct invariants of **s** are defined as follows:

$$\bar{J}_1 = \text{tr } \mathbf{s} = 0 \tag{2-200}$$

$$\bar{J}_2 = \text{tr } \mathbf{s}^2 = \mathbf{s} : \mathbf{s} = s_{ik}s_{ki} \tag{2-201}$$

$$\bar{J}_3 = \text{tr } \mathbf{s}^3 = s_{ij}s_{jk}s_{ki} \tag{2-202}$$

The usual invariants of **s** which appear in the characteristic equation, viz.,

$$\mathbf{s}^3 - J_2\mathbf{s} - J_3\delta = 0 \tag{2-203}$$

are defined as follows:

$$J_2 = \frac{1}{2}\,\text{tr}\mathbf{s}^2 = \frac{1}{2}\,\bar{J}_2 \tag{2-204}$$

$$J_3 = \det \mathbf{s} = \frac{1}{3}\,\bar{J}_3 \tag{2-205}$$

It is simple to show that:

$$\bar{J}_2 = \bar{I}_2 - \frac{1}{3}\,\bar{I}_1^2 \tag{2-206}$$

$$\bar{J}_3 = \bar{I}_3 - \bar{I}_1\bar{I}_2 + \frac{2}{9}\,\bar{I}_1^3 \tag{2-207}$$

and that

$$J_2 = \frac{1}{3}\,I_1^2 - I_2 \tag{2-208}$$

$$J_3 = I_3 - \frac{1}{3}\,I_2I_1 + \frac{2}{27}\,I_1^3 \tag{2-209}$$

$$= I_3 + \frac{1}{3}\,J_2I_1 - \frac{1}{27}\,I_1^3 \tag{2-210}$$

Derivatives of Invariants

The following formula will also prove useful:

$$\bar{I}_1 = \text{tr } \sigma \qquad \frac{\partial \bar{I}_1}{\partial \sigma} = \delta \tag{2-211}$$

$$\bar{I}_2 = \text{tr } \sigma^2 \qquad \frac{\partial \bar{I}_2}{\partial \sigma} = 2\sigma \tag{2-212}$$

$$\bar{I}_3 = \text{tr } \sigma^3 \qquad \frac{\partial \bar{I}_3}{\partial \sigma} = 3\sigma^2 \tag{2-213}$$

$$\bar{K}_1 = \text{tr } \sigma\alpha \qquad \frac{\partial \bar{K}_1}{\partial \sigma} = \alpha \tag{2-214}$$

$$\bar{K}_2 = \text{tr } \sigma\alpha^2 \qquad \frac{\partial \bar{K}_2}{\partial \sigma} = \alpha^2 \tag{2-215}$$

$$\bar{K}_3 = \text{tr } \sigma^2\alpha \qquad \frac{\partial \bar{K}_3}{\partial \sigma} = \sigma\alpha + \alpha\sigma \tag{2-216}$$

$$\bar{K}_4 = \text{tr } \sigma^2\alpha^2 \qquad \frac{\partial \bar{K}_4}{\partial \sigma} = \sigma\alpha^2 + \alpha^2\sigma \tag{2-217}$$

$$I_1 = \text{tr } \sigma \qquad \frac{\partial I_1}{\partial \sigma} = \delta \tag{2-218}$$

$$J_2 = \frac{1}{2} \text{tr } s^2 \qquad \frac{\partial J_2}{\partial \sigma} = s \tag{2-219}$$

$$J_3 = \det s \qquad \frac{\partial J_3}{\partial \sigma} = s^2 - \frac{2}{3} J_2\delta \tag{2-220}$$

Invariants—Expanded Formula

The following formulas are included for convenience to the reader:

$$I_1 = \text{tr}\sigma = \sigma_{ii} = \sigma_{11} + \sigma_{22} + \sigma_{33} \tag{2-221}$$

$$I_2 = (\det \boldsymbol{\sigma})(\mathrm{tr}\,\boldsymbol{\sigma}^{-1}) = \frac{1}{2}\,(I_1^2 - \mathrm{tr}\,\boldsymbol{\sigma}^2) = \frac{1}{2}\,(I_1^2 - \sigma_{ij}\sigma_{ij})$$

$$= (\sigma_{11}\sigma_{22} + \sigma_{22}\sigma_{33} + \sigma_{33}\sigma_{11}) - (\sigma_{12}^2 + \sigma_{23}^2 + \sigma_{31}^2) \tag{2-222}$$

$$I_3 = \det \boldsymbol{\sigma}$$

$$= \sigma_{11}\sigma_{22}\sigma_{33} + 2\sigma_{12}\sigma_{23}\sigma_{31} - (\sigma_{11}\sigma_{23}{}^2 + \sigma_{22}\sigma_{31}{}^2 + \sigma_{33}\sigma_{12}{}^2) \tag{2-223}$$

In terms of the principal values of $\boldsymbol{\sigma}$:

$$I_1 = \sigma_1 + \sigma_2 + \sigma_3 \tag{2-224}$$

$$I_2 = \sigma_1\sigma_2 + \sigma_2\sigma_3 + \sigma_3\sigma_1 \tag{2-225}$$

$$I_3 = \sigma_1\sigma_2\sigma_3 \tag{2-226}$$

$$J_2 = \frac{1}{2}\,\mathrm{tr}\,\mathbf{s}^2 = \frac{1}{2}\,(s_{11}^2 + s_{22}^2 + s_{33}^2) + \sigma_{12}^2 + \sigma_{23}^2 + \sigma_{31}^2$$

$$= -(s_{11}s_{22} + s_{22}s_{33} + s_{33}s_{11}) + \sigma_{12}^2 + \sigma_{23}^2 + \sigma_{31}^2$$

$$= \frac{1}{6}\,[(\sigma_{11} - \sigma_{22})^2 + (\sigma_{22} - \sigma_{33})^2 + (\sigma_{33} - \sigma_{11})^2]$$
$$+ \sigma_{12}^2 + \sigma_{23}^2 + \sigma_{31}^2 \tag{2-227}$$

$$J_3 = \det \mathbf{s} = \frac{1}{3}\,\mathrm{tr}\,\mathbf{s}^3 = \frac{1}{3}\,s_{ij}s_{jk}s_{ki}$$

$$= s_{11}s_{22}s_{33} + 2\,\sigma_{12}\sigma_{23}\sigma_{31} - (s_{11}\sigma_{23}^2 + s_{22}\sigma_{31}^2 + s_{33}\sigma_{12}^2) \tag{2-228}$$

In terms of the principal values of $\boldsymbol{\sigma}$:

$$J_2 = \frac{1}{6}\,[(\sigma_1 - \sigma_2)^2 + (\sigma_2 - \sigma_3)^2 + (\sigma_3 - \sigma_1)^2] \tag{2-229}$$

$$J_3 = \frac{1}{27}\,(2\sigma_1 - \sigma_2 - \sigma_3)(2\sigma_2 - \sigma_3 - \sigma_1)(2\sigma_3 - \sigma_1 - \sigma_2) \tag{2-230}$$

REFERENCES

1. Argyris, J. H., et al., "Recent Developments in Finite Elements Analysis of PCRV," *Proceedings 2nd Int. Conf.*, SMIRT, Berlin, 1973.
2. Baker, R., and Desai, C. S., "Induced Anisotropy During Plastic Straining," *Int. J. Numerical and Analytical Methods in Geomechanics*, Vol. 8, 1984, pp. 167–185.
3. Baladi, G. Y., and Rohani, B., "Elastic-Plastic Model for Saturated Sand," *J. Geotech. Eng. Div.*, ASCE, Vol. 105, No. GT5, April 1979, pp. 465–480.
4. Boehler, J. P., "A Simple Derivation of Representation for Non-Polynomial Constitutive Equations in Some Cases of Anisotropy," *ZAMM*, Vol. 59, 1979, pp. 157–167.
5. Chen, W. F., *Limit Analysis and Soil Plasticity, Developments in Geotechnical Engineering 7*, Elsevier Scientific, Amsterdam, 1975.
6. Coon, M. D., and Evans, R. J., "Recoverable Deformations of Cohesionless Soils," *J. Soil Mech. Found. Eng.*, ASCE, Vol. 97, No. SM2, Feb. 1971, pp. 375–390.
7. Coulomb, C. A., "Essai sur une application des Regles de Maximis et Minimis a quelques Problemes de statique relatifs a l'Architecture," *Memoires par Divers Savants*, 1776 (read in 1773).
8. Cosserat, E., and Cosserat, F., *Theorie des Corps Deformables*, Hermann, Paris, 1909.
9. Dafalias, Y. F., "The Plastic Spin Concept and a Simple Illustration of its Role in Finite Plastic Transformation," *Mechanics of Materials*, Vol. 3, 1984, pp. 223–233.
10. DiMaggio, F. L., and Sandler, I. S., "Material Models for Granular Soils," *J. Eng. Mech. Div.*, ASCE, Vol. 97, No. EM3, June 1971, pp. 935–950.
11. Drucker, D. C., "A Definition of Stable Inelastic Material," *Journal of Applied Mechanics*, ASME, Vol. 26, 1959, pp. 106–112.
12a. Drucker, D. C., Gibson, R. E., and Henkel, D. J., "Soil Mechanics and Work-Hardening Theories of Plasticity," *Proceedings*, ASCE, Vol. 81, 1955, pp. 1–14, also *Transactions*, ASCE, Vol. 122, 1957, pp. 338–346.
12b. Drucker, D. C. and Prager, W., "Soil Mechanics and Plasticity Analysis or Limit Design," *Quart. Appl. Math.* Vol. 10, No. 2, 1952, pp. 157–165.
12c. Duncan, J. M., and Chang, C. Y., "Nonlinear Analysis of Stress and Strain in Soils," *J. Soil Mech. Found. Eng.*, ASCE, Vol. 96, No. SM5, Sept. 1970, pp. 1629–1653.

13. Fardshisheh, F., and Onat, E. J., "Representation of Elastoplastic Bchaviour by Mcans of Statc Variables," *Problems of Plasticity,* Sawczuk, ed., Noordhoff, Leiden, 1974, pp. 89–114.

14. Ghaboussi, J., and Momen, H., "Modelling and Analysis of Cyclic Behavior of Sands," *Soil Mechanics—Transient and Cyclic Loads,* G. N. Pande and O. C. Zienkiewicz (Eds.), Wiley, 1982, pp. 313–342.

15. Green, A. E., and Naghdi, P. M., "A General Theory of Elastic-Plastic Continuum," *Arch. of Rational Mech. and Analysis,* Vol. 18, No. 4, 1965.

16. Hicher, P. Y., "Comportement Mecanique des argiles saturees sur divers chemins de sollicitations monotones et cycliques. Application a une modelisation elastoplastique et visco-plastique," These de Doctorat d'Etat, Universite Paris 6, Paris, France, December 1985.

17. Hill, R., *The Mathematical Theory of Plasticity,* Clarendon Press, Oxford, England, 1950.

18. Huber, M. T., *Czasopismo technize,* Vol. 15, Lvov, Poland, 1904.

19. Krieg, R. D. and Krieg, D. B., "Accuracies of Numerical Solution Methods for the Elastic-Perfectly Plastic Model," *J. Pressure Vessel Technol.,* ASME, Vol. 99, 1977, pp. 510–515.

20. Lade, P. V., and Duncan, J. M., "Elastoplastic Stress-Strain Theory for Cohesionless Soil," *J. Geotech. Eng. Div.,* ASCE, Vol. 101, 1975, pp. 1037–1053.

21. Liu, I. S., "On Representations of Anisotropic Invariants," *Int. J. Eng. Sc.,* Vol. 20, 1982, pp. 1099–1109.

22. Loret, B., "On the Effects of Plastic Rotation in the Finite Deformation of Anisotropic Elastoplastic Materials," *Mechanics of Materials,* Vol. 2, 1983, pp. 287–304.

23. Loret, B., and Prevost, J. H., "Accurate Numerical Solutions for Drucker-Prager Elastic Plastic Models," *Comp. Meth. Appl. Mech. Eng.,* Vol. 54, No. 3, 1986, pp. 259–277.

24. Luong, M. P., "Phenomenes Cycliques dans les Sols Pulverulents," *Revue Francaise de Geotechnique,* Vol. 10, 1980, pp. 39–53.

25. Luong, M. P. and Touati, A., "Sols Grenus sous Fortes Contraintes," *Revue Francaise de Geotechnique,* No. 23, 1983, pp. 51–63.

26. Mandel, J., "Director Vectors and Constitutive Equations for Plastic and Viscoplastic Media," *Problems of Plasticity,* A. Sawczuk (Ed.), Noordhoff, Leiden, 1974, pp. 135–141.

27. Mandel, J., "Equations constitutives et directeurs dans les milieux plastiques et viscoplastiques," *Int. J. Solids and Structures,* Vol. 9, 1973, p. 725.

28. Mandel, J., *Plasticite classique et Viscoplasticite,* CISM Courses and Lectures 97, Springer, New York, 1971.

29. Melan, E., "Zur Platizitat des raumlichenn Kontinuums," *Ing.-Arch.,* Vol. 9, 1938, pp. 116–126.

30. Mroz, Z., "On the Description of Anisotropic Workhardening," *J. Mech. and Phys. of Solids,* Vol. 15, 1967, pp. 163–175.

31. Mroz, Z. and Pietruszak, S. T., "A Constitutive Model for Sand with Anisotropic Hardening Rule," *Int. J. Num. Meth. Geom.,* Vol. 7, 1983, pp. 305–320.

32. Nayak, G. C., and Zienkiewicz, O. C., "Elastic-Plastic Stress Analysis: A Generalization for Various Constitutive Relations Including Strain Softening," *Int. J. Numer. Meths. Engrg.,* Vol. 5, No. 1, 1972, pp. 113–135.

33. Nemat-Nasser, S., "On Dynamic and Static Behavior of Granular Materials," *Soil Mechanics-Transient and Cyclic Loads,* G. N. Pande and O. C. Zienkiewicz (Eds.), Wiley, 1982, pp. 439–458.

34. Nemat-Nasser, S., "On Finite Plastic Flow of Crystalline Solids and Geomaterials," *Journal of Applied Mechanics,* ASME, Vol. 50, No. 4b, 1983, pp. 1114–1126.

35. Nguyen, Q. S., "On the Elastic-Plastic Initial Boundary Value Problem and its Numerical Integration, *Int. J. Numer. Meths. Engrg.,* Vol. 11, 1977, pp. 817–832.

36. Ortiz, M., and Simo, J. C., "An Analysis of a New Class of Integration Algorithms for Elastoplastic Constitutive Relations," *Int. J. Num. Meth. Eng.,* Vol. 23, No. 3, 1986, pp. 353–366.

37. Ortiz, M., and Popov, E. P., "Accuracy and Stability of Integration Algorithms for Elastoplastic Constitutive Relations," *Int. J. Num. Meth. Eng.,* Vol. 21, No. 9, 1985, pp. 1561–1576.

38. Prager, W., *Introduction to Plasticity,* Addison-Wesley, Reading, Mass., 1959.

39. Prager, W., "The Theory of Plasticity: A Survey of Recent Achievements," *Proceedings Inst. Mech. Eng.,* London, England, Vol. 169, 1955, pp. 41–57.

40. Prevost, J. H., "A Simple Plasticity Theory for Frictional Cohesionless Soils," *Soil Dynamics and Earthquake Engineering,* Vol. 4, No. 1, 1985, pp. 9–17.

41. Prevost, J. H., "Two-Surface vs. Multi-Surface Plasticity Theories," *Int. J. Num. Meth. Geom.,* Vol. 6, 1982, pp. 323–338.

42. Prevost, J. H., "Anisotropic Undrained Stress-Strain Behavior of Clays," *J. Geotech. Eng. Div.,* ASCE, Vol. 104, No. GT8, 1978, pp. 1075–1090.

43. Prevost, J. H., "Mathematical Modeling of Monotonic and Cyclic Undrained Clay Behavior," *Int. J. Num. Meth. Geom.,* Vol. 1, No. 2, 1977, pp. 195–216.

44. Richard, R. E., Woods, R. D. and Hall, J. R., *Vibrations of Soils and Foundations,* Prentice-Hall, N.J., 1970.

45. Roscoe, K. H. and Burland, J. B., "On the Generalized Stress-Strain Behavior of Wet Clay," *Engineering Plasticity,* J. Heyman and F. Leckie (Eds.), Cambridge University Press, Cambridge, England, 1968, pp. 535–609.

46. Rowe, P. E., "The Stress-Dilatancy Relation for Static Equilibrium of an Assembly of Particles in Contact," *Proc. Roy. Soc.,* Vol. A269, 1962, pp. 500–527.

47. Schofield, A. N., and Wroth, C. P., *Critical State Soil Mechanics,* McGraw Hill, Inc., London, England, 1968.

48. Schreyer, H. L., Kulak, R. F., and Kramer, J. M., "Accurate Numerical Solutions for Elastic-Plastic Models, *J. Pressure Vessel Technol.,* ASME, Vol. 101, 1979, pp. 226–234.

49. Scott, R. F., "Plasticity and Constitutive Relations in Soil Mechanics," *Journal of Geotechnical Engineering,* ASCE, Vol. 111, No. 5, 1985, pp. 563–605.

50. Simo, J. C., and Ortiz, M., "A Unified Approach to Finite Deformation Elastoplastic Analysis Based on the Use of Hyperelastic Constitutive Equations," *Comp. Meth. Appl. Mech. Eng.,* Vol. 49, 1985, pp. 221–245.

51. Smith, G. F., "On Isotropic Functions of Symmetric Tensors, Skew-Symmetric Tensors and Vectors," *Int. J. Eng. Sc.* Vol. 9, 1971, p. 889.

52. Spencer, A. J. M., "Theory of Invariants," *Continuum Physics,* Vol. 1, A.C. Eringen (Ed.), Academic Press, New York, 1972, pp. 239–353.

53. Tresca, H., "Memoire sur le poinconnage des Metaux," *Memoires des Savants Etrangers,* Vol. 20, 1872, pp. 617–638, (read in 1869).

54. Tresca, H., "Memoire sur l'ecoulement des corps solides soumis a de fortes pressions," *Memoires Presentes par Divers Savants,* Vol. 18, 1868, pp. 773–799 (read in 1864).

55. Valanis, K. C., and Read, H. J. E., "A New Endochronic Plasticity Model for Soils," *Soil Mechanics-Transient and Cyclic Loads,* G. N. Pande and O. C. Zienkiewicz (Eds.), Wiley, 1982, pp. 375–417.

56. Vermeer, P. A., "Formulation and Analysis of Sand Deformation Problems," Ph.D. Thesis, Department of Civil Engineering, Delft University, The Netherlands, 1980.

57. Von Mises, R., "Mechanik der festen Korper im plastisch-deformablen Zustand," *Nachrichten der Koniglicher Gessellschaft der Wissenschaft,* Gottingen, Mathematik-Physik Klass, 1913, pp. 582–592.

58. Wang, C. C., "A New Representation Theorem for Isotropic Functions: An Answer to Professor Smith's Criticism of my Paper on Representation for Isotropic Functions," *Arch. Rat. Mech. Analy.*, Vol. 36, 1970, pp. 166–223.
59. Westergard, H. M., "On the Resistance of Ductile Materials to Combined Stress," *J. Franklin Inst.*, Vol. 189, 1920, pp. 627–640.
60. Yoder, P. J. and Whirley, B. G., "On the Numerical Implementation of Elastoplastic Models," *J. Appl. Mech.*, ASME, Vol. 51, 1984, pp. 283–288.

3

Dynamics of Porous Media

Jean H. Prevost
Department of Civil Engineering
Princeton University
Princeton, New Jersey USA

INTRODUCTION

Soils consist of an assemblage of particles with different sizes and shapes which form a skeleton whose voids are filled with water and air or gas. The word "soil" therefore implies a mixture of assorted mineral grains with various fluids. Hence, soil in general must be looked at as a one-phase (dry soil), two-phase (saturated soil), or multiphase (partially saturated soil) material whose state is to be described by the stresses and displacements (velocities) within each phase. There are still great uncertainties on how to deal analytically with partly saturated soils. Attention is therefore restricted in the following to dry and fully saturated soils. The stresses carried by the soil skeleton are conventionally called "effective stresses" in the soil mechanics literature [Terzaghi, 1943], and those in the fluid phase are called the "pore-fluid pressures."

In a saturated soil, when free drainage conditions prevail, the steady-state pore-fluid pressures depend only on the hydraulic conditions and are independent of the soil skeleton response to external loads. Therefore, in that case, a single-phase continuum description of soil behavior is certainly adequate. Similarly, a single-phase description of soil behavior is also adequate when no drainage (i.e., no flow) conditions prevail. However, in intermediate cases in which some flow can take place, there is an interaction between the skeleton strains and the pore-fluid flow. The solution of these problems requires that soil behavior be analyzed by incorporating the effects of the transient flow of the pore-fluid through the voids, and therefore requires that a two-phase continuum formulation be available for porous media. Such a theory was first developed by Biot [1955, 1956, 1957, 1972, 1977, 1978] for an elastic porous medium. However, it is observed experimentally that the stress-strain strength behavior of the soil skeleton is strongly nonlinear, anisotropic, hysteretic, and path-dependent. An extension of Biot's theory into the nonlinear anelastic range is therefore necessary in order to analyze the transient response of soil deposits. This extension has acquired considerable importance in recent years due to the increased concern with the dynamic behavior of saturated soil deposits and associated liquefaction of saturated sand deposits under seismic loading conditions. Such an extension of Biot's formulation [1980], is presented herein. For that purpose, soil is viewed as a multiphase medium and the modern theories of mixtures developed by Green and Naghdi [1965] and Eringen and Ingram [1965] are used. General mixture results can be shown through formal linearization of the field and constitutive equations, to reduce to Biot linear poroelastic model (see, for example, Bowen [1982]).

The general theoretical framework which forms the basis of mixtures theories was first developed by Truesdell and Toupin [1960] early in 1960. Since then, the theoretical description of multiphase materials has received repeated attention in the literature, and fundamental equations for a dynamical theory of interacting continua have been derived (for example, References 12, 15, 17–19, 24, 25, 31, 32). The most recent findings are summarized in References 1, 2, and 10 which contain many references to relevant works. In the following, the general mixture equations are first summarized and applied to describe the flow of water through saturated anelastic porous soil media. Special attention is given in particular to the physical meaning of the partial stresses which appear in the general field equations, and an effort is made to relate them to physical quantities measurable in the field.

BASIC THEORY

During deformations, the solid particles which form the soil skeleton undergo irreversible motions such as slips at grain boundaries, creations of voids by particles coming out of a packed configuration, and combinations of such irreversible motions. When the particulate nature and the microscopic origin of the phenomena involved are not sought, phenomenological equations then provide an adequate description of the behavior of the various phases which form the soil medium. In multiphase theories, the conceptual model is thus one in which each phase (or constituent) enters through its averaged properties obtained as if the particles were smeared out in space. In other words, the particulate nature of the constituents is described in terms of phenomenological laws as the particulates behave collectively as a continuum. Soil is thus viewed herein as consisting of a solid skeleton interacting with the pore fluids. In order to be able to derive multiphase field and constitutive equations for such a medium, a technique for obtaining local average quantities is therefore necessary. Furthermore, the basic kinematics and balance equations for each constituent and for the mixture as a whole must be defined.

Kinematics

Soil is viewed herein as a mixture consisting of $m(1 \leq m \leq 2)$ deformable media, each of which is regarded as a continuum (for saturated soils $m = 2$), and each following its own motion. It is assumed that at any time t each place $\mathbf{x}$ of the mixture is occupied simultaneously by m different particles x^1, x^2, ..., x^m, one for each constituent. As in single-phase theory, to each constituent is assigned a fixed but otherwise arbitrary reference configuration [Truesdell, 1965], and a motion

$$\mathbf{x} = \mathbf{x}^\alpha(\mathbf{X}^\alpha, t) \qquad \alpha = 1, ..., m \tag{3-1}$$

where $\mathbf{X}^\alpha$ denotes the position of the αth-constituent in its reference configuration, and $\mathbf{x}$ the spatial position occupied at the time t by the particle labeled $\mathbf{X}^\alpha$. For simplicity in the following, both the reference and current configurations of each constituent are referred to rectangular Cartesian axes. Capital and lower case letters are used for the indices on coordinates and tensors referred to the undeformed and deformed configuration, respectively (see, for example, References 14 and 21). The usual continuity and differentiability assumptions are made for the deformation functions $\mathbf{x}^\alpha$, and the following restrictions are imposed

$$\det[x^\alpha_{a,A}] = J^\alpha(\mathbf{X}^\alpha, t) > 0; \quad \det[x^\alpha_{a,b}] > 0 \quad \alpha = 1, ..., m \tag{3-2}$$

for physically possible motions, in which det denotes the determinant and a comma (,) a partial derivative. The velocity and acceleration of $\mathbf{X}^\alpha$ at time t are obtained from Equation 3-1 by time differentiation, viz.

$$\mathbf{v}^\alpha = \mathbf{v}^\alpha(\mathbf{x},t) = \frac{\overset{\alpha}{D}}{Dt}(\mathbf{x}) = \dot{\mathbf{x}}^\alpha$$

$$\mathbf{a}^\alpha = \mathbf{a}^\alpha(\mathbf{x},t) = \frac{\overset{\alpha}{D}}{Dt}(\mathbf{v}^\alpha) = \ddot{\mathbf{x}}^\alpha \tag{3-3}$$

where a superimposed dot indicates differentiation with respect to time holding $\mathbf{X}^\alpha$ fixed (i.e., the material derivative following the motion of the α-constituent),

$$(\bullet)^\alpha = \frac{\overset{\alpha}{D}}{Dt}(\ ) = \frac{\partial}{\partial t}(\ ) + \mathbf{v}^\alpha \cdot \mathbf{\nabla}(\ ) \tag{3-4}$$

Here, and in the following, "∇ and $\bar{\nabla}$" are used to denote spatial and material derivatives, respectively. The deformation gradient for $\mathbf{X}^\alpha$ at time t is defined by

$$\mathbf{F}^\alpha = \mathbf{F}^\alpha(\mathbf{x}^\alpha, t) = \bar{\nabla}\mathbf{x}^\alpha = [F^\alpha_{aA}] = [x^\alpha_{a,A}] \tag{3-5}$$

and the velocity gradient is defined by

$$\mathbf{L}^\alpha = \mathbf{L}^\alpha(\mathbf{x}, t) = \nabla\mathbf{v}^\alpha = \dot{\mathbf{F}}^\alpha \cdot (\mathbf{F}^\alpha)^{-1} = [L^\alpha_{ab}] = [v^\alpha_{a,b}] \tag{3-6}$$

in which, $(\mathbf{F}^\alpha)^{-1}$ denotes the inverse of $\mathbf{F}^\alpha$. The symmetric and skew-symmetric parts of $\mathbf{L}^\alpha$ are referred to as the deformation rate, $\mathbf{d}^\alpha$, and spin tensor, $\mathbf{w}^\alpha$, respectively.

Average Quantities

Average quantities are obtained by integrating microscopic quantities over an averaging volume or area. The averaging procedure is used to obtain a field of macroscopic quantities for each phase. In the macroscopic field, the averaging volume represents and characterizes a physical point. Because the averaging volume is macroscopically infinitesimal, it is denoted by dV. Similarly, the averaging area dA, represents and characterizes a physical point on the surface of dV, and is an infinitesimal element of area in the macroscopic field. The characteristic length, D, of the averaging volume or area is selected such that [Whitaker, 1969] $\ell << D << L$, where ℓ is the microscopic scale of the porous medium and L is the scale of gross inhomogeneities. Typically, $\ell = 50$ microns in sands and $\ell = 1$ micron in clays, whereas $L = 1$cm. The part of dV occupied by the α-phase is denoted by dV^α, and the volume fraction, n^α, of the α-phase is the fraction of dV occupied by the α-phase defined by

$$n^\alpha = n^\alpha(\mathbf{x},t) = \frac{dV^\alpha}{dV} \tag{3-7}$$

Clearly, n^α is constrained by

$$\sum_\alpha n^\alpha = 1 \quad \text{and} \quad 0 \le n^\alpha \le 1$$

Similarly, the part of dA lying in the α-phase is denoted by dA^α, and the areal fraction, $\bar{n}^\alpha$, of the α-phase is the fraction of dA which interests the α-phase defined by

$$\bar{n}^\alpha = \bar{n}^\alpha(\mathbf{x}, t) = \frac{dA^\alpha}{dA}$$

(3-8)

subject to

$$\sum_\alpha \bar{n}^\alpha = 1 \quad \text{and} \quad 0 \le \bar{n}^\alpha \le 1$$

In Reference 20, arguments are presented which support the intuitively appealing identity, $\bar{n}^\alpha = n^\alpha$ [Biot, 1956; Schiffman, 1970], and in the following, the identity is assumed to hold.

A *macroscopic* average mass density function, ρ^α, is associated with each constituent and is defined as the volume average of the *microscopic* density function, ρ_α [Gray and Lee, 1977]

$$\rho^\alpha = \frac{1}{dV} \int_{dV^\alpha} \rho_\alpha dv$$

(3-9)

where dv is the microscopic volume element. In this equation and in subsequent developments, the dependence of macroscopic and microscopic quantities on $\mathbf{x}$ and t is understood. The intrinsic volume average mass density is defined as

$$\bar{\rho}_\alpha = \frac{1}{dV^\alpha} \int_{dV^\alpha} \rho_\alpha dv = \frac{1}{n^\alpha} \rho^\alpha$$

(3-10)

Note that only when the mass density of the α-phase is microscopically constant is the intrinsic volume average mass density function equal to the microscopic mass density. In the following, $\bar{\rho}_\alpha = \rho_\alpha$ and thus $\rho^\alpha = n^\alpha \rho_\alpha$.

The mass density, ρ, of the mixture is defined as

$$\rho = \rho(\mathbf{x},t) = \sum_\alpha \rho^\alpha$$

(3-11)

and the mean (or barycentric) velocity, $\mathbf{v}$, for the mixture is defined as

$$\mathbf{v} = \mathbf{v}(\mathbf{x}, t) = \frac{1}{\rho} \sum_\alpha \rho^\alpha \mathbf{v}^\alpha$$

(3-12)

The velocity gradient for the mixture is then

$$\mathbf{L} = \nabla\mathbf{v} = [L_{ab}] = [v_{a,b}] \tag{3-13}$$

It is of importance to emphasize that the velocity $\mathbf{v}^\alpha$ of the α-constituent (Equation 3-3) is its microscopic (intrinsic or seepage) velocity, and is different from the mean (or superficial) velocity $\bar{\mathbf{v}}^\alpha$, used for instance in Darcy's law [Biot and Willis, 1957], defined as

$$\int_{\partial\mathfrak{R}} \rho_\alpha \bar{\mathbf{v}}^\alpha \cdot \mathbf{n} dA = \int_{\partial\mathfrak{R}_\alpha} \rho_\alpha \mathbf{v}^\alpha \cdot \mathbf{n} dA$$

$$= \int_{\partial\mathfrak{R}} \rho^\alpha \mathbf{v}^\alpha \cdot \mathbf{n} dA \tag{3-14}$$

where $\mathbf{n}$ denotes the unit outward normal to the surface $\partial\mathfrak{R}$, of area A, which encloses the fixed region in space $\mathfrak{R}$, of volume V. Clearly, $\bar{\mathbf{v}}^\alpha = n^\alpha \mathbf{v}^\alpha$.

Balance Laws

All equations are postulated at the current time t, and all field quantities are functions of $\mathbf{x}$ and t. When discussing a constituent of the mixture, it is supposed that it can be isolated from the rest of the mixture, provided that allowance is made for the action upon it of the other constituent(s). The balance laws for the two-phase soil mixture are summarized as follows:

Balance of Mass: No chemical interaction is assumed to take place between the solid soil skeleton and the fluid phase. The balance of mass for each α-constituent then takes the form

$$\frac{\overset{\alpha}{D}}{Dt}(\rho^\alpha) + \rho^\alpha \nabla \cdot \mathbf{v}^\alpha = 0 \tag{3-15}$$

Another version of Equation 3-15 is obtained by recalling that $\rho^\alpha = n^\alpha \rho_\alpha$, and

$$\frac{\overset{\alpha}{D}n^\alpha}{Dt} + n^\alpha \nabla \cdot \mathbf{v}^\alpha = -\frac{n^\alpha}{\rho_\alpha}\frac{\overset{\alpha}{D}}{Dt}(\rho_\alpha) \tag{3-16}$$

in which $\overset{\alpha}{D}(\rho_\alpha)/Dt = 0$ if the grains which constitute the α-phase are incompressible. Equation 3-16 will prove most useful in the following.

Balance of Linear Momentum: Before postulating the balance of momentum laws for each α-constituent it is first necessary to consider the forces acting on this constituent within the region $\mathcal{R}$. In addition to body forces, such as gravity forces, one must also account for the effect on the α-constituent of the mixture outside the region $\mathcal{R}$. This effect is accounted for by introducing a vector field $\bar{t}^{\,\alpha}(\mathbf{n}, \mathbf{x}, t)$ defined on $\partial\mathcal{R}$ and measured per unit area of $\partial\mathcal{R}$, such that

$$\int_{\partial\mathcal{R}} \bar{t}\, dA$$

represents the contact force exerted across $\partial\mathcal{R}$ by the α-constituent outside of $\mathcal{R}$ on the α-constituent in $\mathcal{R}$ [Gurtin, Oliver, and Williams, 1972; Truesdell and Toupin, 1960; Williams, 1973]. This notion of a stress vector is in accord with the stress vector notion for a single-substance continuous medium as introduced in classical continuum mechanics (see, for example, References 14, 21, 27–29). Corresponding to the partial stress vector $\bar{t}^{\,\alpha}$, there exists a partial stress tensor $\boldsymbol{\sigma}^\alpha$ [Hassanizadeh and Gray, 1978; Truesdell and Toupin, 1960; Williams, 1973], such that $\bar{t}^{\,\alpha} = \mathbf{n} \cdot \boldsymbol{\sigma}^\alpha$, where $\mathbf{n}$ denotes the unit outward normal to $\partial\mathcal{R}$. Locally,

$$\bar{t}^{\,\alpha} = \frac{1}{dA} \int_{dA^\alpha} \bar{t}_\alpha \, da \tag{3-17}$$

where $\bar{t}_\alpha$ denotes the intrinsic stress vector of the α-phase. Note that: $\bar{t}^{\,\alpha} = n^\alpha \bar{t}_\alpha$ when $\bar{t}_\alpha$ is microscopically constant. From Equation 3-17 and the preceding definition, it is apparent that the partial stress tensor corresponding to the fluid phase, $\boldsymbol{\sigma}^w$, is equal to n^w times the pore fluid stress, $\boldsymbol{\sigma}_w$, i.e., $\boldsymbol{\sigma}^w = n^w \boldsymbol{\sigma}_w$. However, the partial stress tensor corresponding to the solid phase, $\boldsymbol{\sigma}^s$, is *not* the effective stress, $\boldsymbol{\sigma}'^s$, of classical soil mechanics [Terzaghi, 1943] but rather is

$$\boldsymbol{\sigma}^s = \boldsymbol{\sigma}'^s + n^s \boldsymbol{\sigma}_w = n^s \boldsymbol{\sigma}_s = (1 - n^w)\boldsymbol{\sigma}_s \tag{3-18}$$

for a saturated porous medium, where $n^s \boldsymbol{\sigma}_w$ accounts for the effects of the pore fluid stress on the individual solid grains which constitutes the solid skeleton. In deriving Equation 3-18 it has been assumed that the contact areas between the solid grains are negligibly small [Bazant and Krizek, 1975], so that the pore fluid and associated stress completely surrounds each grain. Each solid grain is also subjected to intergranular-forces that are in excess of the pore fluid stress and characterized by the effective stress $\boldsymbol{\sigma}'^s$. The global stress $\boldsymbol{\sigma}$ which is to appear in the general balance equations for the porous medium, is the sum of the partial stresses,

$$\boldsymbol{\sigma} = \Sigma \boldsymbol{\sigma}^{\alpha}$$

and is equal to:

$$\boldsymbol{\sigma} = \boldsymbol{\sigma}^{s} + \boldsymbol{\sigma}^{w} = \boldsymbol{\sigma}'^{s} + \boldsymbol{\sigma}_{w} \tag{3-19}$$

for a saturated porous medium, as postulated in classical soil mechanics [Terzaghi, 1943].

The local version of the balance of linear momentum equations for each constituent then simply writes (see, for example, References 1 and 10)

$$\rho^{\alpha}\mathbf{a}^{\alpha} = \mathbf{v} \cdot \boldsymbol{\sigma}^{\alpha} + \hat{\mathbf{p}}^{\alpha} + \rho^{\alpha}\mathbf{b} \tag{3-20}$$

where $\mathbf{b}$ = body force per unit mass; $\hat{\mathbf{p}}^{\alpha}$ is the momentum supply to the α-constituent from the rest of the mixture due to other interaction effects, for example due to the relative motions of the constituents, subject to

$$\sum_{\alpha} \hat{\mathbf{p}}^{\alpha} = \mathbf{0}$$

[Atkin and Craine, 1976; Bowen, 1975]. It is further assumed that the mixture consists of nonpolar constituents and that there is no moment of momentum supply between the phases. The balance laws of moment of momentum for each phase then yield that, as in single-substance media, the partial stress tensors must all be symmetric.

Entropy Inequality. In setting up constitutive hypotheses for each constituent, one must ensure that they do not violate the entropy inequality which states that for a mixture in which each constituent has the same temperature θ [Atkin and Craine, 1976].

$$-\sum_{\alpha} \rho^{\alpha}\frac{\overset{\alpha}{D}A^{\alpha}}{Dt} - \sum_{\alpha} \rho^{\alpha}s^{\alpha}\frac{\overset{\alpha}{D}\theta}{Dt}$$

$$+\sum_{\alpha} \hat{\mathbf{p}}^{\alpha} \cdot (\mathbf{v}^{\alpha} - \mathbf{v}) + \sum_{\alpha} \boldsymbol{\sigma}^{\alpha} : \mathbf{d}^{\alpha} - \frac{1}{\theta}\mathbf{q} \cdot \mathrm{grad}\,\theta \geq 0 \tag{3-21}$$

in which $\boldsymbol{\sigma}^{\alpha} : \mathbf{d}^{\alpha} = \mathrm{trace}(\boldsymbol{\sigma}^{\alpha} \cdot \mathbf{d}^{\alpha})$, A^{α} = partial Helmoltz free energy, s^{α} = partial entropy, and $\mathbf{q}$ = heat flux for the mixture.

Constitutive Assumptions

Constitutive equations must be provided for the state variables. This is accomplished as follows:

Equations of State. For all practical applications of interest in soil mechanics, the solid grains may be assumed incompressible, and in the following $\rho_s = $ constant. Equation 3-16 for the solid phase then simplifies to:

$$\frac{\overset{s}{D}n^w}{Dt} = (1 - n^w)\mathbf{v} \cdot \mathbf{v}^s \tag{3-22}$$

where $n^w = (1 - n^s) = $ porosity, and Equations 3-16 and 3-22 may be combined to yield the so-called "storage equation," viz.,

$$\mathbf{v} \cdot (n^w \mathbf{v}^w) + \mathbf{v} \cdot (n^s \mathbf{v}^s) = -\frac{n^w}{\rho_w} \frac{\overset{w}{D}}{Dt} (\rho_w) \tag{3-23}$$

Fluid Phase. The following constitutive equation is assumed to describe the behavior of the fluid phase

$$\boldsymbol{\sigma}_w = -p_w \boldsymbol{\delta} \tag{3-24}$$

where $p_w = $ pore-fluid pressure; i.e., it is assumed that the fluid has no *average* shear viscosity. Further, the fluid flow is assumed barotropic so that the fluid kinetic equation of state is independent of the temperature, viz.,

$$F(p_w, \rho_w) = 0 \tag{3-25}$$

from which it follows that

$$\frac{1}{\rho_w} \frac{\overset{w}{D}}{Dt} (\rho_w) = \frac{1}{\lambda^w} \frac{\overset{w}{D}}{Dt} (p_w) \tag{3-26}$$

where $\lambda^w = \rho_w \partial p_w / \partial \rho_w = $ bulk modulus of the fluid phase. The fluid pressure can thus be determined from Equation 3-23 which now writes:

$$\frac{\overset{w}{D}}{Dt} (p_w) = -\frac{\lambda^w}{n^w} [\mathbf{v} \cdot (n^w \mathbf{v}^w) + \mathbf{v} \cdot (n^s \mathbf{v}^s)] \tag{3-27}$$

For soil media, the compressibility of the fluid phase is often much smaller than the compressibility of the solid skeleton. Therefore, the fluid phase may, in some applications, be regarded as incompressible, and Equation 3-23 reduces in that case to

$$\mathbf{v} \cdot [n^w \mathbf{v}^w] + \mathbf{v} \cdot [n^s \mathbf{v}^s] = 0 \tag{3-28}$$

Solid Porous Skeleton. A rate-type constitutive equation is assumed to describe the behavior of the porous solid skeleton, of the following form:

$$\frac{\overset{s}{D}}{Dt}(\sigma'^s) = \mathbf{D}^s : \mathbf{v}^s_{()} + \mathbf{D}^G : \mathbf{v}^s_{[\,]} = \mathbf{D} : \mathbf{v}\mathbf{v}^s \tag{3-29}$$

where $\mathbf{v}^s_{()}$ and $\mathbf{v}^s_{[\,]}$ = symmetric and skew-symmetric parts of the solid velocity gradient, respectively; $\mathbf{D}^s$ is the material constitutive tensor, a (objective) tensor valued function of, possible σ'^s and the solid deformation gradient; $\mathbf{D}^G$ is the contribution from the rotational component of the stress rate, viz.,

$$D^G_{ijkl} = \frac{1}{2}\,[\sigma'^s_{il}\delta_{jk} + \sigma'^s_{jl}\delta_{ik} - \sigma'^s_{ik}\delta_{jl} - \sigma'^s_{jk}\delta_{il}] \tag{3-30}$$

Many nonlinear material models of interest can be put in the above form (e.g., all nonlinear elastic and many elastic-plastic material models). Appropriate expressions for the effective modulus tensor $\mathbf{D}^s$ for soil media are discussed in Chapter 2. For a linear isotropic elastic porous skeleton:

$$D^s_{ijkl} = \lambda^s\delta_{ij}\delta_{kl} + \mu^s(\delta_{ik}\delta_{jl} + \delta_{il}\delta_{jk}) \tag{3-31}$$

where λ^s, μ^s = effective Lame's moduli; δ_{ij} = Kronecker delta.

Momentum Supplies. Momentum interaction between the solid skeleton and the fluid phase is assumed to consist of diffusive and dilatational contributions, viz.,

$$\hat{\mathbf{p}}^s = -\hat{\mathbf{p}}^w = -\boldsymbol{\zeta} \cdot (\mathbf{v}^s - \mathbf{v}^w) - p_w \mathbf{v} n^w \tag{3-32}$$

where $\boldsymbol{\zeta}$ = symmetric, positive definite second-order tensor. The first term accounts for the momentum transfer due to diffusion phenomena and is sometimes called the "stokes drag" [Bowen, 1975]. The inclusion of such a term is basic to all porous media theories (see, for example,

Reference 4–6]. The second term is called a "buoyancy force" in mixture theories.

Coupled Field Equations

Under the assumptions just described above, the linear momentum equations Equation 3-20 simplify to:

$$\rho^s \mathbf{a}^s = \mathbf{v} \cdot \boldsymbol{\sigma}'^s - n^s \mathbf{v} p_w - \boldsymbol{\zeta} \cdot (\mathbf{v}^s - \mathbf{v}^w) + \rho^s \mathbf{b} \qquad (3\text{-}33a)$$

$$\rho^w \frac{\overset{s}{D}}{Dt}(\mathbf{v}^w) = \rho^w(\mathbf{v}^s - \mathbf{v}^w) \cdot \mathbf{v}\,\mathbf{v}^w - n^w \mathbf{v} p_w$$

$$+ \boldsymbol{\zeta} \cdot (\mathbf{v}^s - \mathbf{v}^w) + \rho^w \mathbf{b} \qquad (3\text{-}33b)$$

when the movement of the solid phase is used as the reference motion. When inertia and convective terms are neglected, Equation 3-33b reduces to Darcy's law [Darcy, 1856] as

$$n^w(\mathbf{v}^w - \mathbf{v}^s) = -(n^w)^2 \boldsymbol{\zeta}^{-1} \cdot (\mathbf{v} p_w - \rho_w \mathbf{b}) \qquad (3\text{-}34)$$

and thus $\mathbf{k} = (n^w)^2 \gamma_w \boldsymbol{\zeta}^{-1}$ = Darcy permeability tensor (symmetric, positive definite), (units: L/T); $\gamma_w = g\rho_w$ = unit weight of the fluid; g = acceleration of gravity.

Applications/Special Cases

Several simplified situations of interest in soil mechanics can be obtained as special cases of the general theory presented previously (Equations 3-22, 3-23, and 3-33), as shown hereafter.

Dynamics without Inertia and Convective Effects in Fluid. If inertia and convective effects in the fluid phase are neglected, Equations 3-33a and 3-33b can be combined to yield:

$$\rho^s \mathbf{a}^s = \mathbf{v} \cdot (\boldsymbol{\sigma}'^s - p_w \boldsymbol{\delta}) + \rho \mathbf{b} \qquad (3\text{-}35)$$

where $\rho = \rho^s + \rho^w$ = total mass density of the mixture. Further, Equation 3-33b simplifies to Equation 3-34. Then taking the divergence of both sides of Equations 3-34 and combining with Equation 3-27, one gets

$$-\mathbf{v}\cdot[n^w(\mathbf{v}^w - \mathbf{v}^s)] = \frac{n^w}{\lambda^w}\frac{\overset{w}{D}}{Dt}(p_w) + \mathbf{v}\cdot\mathbf{v}^s$$

$$= -\mathbf{v}\cdot\left[\frac{1}{\gamma_w}\mathbf{k}\cdot(\mathbf{v}p_w - \rho_w\mathbf{b})\right] \tag{3-36}$$

and finally (rearranging terms):

$$\frac{n^w}{\lambda^w}\frac{D^w}{Dt}(p_w) - \mathbf{v}\cdot\left[\frac{1}{\gamma_w}\mathbf{k}(\mathbf{v}p_w - \rho_w\mathbf{b})\right] + \mathbf{v}\cdot\mathbf{v}^s = 0 \tag{3-37}$$

In many practical cases, the compressibility of the fluid phase is often much smaller than that of the solid soil skeleton. Equation 3-37 can then be further simplified by assuming that the fluid compressibility can be neglected, as:

$$-\mathbf{v}\mathbf{k}\cdot\left[\mathbf{v}\left(\frac{p_w}{\gamma_w}\right) - \frac{1}{g}\mathbf{b}\right] + \mathbf{v}\cdot\mathbf{v}^s = 0 \tag{3-38}$$

The term $-1/g(\mathbf{b})$ can conveniently be expressed in a cartesian reference frame as $\bar{\mathbf{v}}y$ where y is the vertical coordinate (y-axis vertical, oriented upward).

Dynamics with Undrained Conditions. In that case $\mathbf{k} \simeq \mathbf{0}$ relative to the rate of loading, the pore-fluid follows the motion of the solid phase (i.e., $\mathbf{v}^w = \mathbf{v}^s$) and Equation 3-33 simplifies to:

$$\rho\mathbf{a} = \mathbf{v}\cdot(\boldsymbol{\sigma}'^s - p_w\boldsymbol{\delta}) + \rho\mathbf{b} \tag{3-39}$$

where $\mathbf{a} = \mathbf{a}^s(= \mathbf{a}^w) =$ acceleration of the mixture. The storage equation Equation 3-23 also simplifies, and

$$\mathbf{v}\cdot\mathbf{v}^s = -\frac{n^w}{\rho_w}\frac{D}{Dt}(\rho_w) = -\frac{n^w}{\lambda^w}\frac{D}{Dt}(p_w) \tag{3-40}$$

If it is further assumed that the fluid compressibility can be neglected, then (from Equations 3-22 and 3-40)

$$\frac{D}{Dt}(n^w) = 0 \quad\text{and}\quad \mathbf{v}\cdot\mathbf{v}^s = \mathbf{v}\cdot\mathbf{v}^w = 0 \tag{3-41}$$

Note that in general, changes in pore-fluid pressures will arise as a result of the strains in the solid soil skeleton.

Dynamics with Drained Conditions. In that case $\mathbf{k} = \infty$ relative to the rate of loading, and no changes in pore-fluid pressures take place as a result of the strains in the solid soil skeleton. Equation 3-33 then simplifies to:

$$\rho^s \mathbf{a}^s = \mathbf{v} \cdot (\boldsymbol{\sigma}'^s - p_w \boldsymbol{\delta}) + \rho \mathbf{b} \tag{3-42}$$

if the pore-fluid is assumed at steady-state, viz., (from Equation 3-37),

$$-\mathbf{v} \cdot \left[\frac{1}{\gamma_w} \mathbf{k} \cdot (\mathbf{v}p_w - \rho_w \mathbf{b}) \right] = 0 \tag{3-43}$$

If it is further assumed that no fluid flow is taking place (static fluid pressures), then Equation 3-33a simplifies to:

$$\rho^s \mathbf{a}^s = \mathbf{v} \cdot \boldsymbol{\sigma}'^s + \rho'^s \mathbf{b} \tag{3-44}$$

Where $\rho'^s = \rho - \rho_w = n^s(\rho_s - \rho_w) = $ buoyant mass density of the porous soil skeleton, and (from Equation 3-33b)

$$\mathbf{v}p_w = \rho_w \mathbf{b} \tag{3-45}$$

Pseudo-Static Loading Conditions. In that case inertia (and convective) effects are neglected both in the solid and fluid phases. The corresponding field equations for the various cases are simply obtained from above by setting $\mathbf{a}^s = \mathbf{a}^w = \mathbf{0}$, viz.,

"Partly" drained loading conditions (consolidation equation)

$$\mathbf{v} \cdot (\boldsymbol{\sigma}'^s - p_w \boldsymbol{\delta}) + \rho \mathbf{b} = \mathbf{0}$$

$$\frac{n^w}{\lambda^w} \frac{\overset{w}{D}}{Dt}(p_w) - \mathbf{v} \cdot \left[\frac{1}{\gamma_w} \mathbf{k} \cdot (\mathbf{v}p_w - \rho_w \mathbf{b}) \right] + \mathbf{v} \cdot \mathbf{v}^s = 0 \tag{3-46}$$

Undrained loading conditions.

$$\mathbf{v} \cdot (\boldsymbol{\sigma}'^s - p_w \boldsymbol{\delta}) + \rho \mathbf{b} = \mathbf{0} \tag{3-47}$$

with

$$\mathbf{v} \cdot \mathbf{v}^s = \mathbf{v} \cdot \mathbf{v}^w = 0 \qquad \left(\frac{D}{Dt} n^w = 0 \right)$$

if it is further assumed that the fluid compressibility can be neglected.

Drained loading conditions.

$$\mathbf{\nabla} \cdot (\boldsymbol{\sigma}'^{\,s} - p_w \boldsymbol{\delta}) + \rho \mathbf{b} = \mathbf{0}$$

$$-\mathbf{\nabla} \cdot \left[\frac{1}{\gamma_w} \mathbf{k} \cdot (\mathbf{\nabla} p_w - \rho_w \mathbf{b}) \right] = 0 \qquad (3\text{-}48)$$

or if no fluid flow is taking place

$$\mathbf{\nabla} \cdot \boldsymbol{\sigma}'^{\,s} + \rho'^{\,s} \mathbf{b} = \mathbf{0}$$

$$\mathbf{\nabla} p_w = \rho_w \mathbf{b} \qquad (3\text{-}49)$$

FORMULATION

The enormous complexities encountered in solving geotechnical engineering problems (complex geometries, geological strata, material behavior) make analytical closed-form solutions very difficult if not impossible. Recent advances in digital computer technology and in numerical methods have now rendered possible, at least in principle, the solution of any properly posed boundary value problem in mechanics. Consequently, applications of numerical techniques to geotechnical engineering problems have grown at a rapid pace during the last decade. Finite element methods appear to be the most popular procedures employed in geotechnical engineering. A number of excellent books on finite element methods are available (See, for example, Bathe and Wilson [1976] and Zienkiewicz [1977]) to which the reader is referred to for more details on the technique. This section will briefly present some applications of the technique to geotechnical engineering problems. It is convenient to classify the problems encountered in geotechnical engineering according to the nature of the differential equations to be solved, as follows:

- *Hyperbolic*—This class includes transient wave propagation and vibration-type problems (in both saturated and dry soils).
- *Parabolic*—This class includes transient consolidation type problems in saturated soil systems.
- *Elliptic*—This class includes steady state seepage flow problems and load/deformation (and failure) analysis of soil and soil-structure systems in saturated (fully drained or undrained conditions) and dry soil media.

Only the techniques employed for solving hyperbolic problems are briefly reviewed in the following.

The problem of the dynamic response of a two-phase soil system can be represented as the following initial-boundary-value problem: Let Ω denote the domain occupied by the two-phase system and Γ its boundary. Find the soil displacement field $\mathbf{u}^s(\mathbf{x},t)$ and the fluid velocity field $\mathbf{v}^w(\mathbf{x},t)$ in the domain $\Omega \in \mathbf{R}^{NSD}$ (NSD = number of space dimensions), such that

$$\rho^s \mathbf{a}^s = \mathbf{v} \cdot \boldsymbol{\sigma}'^s - n^s \mathbf{v} p_w - \boldsymbol{\zeta} \cdot (\mathbf{v}^s - \mathbf{v}^w) + \rho^s \mathbf{b} \tag{3-50}$$

$$\rho^w \frac{D^s}{Dt}(\mathbf{v}^w) = \rho^w(\mathbf{v}^s - \mathbf{v}^w) \cdot \mathbf{v}\mathbf{v}^w - n^w \mathbf{v} p_w + \boldsymbol{\zeta} \cdot (\mathbf{v}^s - \mathbf{v}^w) + \rho^w \mathbf{b} \tag{3-51}$$

$$\frac{D^s}{Dt}(\boldsymbol{\sigma}'^s) = \mathbf{D} : \mathbf{v}\mathbf{v}^s \tag{3-52}$$

$$\frac{D^s}{Dt}(n^w) = (1 - n^w)\mathbf{v} \cdot \mathbf{v}^s \tag{3-53}$$

$$\frac{D^w}{Dt}(p_w) = -\frac{\lambda^w}{n^w}\left(\mathbf{v} \cdot (n^s \mathbf{v}^s) + \mathbf{v} \cdot (n^w \mathbf{v}^w)\right) \tag{3-54}$$

together with the boundary conditions

$$\mathbf{u}^s = \bar{\mathbf{u}}^s \text{ on } \Gamma_{g^s} \tag{3-55a}$$

$$\mathbf{v}^w = \bar{\mathbf{v}}^w \text{ on } \Gamma_{g^w} \tag{3-55b}$$

$$(\boldsymbol{\sigma}'^s - n^s p_w \boldsymbol{\delta}) \cdot \mathbf{n} = \bar{\mathbf{h}}^s \text{ on } \Gamma_{h^s} \tag{3-56a}$$

$$- n^w p_w \mathbf{n} = \bar{\mathbf{h}}^w \text{ on } \Gamma_{h^w} \tag{3-56b}$$

and the initial conditions

$$\mathbf{u}^s(0) = \mathbf{u}^s_0 \tag{3-57a}$$

$$\dot{\mathbf{u}}^\alpha(0) = \dot{\mathbf{u}}^\alpha_0 \tag{3-57b}$$

Γ_{g^s}, Γ_{h^s} and Γ_{g^w}, Γ_{h^w} are the parts of the boundary on which the displacement and traction for the solid and fluid, respectively, are prescribed. They satisfy the following conditions:

$$\overline{\Gamma_{g^s} \cup \Gamma_{h^s}} = \Gamma; \qquad \Gamma_{g^s} \cap \Gamma_{h^s} = 0 \tag{3-58a}$$

$$\overline{\Gamma_{g^w} \cup \Gamma_{h^w}} = \Gamma; \qquad \Gamma_{g^w} \cap \Gamma_{h^w} = 0 \tag{3-58b}$$

When inertia and convective terms are neglected, Equation 3-50 reduces to Darcy's law:

$$n^w(\mathbf{v}^w - \mathbf{v}^s) = -(n^w)^2 \boldsymbol{\zeta}^{-1} \cdot (\boldsymbol{\nabla} p_w - \rho_w \mathbf{b}) \tag{3-59}$$

and thus $\mathbf{k} = (n^w)^2 \gamma_w \boldsymbol{\zeta}^{-1} =$ Darcy permeability tensor (symmetric, positive definite) where $\gamma_w = g\rho_w =$ unit weight of the fluid and $g =$ acceleration of gravity.

In the case of a *compressible* fluid (Equation 3-54) the fluid pressure is determined (as the effective stress, Equation 3-52) from the computed velocities through time integration of Equation 3-54. In the case of an *incompressible* fluid, Equation 3-54 is no longer available but is replaced by the continuity requirement,

$$\boldsymbol{\nabla} \cdot (n^s \mathbf{v}^s) + \boldsymbol{\nabla} \cdot (n^w \mathbf{v}^w) = 0 \tag{3-60}$$

In order to reduce the number of unknowns, it is convenient in that case to eliminate the fluid pressure from the list of unknowns. For that purpose, a *penalty-function* formulation of the continuity constraint expressed by Equation 3-60 is used to compute the fluid pressure as

$$p_w = -\frac{\lambda^w}{n^w}(\boldsymbol{\nabla} \cdot (n^s \mathbf{v}^s) + \boldsymbol{\nabla} \cdot (n^w \mathbf{v}^w)) \tag{3-61}$$

where λ^w is *not* the effective fluid bulk modulus appearing in Equation 3-54, but rather is a *penalty parameter.* The penalty parameter is to be selected as a large number. This is further discussed later.

When fluid inertia and convection terms are neglected, the problem is expressed as follows: Find the soil displacement field $\mathbf{u}^s(\mathbf{x},t)$ and the fluid pressure field $p_w(\mathbf{x},t)$ in the domain $\Omega \in R^{NSD}$ such that

$$\rho^s \mathbf{a}^s = \boldsymbol{\nabla} \cdot (\boldsymbol{\sigma}'^s - p_w \boldsymbol{\delta}) + \rho \mathbf{b} \tag{3-62a}$$

$$\frac{n_w}{\lambda^w}\frac{D^w}{Dt}p_w - \boldsymbol{\nabla} \cdot \left[\frac{1}{\gamma^w}\mathbf{k} \cdot (\boldsymbol{\nabla} p_w - \rho_w \mathbf{b})\right] + \boldsymbol{\nabla} \cdot \mathbf{v}^s = 0 \tag{3-62b}$$

$$\frac{D^s}{Dt}(\boldsymbol{\sigma}'^s) = \mathbf{D} : {}_{\boldsymbol{\nabla}}\mathbf{v}^s \tag{3-63}$$

$$\frac{D^s}{Dt}n^w = (1 - n^w){}_{\boldsymbol{\nabla}} \cdot \mathbf{v}^s \tag{3-64}$$

$$\mathbf{u}^s = \bar{\mathbf{u}}^s \text{ on } \Gamma_{\mathbf{g}^s} \tag{3-65a}$$

$$p_w = \bar{p}_w \text{ on } \Gamma_{\mathbf{g}^w} \tag{3-65b}$$

$$(\boldsymbol{\sigma}'^s - n^s p_w \boldsymbol{\delta}) \cdot \mathbf{n} = \bar{\mathbf{h}}^s \text{ on } \Gamma_{\mathbf{h}^s} \tag{3-66a}$$

$$-n^w p_w \mathbf{n} = \bar{\mathbf{h}}^w \text{ on } \Gamma_{\mathbf{h}^w} \tag{3-66b}$$

$$\mathbf{u}^s(0) = \mathbf{u}^{s0} \tag{3-67a}$$

$$p_w(0) = p_w^0 \tag{3-67b}$$

$$\dot{\mathbf{u}}^s(0) = \dot{\mathbf{u}}^{s0} \tag{3-67c}$$

where $\rho = \rho^s + \rho^w =$ total mass density of the mixture. This formulation is expressed in terms of the solid displacement and the fluid pressure only ($\mathbf{u}^s$ and p_w). However, the resulting semi-discrete finite element equations form a nonsymmetrical stiffness matrix. Taking advantage of the symmetry of the full system of equations, including fluid inertia and convection terms, allows substantial reduction in computations, which may be greater than those resulting from neglecting fluid inertia and convection. In the following, the full system of equations is used.

Weak Form/Semi-Discrete Finite Element Equations

The weak formulation associated with the initial boundary value problem is obtained by proceeding along standard lines (see, for example, Bathe and Wilson [1976] and Zienkiewicz [1977]). The associated matrix problem is obtained by discretizing the domain occupied by the porous medium into non-overlapping finite elements. Each element is defined by nodal points at which shape functions are prescribed. In general, two sets of shape functions are required for the solid displacement and the fluid velocity field, respectively. However, since attention in the following is restricted to low order (i.e., four-node plane and eight-node brick) finite elements which are the most efficient in nonlinear analysis, the same shape functions are used for both the solid and the fluid. The shape functions for the solid displacement and fluid velocity associated with node A are denoted by N^A in the following. They satisfy the relation $N^A(x^B) = \delta_{AB}$ in which x^B denotes the position vector of node B and δ_{AB} = Kronecker delta. The Galerkin counterpart of the weak formulation is expressed in terms of the shape functions and gives rise to the following system of equations

$$\begin{bmatrix} \mathbf{M}^s & 0 \\ 0 & \mathbf{M}^w \end{bmatrix} \begin{Bmatrix} \mathbf{a}^s \\ \mathbf{a}^w \end{Bmatrix} + \begin{bmatrix} \mathbf{Z} & -\mathbf{Z} \\ -\mathbf{Z}^T & \mathbf{Z} \end{bmatrix} \begin{Bmatrix} \mathbf{v}^s \\ \mathbf{v}^w \end{Bmatrix} + \begin{Bmatrix} \mathbf{n}^s \\ \mathbf{n}^w \end{Bmatrix} = \begin{Bmatrix} \mathbf{f}^{sext} \\ \mathbf{f}^{wext} \end{Bmatrix} \tag{3-68}$$

where $\mathbf{M}^\alpha$, $\mathbf{a}^\alpha$, $\mathbf{v}^\alpha$, $\mathbf{n}^\alpha$, and $\mathbf{f}^{\alpha ext}$ represent the (generalized) mass matrix, acceleration, velocity, and force vectors, respectively ($\alpha = s,w$ for the soil and fluid phases, respectively). The element contributions to node A from node B for direction i and j to the matrices appearing in Equation 3-68 are defined in the following. The terms are integrated over the spatial domain occupied by the element Ω^e. For computational simplicity a diagonal "lumped" mass matrix is used,

$$(m_{ij}^{AB})^\alpha = \delta_{ij}\delta^{AB} \int_{\Omega^e} \rho^\alpha N^A \, d\Omega \tag{3-69}$$

The momentum transfer terms give rise to the damping matrix $\mathbf{Z}$:

$$(Z_{ij}^{AB}) = \int_{\Omega^e} N^A \xi_{ij} N^B \, d\Omega \tag{3-70}$$

The external force $(\mathbf{f}^\alpha)^{ext}$ (i.e., body force, surface traction) is computed as follows:

$$(f_i^A)^{\alpha ext} = \int_{\Omega^e} \rho^\alpha b_i N^A \, d\Omega + \text{boundary terms} \tag{3-71}$$

The internal stress forces $\mathbf{n}^\alpha$ are computed as follows:

$$(n_i^A)^s = \int_{\Omega^e} (\sigma_{ij}^{'s} - n^s p_w \delta_{ij}) N_{,j}^A \, d\Omega \tag{3-72}$$

and

$$(n_i^A)^w = \int_{\Omega^e} \rho^w N^A (v_j^w - v_j^s) v_{i,j}^w \, d\Omega - \int_{\Omega^e} N_{,i}^A n^w p_w \, d\Omega \tag{3-73}$$

Time Integration

Time integration of the semi-discrete finite element equations is accomplished by a finite difference time stepping algorithm. In general im-

plicit and explicit integration procedures are available. Explicit procedures are computationally the simplest since they do not require (for a
diagonal mass matrix) equation solving to advance the solution. However, stability restricts the size of the allowable time step. On the other
hand, unconditional stability can usually be achieved in implicit procedures, which require solution of a system of equations at each time step.
For the problem at hand, a purely explicit procedure is not usually appropriate because of the unreasonably stringent time-step restriction resulting from the presence of the very stiff fluid in the mixture (even for
highly nonlinear solid material models). Recently developed methods
combine the attractive features of explicit and implicit integration. The
method used here falls under the category of "split operator methods."
Different portions of the system of equations are treated implicitly and
explicitly, reducing the system of equations to be solved. The specific
split to be made is obviously problem dependent, and the appropriate im-
plicit/explicit splits for the problem at hand are detailed in Prevost [1982,
1985].

The discretized equations of motion (Equation 3-68) can be written
symbolically as follows:

$$\mathbf{Ma} + \mathbf{Cv} + \mathbf{n}(\mathbf{d},\mathbf{v}) = \mathbf{f}^{\text{ext}} \tag{3-74}$$

where $\mathbf{M}$, $\mathbf{C}$, represent the generalized mass and damping matrices, $\mathbf{n}$
and $\mathbf{f}^{\text{ext}}$ represent generalized internal stress and external forces and $\mathbf{a}$, $\mathbf{v}$,
and $\mathbf{d}$ represent the generalized acceleration, velocity, and displacement.
Time integration is performed by using the implicit-explicit algorithm of
Hughes, et al. [1978, 1979] which consists of satisfying the following
equations:

$$\mathbf{Ma}_{n+1} + \mathbf{C}^{\mathrm{I}}\mathbf{v}_{n+1} + \mathbf{C}^{\mathrm{E}}\widetilde{\mathbf{v}}_{n+1} + \mathbf{n}^{\mathrm{I}}(\mathbf{d}_{n+1}, \mathbf{v}_{n+1})$$
$$+ \mathbf{n}^{\mathrm{E}}(\widetilde{\mathbf{d}}_{n+1}, \widetilde{\mathbf{v}}_{n+1}) = \mathbf{f}^{\text{ext}}_{n+1} \tag{3-75}$$

$$\mathbf{d}_{n+1} = \widetilde{\mathbf{d}}_{n+1} + \gamma\Delta t^2 \mathbf{a}_{n+1} \tag{3-76a}$$

$$\mathbf{v}_{n+1} = \widetilde{\mathbf{v}}_{n+1} + \beta\Delta t \mathbf{a}_{n+1} \tag{3-76b}$$

where

$$\widetilde{\mathbf{d}}_{n+1} = \mathbf{d}_n + \Delta t\, \mathbf{v}_n + (1 - 2\beta)\frac{\Delta t^2}{2}\, \mathbf{a}_n \tag{3-77a}$$

$$\widetilde{\mathbf{v}}_{n+1} = \mathbf{v}_n + (1 - \gamma)\Delta t \mathbf{a}_n \tag{3-77b}$$

The superscripts I and E refer to the parts of $\mathbf{C}$ and $\mathbf{n}$ which are treated implicitly and explicitly, respectively. Δt = time step; $\mathbf{f}_n^{ext} = \mathbf{f}^{ext}(t_n)$; $\mathbf{d}_n$, $\mathbf{v}_n$, and $\mathbf{a}_n$ are the approximations to $\mathbf{d}(t_n)$, $\mathbf{v}(t_n)$ and $\mathbf{a}(t_n)$; γ and β are algorithmic parameters which control accuracy and stability of the method. It may be recognized that when all terms are treated implicitly the procedure corresponds to the Newmark method [1959]. The quantities $\tilde{\mathbf{d}}_{n+1}$ and $\tilde{\mathbf{v}}_{n+1}$ are "predictor" values, while $\mathbf{d}_{n+1}$ and $\mathbf{v}_{n+1}$ are "corrector" values. From Equations 3-75–3-77b it is apparent that the calculations are rendered partly explicit by evaluating part of the viscous term, $\mathbf{C}^E\tilde{\mathbf{v}}_{n+1}$, and the force $\mathbf{n}^E$ in terms of predictor values based on data known from the previous step. Calculations commence with the given initial data ($\mathbf{d}_0$ and $\mathbf{v}_0$) and $\mathbf{a}_0$ which is defined by

$$\mathbf{M}\mathbf{a}_0 = \mathbf{f}_0^{ext} - \mathbf{C}\mathbf{v}_0 - \mathbf{n}(\mathbf{d}_0,\mathbf{v}_0) \tag{3-78}$$

Since $\mathbf{M}$ is diagonal, the solution of Equation 3-78 is trivial.

Implementation

At each time step Equations 3-75–3-77b constitute a nonlinear algebraic problem which is solved by an iterative procedure. An "effective static problem" is formed in terms of $\mathbf{a}_{n+1}$, which is then linearized. Within each time step the calculations are performed iteratively as summarized in Flowchart 3-1 for a Newton-Raphson-type implementation. In Flowchart 3-1,

$$\mathbf{K}^I = \partial\mathbf{n}^I/\partial\mathbf{d} \tag{3-79a}$$

$$\mathbf{C}^I = \mathbf{C} + \partial\mathbf{n}^I/\partial\mathbf{v} \tag{3-79b}$$

denote the parts of the tangent stiffness and damping matrices, respectively, to be treated implicitly. Other iterative techniques are discussed in the next section. Implicit treatment of nonlinear terms usually requires matrix reformation and factorization at each time step (and for every iteration to be performed if Newton-Raphson iterations are used). In general, it is therefore desirable to treat nonlinearities explicitly; γ and β in Equations 3-76a and b and 3-77a and b are then selected to achieve unconditional stability in the implicit solution to the problem. Then the maximum stable time step for the problem is determined from a Courant condition which must be satisfied in the explicit element group. However, if any portion of the explicit element group requires extremely small time steps, it becomes more efficient to treat that portion implicitly. The appropriate implicit-explicit split for the hyperbolic problem is discussed in Prevost [1982, 1985], and is summarized as follows.

Flowchart 3-1
Newton-Raphson Iterations

1. Set iteration counter i = 0
2. Predictor phase:

$$\mathbf{d}^{(i)}_{n+1} = \widetilde{\mathbf{d}}_{n+1}$$

$$\mathbf{v}^{(i)}_{n+1} = \widetilde{\mathbf{v}}_{n+1}$$

$$\mathbf{a}^{(i)}_{n+1} = 0$$

3. Form out-of-balance force:

$$\Delta\mathbf{f}^{(i)} = \mathbf{f}^{\text{ext}}_{n+1} - \mathbf{M}\mathbf{a}^{(i)}_{n+1} - \mathbf{C}\mathbf{v}^{(i)}_{n+1} - \mathbf{n}(\mathbf{d}^{(i)}_{n+1}, \mathbf{v}^{(i)}_{n+1})$$

4. Form effective mass: (reform and factorize only if required)

$$\mathbf{M}^* = \mathbf{M} + \Delta t\gamma\mathbf{C}^{\mathrm{I}} + \Delta t^2\beta\mathbf{K}^{\mathrm{I}}$$

5. Solution phase:

$$\mathbf{M}^*\Delta\mathbf{a}^{(i)} = \Delta\mathbf{f}^{(i)}$$

6. Corrector phase:

$$\mathbf{a}^{(i+1)}_{n+1} = \mathbf{a}^{(i)}_{n+1} + \Delta\mathbf{a}^{(i)}$$

$$\mathbf{v}^{(i+1)}_{n+1} = \widetilde{\mathbf{v}}_{n+1} + \Delta t\gamma\mathbf{a}^{(i+1)}_{n+1}$$

$$\mathbf{d}^{(i+1)}_{n+1} = \widetilde{\mathbf{d}}_{n+1} + \Delta t^2\beta\mathbf{a}^{(i+1)}_{n+1}$$

7. Convergence check: (only if i > 0)

$$\text{IF}\left[\frac{||\Delta\mathbf{f}^{(i)}||}{||\Delta\mathbf{f}^{(0)}||} \leq \text{TOL}^* \text{ . and . } \frac{||\Delta\mathbf{a}^{(i)}||}{||\Delta\mathbf{a}^{(0)}||} Q \leq (1-Q)\text{TOL}^*\right] \quad \textit{GOTO } 8$$

where

$$Q = \max\left[\frac{||\Delta\mathbf{a}^{(i)}||}{||\Delta\mathbf{a}^{(i-1)}||}, \frac{||\Delta\mathbf{a}^{(i-1)}||}{||\Delta\mathbf{a}^{(i-2)}||}\right]$$

OTHERWISE i = >i + 1 and *GOTO* 3

8. n = >n + 1

---

* Typically, TOL = 10^{-3}

Incompressible Fluid. In this case, $\mathbf{C}^\mathrm{I}$ is selected to contain both the momentum transfer and the penalty terms contribution to the equations of motion, viz.,

$$\mathbf{C}^\mathrm{I} = \begin{bmatrix} \mathbf{Z} + \mathbf{C}^\mathrm{ss} & -\mathbf{Z} + \mathbf{C}^\mathrm{sw} \\ -\mathbf{Z}^\mathrm{T} + \mathbf{C}^\mathrm{ws} & \mathbf{Z} + \mathbf{C}^\mathrm{ww} \end{bmatrix} \tag{3-80}$$

where $\mathbf{C}^{\alpha\beta}$ $(\alpha,\beta = \mathrm{s,w})$ are damping matrices arising from the penalty treatment of the fluid contribution as

$$(C_{ij}^{AB})^{\alpha\beta} = \int_{\Omega^e} \lambda^\mathrm{w} \frac{n^\alpha n^\beta}{n^\mathrm{w}} N_{,i}^A N_{,j}^B \, d\Omega \tag{3-81}$$

The convective fluid force (see Equation 3-73) is usually small and therefore is treated explicitly with no resulting computational difficulty. Note that since $\mathbf{C}^\mathrm{sw} = (\mathbf{C}^\mathrm{ws})^\mathrm{T}$, the resulting $\mathbf{C}^\mathrm{I}$ is symmetric for the choice adopted here.

As for the solid stress force (Equation 3-72) contribution to the equations of motion, three options are possible: "implicit," "explicit," or "implicit-explicit" treatment. The choice is to be made as follows:

Wave-Propagation-Type Calculations. Very short time scale (and high frequency) solutions are sought and an explicit treatment of the solid effective stress contribution is usually found most appropriate in that case. The time-step size restriction resulting from stability considerations is of the same order as the one resulting from accuracy considerations for nonlinear material models. Further, the specific implicit treatment adopted for the fluid contribution allows the calculations to be carried at a time step usually close to the time step corresponding to the propagation of the solid compressional wave.

Vibration-Type Calculations. Since the frequencies to be captured are usually much smaller than the foregoing, an implicit treatment of the solid effective stress contribution is usually most convenient in that case since it allows the time step to be selected following accuracy considerations only. Unconditional stability is to be achieved by proper selection of the algorithmic parameters as discussed later. However, for nonlinear analyses, a purely implicit treatment requires a matrix to reform/factorize at each time step (and for every iteration to be performed, in general), thus producing a considerable computational burden. In that case, it is convenient to adopt an implicit-explicit treatment of the effective stress contribution as follows: The linear part of the stiffness is treated implicitly while the remaining nonlinear part is treated explicitly. For that purpose, a solid stiffness operator is defined from Equation 3-72 through linearization as

$$\mathbf{K}_s = \begin{bmatrix} \mathbf{K}^s + \mathbf{K}^G & \mathbf{0} \\ \mathbf{0} & \mathbf{0} \end{bmatrix} \tag{3-82}$$

where $\mathbf{K}^s$ is the material tangent part, and $\mathbf{K}^G$ is the so-called "initial stress" or geometric part, formed from the tensors $\mathbf{D}^s$ and $\mathbf{D}^G$ (Equation 3-78) in the usual manner.

In the implicit-explicit procedure,

$$\mathbf{K}^I = \begin{bmatrix} \mathbf{K}_E^s & \mathbf{0} \\ \mathbf{0} & \mathbf{0} \end{bmatrix} \tag{3-83}$$

where $\mathbf{K}_E^s$ is the linear elastic contribution to the material tangent stiffness. Such a choice does not always lead to unconditional stability. The difficulty is not usually associated with the explicit treatment of $\mathbf{K}^G$ (which contains terms of the stress order and therefore usually has a negligible impact on stability), but rather from the explicit treatment of the nonlinear term $(\mathbf{K}^s - \mathbf{K}_E^s)$ for materials with a locking tendency.

In that case, care must be exercised in selecting a time step smaller than the one associated with the fastest expected wave speed corresponding to the subsequent stress histories to be followed by the material elements.

Diffusion-Type Calculations. It is sometimes desirable to capture the purely diffusive part ("consolidation" part) of the solution "dynamically." Such a necessity arises in situations in which both short and long time solutions to a dynamical problem are sought (such as in seismic- or blast-induced liquefaction simulations). As shown in Prevost [1982, 1985], by switching to an appropriate choice of the Newmark parameters, $\gamma = 3/2$ and $\beta = 1$, and by using the implicit-explicit option just described, all dynamic transients can be damped out, and purely diffusive (consolidation) solutions can be obtained "dynamically" by solving the dynamic equations.

Remarks

- The penalty treatment of the storage equation constraint requires that λ^w be selected as a "large number" capable of predominating the other moduli. It should be picked according to the relation

$$\lambda^w = C \ \mathrm{Max} \left[\frac{(n^w)^2 \gamma_w}{k}, \ \frac{1}{2} \Delta t(\lambda^s + 2 \ \mu^s) \right] \tag{3-84}$$

where C is a constant which depends only on the computer word length. Numerical studies reveal that for floating-point word lengths of 60 to 64 bits, $C \approx 10^7$.

- It has been determined, on the basis of numerical experiments, that it is better to use reduced integration of the penalty terms, and of the solid volumetric stiffness ($\mathbf{v} \cdot \mathbf{v}^s$ contribution) to alleviate potential mesh locking phenomena.

Compressible Fluid. In that case, $\mathbf{C}^I$ is selected to contain the momentum transfer term contribution to the equations of motion, viz.,

$$\mathbf{C}^I = \begin{bmatrix} \mathbf{Z} & -\mathbf{Z} \\ -\mathbf{Z}^T & \mathbf{Z} \end{bmatrix} \tag{3-85}$$

and the convective fluid force is treated explicitly as previously.

Again, the fluid pressure contribution is treated implicitly. For that purpose, a fluid stiffness operator is defined from Equations 3-54 and 3-72 through linearization as

$$\mathbf{K}_w = \begin{bmatrix} \mathbf{C}^{ss} & \mathbf{C}^{sw} \\ \mathbf{C}^{ws} & \mathbf{C}^{ww} \end{bmatrix} \tag{3-86}$$

where $\mathbf{C}^{\alpha\beta}(\alpha,\beta = s,w)$ are the same matrices as previously defined in Equation 3-80. However, note that in that case, the $\mathbf{C}^{\alpha\beta}$ matrices contribute to the stiffness matrix rather than to the damping matrix.

The same options as described previously for the solid stress force contribution to the equations are used. Note that in the implicit-explicit procedure,

$$\mathbf{K}^I = \begin{bmatrix} \mathbf{K}_E^s + \mathbf{C}^{ss} & \mathbf{C}^{sw} \\ \mathbf{C}^{ws} & \mathbf{C}^{ww} \end{bmatrix} \tag{3-87}$$

which again is a symmetric matrix.

Stability Conditions. The resulting stability conditions are summarized as follows (see Hughes, [1979] and Newmark [1959]). In all cases $\gamma \geq 1/2$:

1. *Implicit treatment:* Unconditional stability is achieved if $\beta \geq \gamma/2$ and it is recommended [Hilber, 1976] that

$$\beta = \left(\gamma + \frac{1}{2}\right)^2 / 4 \tag{3-88}$$

to maximize high-frequency numerical dissipation.

2. *Implicit-explicit treatment:* The time step restriction is

$$\omega \Delta t < 2/\left(\gamma + \frac{1}{2}\right) \tag{3-89}$$

to maximize high-frequency numerical dissipation; ω = highest natural frequency associated with the explicit part of the stiffness operator. The maximum expected frequency may be bounded by the frequency of the smallest element, viz., for a rectangular four-node bilinear element,

$$\omega = 2\,\frac{C}{L} \tag{3-90}$$

where L = smallest dimension of the element

$$C = \text{wave speed } (C = \sqrt{(\lambda^s + 2\mu^s)/\rho^s} \text{ for the linear model}).$$

For vibration calculations optimum accuracy is obtained with $\gamma = 1/2$. Increasing γ serves to dissipate high frequency numerical noise. The purely diffusive part (consolidation part) of the problem can be captured "dynamically" using the hyperbolic system developed by setting $\gamma = 3/2$ and $\beta = 1$ and using implicit-explicit integration. This choice damps out all dynamic transients.

Iterative Techniques

The nonlinear discretized system of equations developed earlier is solved at each time step using an iterative or Newton method. At the beginning of each step a load is applied and an initial solution is determined. However, since the coefficient matrix is formed using constants representing material properties of the structure at a fixed configuration and since the material properties change during the load step, equilibrium is not necessarily satisfied with the first guess at the solution, leaving an out-of-balance force. In subsequent iterations, the out-of-balance force is calculated and corrections to the solution are computed. For that purpose, the coefficient matrix may be updated to more closely represent the current material properties. Iterations are repeated until the last approximation has converged satisfactorily to the correct solution. A large amount of computational effort is expended in the iterative phase of the solution of the nonlinear system of equations. Consequently, the iterative procedure used must be both reliable and efficient for the problem to be solved. This section investigates the efficiency of various iterative procedures.

Three iterative procedures are examined here: Newton-Raphson, modified Newton-Raphson and quasi-Newton methods. Newton-Raphson involves reforming and factoring the coefficient matrix for each iteration at each time step, a computationally expensive process. However, the reformed matrix, reflecting material properties from the latest approximation to the solution, in general produces a better approximation in the next iteration. The procedure has been used successfully for both hardening and softening systems. In the modified Newton-Raphson procedure the coefficient matrix is not reformed during the iterations. Typically it is formed and factored at the beginning of each time step, but other strategies may use only the matrix formed at the first step, or require a reform after a given number of steps. Iterations are then computationally inexpensive, but convergence may require many iterations or may not be achieved. Modified Newton-Raphson iterations are not always able to solve hardening systems. Quasi-Newton methods have been used more recently to solve nonlinear systems and to minimize nonquadratic functionals. These methods involve rank one or rank two modifications (see Brodlie et al. [1973]) of the inverse of the coefficient matrix at each iteration. The modifications are cheaper to accomplish than a full reform and refactor of the matrix and yet convey enough information about the system properties to reduce the number of iterations required and ensure convergence.

Newton-Raphson Iterations. The Newton-Raphson method requires the formation and factorization of the tangent stiffness matrix corresponding to the current deformation state at each iteration i. Here the effective stiffness $\mathbf{M}^{*(i)}$ is computed as:

$$\mathbf{M}^{*(i)} = \mathbf{M} + \gamma\Delta t\mathbf{C} + \gamma\Delta t\mathbf{C}_T(\mathbf{d}^{(i)},\mathbf{v}^{(i)}) + \beta\Delta t^2\mathbf{K}_T(\mathbf{d}^{(i)},\mathbf{v}^{(i)}) \qquad (3\text{-}91)$$

where

$$\mathbf{K}_T = \partial\mathbf{n}/\partial\mathbf{d} \qquad (3\text{-}92a)$$

$$\mathbf{C}_T = \partial\mathbf{n}/\partial\mathbf{v} \qquad (3\text{-}92b)$$

At each iteration $\mathbf{K}_T$ and $\mathbf{C}_T$ (and thus $\mathbf{M}^{*(i)}$) are updated and expressed in terms of the current (i th) approximation of the displacement and velocity. The updated solution is calculated as

$$\mathbf{a}^{(i+1)} = \mathbf{a}^{(i)} + \Delta\mathbf{a}^{(i)} \qquad (3\text{-}93)$$

where

$$\Delta \mathbf{a}^{(i)} = [\mathbf{M}^{*(i)}]^{-1} \Delta \mathbf{f}^{(i)} \tag{3-94}$$

Thus each solution increment is tangential to the curve at the last iteration as depicted in Figure 3-1. Iterations proceed as described in Flowchart 3-1.

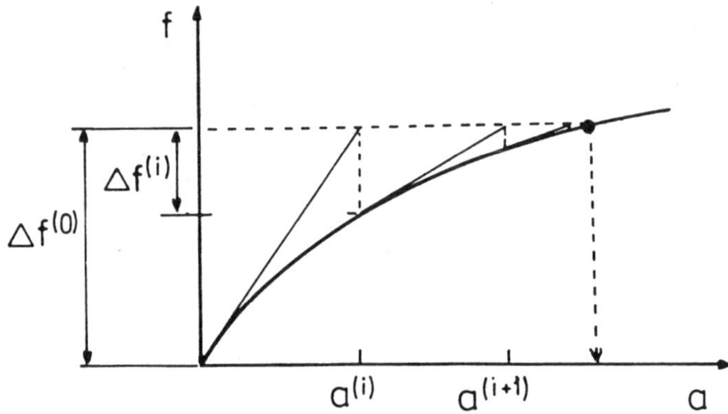

Figure 3-1. Newton-Raphson iterations.

Provided that the first iteration falls within the domain of attraction to the correct solution for Newton's method, the procedure converges superlinearly. If the effective stiffness satisfies a Lipschitz condition at $\mathbf{a}^*$ (where $\Delta \mathbf{a} = 0$), (that is $\|\mathbf{M}^*(\mathbf{a}) - \mathbf{M}^*(\mathbf{a}^*)\| \leq k\|\mathbf{a} - \mathbf{a}^*\|$ for constant k), convergence is second order [Matthies and Strang, 1979]. The disadvantage of this procedure is the great number of calculations involved. The cost of reforming and factoring the stiffness matrix at each iteration for a large system with both fluid and solid degrees of freedom for each time step of the input signal of an earthquake is very high.

Modified Newton-Raphson Iterations. For modified Newton-Raphson iterations, the tangent stiffness matrix is formed and factored only once and the update to the solution is calculated as:

$$\Delta \mathbf{a}^{(i+1)} = [\mathbf{M}^{*(0)}]^{-1} \Delta \mathbf{f}^{(i)} \tag{3-95}$$

For the iterations the initial stiffness is used and it is only necessary to calculate the out-of-balance loads and back substitute. When modified Newton iterations converge in a reasonable number of iterations, this procedure is computationally efficient. As the coefficient matrix is not updated based on information from the last iteration, it entails slow convergence for suddenly softening systems as illustrated in Figure 3-2. In other cases the procedure may diverge. For example, in the stiffening single degree-of-freedom system illustrated in Figure 3-3, at each iteration a step is taken in the direction of the initial stiffness. The system can never reach the required solution of a_{n+1} corresponding to a load increase of $\Delta f^{(0)}$. It can therefore be expected that modified Newton-Raphson iterations will perform poorly in problems involving significant compaction of geomaterials when the solid bulk modulus typically increases, and the system hardens. When the input changes from loading to unloading, materials which have yielded plastically typically harden considerably, recovering the original or elastic stiffness. In that case modified Newton-Raphson iterations are also expected to perform poorly (and may fail to converge) as shown in Figure 3-4. Seismic excitation involves persistent changes from loading to unloading, and so this method should be applied with caution to problems of nonlinear response to earthquakes.

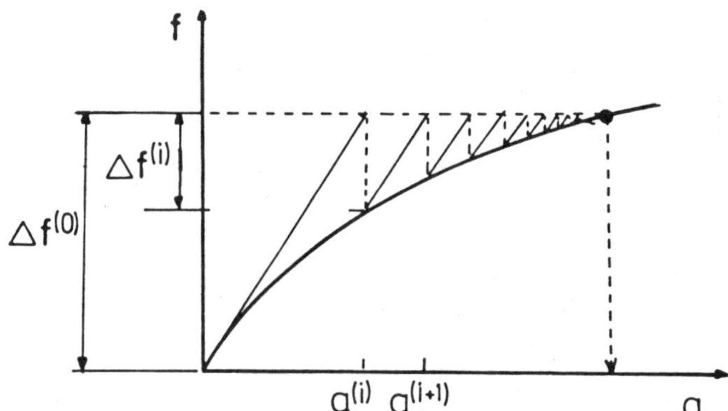

Figure 3-2. Modified Newton-Raphson iterations.

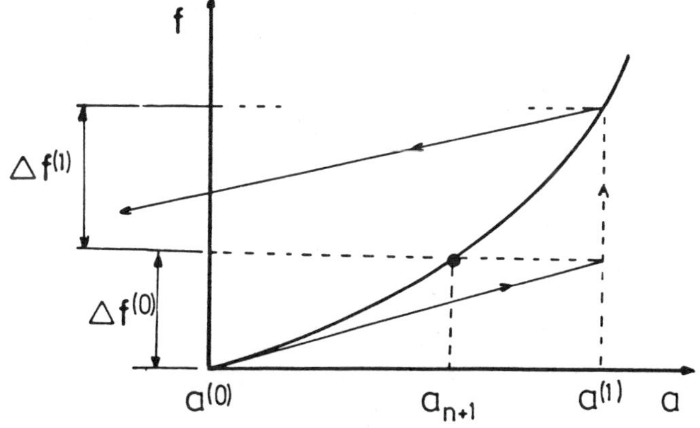

Figure 3-3. Modified Newton-Raphson iterations with divergence for a system with hardening.

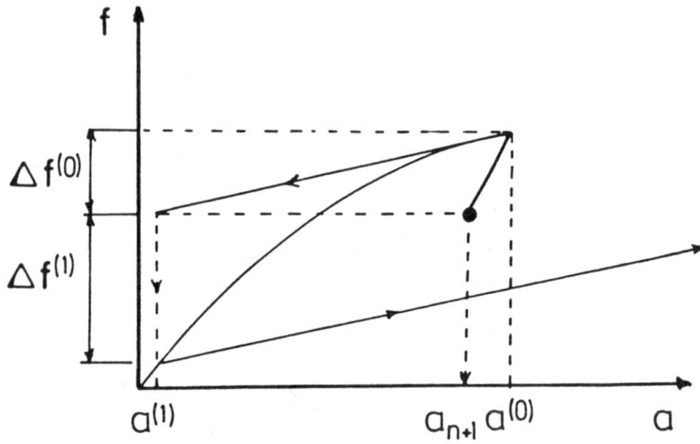

Figure 3-4. Modified Newton-Raphson iterations with divergence for nonlinear materials on unloading.

Quasi-Newton Iterations. Two quasi-Newton iterative methods are presented below: BFGS and Broyden. The quasi-Newton update of the effective stiffness matrix satisfies the quasi-Newton equation or the secant condition:

$$\mathbf{M}^{*(i)}(\mathbf{a}^{(i)} - \mathbf{a}^{(i-1)}) = \Delta\mathbf{f}^{(i)} - \Delta\mathbf{f}^{(i-1)} \tag{3-96}$$

For a single degree-of-freedom system this condition leads to the secant method, shown schematically in Figure 3-5.

The quasi-Newton or matrix update methods form a subclass of the Newton methods, a strategy which is intermediary to Newton-Raphson

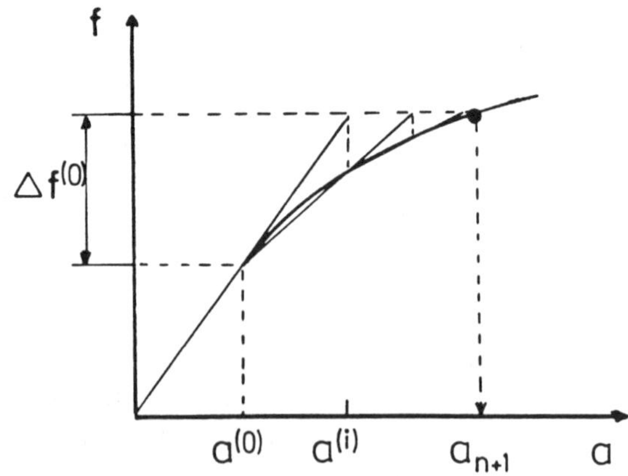

Figure 3-5. Performance of a quasi-Newton method in a one-dimensional example.

and the modified Newton methods. The coefficient matrix is modified when useful without actually reforming or refactoring the matrix. The matrix is corrected by the addition of a matrix of rank one or rank two. In practice, this modification is performed on the inverse matrix and can be accomplished by vector multiplications as shown later, so that the effective stiffness matrix is formed and factored only once for each time step. For each iteration quasi-Newton methods require fewer operations to update the coefficient matrix than the Newton-Raphson method. The convergence, however, is superlinear rather than quadratic in this case [Dennis and More, 1977]. As opposed to the modified Newton update, the quasi-Newton update can solve systems subject to loading and unloading and material hardening.

Both procedures follow the same general outline. At each iteration the effective stiffness is updated in a manner particular to the procedure, as described hereafter. The direction of the incremental solution is obtained as:

$$\mathbf{x}^{(i)} = [\mathbf{M}^{*(i)}]^{-1}\Delta\mathbf{f}^{(i)} \tag{3-97}$$

Typically an acceleration factor s is then found (by an iterative procedure or line search) which minimizes the component of the out-of-balance force in the direction of the incremental solution

$$\mathbf{x}^{(i)\mathrm{T}} \cdot \Delta\mathbf{f}(\mathbf{a}^{(i)} + s^{(i)}\mathbf{x}^{(i)}) = 0 \tag{3-98}$$

The new solution is then computed as

$$\mathbf{a}^{(i+1)} = \mathbf{a}^{(i)} + \Delta\mathbf{a}^{(i)} \tag{3-99}$$

where

$$\Delta\mathbf{a}^{(i)} = s^{(i)}\mathbf{x}^{(i)} \tag{3-100}$$

BFGS Iterations. The BFGS matrix update, discovered independently in the same year [1970] by Broyden, Fletcher, Goldfarb, and Shanno updates the coefficient matrix in a way that preserves both symmetry and positive-definiteness from one iteration to the next [Dennis and More, 1977]. The corresponding update to the inverse matrix is expressed as follows:

$$[\mathbf{M}^{*(i)}]^{-1} = (\mathbf{I} - \rho^{(i)}\boldsymbol{\delta}^{(i)}\boldsymbol{\gamma}^{(i)\mathrm{T}})[\mathbf{M}^{*(i-1)}]^{-1}(\mathbf{I} - \rho^{(i)}\boldsymbol{\gamma}^{(i)}\boldsymbol{\delta}^{(i)\mathrm{T}})$$
$$+ \rho^{(i)}\boldsymbol{\delta}^{(i)}\boldsymbol{\delta}^{(i)\mathrm{T}} \tag{3-101}$$

where

$$\delta^{(i)} = \mathbf{a}^{(i)} - \mathbf{a}^{(i-1)} \tag{3-102}$$

$$\gamma^{(i)} = \Delta\mathbf{f}^{(i)} - \Delta\mathbf{f}^{(i-1)} \tag{3-103}$$

$$\rho^{(i)} = \frac{1}{\delta^{(i)\mathrm{T}} \cdot \gamma^{(i)}} \tag{3-104}$$

In practice it is not necessary to reform the coefficient matrix. Flowchart 3-2 outlines the details of the update procedure. The solution increment is solved in terms of the inverse of the initial coefficient matrix and the matrix update is accomplished by vector multiplications.

BFGS Product Form. When the stiffness matrix is symmetric and positive definite, it is possible to express the BFGS update in a product form [Brodlie, et al., 1973; Bathe and Cimento, 1980; Matthies and Stang, 1979]:

$$[\mathbf{M}^{*(i)}]^{-1} = (\mathbf{I} + \mathbf{u}^{(i)}\mathbf{v}^{(i)\mathrm{T}})[\mathbf{M}^{*(i-1)}]^{-1}(\mathbf{I} + \mathbf{v}^{(i)}\mathbf{u}^{(i)\mathrm{T}}) \tag{3-105}$$

where

$$\mathbf{u}^{(i)} = \frac{1}{\delta^{(i)\mathrm{T}} \cdot \gamma^{(i)}} \, \delta^{(i)} \tag{3-106}$$

$$\mathbf{v}^{(i)} = \left[\frac{\delta^{(i)\mathrm{T}} \cdot \gamma^{(i)}}{\delta^{(i)\mathrm{T}} \mathbf{M}^{*(i-1)} \delta^{(i)}}\right]^{1/2} \mathbf{M}^{*(i-1)}\delta^{(i)} - \gamma^{(i)} \tag{3-107}$$

It is not necessary to form the matrix $\mathbf{M}^{*(i)}$, since $\mathbf{v}^{(i)}$ can be expressed as

$$\mathbf{v}^{(i)} = \Delta\mathbf{f}^{(i-1)} \left[1 - s^{(i-1)} \left[\frac{\mathbf{x}^{(i-1)\mathrm{T}} \cdot \gamma^{(i)}}{\delta^{(i)\mathrm{T}} \cdot \Delta\mathbf{f}^{(i-1)}}\right]^{1/2}\right] - \Delta\mathbf{f}^{(i)} \tag{3-108}$$

and the solution is calculated as

$$\mathbf{x}^{(i)} = [\Pi(\mathbf{I} + \mathbf{w}^{(j)}\mathbf{v}^{(j)\mathrm{T}})][\mathbf{M}^{*(0)}]^{-1}[\Pi(\mathbf{I} + \mathbf{v}^{(j)}\mathbf{w}^{(j)\mathrm{T}})]\Delta\mathbf{f}^{(i)} \tag{3-109}$$

involving formation and factorization only of the initial coefficient matrix. Flowchart 3-3 outlines the details of the update procedure. In this particular form of the BFGS algorithm the matrix update is easily controlled. As the two factors involved in the update are transpositions of each other, the inverse of the coefficient matrix remains symmetric and

Flowchart 3-2
BFGS Iterations

1. Form out-of-balance force

$$\mathbf{q}^{(i)} = \Delta \mathbf{f}^{(i)} = \mathbf{f}^{ext} - \mathbf{M}\mathbf{a}^{(i)} - \mathbf{C}\mathbf{v}^{(i)} - \mathbf{n}(\mathbf{d}^{(i)},\mathbf{v}^{(i)})$$

2. Compute and store

$$\boldsymbol{\delta}^{(i)} = \mathbf{a}^{(i)} - \mathbf{a}^{(i-1)}$$

$$\boldsymbol{\gamma}^{(i)} = \Delta \mathbf{f}^{(i)} - \Delta \mathbf{f}^{(i-1)}$$

$$\rho^{(i)} = \frac{1}{\boldsymbol{\delta}^{(i)T} \cdot \boldsymbol{\gamma}^{(i)}}$$

3. For $j = (i-1),\ldots,0$

   Compute and store

$$\alpha^{(j)} = \rho^{(j+1)}\boldsymbol{\delta}^{(j+1)T} \cdot \mathbf{q}^{(j+1)}$$

   Compute

$$\mathbf{q}^{(j)} = \mathbf{q}^{(j+1)} - \alpha^{(j)}\boldsymbol{\gamma}^{(j+1)}$$

4. Solve

$$\mathbf{x}^{(0)} = [\mathbf{M}^{*(0)}]^{-1}\mathbf{q}^{(0)}$$

5. Compute new search direction. For $j = 1,2,\ldots,i$ Compute

$$\beta^{(j)} = \rho^{(j)}\boldsymbol{\gamma}^{(j)T} \cdot \mathbf{x}^{(j-1)}$$

$$\mathbf{x}^{(j)} = \mathbf{x}^{(j-1)} + (\alpha^{(j-1)} - \beta^{(j)})\boldsymbol{\delta}^{(j)}$$

6. Line search for acceleration factor $s^{(i)}$

$$\mathbf{x}^{(i)T} \cdot \Delta \mathbf{f}(\mathbf{a}^{(i)} + s^{(i)}\mathbf{a}^{(i)}) = 0$$

7. Update

$$\Delta \mathbf{a}^{(i)} = s^{(i)}\mathbf{x}^{(i)}$$

positive definite. The determinant of the updated inverse matrix is easily calculated:

$$\det[[\mathbf{M}^{*(i+1)}]^{-1}] = \det[[\mathbf{M}^{*(i)}]^{-1}](1 + \mathbf{v}^{(j)T} \cdot \mathbf{w}^{(j)}) \qquad (3\text{-}110)$$

When this value becomes small, the updated matrix is nearly singular and the "condition number" becomes large. Matrix updates are omitted in this case.

Flowchart 3-3
BFGS Iterations, Product Form

1. Form out-of-balance force

$$\Delta\mathbf{f}^{(i)} = \mathbf{f}^{ext} - \mathbf{M}\mathbf{a}^{(i)} - \mathbf{C}\mathbf{v}^{(i)} - \mathbf{n}(\mathbf{d}^{(i)},\mathbf{v}^{(i)})$$

2. Compute

$$\boldsymbol{\delta}^{(i)} = \mathbf{a}^{(i)} - \mathbf{a}^{(i-1)}$$

$$\boldsymbol{\gamma}^{(i)} = \Delta\mathbf{f}^{(i)} - \Delta\mathbf{f}^{(i-1)}$$

3. Compute and store

$$\mathbf{v}^{(i)} = \left[1 + s^{(i-1)}\left[\frac{\mathbf{d}^{(i-1)T}\cdot\boldsymbol{\gamma}^{(i)}}{\boldsymbol{\delta}^{(i)T}\cdot\Delta\mathbf{f}^{(i-1)}}\right]^{1/2}\right]\Delta\mathbf{f}^{(i-1)} - \Delta\mathbf{f}^{(i)}$$

$$\mathbf{w}^{(i)} = \frac{1}{\boldsymbol{\delta}^{(i)T}\cdot\boldsymbol{\gamma}^{(i)}}\boldsymbol{\delta}^{(i)}$$

4. For $j = (i - 1),\dots,0$ compute

$$\Delta\mathbf{f}^{(j)} = \Delta\mathbf{f}^{(j+1)} + (\mathbf{w}^{(j)T}\cdot\Delta\mathbf{f}^{(j+1)})\mathbf{v}^{(j)}$$

5. Solve

$$\mathbf{x}^{(0)} = [\mathbf{M}^{*(0)}]^{-1}\Delta\mathbf{f}^{(0)}$$

6. Compute new search direction. For $j = 1,2,\dots,i$ compute

$$\mathbf{x}^{(j)} = \mathbf{x}^{(j-1)} + (\mathbf{v}^{(j)T}\cdot\mathbf{x}^{(j-1)})\mathbf{w}^{(j)}$$

7. Line search for acceleration factor $s^{(i)}$

$$\mathbf{x}^{(i)T}\cdot\Delta\mathbf{f}(\mathbf{a}^{(i)} + s^{(i)}\mathbf{a}^{(i)}) = 0$$

8. Update

$$\Delta\mathbf{a}^{(i)} = s^{(i)}\mathbf{x}^{(i)}$$

The product form is not a valid expression for the BFGS update when the stiffness is not symmetric and positive definite. However, the computation cost is similar to the original BFGS update expression and it can be a useful expression when it is important to monitor the changing value of the determinant or condition number of the stiffness.

Broyden Iterations. When the coefficient matrix is not symmetric or positive definite, there is no reason to impose the constraint of preserving these characteristics. An earlier update, proposed by Broyden and known as the Broyden update [Dennis and More, 1977; Engelman et al., 1981] requires that in addition to satisfying the secant condition, $\mathbf{M}^{*(i+1)}$ must coincide with $\mathbf{M}^{*(i)}$ on the orthogonal complement of $\mathbf{x}^{(i)}$ [Dennis and More, 1977] resulting in the matrix update formula

$$\mathbf{M}^{*(i+1)} = \mathbf{M}^{*(i)} + \frac{(\boldsymbol{\gamma}^{(i)} - \mathbf{M}^{*(i)}\boldsymbol{\delta}^{(i)})\boldsymbol{\delta}^{(i)\mathrm{T}}}{\boldsymbol{\delta}^{(i)\mathrm{T}}\boldsymbol{\delta}^{(i)}} \tag{3-111}$$

or in inverse form

$$[\mathbf{M}^{*(i)}]^{-1} = [\mathbf{M}^{*(i-1)}]^{-1} + \frac{(\boldsymbol{\delta}^{(i)} - [\mathbf{M}^{*(i-1)}]^{-1}\boldsymbol{\gamma}^{(i)})\boldsymbol{\delta}^{(i)\mathrm{T}}}{\boldsymbol{\delta}^{(i)\mathrm{T}}[\mathbf{M}^{*(i-1)}]^{-1}\boldsymbol{\delta}^{(i)}}\,[\mathbf{M}^{*(i-1)}]^{-1} \tag{3-112}$$

Details of the Broyden procedure are presented in Flowchart 3-4. Note that the matrix updates require only post multiplication rather than pre and post multiplication of the effective stiffness as in the BFGS procedure. However, the Broyden method of matrix updating does not necessarily produce a stiffness which converges to the Jacobian for a given load and acceleration increment (see Dennis and More [1977]).

Examples. The iterative procedures are tested for the example of consolidation of a sphere of saturated soil [Lacy, 1986]. This axisymmetric problem is modeled two-dimensionally, using one quadrant of a sphere of radius a = 4 (see Figure 3-6). At time t = 0, a uniform load of 10 is applied to the outside of the sphere, causing an initial uniform jump in fluid pressure ($p_w = 10$). Then consolidation takes place. The following material properties are assumed: elastic shear modulus $G_0 = 1000.0$, elastic bulk modulus $B_0 = 2G_0/3$, porosity $n^w = 0.3$, permeability $k_1 = k_2 = 1.6 \times 10^{-5}$, solid density $\rho_s = 0.02$, and fluid density $\rho_w = 0.01$. The fluid is assumed incompressible and the fluid pressure is treated with a penalty formulation as described previously. Figure 3-7 shows the assumed pressure-volume behavior of the soil skeleton. The curve follows an S-shape, typical for soil under compaction, with the slope of the curve (i.e., the bulk modulus), first decreasing and then in-

---

Flowchart 3-4
Broyden Iterations

1. Form out-of-balance force

$$\Delta \mathbf{f}^{(i)} = \mathbf{f}^{\text{ext}} - \mathbf{M}\mathbf{a}^{(i)} - \mathbf{C}\mathbf{v}^{(i)} - \mathbf{n}(\mathbf{d}^{(i)},\mathbf{v}^{(i)})$$

2. Solve

$$\mathbf{b}^{(1)} = [\mathbf{M}^{*(0)}]^{-1}\Delta\mathbf{f}^{(i)}$$

3. For $j = 1,2,\ldots,(i-1)$ compute

$$\mathbf{b}^{(j+1)} = \mathbf{b}^{(j)} + \rho^{(j)}(\boldsymbol{\delta}^{(j)} - \mathbf{r}^{(j)})\boldsymbol{\delta}^{(j)T}\cdot\mathbf{b}^{(j)}$$

4. Compute and store

$$\mathbf{r}^{(i)} = [\mathbf{M}^{*(i-1)}]^{-1}\boldsymbol{\gamma}^{(i)} = \mathbf{b}^{(i)} - \mathbf{x}^{(i-1)}$$

$$\boldsymbol{\delta}^{(i)} = \mathbf{a}^{(i)} - \mathbf{a}^{(i-1)} = -s^{(i-1)}\mathbf{x}^{(i-1)}$$

$$\rho^{(i)} = \frac{1}{\boldsymbol{\delta}^{(i)T}\cdot\mathbf{r}^{(i)}}$$

5. Calculate search direction

$$\mathbf{x}^{(i)} = \mathbf{b}^{(i)} + \rho^{(i)}(\boldsymbol{\delta}^{(i)} - \mathbf{r}^{(i)})\boldsymbol{\delta}^{(i)T}\cdot\mathbf{b}^{(i)}$$

6. Line search for acceleration factor $s^{(i)}$

$$\mathbf{x}^{(i)T}\cdot\Delta\mathbf{f}(\mathbf{a}^{(i)} + s^{(i)}\mathbf{a}^{(i)}) = 0$$

7. Update

$$\Delta\mathbf{a}^{(i)} = s^{(i)}\mathbf{x}^{(i)}$$

---

creasing. The elastoplastic constitutive theory presented in Lacy [1986] is used to model the non-linear stress-strain behavior of the material. Figure 3-8A shows the fluid pressure at the center of the sphere over the period of consolidation. The pressure initially increases and then decreases as the fluid drains out of the sphere (known as the Mandel-Cryer effect [Mandel, 1953; Cryer, 1963]. Figure 3-8B shows the displacement over time of a node (no. 65) on the periphery of the sphere.

The problem is solved using two solution techniques. First the two-phase equations are solved as a parabolic problem (see Equations 3-13–3-18), in terms of three unknowns: soil displacement in two perpendicular directions and the pore fluid pressure. This gives rise to a nonsymmetric coefficient matrix. The same problem is also solved using

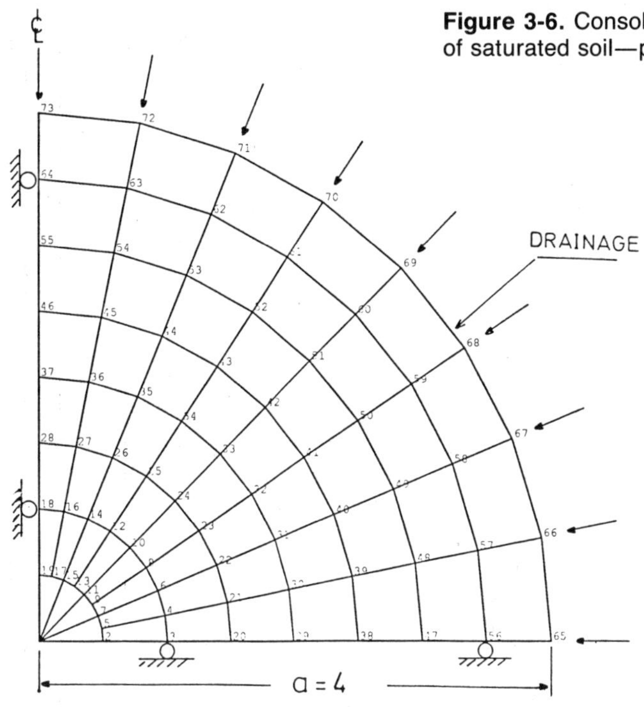

Figure 3-6. Consolidation of a sphere of saturated soil—problem geometry.

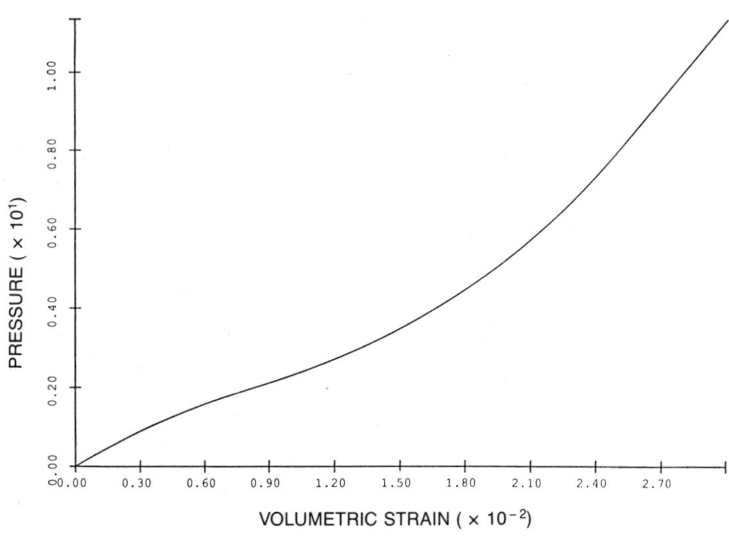

Figure 3-7. Pressure vs. volumetric strain.

the hyperbolic formulation (see Equations 3-1–3-6), in terms of four unknowns: soil displacement in two perpendicular directions and fluid velocities in two perpendicular directions. As mentioned earlier setting the Newmark parameters $\gamma = 1.5$ and $\beta = 1$ allows consolidation to be captured dynamically. This formulation gives rise to a symmetric coefficient matrix. At each of 50 time steps of equal duration, iterations were performed until convergence of 10^{-3} was reached. All calculations were performed on a Microvax II computer.

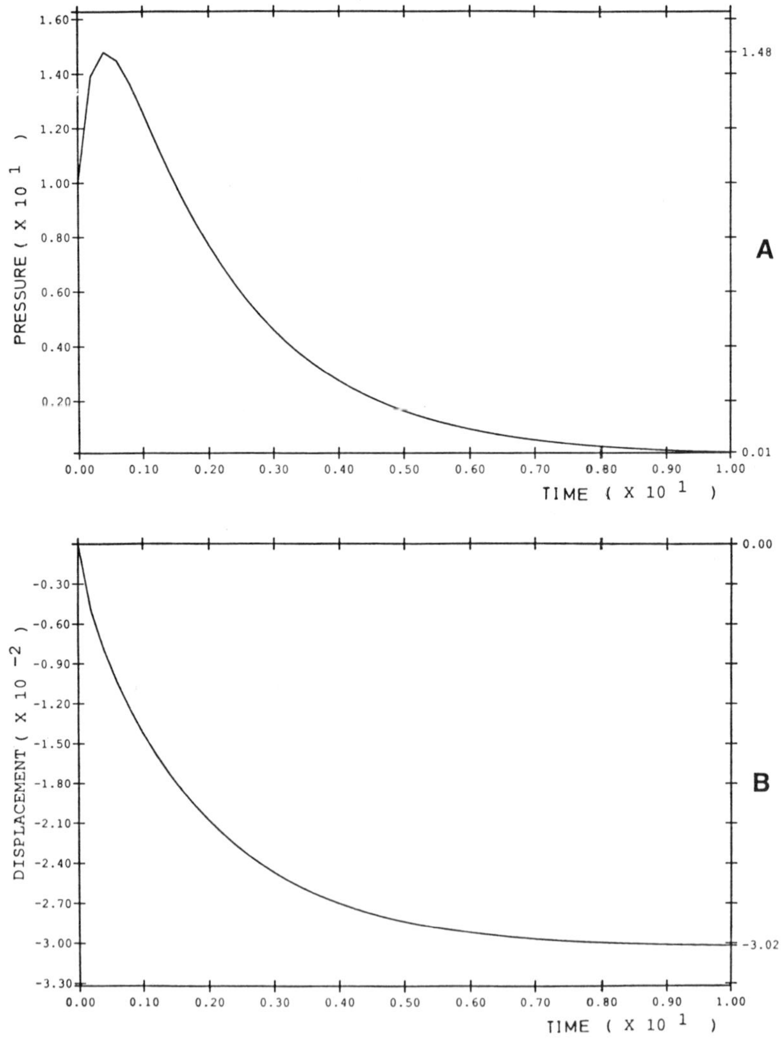

Figure 3-8. (A) Fluid pressure at sphere center (Node 1). (B) Displacement of sphere periphery (Node 65).

Table 3-1 reports the total number of iterations and the number of seconds required for solving the problem for each procedure for the parabolic formulation. Each method was used first requiring the coefficient matrix to be reformed and factored at the beginning of each time step and second using the coefficient matrix from the first time step throughout the calculations. Each procedure required more iterations when the matrix was not reformed, and except for modified Newton-Raphson the total solution time increased. The BFGS procedure performed badly, diverging in the third time step in each case. The BFGS product form, which is only valid for symmetric matrices and should not legitimately be used for this nonsymmetric formulation, did converge to solution, but required the most iterations and a lengthy solution time. For this problem the Broyden procedure performed best. It is interesting to note that the Broyden method, which is coupled with a line search procedure, took fewer iterations than the full Newton-Raphson method but took more time per iteration. The modified Newton-Raphson procedure with no line search took over twice as many iterations to converge as the Broyden method, but only required 15% to 20% more solution time.

Table 3-2 reports the details of solution of the hyperbolic (symmetric) problem. First, it should be noted that except for the Broyden iterations, the solution time for this symmetric formulation (with four degrees of freedom per node) is less than that required for the nonsymmetric parabolic formulation (with three degrees of freedom per node). All three quasi-Newton procedures performed comparably and their performances improved when the coefficient matrix was not reformed during the course of solution. The time per iteration for each method is similar. The full Newton-Raphson procedure required almost twice as many iterations as the quasi-Newton methods and larger solution times. In this case, the

Table 3-1
Consolidation of a Sphere-Parabolic Solution

	Iters.	Sec
Coefficient Matrix Reformed at each Time Step		
Newton-Raphson	127	2,683.78
Modified Newton-Raphson	153	2,182.55
BFGS	Diverges in first time step	
BFGS Product Form	168	3,113.77
Broyden	69	1,836.28
No Reformation of Coefficient Matrix		
Modified Newton-Raphson	186	1,950.42
BFGS	Diverges in first time step	
BFGS Product Form	314	4,728.38
Broyden	100	1,695.97

peformance of the modified Newton-Raphson method did not improve when the stiffness was updated at each time step, and it required only about 5% more solution time than the quasi-Newton methods.

Table 3-2
Consolidation of a Sphere-Hyperbolic Solution

	Iters.	Sec
Coefficient Matrix Reformed at each Time Step		
Newton-Raphson	116	2,586.15
Modified Newton-Raphson	124	2,017.47
BFGS	66	1,943.90
BFGS Product Form	66	1,947.45
Broyden	67	1,951.13
No Reformation of Coefficient Matrix		
Modified Newton-Raphson	139	1,663.73
BFGS	73	1,566.43
BFGS Product Form	73	1,540.10
Broyden	74	2,070.67

EXAMPLES

In this section, numerical results for dynamic transient phenomena in saturated porous media are presented. In the calculations reported hereafter, the four-node bilinear isoparametric element (see, for example, Bathe and Wilson [1976] and Zienkiewicz [1977] for detailed descriptions) was used with the standard selective integration scheme [Malkus and Hughes, 1978]. Also, the algorithmic parameters were selected such that $\gamma \geqslant \frac{1}{2}$ and $\beta = (\gamma + \frac{1}{2})^2/4$ to maximize high-frequency numerical dissipation.

One-Dimensional Wave Propagation

Biot [1956] first analyzed the propagation of plane progressive waves in a fluid-saturated elastic porous medium. Further derivations on the subject matter can be found in Bowen and Lockett [1983], Bowen and Reinicke [1978], Bowen and Chen [1975]. One solid rotational and two dilatational diffusive waves will in general propagate through the porous medium.

In order to illustrate the essential features of the phenomena associated with the propagation of dilatational waves, and the performance of the proposed numerical schemes, numerical results for a simple one-dimensional wave propagation are presented. In the analyses to follow, the initial-value problem consists of a porous medium column, fixed at one end and subjected to an impulsive load at the other end. The first 10 m are assumed dry, and the remains saturated. The movements of both solid and fluid phases are constrained to take place in the vertical direction only. The mesh is shown in Figure 3-9 and consists of three groups of elements:

Group 1: $0 \leq x \leq 10$ consists of 10 equally spaced elements

Group 2: $10 \leq x \leq 100$ consists of 90 equally spaced elements

Group 3: $100 \leq x \leq 800$ consists of 100 equally spaced elements.

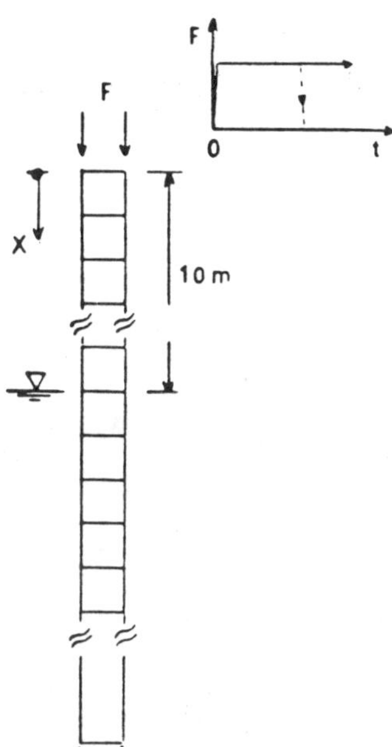

Figure 3-9. One-dimensional wave propagation: problem geometry—finite element mesh.

The total length of the mesh was selected such that within the space-time window of interest, no reflections from the fixed end of the column would take place in order not to obscure the interpretation of the results. Upon impact at the free end, a square compressive wave is sent through the column, where it hits the water table and gets partly reflected, partly transmitted through the saturated porous medium.

Elastic Porous Skeleton Filled with a Compressible Fluid. The material properties are selected as follows:

- *Solid skeleton:*

$$\lambda^s + 2\mu^s = 10^7 \text{ N/m}^2 \qquad \text{(Young's modulus)}$$

$$\rho_s = 2 \times 10^3 \text{ kg/m}^3 \qquad \text{(Mass density; solid grains)}$$

$$n^w = 0.50 \qquad \text{(Porosity)}$$

- *Fluid:*

$$\lambda^w = 5.0 \times 10^8 \text{ N/m}^2$$

$$\rho_w = 1.0 \times 10^3 \text{ kg/m}^3 \qquad \text{(Mass density; fluid phase)}$$

The porous medium column is subjected to a ramp load at its free end. In order to obtain simple (and easy to interpret) pictures, absorbing dampers have been placed at the free end to prevent reflections.

From Biot [1956], Bowen and Lockett [1983], Bowen and Reinicke [1978], Bowen and Chen [1975], the propagation speeds C_{P_1} and C_{P_2} associated with the characteristic manifolds corresponding to Equations 3-50 and 3-51 are

$$C_{P_2} = 162.63 \text{ m/sec}; \qquad C_{P_1} = 856.48 \text{ m/sec}$$

and the "frozen" mixture speed is:

$$C_0 = 820.57 \text{ m/sec}$$

Upon impact, two dilatational waves (both dispersive and diffusive) will in general propagate through the saturated porous medium with wave speeds C_1 and C_2 such that:

$$0 \le C_2 \le C_{P_2} < C_0 \le C_1 \le C_{P_1}$$

In the following, the permeability is selected as

$$k = 5.00 \times 10^{-2} \text{ m/sec}$$

such that the drag coefficient

$$\zeta = 5 \times 10^3 \text{ kg/m}^3/\text{sec}$$

and the characteristic frequency is

$$\omega_0 = \zeta \left(\frac{1}{\rho^s} + \frac{1}{\rho^w} \right) = 15 \text{ rad/sec}$$

In the calculations reported hereafter, the algorithmic parameters are selected such that $\alpha = 0.65$ and $\beta = 0.33$, and the solid effective stress is treated explicitly. The time integration is performed with $\Delta t = 8.5 \times 10^{-3}$ sec close to the critical time step ($\Delta t = 8.7 \times 10^{-3}$ sec). Figure 3-10 shows the computed solid (Figure 3-10A) and fluid (Figure 3-10B) velocities as recorded at distances x = 5.0; 10.0; 15.0; 20.0; 25.0; 30.0; 35.0; 40.0; 45.0; and 50.0 m away from the free-end. Note the distinct two wave fronts structure in the responses recorded in the near field of the saturated medium. The second wave is more diffusive than the first one and eventually disappears in the far field, where only the wave of the first kind prevails. Note the sharpness of the computed results even though the calculations are carried at the relatively large time step associated with the solid wave speed 100 m/sec.

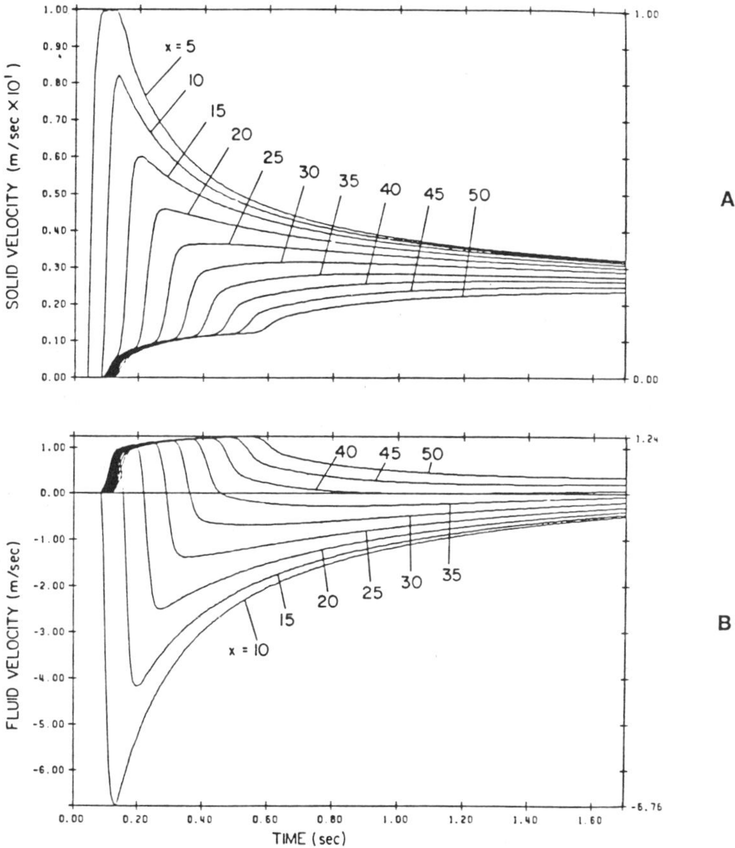

Figure 3-10. Computed velocity time histories—compressible fluid: (A) Solid. (B) Fluid.

Figure 3-11 shows the computed changes in effective axial stress (Figure 3-11A) and pore fluid pressure (Figure 3-11B), at distances $x = 15.0$; 20.0; 25.0 … ,; and 50.0 m away from the free end. Note the effect of diffusion taking place after passage of the waves.

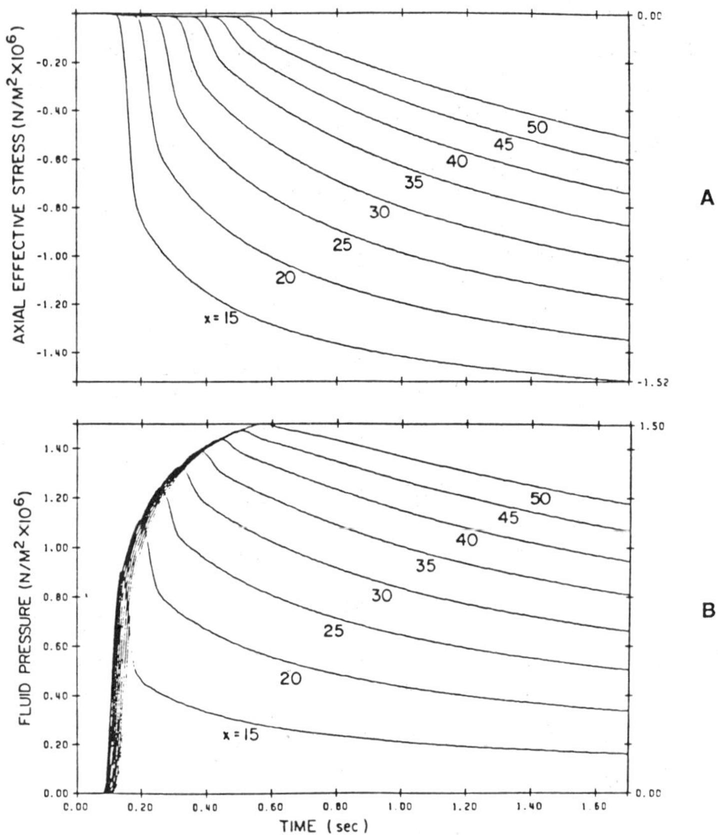

Figure 3-11. Computed stress time histories—compressible fluid: (A) Axial effective stress. (B) Fluid pressure.

Elastic Porous Skeleton Filled with an Incompressible Fluid. The problem is the same as just analyzed except that the fluid is now assumed incompressible, and a penalty formulation is adopted. In the following, the penalty parameter is selected as

$$\lambda^w = 10^{12}$$

The same algorithmic parameters and the same time step as in the previous example are used. Figure 3-12 shows the computed solid (Figure

3-12A) and fluid (Figure 3-12B) velocities as recorded at distances
x = 5.0; 10.0; ... ; and 50.0 m away from the free-end. Note that due to
the assumption of an incompressible fluid, the first wave-like motion has
disappeared and only the wave of the second kind remains. It is the wave
of the second kind which is of most interest in engineering applications
since it is the one that triggers the large changes in effective normal
stresses (see Figure 3-11A). Figure 3-13 shows the computed changes in
effective axial stress (Figure 3-13A) and pore fluid pressure (Figure
3-13B), at distances x = 15.0; 20.0; ... ; and 50.0 m away from the free
end. As expected, a jump in fluid-pressure is recorded at every point
down the saturated porous medium at time t = 0.1 sec when the solid
wave first hits the water surface. No changes in effective stresses occur
before arrival of the wave of the second kind. As previously, changes in
effective stress and pore fluid pressure are induced by the passage of the
solid stress wave and diffusion takes place thereafter.

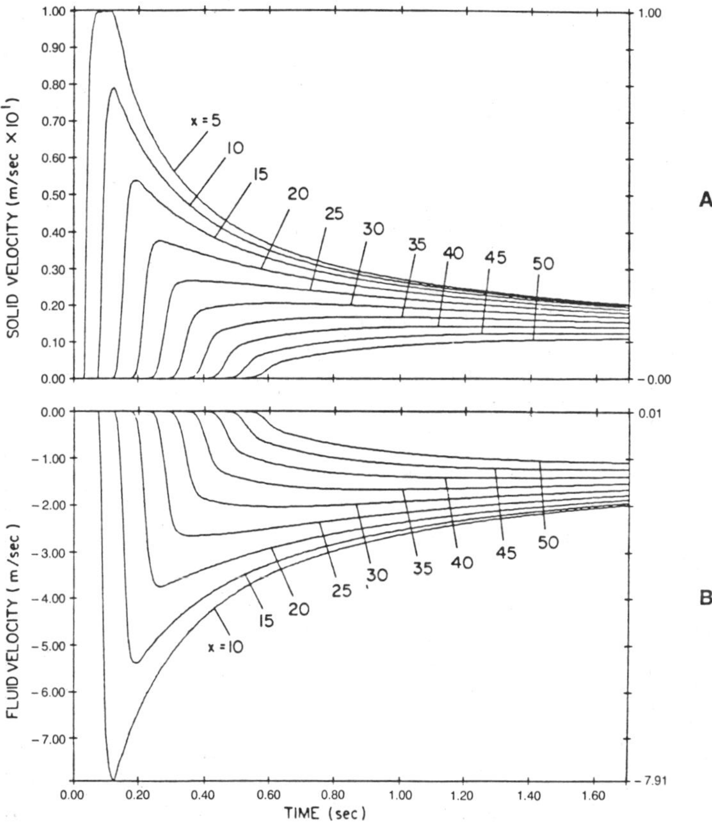

Figure 3-12. Computed velocity time histories—incompressible fluid: (A) Solid.
(B) Fluid.

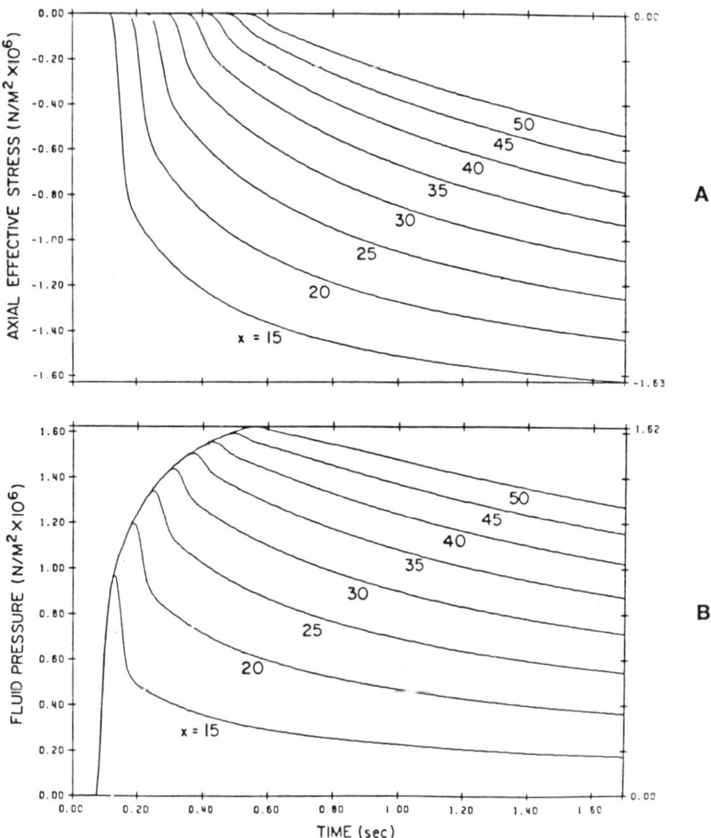

Figure 3-13. Computed stress time histories—incompressible fluid: (A) Axial effective stress. (B) Fluid pressure.

Seismic Site Response Analysis

A most important and essential step in seismic analysis of dynamic soil-structure interaction is the solution of the free-field ground motion problem (often called the site response problem). This step is required to compute compatible motions at the boundaries of the discrete soil model in direct methods of analysis. Also, it is required for the convolution analysis to define appropriate motions at the bottom (or outcropping), which yield the (usually specified) free-field motions at the surface of the soil deposit.

To compute the site response, assumptions must be made regarding the types of waves propagated during the earthquake. The most common assumptions are that the soil is horizontally stratified and that the excitation consists of vertically propagating dilatational (P-waves) and shear (S-waves). The dilatational waves only produce vertical accelerations, and the problem then becomes one-dimensional. However, due to the presence of coupling between shear and volumetric deformations in soil media, the shear waves will, in general, produce both horizontal and vertical motions. These effects are usually disregarded, and the shear wave propagation is commonly analyzed as a one-dimensional problem (see, for example, Schnabel et al. [1972]). Such an assumption is certainly valid for saturated soil media if no drainage of the pore fluid can take place within the time frame of the seismic excitation. However, for dry soil deposits and for saturated soil deposits of moderate permeabilities in which drainage can take place, horizontal motions will in general be accompanied with vertical motions and such effects should be accounted for in the analysis. Further, in cases in which potential liquefaction is of concern, a complete effective stress analysis which models directly the nonlinear hysteretic stress-strain response of soils should be conducted. In such an analysis, the build ups in pore water pressures and their dissipation with time are computed, and their effects on the dynamic response are taken into account. Effective stress analysis based on the solution of *uncoupled* equations for the porous solid soil skeleton and the pore water fluid have been proposed (see, for example, Lee, Finn, Liam [1975]; Martin and Seed [1979] and Streeter et al. [1974]). However, a rational and complete analysis should be based on the solution of the *fully coupled* solid soil skeleton/pore water fluid, as proposed by Dikmen and Ghaboussi [1977, 1978, 1979, 1981, 1984]. This becomes an essential feature when the analysis capabilities are not to be restricted to site response analysis only, but are to be applicable also to the solution of the complete soil-structure interaction problem. A very important point first raised by Roesset [1981] which must then be considered, is that in order to conduct *consistent* analyses, the determination of the compatible motions in the site response analysis must be performed with the *same* mathematical model used for the study of the complete soil-structure system. This will ensure compatibility of formulations, field and constitutive equations, and time integration procedures. Performing the site response analysis with a *special purpose* program such as LASS [Ghaboussi and Dikmen, 1977, 1979] and the main analyses with *another more general* multidimensional computer program, introduces unnecessary and undesirable uncertainties into the analysis.

In the following a procedure is presented [Prevost, 1986] which allows site response analyses to be performed with any general *multidimensional* finite element analysis package. The procedure is general, and is easily and concisely implemented in any multi-purpose finite element analysis package. As an illustration of the procedure, the results of an effective stress analysis for the seismic response and liquefaction of a horizontally layered saturated sand deposit are presented.

Method of Analysis. For a proper simulation of the free-field conditions in a horizontally stratified ground, the nodes located on the same horizontal plane must remain equally spaced and must undergo parallel motions. The difficulty in simulating such conditions by using two-dimensional (plane-strain) or three-dimensional finite elements stems from the presence of elemental rocking rigid body motions (whose presence is necessary in order to satisfy completeness requirements for the elemental shape functions). For instance, consider a column of two-dimensional plane elements such as is shown in Figure 3-14. To each node are assigned two translational degrees of freedom, in the horizontal and vertical directions, respectively. It is well known that if a plane horizontal shear wave propagation is initiated at the base of the column, it will not propagate vertically as a plane horizontal shear wave, unless the elements' rocking motions are eliminated by constraining the nodes from moving vertically (i.e., roller-type boundary conditions are used for all nodes). However, to resort to such a procedure is not appropriate for free-field simulations in anisotropic soil media, where vertically propagating shear waves will, in general, produce both horizontal and vertical motions. Further, the free-field requirement of having the nodes on the same horizontal planes undergo parallel motions, would also require the nodes to be artificially tied together (e.g., by using stiff truss elements).

The free-field simulation requirements, namely that the nodal planes must remain horizontal and must undergo parallel motions, are *exactly* specified by assigning the *same equation number* to each nodal degree of freedom on the same horizontal plane (e.g., by having nodes A and B in Figure 3-14 share the same equation numbers (1 and 2) in the horizontal and vertical directions, respectively). The procedure is *exact* in satisfying the free-field conditions. Further, it is trivially implemented. A similar procedure has been used previously in Zienkiewicz and Scott [1972] to simulate cyclic symmetry in structures.

Example. The horizontally-layered ground is divided into a number of two-dimensional plane elements as shown in Figure 3-14. The system is to be subjected to base acceleration. In the analysis to follow, the water table is assumed to be located at the ground surface, and the site to consist of a saturated cohesionless soil deposit. To each node are assigned four translational degrees of freedom: two for the solid phase (soil skeleton) and two for the fluid phase (pore water). In the free-field simulation, the solid nodal planes must remain horizontal and must undergo parallel motions. Further, the fluid phase must only move (relative to the solid phase) in the vertical direction (i.e., no lateral drainage is allowed). Again, this is *exactly* specified by assigning the same equation number to each nodal degree of freedom (i.e., both for the solid and fluid phases in the horizontal and vertical directions, respectively), on the same horizon-

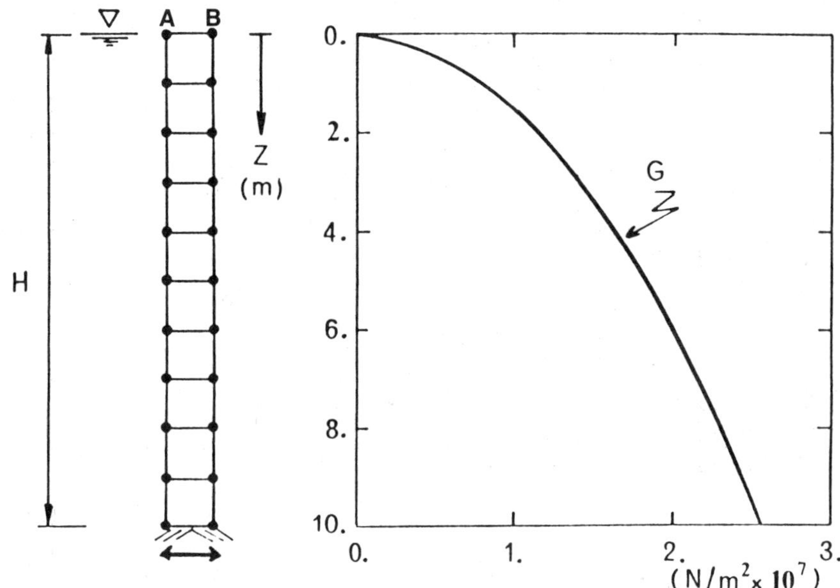

Figure 3-14. Seismic site response soil profile used in the analysis—finite element mesh.

tal planes. Therefore, for the mesh shown in Figure 3-14, which consists of ten four-noded bilinear isoparametric elements with four degrees of freedom at each node, only forty equations are required.

Figure 3-14 shows the particular soil deposit to be analyzed and the finite element mesh used. The soil deposit is 10 m deep with the water table at the ground surface. The deposit is discretized by using 10 four-node plane finite elements and is to be subjected as a horizontal base acceleration to the Pacoima acceleration time history recorded during the San Fernando Earthquake (California) on February 9, 1971, with the maximum acceleration scaled to 0.12 g. The first 12 seconds of the input ground shaking (scaled) are shown in Figure 3-15A, and its Fourier transform is shown in Figure 3-15B. The signal is rich in frequencies from .5 to 5Hz. No drainage is assumed to take place through the rigid bottom boundary, and the ground shaking is applied as a horizontal input acceleration at the bottom boundary nodes. The particular loose sand which forms the deposit to be analyzed is the one reported in Chapter 2. The nonlinear and hysteretic stress-strain behavior of the sand is modeled by using the effective-stress elastic-plastic model also reported in Chapter 2.

For the deposit shown in Figure 3-14, it is assumed that the initial soil moduli (elastic and plastic) increase with the square-root of the depth, as illustrated in Figure 3-14 for the initial elastic shear modulus. The material properties are assumed as follows:

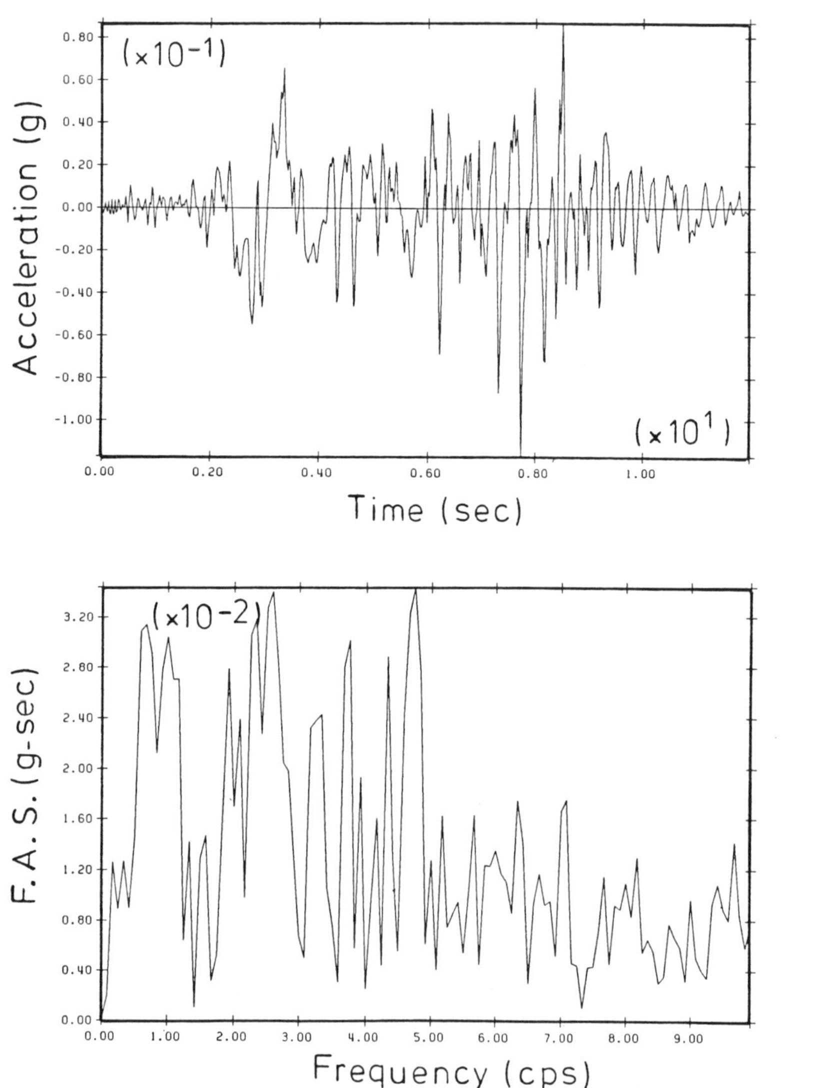

Figure 3-15. Input solid horizontal acceleration (Pacoima record scaled to 0.12 g): (A) Time history. (B) Fourier transforms.

- *Soil Skeleton*

$$\rho_s = 2.65 \times 10^3 \text{ kg/m}^3 \qquad \text{(mass density; solid grains)}$$
$$n^w = 0.43 \qquad \text{(porosity)}$$
$$k_{11} = k_{22} = k; \quad k_{12} = 0 \qquad \text{(coefficient of permeability)}$$

- *Fluid*

$$\lambda^w = 10^{10} \text{ N/m}^2 \qquad \text{(bulk modulus)}$$
$$\mu^w = 0.0 \qquad \text{(shear viscosity)}$$
$$\rho_w = 10^3 \text{ kg/m}^3 \qquad \text{(mass density; fluid phase)}$$

The total mass density (solid and fluid) is defined through

$$\rho = (1 - n^w)\rho_s + n^w\rho_w = 1.94 \times 10^3 \text{ kg/m}^3$$

and the buoyant mass density is computed as:

$$\rho_b = \rho - \rho_w = 0.94 \times 10^3 \text{ kg/m}^3$$

The unit weights (total, buoyant, . . .) γ are defined by $\gamma = \rho g$, where $g = 9.81$ m/sec^2 = acceleration of gravity.

Figure 3-16 shows the computed horizontal acceleration time histories at the top of the soil column for various values of the permeabilities: $k = \infty$ (fully drained), $k = 10^{-2}$m/sec and $k = 10^{-4}$m/sec, respectively, together with the input base acceleration time history. Note the strong amplification of the signal computed at the surface for drained (or partly drained) conditions. However, note that as a result of the decrease in permeability, major pore water pressure increases occur, and the deposit eventually liquefies for $k = 10^{-4}$ m/sec after 8 sec of shaking. This is illustrated in Figure 3-17 which shows the time histories at various depths of excess pore water pressures, vertical effective stresses and shear stresses normalized with respect to the *in-situ* initial vertical effective stresses. It can be seen that major pore water pressure increases occur during the strong phase of the shaking (after about 8 sec of shaking), and that the top 3 m of the deposit thereafter, liquefy.

It is also of interest to note that as a result of the coupling between shear and volumetric deformations, vertical motions are in general expected to be generated by the passage of the shear waves through the deposit. This is illustrated in Figure 3-18, which shows the computed vertical acceleration time histories at the top of the deposit for a fully drained case ($k = \infty$). The peak vertical acceleration in that case is computed to be 0.075 g.

Nonlinear Response Analysis of an Earth Dam

The techniques described in the previous sections are used to analyze the dynamic response of the Santa Felicia dam to the San Fernando earthquake [Lacy and Prevost, 1986]. This example is chosen for several rea-

 Geotechnical Modeling and Applications

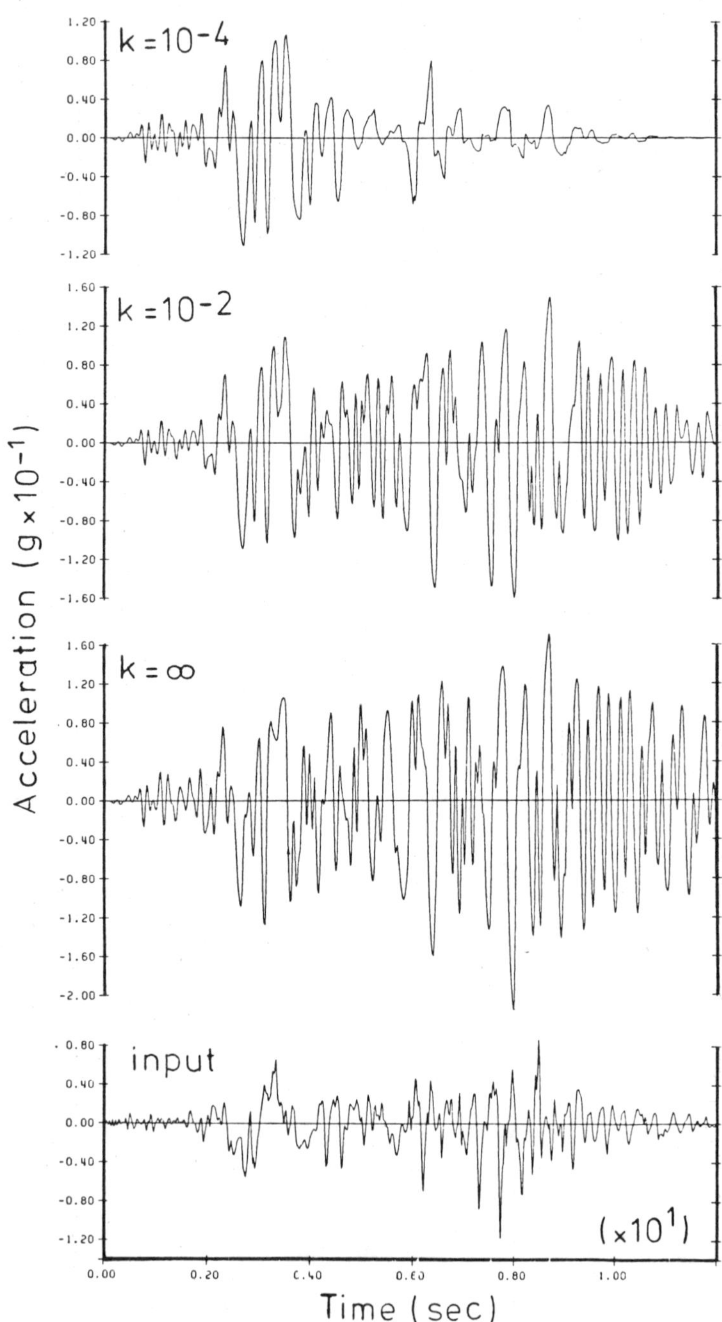

Figure 3-16. Computed horizontal surface acceleration time histories.

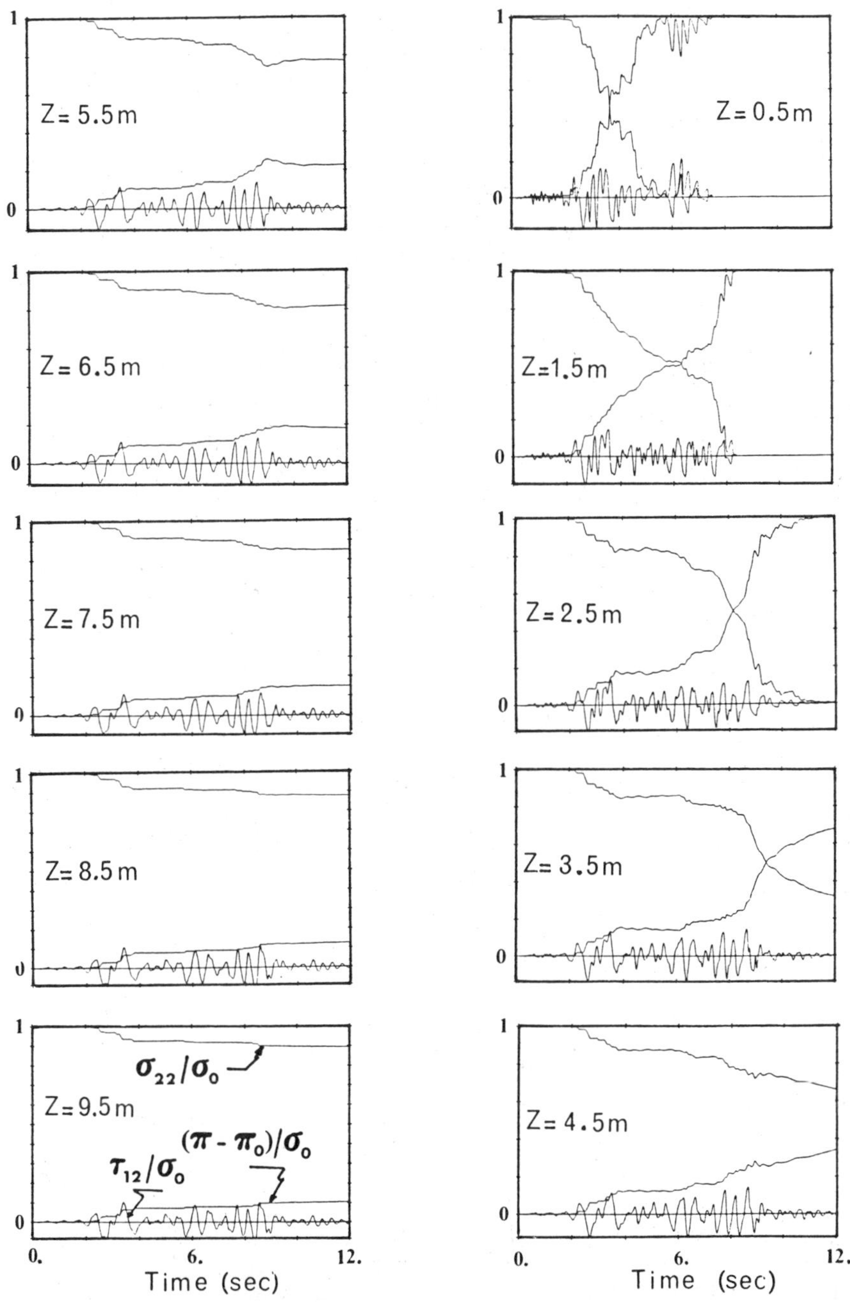

Figure 3-17. Computed excess pore water pressures, vertical effective stresses, and shear stresses time histories (in this example $k = 10^{-4}$ m/sec).

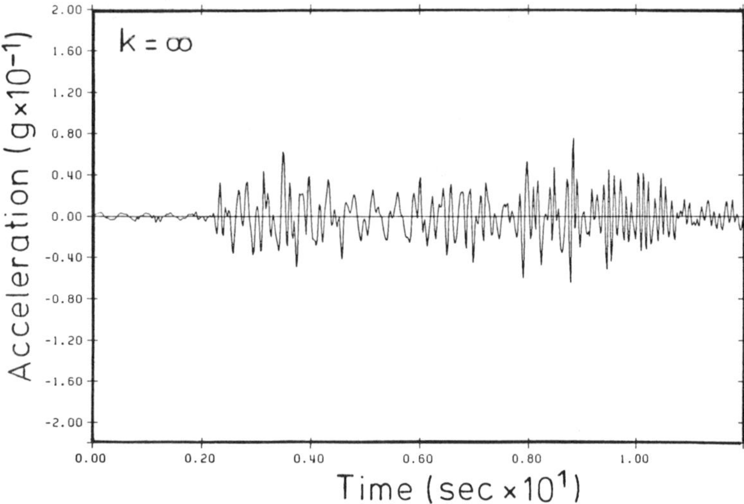

Figure 3-18. Computed vertical surface acceleration time history (in this example k = ∞(drained).

sons. First, the response of an earth dam to an earthquake is a complicated problem which uses a wide range of the numerical schemes presented in this work. The dam is large and nonhomogeneous with a complicated geometry that favors solution by the finite element method. The soil is inelastic and saturated beneath the free surface (a priori unknown). The nonlinearities in the material recommend the use of a sophisticated constitutive model. Also, the earth dam and earthquake in this example are well documented. There are available detailed reports of the material properties of the Santa Felicia dam [Abdel-Ghaffar and Scott, 1978]. Recorded accelerations of the San Fernando earthquake are available at both the abutment and crest of the dam. Using the abutment record as input, the predicted crest acceleration may be compared with the recorded response.

Problem Geometry/Finite Element Mesh. The Santa Felicia dam is a modern rolled-fill earth embankment built in 1954–1955 and located in Ventura County, California, about 65 km northwest of Los Angeles. The dam is 273 feet high above its lowest foundation. The crest is 30 feet wide and is 1,275 feet long. The upstream and downstream faces slope at 2.25:1 and 2:1, respectively, The dam has an impervious core that rises from bedrock with slopes of 0.33:1. Figure 3-19 shows the cross section at midlength, a longitudinal cross section, and a plan view of the dam.

Figure 3-20 shows the two-dimensional finite element model, representing the maximum cross section of the dam, used for the dynamic analysis. The mesh consists of 156 nodes. Each node is assigned 4 possible degrees of freedom: two perpendicular directions of soil displacement and two perpendicular directions of fluid velocity. For purposes of this analysis it was assumed that the reservoir was full to 240 feet. The free surface of fluid was located using the procedure described in Lacy [1986], and Lacy and Prevost [1986]. Then the dam was divided into 66 saturated elements and 67 dry elements with nodes having no fluid degrees of freedom. A plane strain assumption was used.

The canyon walls are made of rock. These walls were assumed rigid in all cases, and the nodes located along them were constrained to move in

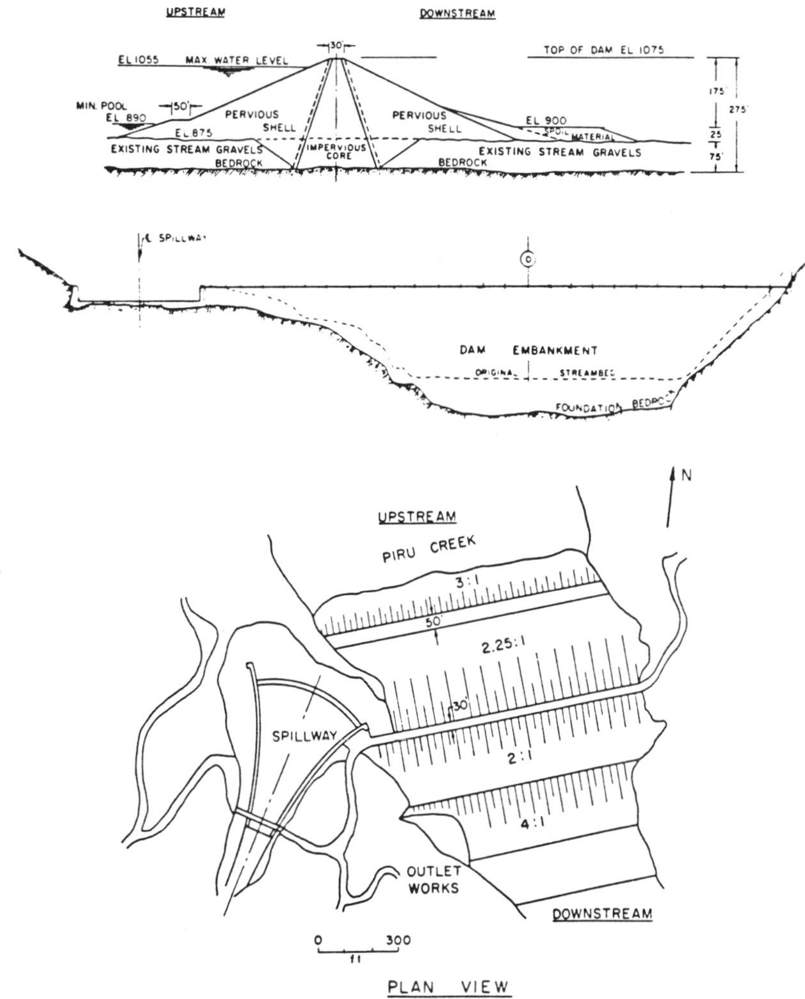

Figure 3-19. Santa Felicia earth dam structural details.

phase and with the same amplitude. Given the size of the dam, variation of the earthquake signal along the canyon boundary is likely, and the assumption of uniformity probably decreases the accuracy of the solution.

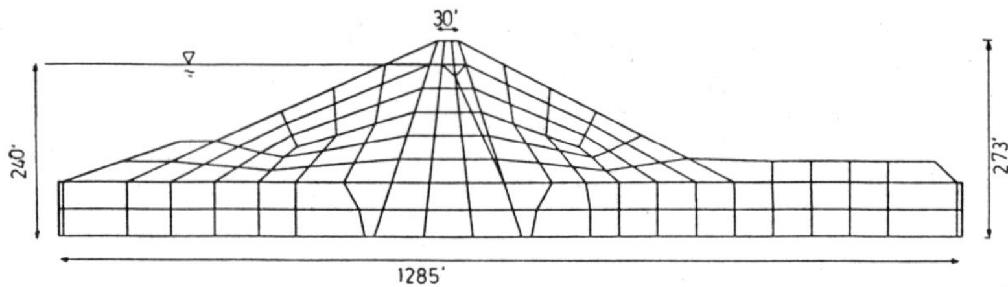

Figure 3-20. Santa Felicia earth dam finite element mesh.

Material Properties. The Santa Felicia dam is made of an impervious clay core covered by a sand and gravel shell resting on a stiff foundation layer of gravel and sand down to bedrock. The material properties were extracted from tests reported in Abdel-Ghaffar and Scott [1978, 1981]. Table 3-3 presents solid mass density (ρ_s), porosity (n^w), permeability (k), and friction angle (ϕ) for each material. The low strain elastic shear moduli for each material at each depth were back calculated from shear wave velocity measurements taken at the site [Abdel-Ghaffar and Scott, 1978]. These are listed in Table 3–4. The elastic bulk modulus is assumed to be $B_0 = 2/3G_0$. The shear and bulk moduli vary with pressure

$$ G = G_0 \left[\frac{p}{p_0}\right]^n ; \qquad B = B_0 \left[\frac{p}{p_0}\right]^n $$

where n = 0.5 for sand
      n = 1.0 for clay
and p_0 = the computed overburden pressure for the element group.

The pore fluid is assumed compressible, with a bulk modulus of $\lambda^w = 4.464 \times 10^7$ lb/ft^2.

The constitutive theory presented in Lacy [1986] and Prevost [1985] was used to model the nonlinear hysteretic stress-strain behavior of the dam materials. The material plasticity is pressure-sensitive and modeled by nested conical yield surfaces. A hyperbolic shear stress-strain curve was generated for each element group about the reference pressure for those elements. The reference strain, γ_r, was assumed equal to 0.015 in tension and -0.015 in compression. Figure 3-21 shows the curve produced for saturated clay elements at 148.5 ft below the dam crest. Ten conical yield surfaces were used to approximate the shear stress strain behavior. A purely kinematic hardening rule was assumed, forcing the

Table 3-3
Santa Felicia Dam Materials

Property	Core	Shell	Foundation
ρ_s (slugs/ft^3)	5.194	5.213	5.213
n^w	0.318	0.255	0.309
k (ft/day)	0.001	20.0	150.0
ϕ (degrees)	31.0	40.0	39.0

Table 3-4
Low Strain Shear Moduli for Santa Felicia Dam*

Depth Below Crest (ft)	Core	G_0 (lb/ft^2) Shell	Foundation
254.25	—	4,763,929	39,726,600
216.75	—	4,763,929	39,726,600
181.5	3,972,660	4,763,929	—
148.5	3,875,524	4,719,510	—
115.5	3,603,646	4,440,157	—
82.5	3,155,159	3,882,862	—
49.5	2,316,342	3,185,766	—
16.5	1,937,243	2,441,492	—

* Curve produced for saturated clay elements at 148.5 ft below the dam crest.

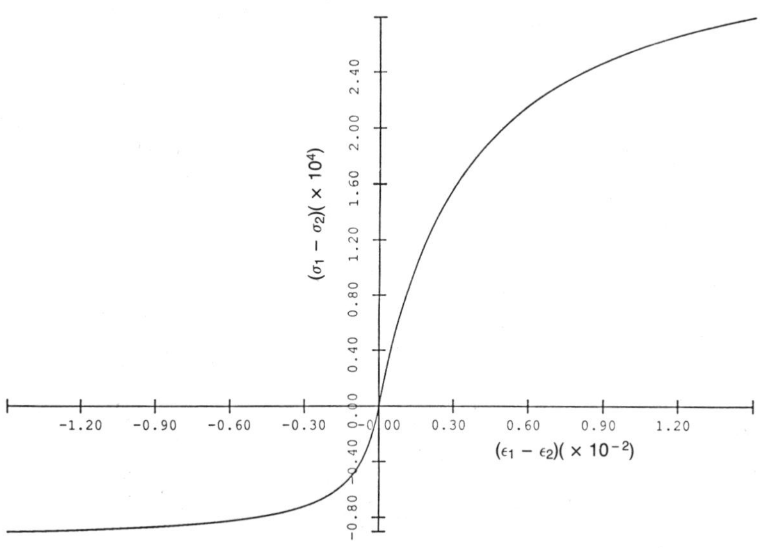

Figure 3-21. Stress-strain relation: clay core at depth H/3.

size of each yield surface to remain constant. One yield surface was used to model the hydrostatic compression vs. volumetric strain relation.

Input Ground Motion. The model dam was subjected to the input ground motion recorded at the Santa Felicia dam site, near the outlet works, during the 1971 San Fernando earthquake ($M_L = 6.3$). The first 16 seconds of the recorded acceleration was used, with data at .02 second intervals and peak acceleration of 0.22 g in the upstream-downstream di-

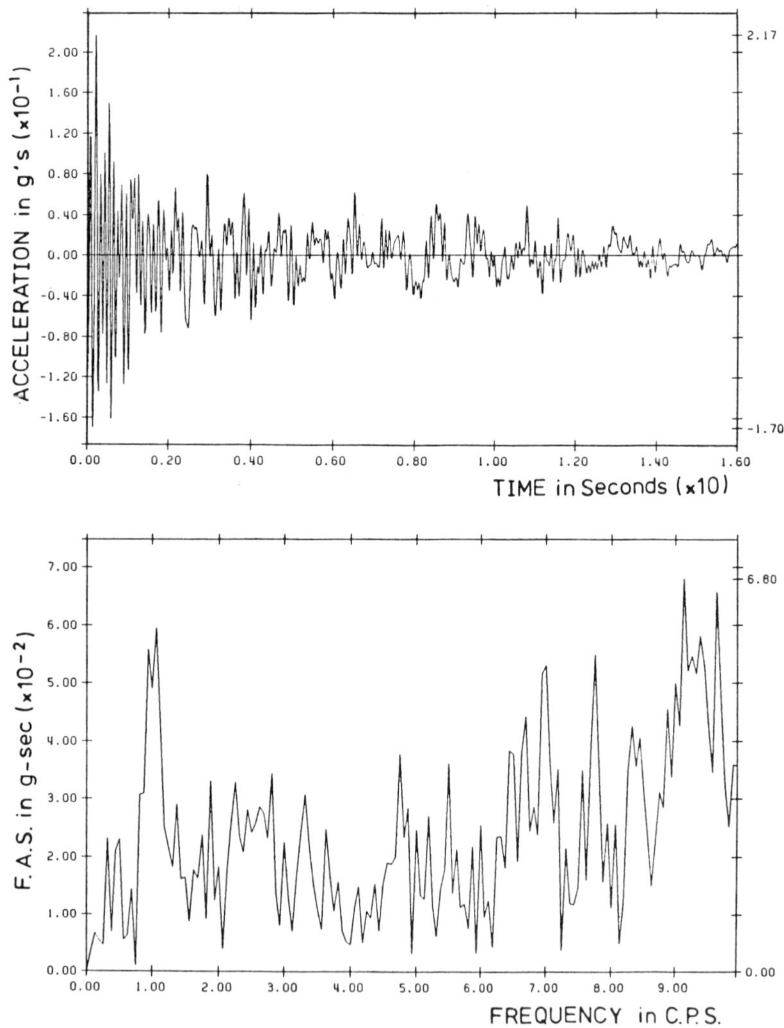

Figure 3-22A. Input ground motion: upstream-downstream direction.

rection. Figure 3-22 shows the input acceleration time histories in the upstream-downstream and vertical directions and their respective Fourier transforms. It should be noted that the peaks at about 10 Hz in the Fourier amplitude spectra were caused by the vibrations of a concrete standpipe underlying the floor of the valve house upon which the strong motion instrument was fixed. The two components of ground motion were applied simultaneously as prescribed uniform input accelerations at the solid canyon boundary nodes.

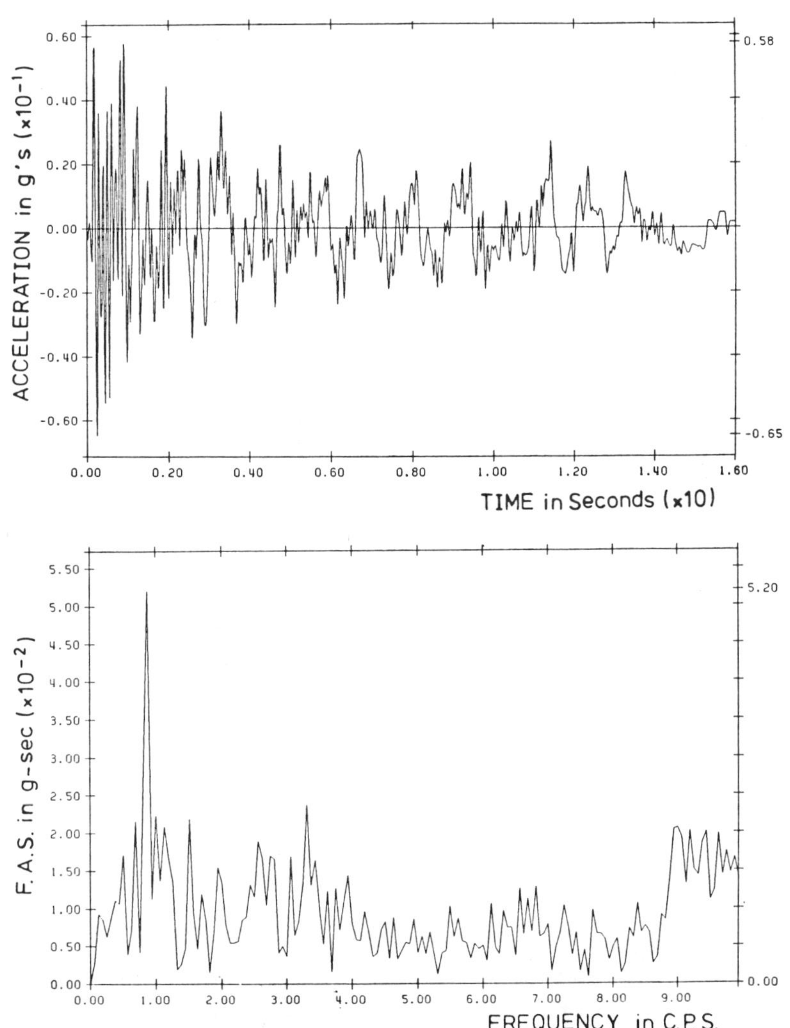

Figure 3-22B. Input ground motion: vertical direction.

Implementation. The solution was performed in two steps. First, gravity loads and loads representing the fluid on the upstream face of the dam were applied and the dam consolidated during ten load steps at three-day increments. The consolidation phase was calculated dynamically by setting the Newmark parameters in the integration scheme to $\gamma = 1.5$ and $\beta = 1.0$ as described earlier. When consolidation was completed, the nodal velocities and accelerations were zeroed, the time was reset to zero and the recorded earthquake signal was applied at the canyon boundary. The Newmark parameters in the integration scheme were chosen as $\gamma = 0.55$ and $\beta = (\gamma + \frac{1}{2})^2/4 = 0.28$. This choice of γ introduces a slight numerical damping ($\gamma = 0.50$ corresponds to no numerical damping) and the selected value for β maximizes high frequency numerical dissipation. No additional viscous physical damping was introduced. BFGS iterations were performed at each time step, with reformation of the coefficient matrix at the beginning of each time step. A maximum of 20 iterations was allowed and the convergence tolerance was set to 10^{-3}. All calculations were performed on a Cray computer.

Results. Figures 3-23A and 3-23B compare the computed acceleration histories at the dam crest of the two-phase model with acceleration histories computed with a one-phase soil model [Prevost et al., 1985a and 1985b]. Acceleration histories are shown in the upstream-downstream and vertical directions along with their respective Fourier transforms. In the upstream-downstream direction, maximum acceleration of the two-phase model (.216g) is substantially lower than the maximum for the one-phase model (.278g). As expected, introduction of pore-fluid soil skeleton interaction has damped the system response. The frequency content of the two signals is only slightly lower for the two-phase system. In the vertical direction, the response of the two-phase system is greater than that of the one-phase system. It is interesting to note that in a previous study [Prevost et al. 1985a and 1985b] of a three-dimensional nonlinear one-phase model, the response in the vertical direction also unexpectedly exceeded that in the two-dimensional one-phase model.

Figures 3-24A and 3-24B compare the results of the two phase calculations to the recorded response at the dam crest during the San Fernando earthquake. The upstream-downstream signals are remarkably similar. The maximum computed acceleration (.216g) differs from the maximum recorded acceleration (.207g) by only 4%. The model also closely predicts the frequency content of the signal. In the vertical direction the magnitude of the computed acceleration greatly overestimates that of the recorded signal. However, the range of frequencies for the two signals is similar.

Figure 3-25 shows the computed velocity response spectra for the measured, one-phase and two-phase acceleration time histories in the up-

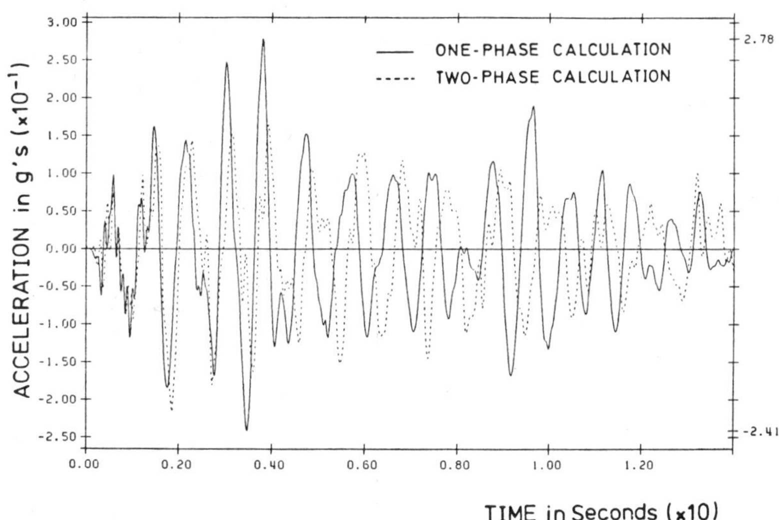

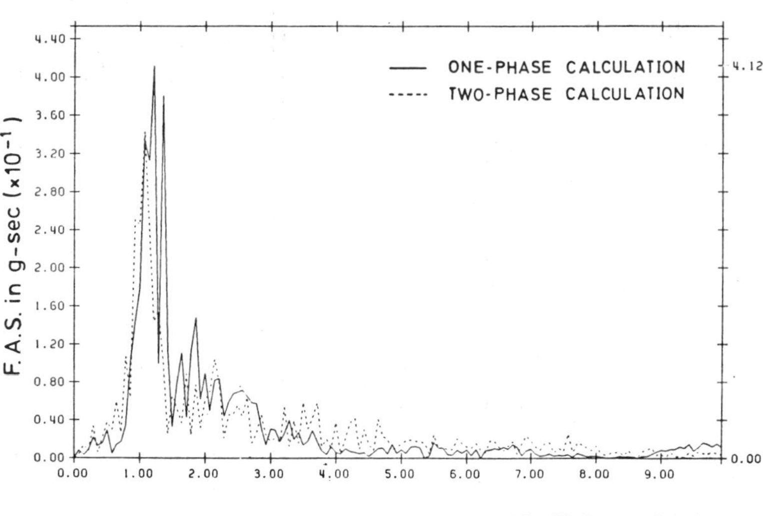

Figure 3-23A. Acceleration of dam crest—one-phase vs. two-phase calculations: upstream-downstream direction.

stream-downstream directions. The frequency content of both one-phase and two-phase computed accelerations is in close agreement with the recorded motion.

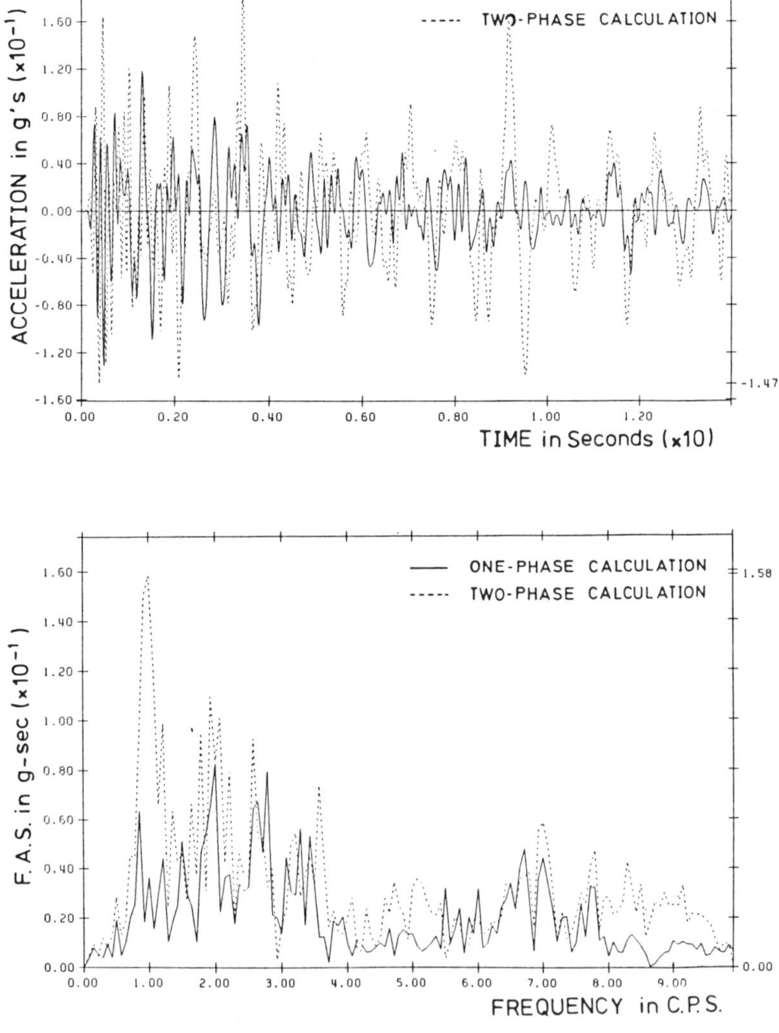

Figure 3-23B. Acceleration of dam crest—one-phase vs. two-phase calculations: vertical direction.

Figures 3-26 and 3-27 show the correlation coefficient between re-corded and computed accelerations for both one-phase and two-phase models. In Figure 3-26 the correlation is shown for a gradually increas-

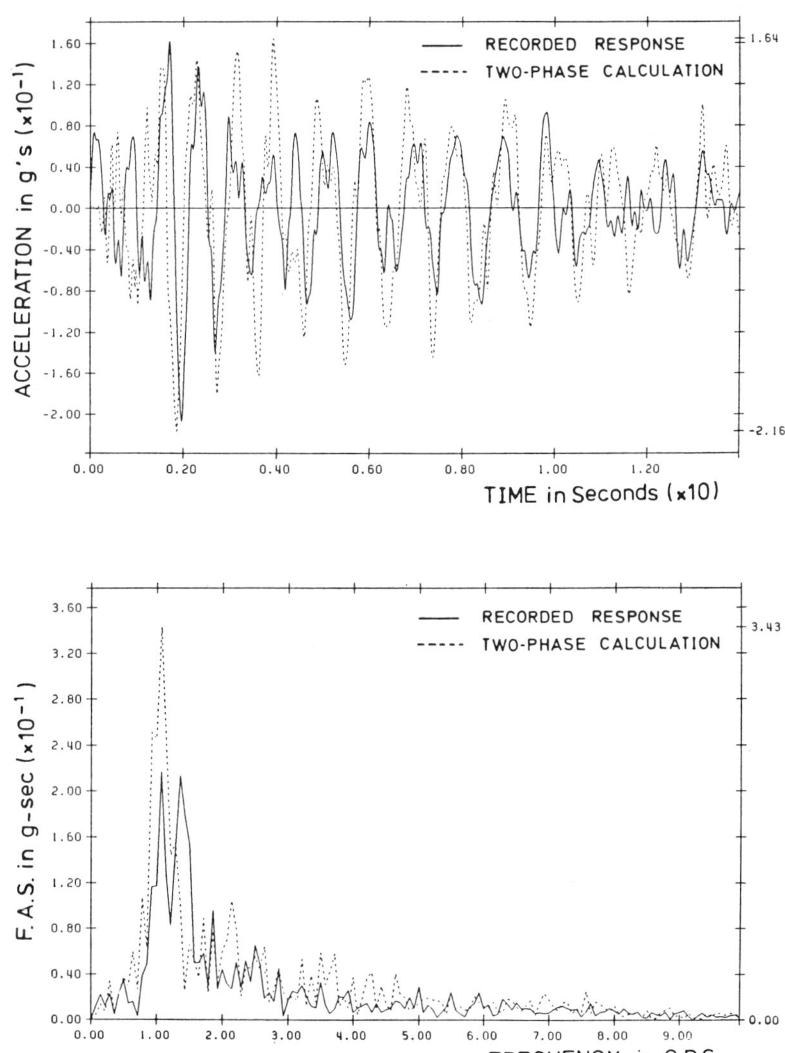

Figure 3-24A. Acceleration of dam crest—two-phase calculations vs. recorded response: upstream-downstream direction.

ing time window. From Figure 3-26, it is clear that the early stages of both sets of computed results, poorly correlate with the measured response. However, as the time window is increased, the two-phase results

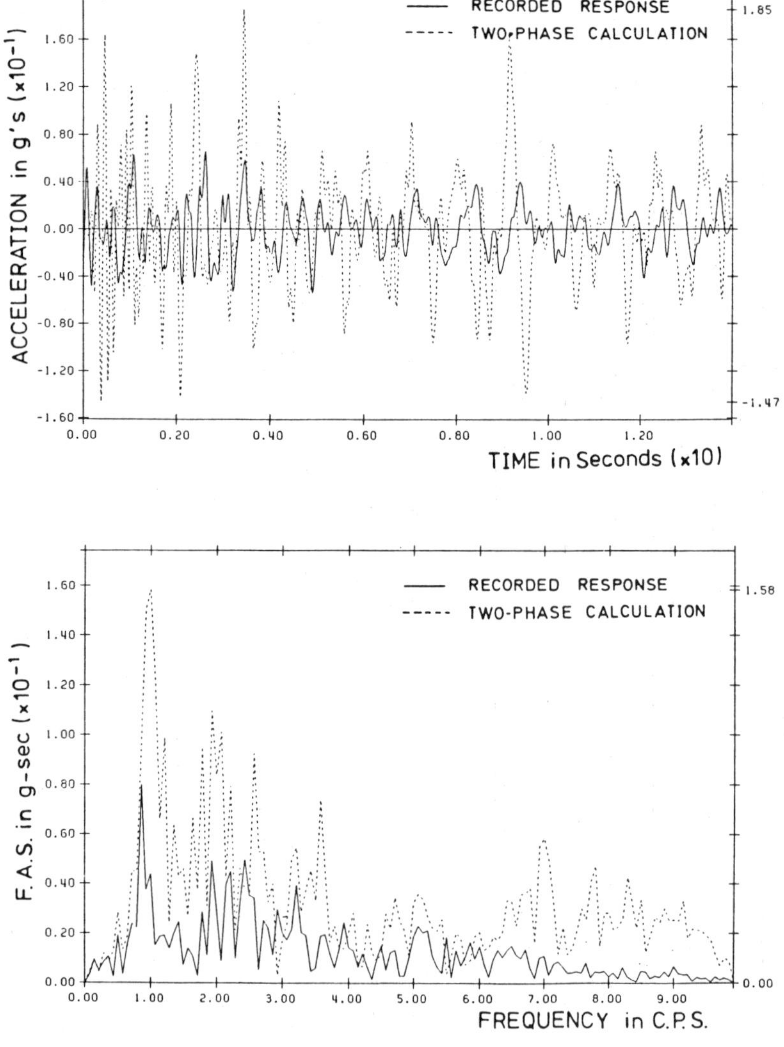

Figure 3-24B. Acceleration of dam crest—two-phase calculations vs. recorded response: vertical direction.

gradually improve to reach a correlation value of 0.54. On the other hand, the one-phase results reach a maximum correlation of 0.24 (for 4 seconds of motions), but as the time window is increased the correlation

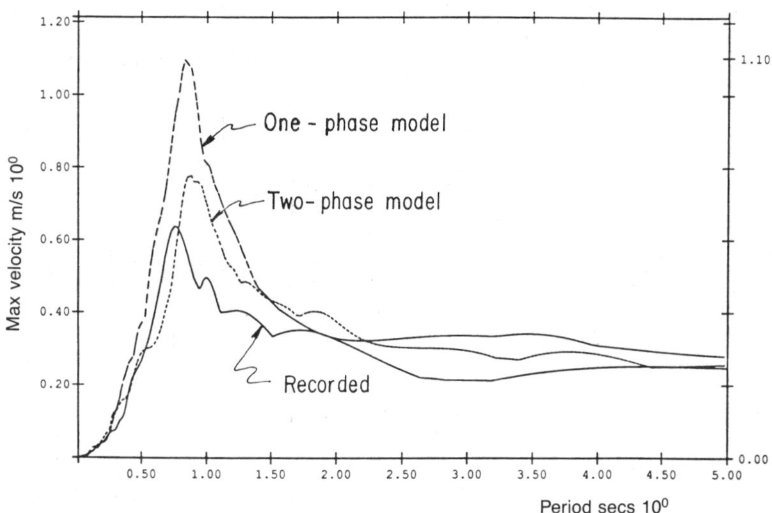

Figure 3-25. Acceleration of dam crest in the upstream-downstream direction— velocity response spectra.

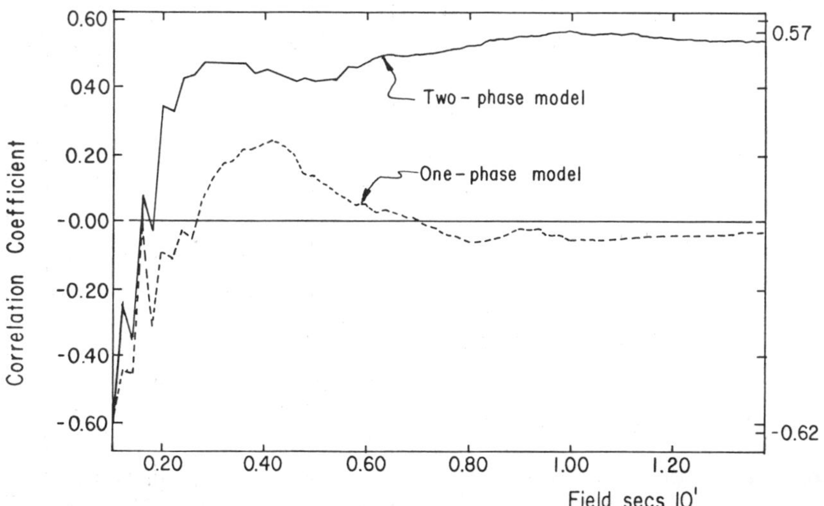

Figure 3-26. Acceleration of dam crest in the upstream-downstream direction— correlation coefficient for increasing time window.

drops to almost zero (i.e., no correlation). In Figure 3-27 the correlation calculations were performed for the entire time history by shifting the calculated motions relative to the recorded motion by ϵ-seconds. The results shown in Figure 3-27 indicate that the two-phase model is virtually in phase with the measured motion and can hardly be improved upon by shifting. On the other hand, the one-phase model could be improved by shifts of 0.25 sec or 0.7 sec.

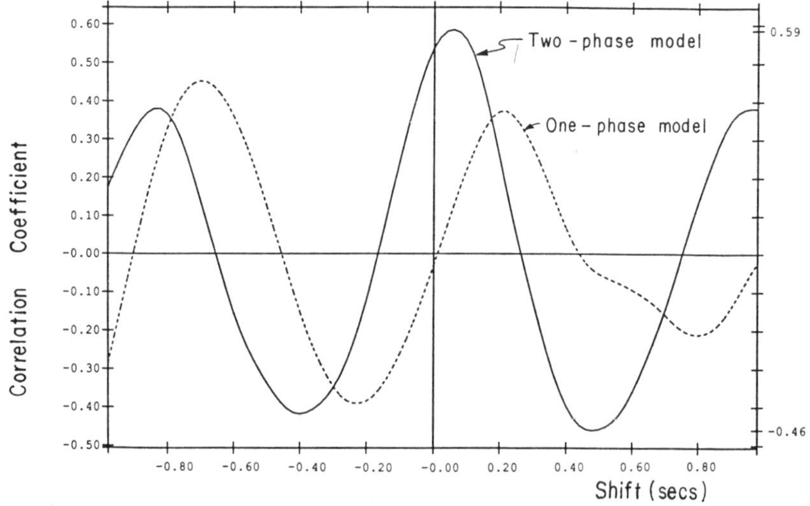

Figure 3-27. Acceleration of dam crest in the upstream-downstream direction—correlation coefficient with time shifting.

CONCLUDING REMARKS

Numerical methods have been proposed which make an accurate analysis of the dynamic response of two-phase soil systems possible. As an example, the elastoplastic response of a nonhomogeneous earth dam has been examined. The calculated response compares well with the recorded response of the dam at the crest during an earthquake.

Although the proposed methodology is sound, it is still far from having reached a mature stage. Nonlinear calculations are still difficult to complete successfully. This is illustrated in Figure 3-28 which shows a comparison of the computed acceleration time histories using BFGS and modified Newton-Raphson iterations with reforms at the beginning of each time step. Clearly the two computed responses are identical in both the upstream-downstream and vertical directions up to time = 12 sec af-

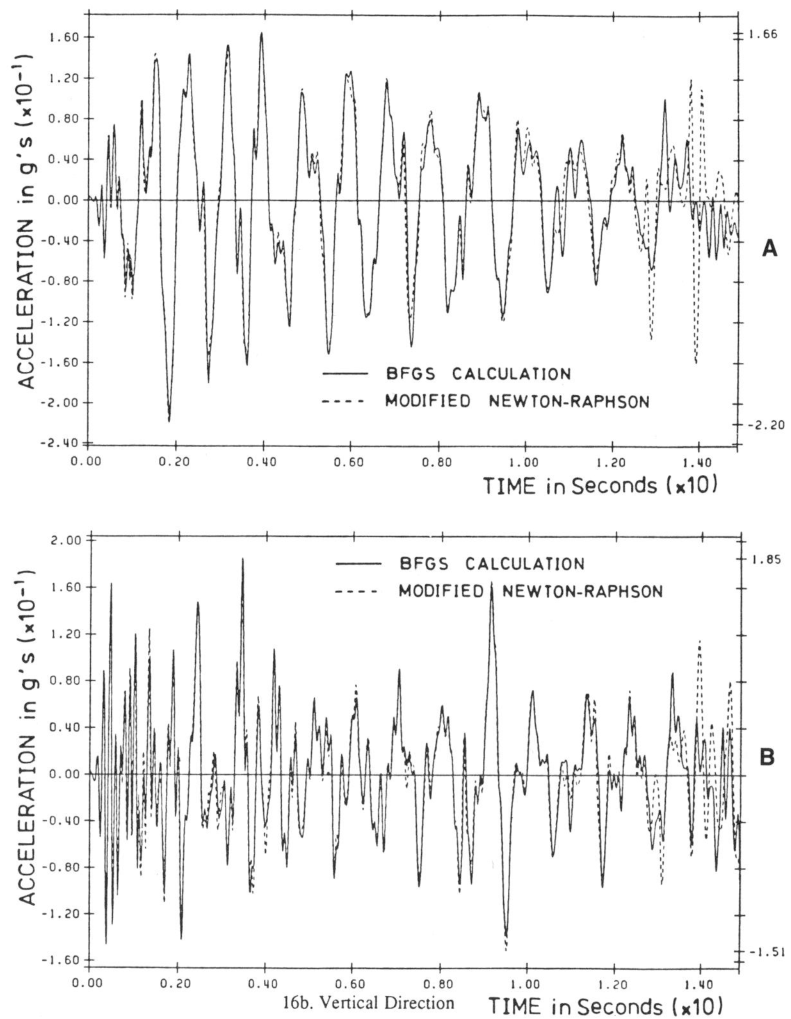

Figure 3-28. Acceleration of dam crest—BFGS vs. modified Newton-Raphson iterations: (A) Upstream- downstream direction. (B) Vertical direction.

ter which they differ grossly. Therefore, although very stringent and conservative convergence criteria had been used (see Flowchart 3-1), the modified Newton-Raphson iterations failed to provide the right answer after time = 12.0 sec (although the solution kept passing the convergence test!). Further research should therefore be devoted to devising better, sturdier, and more reliable algorithmic strategies before such analyses can become "routine" and made widely available.

REFERENCES

1. Abdel-Ghaffar, A. M., and Scott, R. F., "An Investigation of the Dynamic Characteristics of an Earth Dam," Report No. EERL 78-02, California Inst. of Technology, Pasadena, Calif., August 1978.

2. Abdel-Ghaffar, A. M., and Scott, R. F., "Vibration Tests of a Full-Scale Earth Dam," *Journal of the Geotechnical Engineering Division*, ASCE, Vol. 107, No. GT3, March 1981, pp. 241–269.

3. Atkin, R. J., and Craine, R. E., "Continuum Theories of Mixtures: Basic Theory and Historical Development," *Quarterly J. of Mechanics and Applied Mathematics*, Vol. 29, 1976, pp. 209–244.

4. Atkin, R. J., and Craine, R. E., "Continuum Theories of Mixtures Applications," *Journal of the Institute of Mathematics and its Applications*, Vol. 17, 1976, pp. 153–207.

5. Bathe, K. J., and Wilson, E. L., *Numerical Methods in Finite Element Analysis*, Prentice Hall, Englewood Cliffs, N.J., 1976.

6. Bathe, K. J., and Cimento, A. P., "Some Practical Procedures for the Solution of Nonlinear Finite Element Equations," *J. Comp. Meth. Appl. Mech. Eng.* Vol. 22, 1980, pp. 59–85.

7. Bazant, Z. P., and R. J. Krizek, "Saturated Sand as an Inelastic Two-Phase Medium," *J. of the Eng. Mech. Div.*, ASCE, Vol. 101, No. EM4, 1975, pp. 317–332.

8. Biot, M. A., "Theory of Elasticity and Consolidation for a Porous Anisotropic Solid," *J. of Applied Physics*, Vol. 26, 1955, pp. 182–185.

9. Biot, M. A., "Theory of Propagation of Elastic Waves in a Fluid-Saturated Porous Solid, I. Low Frequency Range," *J. of the Acoustical Society of America*, Vol. 28, 1956, pp. 168–191.

10. Biot, M. A., "Theory of Finite Deformations of Porous Solids," *Indiana University Mathematics Journal*, Vol. 21, No. 7, 1972, pp. 597–620.

11. Biot, M. A., "Variational Lagrangian Thermodynamics of Non-Isothermal Finite Strain Mechanics of Porous Solids and Thermomolecular Diffusion," *International Journal of Solids and Structures*, Vol. 13, 1977, pp. 579–597.

12. Biot, M. A., "Variational Irreversible Thermodynamics of Heat and Mass Transfer in Porous Solids: New Concepts and Methods," *Quarterly of Applied Mathematics*, Vol. 36, 1978, pp. 19–38.

13. Biot, M. A., and Willis, D. G., "The Elastic Coefficients of a Theory of Consolidation," *J. of Applied Mechanics*, ASME, Vol. 79, 1957, pp. 594–601.

14. Bowen, R. M., "Theory of Mixtures," in *Continuum Physics*, Vol. 3, A. C. Eringen (Ed.), Academic Press, New York, 1975, pp. 1–127.

15. Bowen, R. M., "Compressible Porous Media Models by Use of the Theory of Mixtures," *Int. Journal of Eng. Science,* Vol. 20, No. 6, 1982, pp. 697–735.

16. Bowen, R. M., and Lockett, R. R., "Inertial Effects is Poroelasticity," *J. Appl. Mech.,* ASME, Vol. 50, 1983, pp. 334–342.

17. Bowen, R. M., and Reinicke, K. M., "Plane Progressive Waves in a Binary Mixture of Linear Elastic Materials," *J. Applied Mechanics,* ASME, Vol. 45, 1978, pp. 493–499.

18. Bowen, R. M., and Chen, P. J., "Waves in a Binary Mixture of Linear Elastic Materials," *J. Mecanique,* Vol. 14, No. 2, 1975, pp. 237–266.

19. Brodlie, K. W., Gourlay, A. R., and Greenstadt, J., "Rank-One and Rank-Two Corrections to Positive Definite Matrices Expressed in Product Form," *J. Inst. Maths Applics.,* Vol. 11, 1973, pp. 73–82.

20. Crochet, M. J., and Naghdi, P. M., "On Constitutive Equations for Flow of Fluid Through an Elastic Solid," *Int. J. of Engineering Science,* Vol. 4, 1966, pp. 383–401.

21. Cryer, C. W., "A Comparison of the Three-Dimensional Consolidation Theories of Biot and Terzaghi," *Quarterly J. of Mechanics and Applied Mathematics,* Vol. 16, 1963, pp. 401–412.

22. Darcy, H. J., "Les Fontaines Publiques de la ville de Dijon," V. Dalmont, Paris, 1856.

23. Dennis, J. E., Jr., and More, J. J., "Quasi-Newton Methods, Motivation and Theory," *SIAM Review,* Vol. 19, No. 1, January 1977, pp. 46–89.

24. Dikmen, S. U., and J. Ghaboussi, "Effective Stress Analysis of Seismic Response and Liquefaction: Theory," *J. Geotech. Eng.,* ASCE, Vol. 110, No. 5, May 1984, pp. 628–644.

25. Engelman, M. S., Strang, G., and Bathe, K.-J., "The Application of Quasi-Newton Methods in Fluid Mechanics," *Int. J. Num. Meth. Eng.* Vol. 17, 1981, pp. 707–718.

26. Eringen, A. C., *Nonlinear Theory of Continuous Media,* McGraw-Hill, New York, 1962.

27. Eringen, A. C., and Ingram, J. D., "A Continuum Theory of Chemically Reacting Media I and II," *Int. J. of Engineering Science,* Vol. 3, 1965, pp. 197–212, and Vol. 5, 1967, pp. 289–322.

28. Ghaboussi, J., and Dikmen, S. U., "LASS-II, Computer Program for Analysis of Seismic Response and Liquefaction of Horizontally Layered Sands," *Report No. UILU-ENG-77-2010,* Dept. of Civil Eng., Univ. of Illinois at Urbana-Champaign, Urbana, IL, 1977.

29. Ghaboussi, J., and Dikmen, S. U., "Liquefaction Analysis of Horizontally Layered Sands," *J. Geotech. Eng. Div.,* ASCE, Vol. 104, No. GT3, March 1978, pp. 341–356.

30. Ghaboussi, J., and Dikmen, S. U., "LASS-III, Computer Program for Seismic Response and Liquefaction of Layered Ground under Multi-Directional Shaking," *Report No. UILU-ENG-79-2012*, Dept. of Civil Eng., Univ. of Illinois at Urbana-Champaign, Urbana, IL, 1979.

31. Ghaboussi, J., and Dikmen, S. U., "Liquefaction Analysis for Multidirectional Shaking," *J. Geotech. Eng. Div.*, ASCE, Vol. 107, No. GT5, May 1981, pp. 605–627.

32. Gray, W. G., and Lee, P. C. Y., "On the Theorems for Local Volume Averaging of Multiphase Systems," *Int. Journal of Multiphase Flow*, Vol. 3, 1977, pp. 333–340.

33. Green, A. E., and Naghdi, P. M., "A Dynamical Theory of Interacting Continua," *Int. J. of Eng. Science*, Vol. 3, 1965, pp. 231–241.

34. Green, A. E., and Steel, T. R., "Constitutive Equations for Interacting Continua," *Int. J. of Eng. Science*, Vol. 4, 1966, pp. 483–500.

35. Gurtin, M. E., Oliver, M. L., and Williams, W. O., "On Balance of Forces for Mixtures," *Quarterly of Applied Mathematics*, Vol. 30, 1972, pp. 527–530.

36. Hassanizadeh, M., and Gray, W. G., "General Balance Equations for Multiphase Systems," Princeton University, October, 1978.

37. Hilber, H. M., "Analysis and Design of Numerical Integration Methods in Structural Dynamics," *Rep. No. EERC 76-29*, Earthquake Engineering Research Center, University of California, Berkeley, CA, November 1976.

38. Hughes, T. J. R., and Liu, W. K., "Implicit-Explicit Finite Elements in Nonlinear Transient Analysis," *Comp. Meth. Appl. Mech. Eng.*, ASME, Vol. 45, 1978, pp. 371–374.

39. Hughes, T. J. R., and Liu, W. K., "Implicit-Explicit Finite Elements in Transient Analysis: Implementation and Numerical Examples," *J. Appl. Mech.*, ASME, Vol. 45, 1978, pp. 395–398.

40. Hughes, T. J. R., Pister, K. S., and Taylor, R. L., "Implicit-Explicit Finite Elements in Nonlinear Transient Analysis," *Comp. Meth. Appl. Mech. Eng.*, Vol. 17/18, 1979, pp. 159–182.

41. Lacy, S. J., "Numerical Procedures for Nonlinear Transient Analysis of Two-Phase Soil Systems," *PhD Thesis*, Department of Civil Engineering, Princeton University, Princeton, NJ, Oct. 1986.

42. Lacy, S. J., and Prevost, J. H., "Flow Through Porous Media: A Procedure for Locating the Free Surface," *Int. J. Num. Analyt. Meth. Geomechanics*, 1986 (in process).

43. Lacy, S. J., and Prevost, J. H., "Nonlinear Seismic Response Analysis of Earth Dams," *Soil Dynamics and Earthquake Engineering*, (1986) (to appear).

44. Lee, M. K. W., and Finn, N. D., Liam, "DESRA-1: Program for the Dynamic Effective Stress Response Analysis of Soil Deposits Including Liquefaction Evaluation," *Soil Mechanics Series Report No. 38,* Dept. of Civil Eng., Univ. of British Columbia, Vancouver, B.C., 1975.
45. Malkus, D. S., and Hughes, T. J. R., "Mixed Finite Elements Methods—Reduced and Selective Integration Techniques," *Comp. Meth. Appl. Mech. Eng.,* Vol. 15, No. 1, 1978, pp. 63–81.
46. Malvern, L. E., *Introduction to the Mechanics of a Continuous Medium,* Prentice-Hall, 1969.
47. Mandel, J., "Consolidation des sols (Etude Mathematique)," *Geotechnique,* Vol. 3, 1953, pp. 287–299.
48. Martin, P. O., and Seed, H. B., "Simplified Procedure for Effective Stress Analysis of Ground Response," *J. Geotech. Eng. Div.,* ASCE, Vol. 105, No. GT6, 1979, pp. 739–758.
49. Matthies, H., and Strang, G., "The Solution of Nonlinear Finite Element Equations," *Int. J. Num. Meth. Engng.,* Vol. 14, 1979, pp. 1613–1626.
50. Newmark, N. M., "A Method of Computation for Structural Dynamics," *J. Eng. Mech. Div.,* ASCE, Vol. 85, No. EM3, 1959, pp. 67–94.
51. Prevost, J. H., "Mechanics of Continuous Porous Media," *Int. J. of Engineering Science,* Vol. 18, No. 5, 1980, pp. 787–800.
52. Prevost, J. H., "Nonlinear Transient Phenomena in Saturated Porous Media," *Comp. Meth. Appl. Mech. Eng.,* Vol. 30, 1982, pp. 3–18.
53. Prevost, J. H., "Wave Propagation in Fluid-Saturated Porous Media: An Efficient Finite Element Procedure," *Int. J. Soil Dyn. Earthq. Eng.,* Vol. 4, No. 4, 1985, pp. 183–202.
54. Prevost, J. H., "A Simple Plasticity Theory for Cohesionless Frictional Soils," *Int. J. Soil Dyn. Earth. Eng.,* Vol. 4, No. 1, 1985, pp. 9–17.
55. Prevost, J. H., "Effective Stress Analysis of Seismic Site Response," *Int. J. Num. Analyt. Meth. Geomechanics,* Vol. 10, No. 6, 1986, pp. 653–665.
56. Prevost, J. H., Abdel-Ghaffar, A. M., and Lacy, S. J., "Nonlinear Dynamic Analyses of an Earth Dam," *Journal of Geotechnical Engineering,* ASCE, Vol. 111, No. 7, July 1985, pp. 882–897.
57. Prevost, J. H., Abdel-Ghaffar, A. M., and Lacy, S. J., "Analyse Dynamique Non-Lineaire et Tridimensionnelle d'un Barrage en Terre," *Revue Francaise de Geotechnique,* Vol. 33, 1985, pp. 19–36.
58. Roesset, J. M., "A Review of Soil-Structure Interaction," in *Soil Structure Interaction: The Status of Current Analysis Methods and*

Research, prepared for U.S. Nuclear Regulatory Commission, by J. J. Johnson, NUREG/CR - 1780, UCRL-53011, Lawrence Livermore Laboratory, Livermore, CA, 1981.

59. Schiffman, R.L., "Stress Components of a Porous Medium," *J. of Geophysical Research,* Vol. 75, 1970, pp. 4035–4038.

60. Schnabel, P. B., Lysmer, J., and Seed, H. B., "SHAKE—A Computer Program for Earthquake Response Analysis of Horizontally Layered Sites," *Report EERC 72-12,* Univ. of California, Berkeley, 1972.

61. Steel, T. R., "Applications of a Theory of Interacting Continua," *Quarterly Journal of Mechanics and Applied Mathematics,* Vol. 20, 1967, pp. 57–72.

62. Streeter, V. L., Wylie, E. B., and Richart, F. E., "Soil Motion Computation by Characteristics Method," *J. Geotech. Eng. Div.,* ASCE, Vol. 100, No. GT3, March 1974, pp.

63. Tabaddor, F., and Little, R. W., "Interacting Continuous Medium Composed of an Elastic Solid and an Incompressible Newtonian Fluid," *Int. J. of Eng. Science,* Vol. 7, 1971, pp. 825–841.

64. Terzaghi, K., *Theoretical Soil Mechanics,* John Wiley and Sons, N.Y., 1943.

65. Truesdell, C., *The Elements of Continuum Mechanics,* Springer-Verlag, New York, 1965.

66. Truesdell, C., and Toupin, R., *The Classical Field Theories, Handbuch der Physik,* S. Flugge (Ed.), Springer-Verlag, Berlin, Vol. III/1, 1960.

67. Truesdell, C., and Noll, W., *The Non-Linear Field Theories of Mechanics, Handbuch der Physik,* S. Flugge (Ed.), Springer-Verlag, Berlin, Vol. III/3, 1965.

68. Whitaker, S., "Advances in Theory of Fluid Motion in Porous Media," *Industrial and Engineering Chemistry,* Vol. 61, No. 12, 1969, pp. 14–28.

69. Williams, W. O., "On the Theory of Mixtures," *Archives for Rational Mechanics and Analysis,* Vol. 51, 1973, pp. 239–260.

70. Williams, W. O., "Constitutive Equations for Flow of an Incompressible Viscous Fluid Through a Porous Medium," *Quarterly of Applied Mathematics,* Vol. 36, October, 1978, pp. 255–267.

71. Zienkiewicz, O. C., *The Finite Element Method,* 3rd Edition, McGraw-Hill, London, 1977.

72. Zienkiewicz, O. C., and Scott, F. C., "On the Principle of Repeatability and Its Application in Analysis of Turbine and Pump Impellers," *Int. J. Num. Meth. Eng.,* Vol. 4, No. 3, 1972, pp. 445–448.

4

Static and Cyclic Response of Interfaces for Analysis and Design of Soil-Structure Interaction Problems

C. S. Desai
Department of Civil Engineering and Engineering Mechanics
University of Arizona
Tucson, Arizona USA

The response of structure-foundation systems subjected to static and dynamic loadings is influenced significantly by the existence of and behavior of the interfaces or junctions between the structure and foundation soils. Often, designs of such systems are performed by assuming complete bonding at the interfaces. Although this assumption simplifies the design procedure, it can introduce errors in the prediction of the stresses and deformations. This is mainly due to the fact that relative motions such as sliding, separation, or debonding and rebonding during the loading influence the stress-deformation response. Inclusion of the relative motions and resulting mechanisms of load transfer from the structure to soil and vice versa can introduce strong nonlinearities. Hence, analytical closed-form solutions become difficult, and recourse has to be made to numerical techniques such as the finite element, boundary element and finite difference methods.

Figure 4-1 shows typical problems of buildings, retaining walls, and piles where interfaces exist at the junctions of the structures and the foundation soils. Under static or dynamic loads, an interface can experience various modes of deformation, depicted schematically in Figure 4-2. During the no-slip or stick mode, the shear stress (τ) has not reached the failure or ultimate state, and the normal stress (σ_n) is compressive (positive). When the shear stress exceeds a failure stress defined by criteria such as Mohr-Coulomb, slip occurs; here σ_n is still compressive.

The interface is usually under compression in the initial or in situ state. During the loading, the compressive stress (σ_n) may reduce gradually, become zero and transit to the tensile regime (Figure 4-3). Depending upon the tensile strength (σ_t) of the interface (when $\sigma_n \leq 0$), the interface may experience debonding or separation during the no-slip or slip modes. During the subsequent loading, σ_n may become compressive again, and a rebonded state is established, and reloading can occur (Figure 4-3).

Another important mode, not shown in Figure 4-2, is interpenetration of one material into its neighbor. For the soil-structure interaction problems it is often necessary to control this so as to satisfy physical conditions between the two materials. For instance, for a soil-steel interface the interpenetration may be almost nonexistent.

REVIEW OF DEVELOPMENTS

Although the importance of the interface behavior and resulting soil-structure interaction effects have been recognized for a long time, development of realistic models for simulation of various modes of deformation has been undertaken only recently. This is mainly due to the mathematical difficulties in modelling the contact problem, and the lack of adequate (laboratory) testing equipment.

A number of models have been reported in the literature for simulating interface behavior under static and dynamic loadings; very often, models

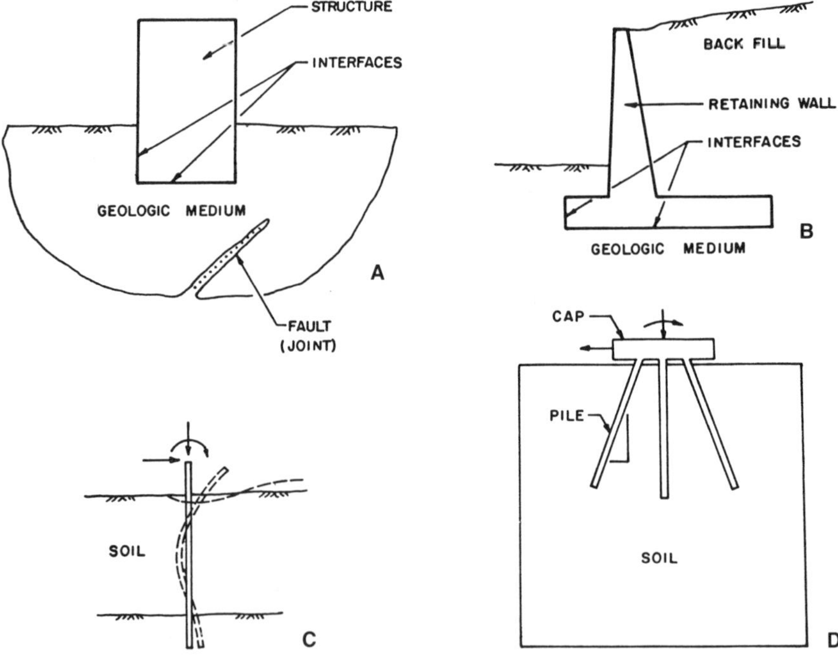

Figure 4-1. Soil-structure interaction problems: (A) Structure-foundation system. (B) Retaining wall. (C) Single pile. (D) Pile Group.

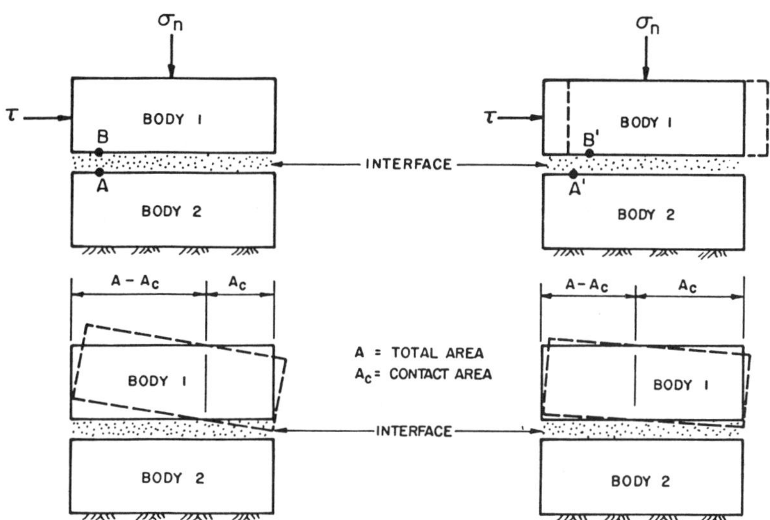

Figure 4-2. Schematic of modes of deformation at interface (two-dimensional): (A) Stick or no slip. (B) Slip or sliding. (C) Debonding. (D) Rebonding.

developed in other disciplines such as for rock joints have been adopted with empirical modifications to suit the specific requirements of the interface problem. It is not possible to provide a detailed review herein; comprehensive reviews are available elsewhere [Desai, 1977, 1979, 1981; Idriss et al., 1979; Wolf, 1985]. A short review of the major developments relevant to the presentation herein follows.

The problem of interfaces belongs to the general class of the contact and friction problems of mechanics which occur in many disciplines of engineering and physics. Thus the contact between two dissimilar materials, in metals, composite, joints (in rocks) and interfaces, involves similar physical and mathematical considerations. The early works have provided a basis for the solution of (metal) friction problems [Bowden and Tabor, 1964]. For the analysis of dams and embankments, Newmark [1965] considered interface behavior as the sliding of a rigid block; a similar model was adopted by Crandall, Lee, and Williams [1974].

Static Analysis

Ngo and Scordelis [1967] presented a linkage element for simulating cracks in concrete using springs to represent normal and shear behavior. Goodman, Taylor, and Brekke [1968] proposed a model for rock joints by expressing the behavior based on the relative displacements between the

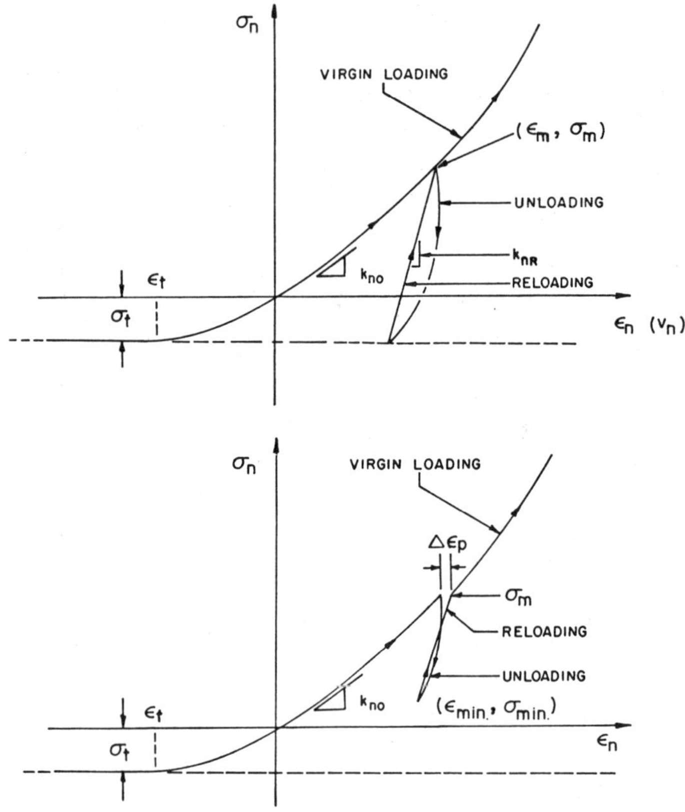

Figure 4-3. Schematic of stress-strain response for normal behavior: (A) Virgin loading and full unloading with tensile stress condition. (B) Partial unloading.

two neighboring parent rocks. Zienkiewicz et al. [1970] developed a similar model with the isoparametric finite element. Ghaboussi, Wilson, and Isenberg [1973] presented a model for joints in which the relative displacements are treated as independent degrees-of-freedom. These models for rock joints were adopted to simulate the interface behavior by assuming zero-thickness for the interface element [Clough and Duncan, 1971; Desai, 1974, 1976; Desai and Holloway, 1972; Desai et al., 1974; Lightner and Desai, 1979]. Desai and coworkers formalized the use of this model including extension for axisymmetric problems [Desai, 1974; Desai and Holloway, 1972]. Desai and Appel [1976], Desai, Phan and Perumpral [1982] developed this model for static analysis of three-dimensional nonlinear soil-structure interaction problems.

Katona [1983] derived an interface model based on the virtual work principle with special constraints relevant to the interface. Herrmann [1978] presented a model with constraints and an algorithm to include sliding and debonding.

Dynamic Analysis

Isenberg, Lee, and Agbabian [1973] used the element proposed by Ghaboussi et al. [1973] for three-dimensional analysis of a structure subjected to blast loadings. Wolf [1976, 1985] observed that the torsional effects due to traveling shear waves induce rocking perpendicular to the direction of excitation when lift off or slip occurs.

Kausel et al. [1979] discussed the importance of sliding or relative slip and proposed a model to account for translational and rotational models. Vaughn and Isenberg [1983] gave comprehensive consideration to the interface behavior and used a model that involved a finite-sized zone to represent the interface. They considered slip, debonding, and rebonding with an iterative procedure. Toki et al. [1981] used the Goodman et al. [1968] model for dynamic analysis of pier foundations.

The author and coworkers [Desai et al., 1985; Desai and Siriwardane, 1982; Desai et al., 1984] have developed new and alternative concepts to characterize interface behavior under static and cyclic loadings and unique laboratory test devices to simulate the normal and shear responses. The main objective here is to present these developments and typical applications to static and dynamic soil-structure interaction problems.

Models

The response of the interface involves two main components, shear and normal. Although both are coupled, often they are assumed to be decoupled with the following constitutive relation:

$$\{\sigma\} = [C]\, \{\epsilon\} \tag{4-1}$$

where $\{\sigma\}$ = vector of stress components
$\{\epsilon\}$ = vector of relative displacements or strains
$[C]$ = constitutive matrix given by

$$[C] = \begin{bmatrix} k_n & 0 \\ 0 & k_s \end{bmatrix} \tag{4-2}$$

k_n, k_s = normal and shear stiffnesses or moduli, respectively

The shear response can be obtained from direct shear or torsional shear (laboratory) tests involving testing of interfaces between the structural and geologic materials. However, it is difficult to perform tests that can allow for definition of the normal response. As a result, most previous models have defined the shear stiffness k_s from results of shear tests in terms of the $\tau - u_r$, where τ = shear stress and u_r = relative shear displacement, with sliding identified by using the peak or failure stress with criteria such as Mohr-Coulomb. However, for the normal behavior, arbitrary values of the normal stiffness are often chosen before and after debonding ($\sigma_n \leq 0$). During no-slip and sliding ($\sigma_n > 0$), a very high value for k_n ($\cong 10^8$ units) is chosen, whereas it is reduced to a very small value ($\cong 100$ units) when debonding occurs. It is believed that this approach may not be realistic for many soil-structure interaction problems, and can result in erratic and unrealistic (normal) stresses at the interface [Desai et al., 1974]. As a consequence, most results (from numerical calculations) for the normal stresses on the structures reported in the literature are usually not in the interface but usually in the geologic materials in the vicinity [Clough and Duncan, 1971; Desai et al., 1974].

The interface introduces a discontinuity in a continuum system. The models developed essentially introduce force and/or displacement constraints to simulate the interface, while the continuum theories are used for the overall system. However, it is more appropriate to develop models that can allow for the entire shear and normal responses including virgin loading, unloading, and reloading.

Various Modes Described Earlier. The author and coworkers [Desai, 1981; Desai et al., 1985; Desai et al., 1984] have developed the thin-layer element, laboratory testing equipment, and procedures to evolve such models. The remainder of this chapter describes these approaches with typical applications.

PROPOSED MODELS

In many soil-structure interaction problems it is appropriate to assume a thin soil zone in the vicinity of the structure to participate in the interface action. For example, from field pile load tests under cyclic loading, Matlock, Bogard and Cheang [1983] observed a thin zone containing a large number of sub layers that might have formed during the shear transfer and failure.

The thin-layer element is shown in Figure 4-4. The thickness of the thin layer is an important parameter. From a parametric study involving finite element calculations and laboratory shear tests on sand-concrete interfaces, it was found that the thickness t can be chosen in the following range [Desai et al., 1984]:

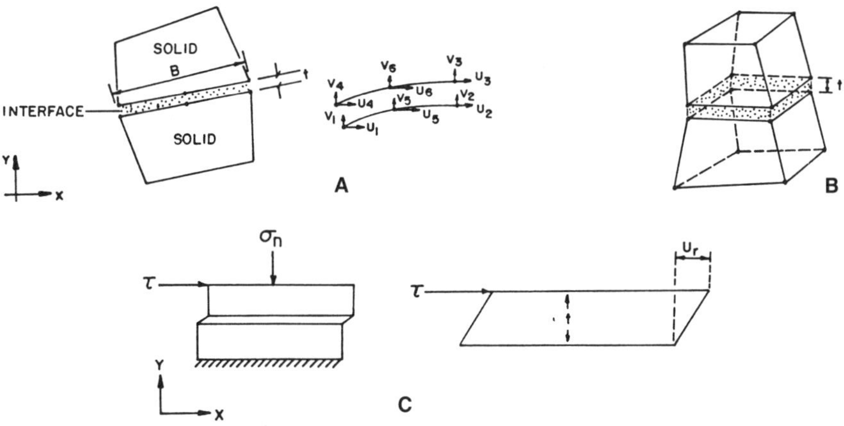

Figure 4-4. Thin-layer interface and solid elements: (A) Two-dimensional. (B) Three-dimensional. (C) Behavior at interface.

$$0.01 \leq \frac{t}{B} \leq 0.1 \tag{4-3}$$

where B = average width of the adjoining soil and structural elements

Shear Behavior

The shear stiffness (k_s) or shear modulus G for the interface can be obtained based on observed $\tau - u_r$ relations as

$$G \cong \frac{\partial \tau}{\partial u_r} \times t \tag{4-4}$$

and

$$k_s = \frac{\partial \tau}{\partial u_r} \tag{4-5}$$

For the cyclic behavior, to allow for virgin loading, unloading, and reloading, a modified form of the Ramberg-Osgood (R-O) model is used, Figure 4-5, [Desai et al., 1985; Desai and Siriwardane, 1982; Drumm, 1983; Drumm and Desai, 1986; Idriss et al., 1978] in which the $u_r - \tau$ relation is expressed as

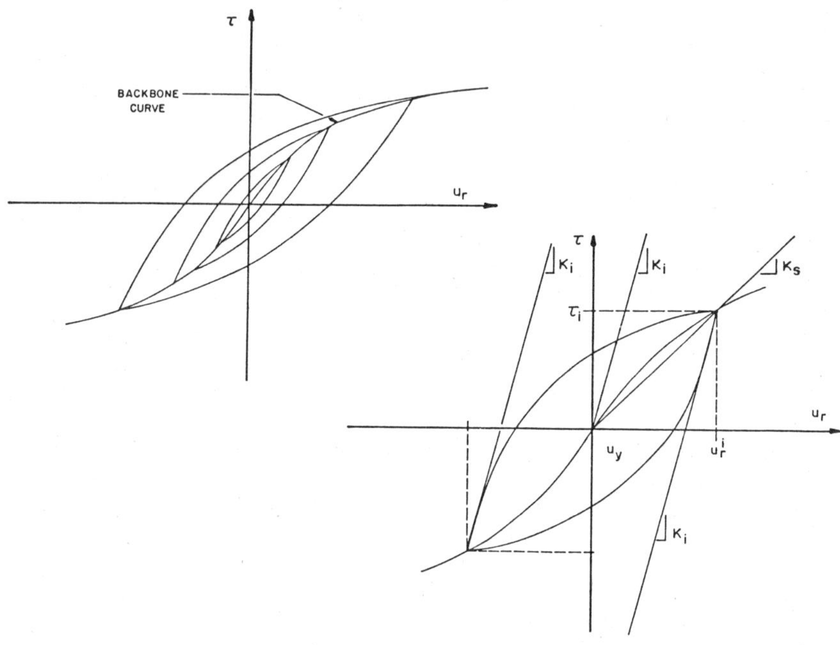

Figure 4-5. Ramberg-Osgood model for cyclic shear stress-displacement relationship.

$$u_r = u_y \left(\frac{\tau}{\delta \, k_{si} \, u_y} \right) \left(1 + \frac{2\alpha}{2^R} \left| \frac{\tau}{\delta \, k_{si} \, u_y} \right|^{R-1} \right) \tag{4-6a}$$

$$u_r \pm u_i = u_y \left(\frac{\tau \pm \tau_i}{\delta_s \, k_{si} \, u_y} \right) \left(1 + \alpha \left| \frac{\tau \pm \tau_i}{\delta \, k_{si} \, u_y} \right|^{R-1} \right) \tag{4-6b}$$

where Equation 4-6a relates to the backbone curve and Equation 4-6b to the cyclic loading, unloading, reloading response

and $\;\;\tau_i$, u_i = values of τ and u_r at the point of unloading

$\qquad$ u_y = reference (relative) displacement

$\qquad$ k_{si} = initial shear stiffness

$\qquad$ δ_s = cyclic index

$\quad$ α, R = positive material parameters

To find the constants involved in Equations 4-6a and 4-6b, a comprehensive series of tests were performed by using a cyclic multi-degree-of-freedom (CYMDOF) shear device (Figure 4-6). Details of the development and computations of the constants for the shear model for a sand (Ottawa)/concrete interface, as affected by factors such as the states and amplitudes of shear and normal stresses, frequency (ω), initial density (D_r), and number of loading cycles (N), are given by Desai, Drumm and Zaman [1985], Drumm [1983], and Drumm and Desai [1986].

Normal Behavior

When the interface is in the no-slip mode and loading is such that the normal stress, σ_n, increases from the initial compressive, virgin loading occurs (Figure 4-3). During this state, the normal stiffness can increase, as defined by the slope of the $\sigma_n - v_n$ (or strain, $\epsilon_n = v_n/t$) curve, where v_n = normal displacement, and t = thickness of the interface.

As indicated earlier, the interface may experience gradual reduction (unloading) from high compressive σ_n, which may reach zero and then become tensile. If the interface has no tensile strength, debonding occurs when $\sigma_n = 0$; otherwise, when $\sigma_n \geq \sigma_t$, the normal stiffness tends toward a small value. During subsequent (re)loading, the normal stress may become compressive again, the contact is reestablished, and $\sigma_n > 0$.

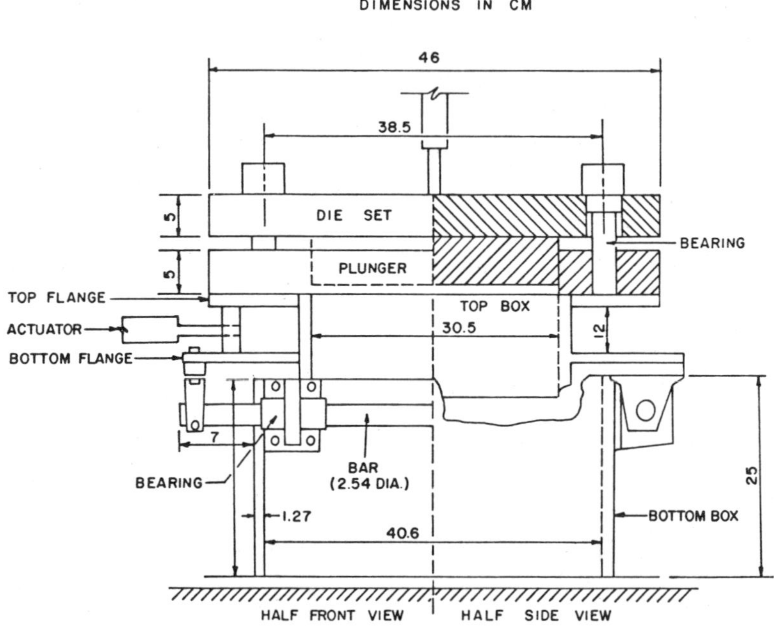

Figure 4-6. Details of translational box in CYMDOF.

Based on a series of cyclic normal, shear, and combined normal and shear tests using the CYMDOF device for the sand-concrete interface, the following functions were proposed for the virgin loading, unloading, and reloading [Desai and Nagaraj, 1986; Nagaraj, 1986].

Virgin Loading. This behavior is simulated by using the following exponential function:

$$\sigma_n = a_o \left(e^{a_1(\epsilon_n - \epsilon_p)} - 1.0 \right), \text{ for } \epsilon_n > \epsilon_m \text{ and } \sigma > \sigma_m \tag{4-7a}$$

$$\sigma_n = -\sigma_t, \text{ for } \epsilon_n < \epsilon_t = \frac{1}{a_1} \ell n \left(\frac{-\sigma_t}{\sigma_o} + 1 \right) \tag{4-7b}$$

and the tangent normal stiffness is given by

$$k_{nt}^v = \frac{\partial \sigma_n}{\partial \epsilon_n} = k_{no} \, e^{a_1(\epsilon_n - \epsilon_p)} \tag{4-8}$$

where $\epsilon_p = \sum_1^N \Delta\epsilon_p$, $\Delta\epsilon_p =$ difference between irreversible strain for partial unloading and full unloading

$N =$ number of loading cycles

σ_m and $\epsilon_m =$ previous maximum stress and strain at unloading, respectively

$\sigma_t =$ tensile strength of interface

$k_{no} = a_o a_1$, initial normal stiffness modulus

a_o and $a_1 =$ constants determined from laboratory tests

Unloading. The unloading response is simulated by using a hyperbolic relation such as

$$\sigma_n = \sigma_m - \frac{\epsilon_m - \epsilon_n}{a_2 + P_1 \, (\epsilon_m - \epsilon_n)}, \text{ for } \epsilon_m - \frac{\sigma_m + \sigma_t}{k_{nR}^N} < \epsilon_n < \epsilon_m \tag{4-9a}$$

$$\sigma_n = -\sigma_t, \text{ for } \epsilon_n < \left(\epsilon_m - \frac{\sigma_m + \sigma_t}{k_{nR}^N} \right) \tag{4-9b}$$

$$P_1 = \frac{(1 - a_2 \, k_{nR}^N)}{(\sigma_m + \sigma_t)} \tag{4-9c}$$

Here, k_{nR}^N, which is defined later, is used to identify the range of strains within which unloading occurs. The tangent normal stiffness is given by

$$k_{nt}^u = \frac{\partial \sigma_n}{\partial \epsilon_n} = \frac{a_2}{[a_2 + P_1 \, (\epsilon_m - \epsilon_n)]^2} \tag{4-10}$$

where a_2 = material parameter
P_1 = related to the asymptotic value of σ_n during unloading

From the laboratory tests it is found that the value of a_2, which denotes the slope of the unloading curve at the point of unloading, remains essentially constant with N for given values of σ_o and σ_a; however, it is dependent on the magnitudes of σ_o and σ_a. Hence, it is expressed as follows in terms of the initial reloading modulus, k_{nR}^1, which is constant for a given test, and a_f, which is dependent on σ_o and σ_a as

$$a_2 = \frac{1}{a_f \, k_{nR}^1} \tag{4-11a}$$

Here a_f is a factor that gives the stiffness modulus at the point of unloading when it is multiplied by k_{nR}^1, and is expressed as

$$a_f = a_{fo} \, e^{-\bar{\sigma}\alpha_o} \tag{4-11b}$$

where a_{fo} and α_o = material constants and $\bar{\sigma}$ is a nondimensional parameter:

$$\bar{\sigma} = \frac{\sigma_o - \sigma_a}{\sigma_a} \tag{4-11c}$$

Reloading. If during (Figure 4-3b) or at the end of unloading (Figure 4-3a) the stress level is increased, reloading occurs. From the test data, the reloading response is found to be essentially linear; however, the value of its slope (normal stiffness modulus) is dependent on the state of stress at which reloading occurred. Hence, the stress-strain relation is expressed in terms of σ_n and ϵ_n at any point during reloading:

$$\sigma_n = k_{nt}^r \, (\epsilon_n - \epsilon_o) + \sigma_o \tag{4-12a}$$

$$k_{nt}^r = \max \, (k_{nR}^N, \, k_{nt}^u) \tag{4-12b}$$

$$\epsilon_o = \max \left\{ \left(\epsilon_m - \frac{\sigma_m + \sigma_t}{k_{nR}^N} \right), \ \epsilon_t, \ \epsilon_{min} \right\} \tag{4-12c}$$

$$\sigma_o = \max \left(-\sigma_t, \ \sigma_{min} \right) \tag{4-12d}$$

where σ_{min} and ϵ_{min} = minimum stress and strain when reloading occurs

k_{nt}^u = tangent stiffness modulus of unloading curve at the time of reloading; that is, for $\epsilon_n = \epsilon_{min}$

k_{nR}^N = reloading stiffness modulus for cycle N

For the static loading, k_{nR}^N remains essentially constant and equals the initial reloading modulus k_{nR}^1.

The effects of loading cycles and σ_o and σ_a on k_{nR}^N are incorporated by using the observed variation of k_{nR}^N/k_{nR}^1. Based on these results, the effect of N is expressed as

$$\frac{k_{nR}^N}{k_{nR}^1} = 1.0 + \delta m \log N \tag{4-13a}$$

and is given by

$$m = \beta_o + \beta_1 \bar{\sigma} \tag{4-13b}$$

in which β_o and β_1 = constants.

Effect of $\bar{\sigma}$ on k_{nR}^1 is given by

$$\frac{k_{nR}^1}{k_{nRo}^1} = 1.0 + c_1 \bar{\sigma} + c_2 \bar{\sigma}^2 \tag{4-14}$$

where $k_{nRo}^1 = k_{nR}^1$ for $\bar{\sigma} = 0$, which refers to reloading modulus from static load

$c_1, \ c_2$ = constants.

The constants involved in the model for normal behavior, for Ottawa sand/concrete interface, are obtained from a series of laboratory tests using the CYMDOF device. These tests include cyclic normal starting from a given initial normal stress and combined cyclic normal and shear loading. The tests are designed such that the interface undergoes normal stress from a high value to a near zero value, thus permitting definition of gradual variation of normal stiffness from high to low values, Equations 4-7, 4-9, and 4-12.

The constants for the shear behavior with the R-O model and for the normal behavior are given in Tables 4-1 and 4-2, respectively.

Table 4-1
Summary of Ramberg-Osgood Parameters for Sand-Concrete Interface

Initial stiffness, K_i
$$K_i = (20.0 + 0.67\,D_r)\sigma_n$$

Parameter R
$$R = 1.4 + (1.37 \times 10^{-3} \times D_r)\sigma_n$$

Reference displacement, u_y
$$u_y = 0.01 \text{ in.}$$

Parameter α

$D_r = 80\%$	$D_r = 15\%$
$\alpha = 2.55 + 0.15\sigma_n$	$\alpha = 1.60 + 0.027\sigma_n$

Linear interpolation for intermediate densities

Cyclic parameter, t

$D_r = 80\%$	$D_r = 15\%$
$t = 0.025$	$t = 0.35 - u_r^m/(0.18 + 2.3u_r^m)$

Linear interpolation for intermediate densities

From Drumm [1983]; Drumm and Desai [1986].
D_r in percent, σ_n in psi, and u_y in in.

Table 4-2
Constants for Model for Normal Behavior

From Specific Tests
(1 psi = 6.89 kPa)

Test			Virgin Loading		Unloading			Reloading	
σ_o psi	σ_a psi	$\bar{\sigma}$ psi	a_o psi	a_1	k_{nR}^1 psi	a_f	$1/a_2$ psi	δ	m
15	10	0.50			3,700	1.82	6,700	1.0	0.21
15	15	0.00			3,100	4.50	14,000	1.0	0.30
20	10	0.50			4,500	1.76	7,900	1.0	0.16
20	15	0.25	34,000	0.0044	3,400	2.00	6,700	1.0	0.23
20	20	0.00			3,100	4.50	14,000	1.0	0.30
25	10	0.60			5,150	1.73	8,900	1.0	0.14
25	15	0.40			4,000	1.80	7,100	1.0	0.19
25	20	0.20			3,300	1.79	6,250	1.0	0.24
Static	—	—			3,100	1.50	—	0.0	—

Effect of σ_o, σ_a, and N

Variation	Equations
Variation of a_f with $\bar{\sigma}$, Equation 4-11	$a_{fo} = 4.5$
	$\alpha_o = 0.10$
Variation of reloading modulus with $\bar{\sigma}$	Equation 4-13
	$\beta_o = 0.30$
	$\beta_1 = -0.27$
	Equation 4-14
	$c_1 = -0.10$
	$c_2 = 2.0$

From Desai and Nagaraj [1986]; Nagaraj [1986].

Combined Normal and Shear Behavior

The incremental form of Equation 4-1 can be written as

$$\{d\sigma\} = [C^t] \{d\epsilon\} \tag{4-15}$$

where d denotes increment and superscript t denotes tangent moduli k_s^t and k_n^t, which are obtained by appropriate differentiation of Equations 4-6, 4-7, 4-9, and 4-12. It is possible to use relative displacements du and dv, or increments of corresponding strain quantities defined as $\epsilon_s = u_r/t$ and $\epsilon_n = v_n/t$, with the assumption that the shear and normal strains are uniform over the thin-layer element [Desai and Nagaraj, 1986; Nagaraj, 1986].

Equation 4-15 can now be introduced in the incremental finite element equations:

$$[k^t] \{\Delta q\} = \{\Delta Q\} \tag{4-16}$$

where $\{\Delta q\}$, $\{\Delta Q\}$ = vectors of incremental displacements and loads, respectively

$[k^t]$ = tangent stiffness matrix given by

$$[k^t] = \int_V [B]^T [C^t] [B] \, dV$$

where   [B] = transformation matrix
                          V = volume

For dynamic analysis, Equation 4-16 would include terms related to mass and damping matrices, and the resulting matrix equations would be solved in the time domain [Nagaraj, 1986; Zaman et al., 1984].

Generalized Models

It may be more appropriate and realistic to develop models that combine both the normal and shear responses as well as allow for factors such as dilatation and softening. Such models, for the interfaces and joints, have been proposed and are being developed as special cases of the hierarchical plasticity-based concept for solid (geologic) materials [Desai et al., 1987; Desai et al., 1986; Fishman and Desai, 1987].

VERIFICATION AND APPLICATION

The proposed interface element and constitutive models are verified by

- Comparisons of the model predictions with results from laboratory stress-strain tests [Desai et al., 1985; Desai and Nagaraj, 1986; Drumm, 1983; Drumm and Desai, 1986; Nagaraj, 1986; Zaman et al., 1984]
- By comparisons of finite element predictions with observed behavior of various practical problems in geotechnical engineering [Desai, Muqtadir, and Scheele, 1986; Desai et al., 1984; Nagaraj, 1986; Zaman et al., 1984].

A few examples follow.

LABORATORY TESTS

Static Direct Shear Test

Figure 4-7A shows static direct shear tests between a sand and concrete [Desai, 1976]. The finite element mesh for the test set up is shown in Figure 4-7B. The thin-layer element (Figure 4-4) was used to simulate the interface behavior. In the finite element analysis, the constant normal stress σ_n was applied and then the shear stress τ was applied incrementally as in the laboratory tests.

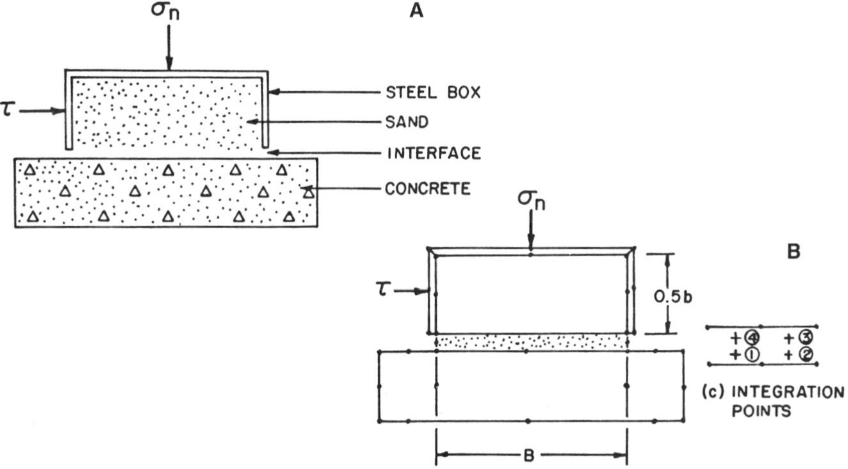

Figure 4-7. Analysis for direct shear test for choice of thickness. (A) Direct shear test for concrete-soil interface. (B) Finite element mesh.

Figure 4-8 shows observations for two tests with $\sigma_n = 4.77$ and 9.55 kg/cm^2. Typical results from the finite element calculations were compared with the observations in Figure 4-8; this led to the criterion, Equation 4-3, for the choice of the thickness t of the interface.

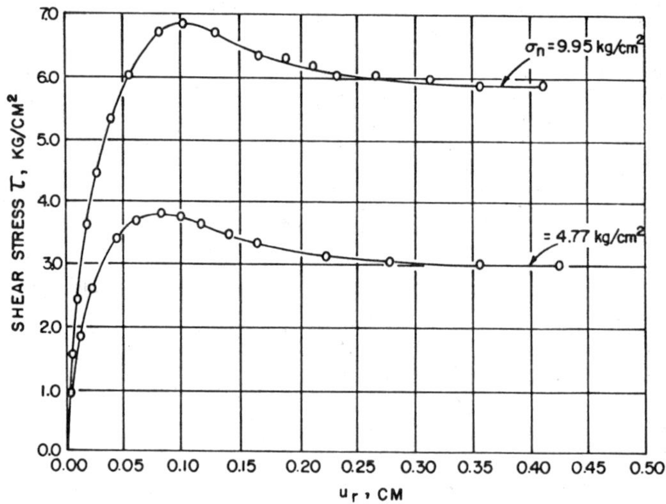

Figure 4-8. Typical direct shear test results (D_r = 80%) [Desal, 1976; Desal, et al., 1984].

Cyclic Shear Tests

Figure 4-9A shows finite element mesh for the cyclic shear tests. Figure 4-9B shows comparisons between predictions and observations of shear stress (τ) vs. relative displacement (u_r) from cyclic displacement controlled tests (amplitude $u_a = 0.05$ in. (0.13 cm), frequency $\omega_x = 1.0$ Hz) for Ottawa sand/concrete interface with relative density $D_r = 80\%$ and $\sigma_n = 28$ psi (190 kPa) [Desai et al., 1985; Drumm, 1983].

Cyclic Combined Normal and Shear Tests

Figure 4-10 shows mesh for a typical combined test in which displacement-controlled shear load with amplitude $u_a = 0.05$ in. and frequency $\omega_x = 1.0$ Hz, and normal stress-controlled test with amplitude $\sigma_a = 15$ psi (103 kPa) initial stress $\sigma_o = 20$ psi (138 kPa), and $\omega_z = 0.1$ Hz. Here the initial normal stress σ_o was first applied and then the cyclic load with $\sigma_a = 15$ psi (103 kPa) was applied. Figure 4-11 shows comparisons between predicted and observed responses in terms of τ vs. time [Desai and Nagaraj, 1986; Nagaraj, 1986].

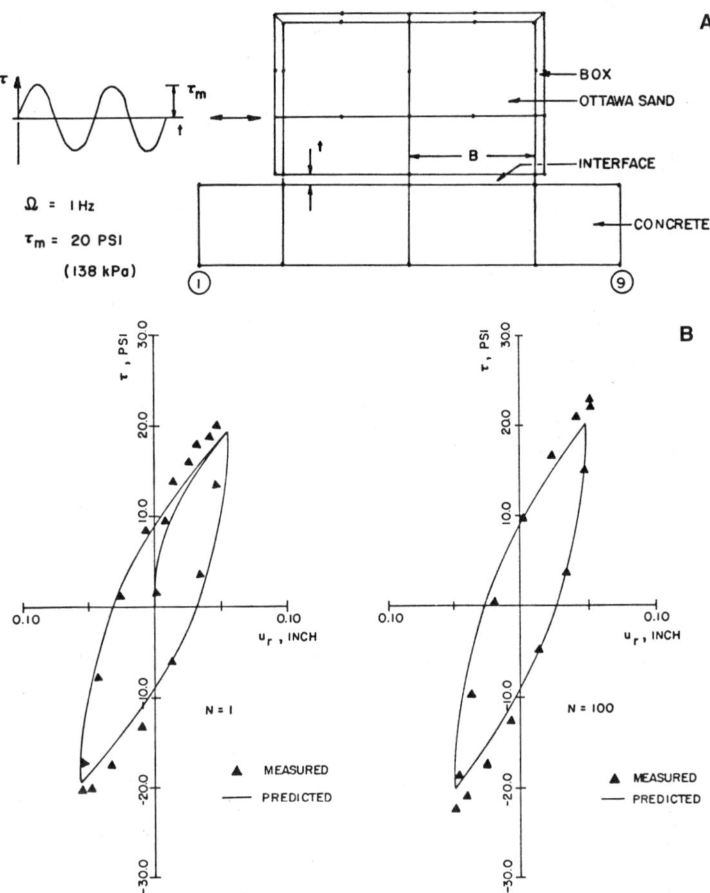

Figure 4-9. Mesh and comparisons for cyclic shear test (D_r = 80%, σ_n = 28 psi): (A) Mesh. (B) Comparisons at cycles N = 1 and 2.

PRACTICAL PROBLEMS

The thin-layer element concept with the models defined from static and cyclic shear tests was introduced in static and dynamic nonlinear finite element procedures. Then behavior of number of geotechnical problems simulated in the laboratory and in the field were predicted. The predictions were also compared with closed-form solutions for a number of problems. These problems included footings on clays and sands, single piles in clays and sands, pile groups in sand, tied-back retaining walls, buried pipes in soils, track support structures, anchors in sand, and model nuclear power plant (SIMQUAKE) structures in sand. Typical results follow.

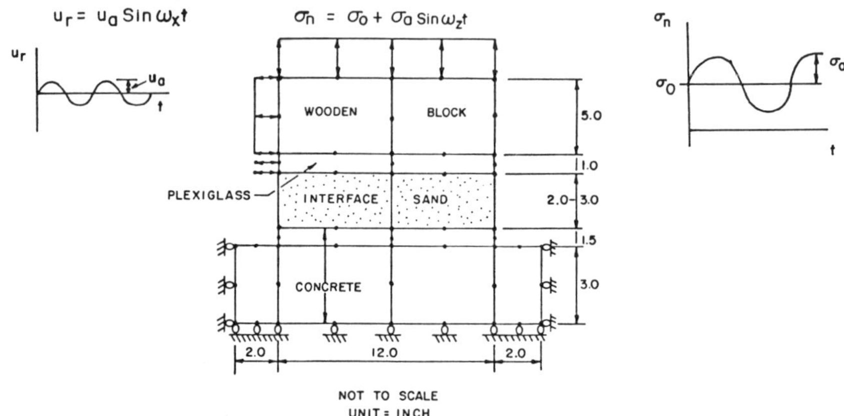

Figure 4-10. Mesh for combined cyclic normal and shear loading.

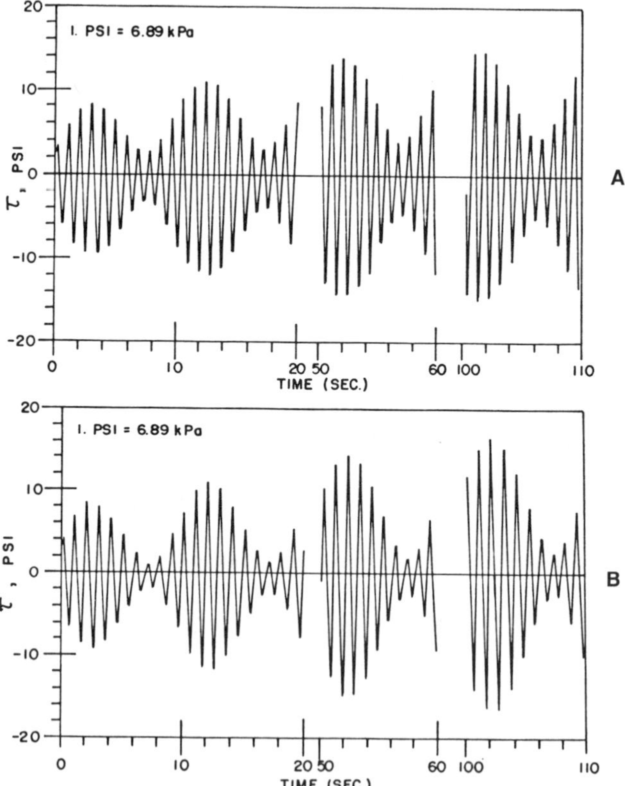

Figure 4-11. Comparisons between observed and predicted responses for cyclic combined normal and shear tests (σ_0 = 20 psi, σ_a = 15 psi, u_a = 0.05 in.): (A) Observed. (B) Finite element predictions.

Buried Pipe

Figure 4-12A shows an idealized buried steel pipe in linear elastic soil [Desai, et al., 1984; Katona, 1983]. The thin-layer element was used by assuming three interface conditions: Frictionless slip with tan $\phi = 0.001$, stick or bonded with tan $\phi = 2.0$, and frictional slip with tan $\phi = 0.25$; here ϕ = angle of friction between the sand and the steel. The following material properties were used:

$Soil:$ E_s (modulus of elasticity) = 3,000 psi (21 MPa)
   ν_s (Poisson's ratio) = 0.333
$Pipe:$ $E_p = 2.2 \times 10^6$ psi (15 GPa), $\nu_p = 0.25$
$Interface:$ G_i (shear modulus) = 128.8 × t lb/in.^{2} (3.56 × t GPa)
   $\nu_i = 0.333$
   $\tau = 128.8\ (u_r) - 250\ u_r^2$
   $t = 0.01\ (\pi R/8)$

where R = radius of the pipe

Figure 4-12B shows the finite element mesh. The soil is assumed weightless and the overburden pressure is applied in a number of incre-

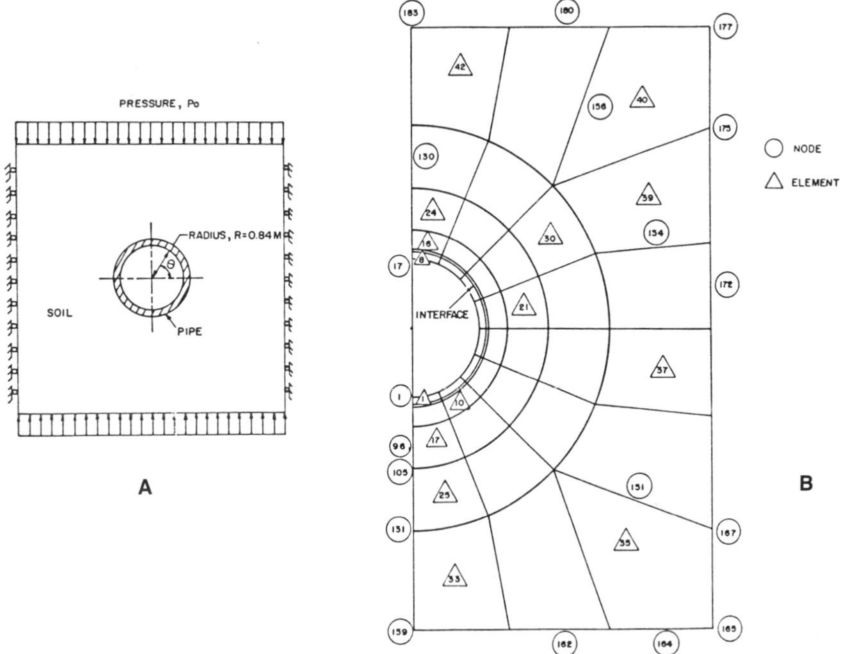

Figure 4-12. Buried pipe and mesh: (A) Pipe [Katona, 1983]. (B) Mesh.

ments. Figures 4-13A, B, and C show comparisons between the predictions using the thin-layer element, Katona's element [1983], and the closed-form solution [Burns and Richards, 1964]. It can be seen that the proposed element provides very good correlation with the closed-form solutions and Katona's results.

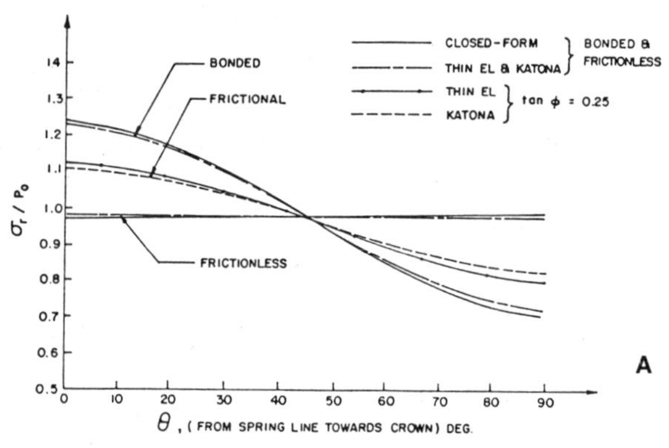

Figure 4-13. Comparison of numerical and closed-form solutions. (A) σ_r/P_o vs. θ (B) τ/P_o vs. θ (C) Displacements along crown.

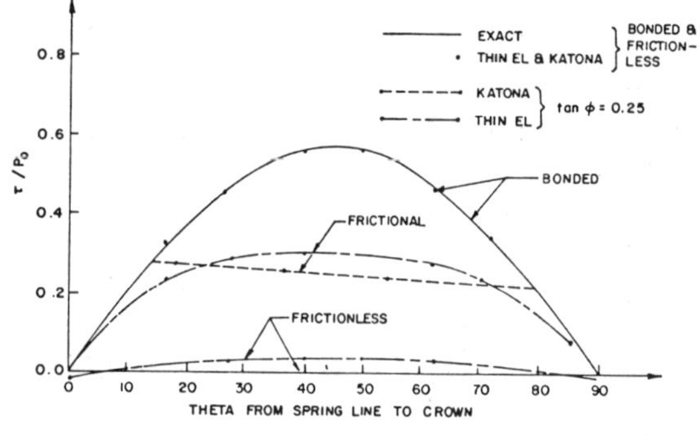

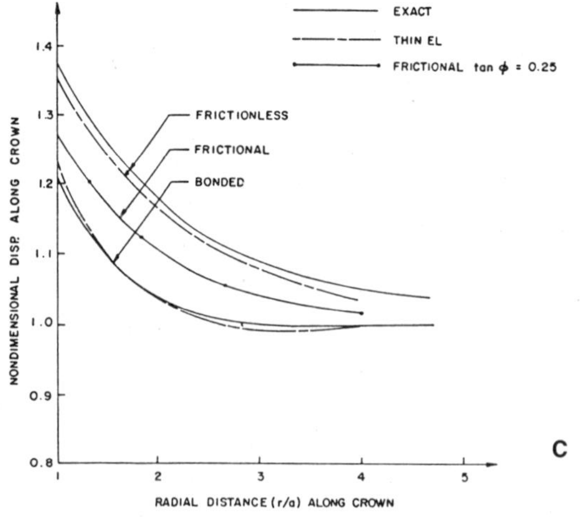

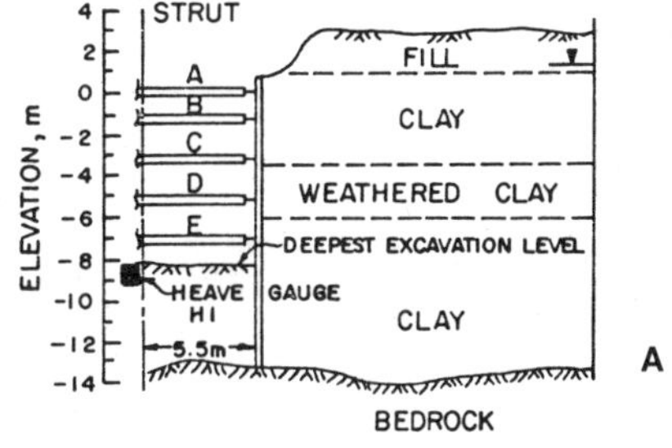

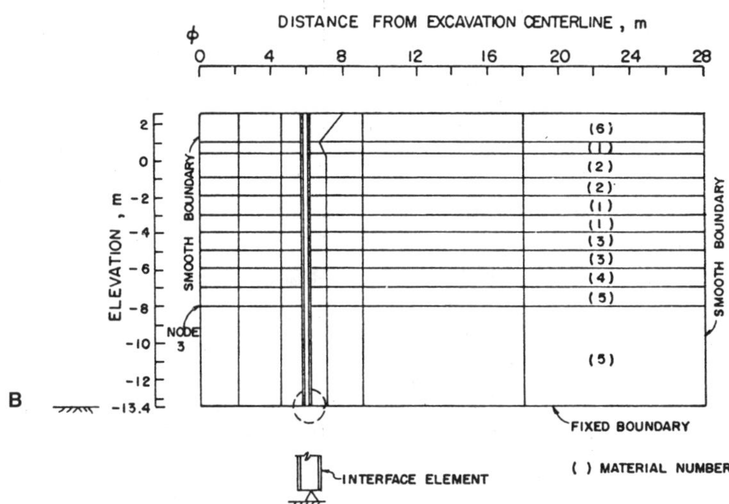

Figure 4-14. Retaining wall and mesh: (A) Wall section [Clausen, 1971]. (B) Finite element mesh for Vaterland 1 wall.

Tied-Back Wall

Figures 4-14A and B show the tied-back retaining wall [Clausen, 1971; Desai et al., 1984; Mana, 1978] and the finite element mesh, respectively. The wall was constructed in eight stages including the in situ condition, excavation, and installation of the ties A, B, C, and D. The material properties for various materials and the thin-layer interface are shown in Table 4-3 [Clausen, 1971; Desai, 1984; Mana, 1978]. The cohesive soil was characterized by using the Von Mises criterion with incremental plasticity. The mixed finite-element procedure was used [Desai et al., 1984] and included simulation of the sequential constructions, with computation of stresses, pressures, and displacements after each stage.

Table 4-3
Soil Properties for Strutted Wall

Material Number	E (t/m²)	ν	c (t/m²)	ϕ	γ (t/m³)	K_o
1	800.0	0.4933	4.0	0	1.95	0.65
2	760.0	0.4933	3.8	0	1.95	0.65
3	900.0	0.4933	4.5	0	1.95	0.65
4	500.0	0.4967	5.0	0	1.95	0.65
5	410.0	0.4967	4.1	0	1.95	0.65
6	368.0	0.3000	1.42	0	1.95	0.65

From Clausen [1985]; Desai et al. [1984]; and Mana [1978].

Wall properties: $E = 2.76 \times 10^6$ t/m², $\nu = 0.33$
Interface properties: $E = 450.0$ t/m², $\nu = 0.33$, $G_i = 0.10$ t/m², t = 0.05 m

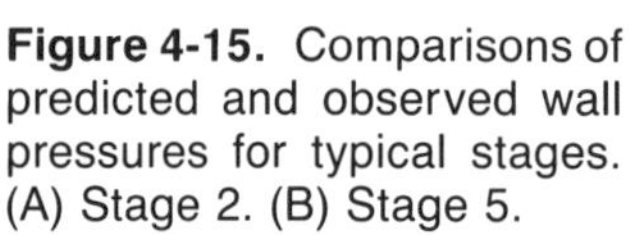

Struct properties
1t = 1,000 kg

A,D: Area = 1.0 m², $E = 9.2.0$ t/m²

B,C,E: Area = 1.0 m², $E = 4560.0$ t/m²

Typical comparisons between predictions and measurements for the wall pressures; that is, normal stress in the interface, are shown in Figures 4-15A and B and for Stages 2 and 5, respectively. It can be seen that the correlation is satisfactory and also that the thin-layer model is capable of providing consistent stresses in the interfaces.

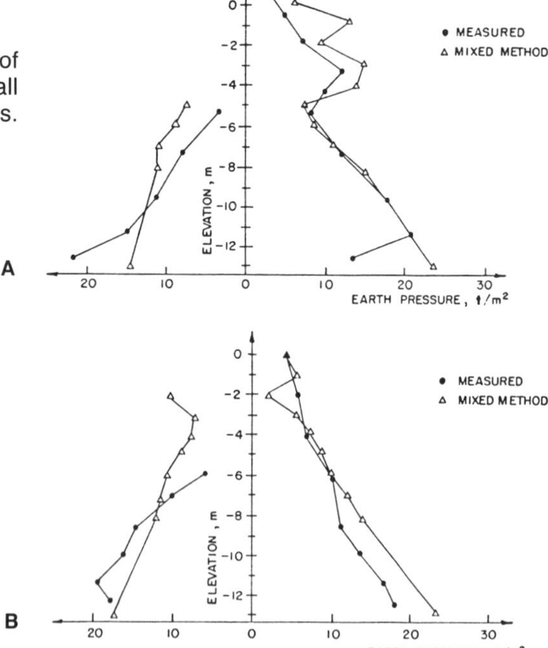

Figure 4-15. Comparisons of predicted and observed wall pressures for typical stages. (A) Stage 2. (B) Stage 5.

Anchors in Sand

A series of anchors (Figure 4-16) were tested by Scheele [1981] in the field in a sand called Munich sand in W. Germany. Cubical specimens of the sand were tested in the laboratory using the multiaxial testing device and a plasticity-based isotropic hardening model [Desai et al., 1986] was used to characterize its behavior. The interface of the grouted anchor was simulated by using three-dimensional thin-layer interface element (Figure 4-4). The material properties for the sand, anchor and interface are given in Desai, Muqtadir, and Scheele [1986].

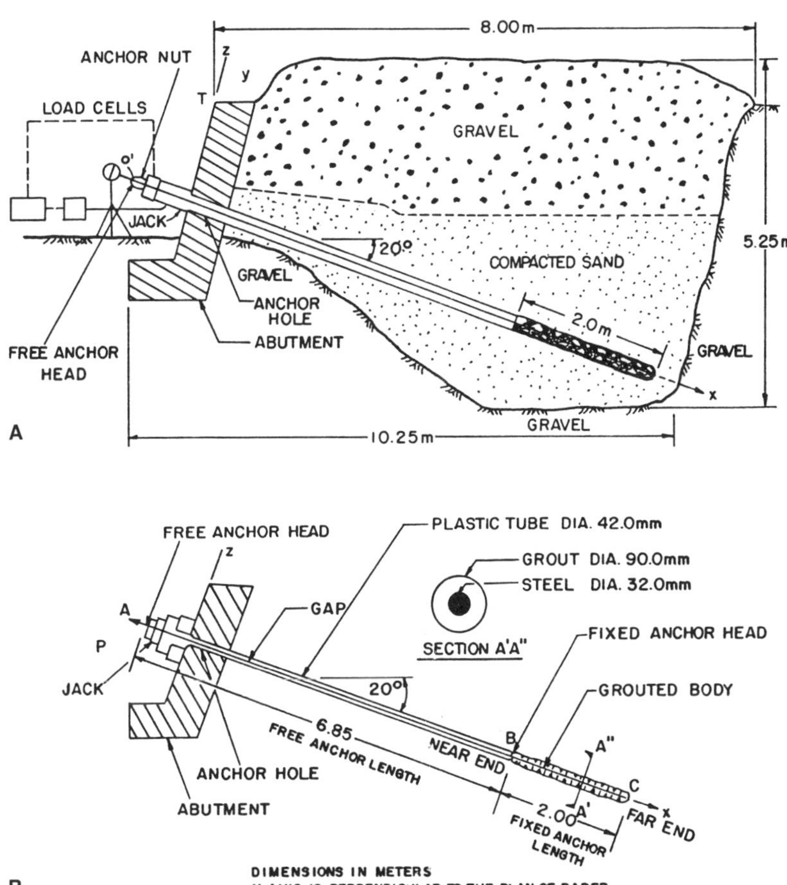

Figure 4-16. Anchor in sand [Scheele, 1981]: (A) Elevation of anchor-soil system (schematic). (B) Details of components of anchor.

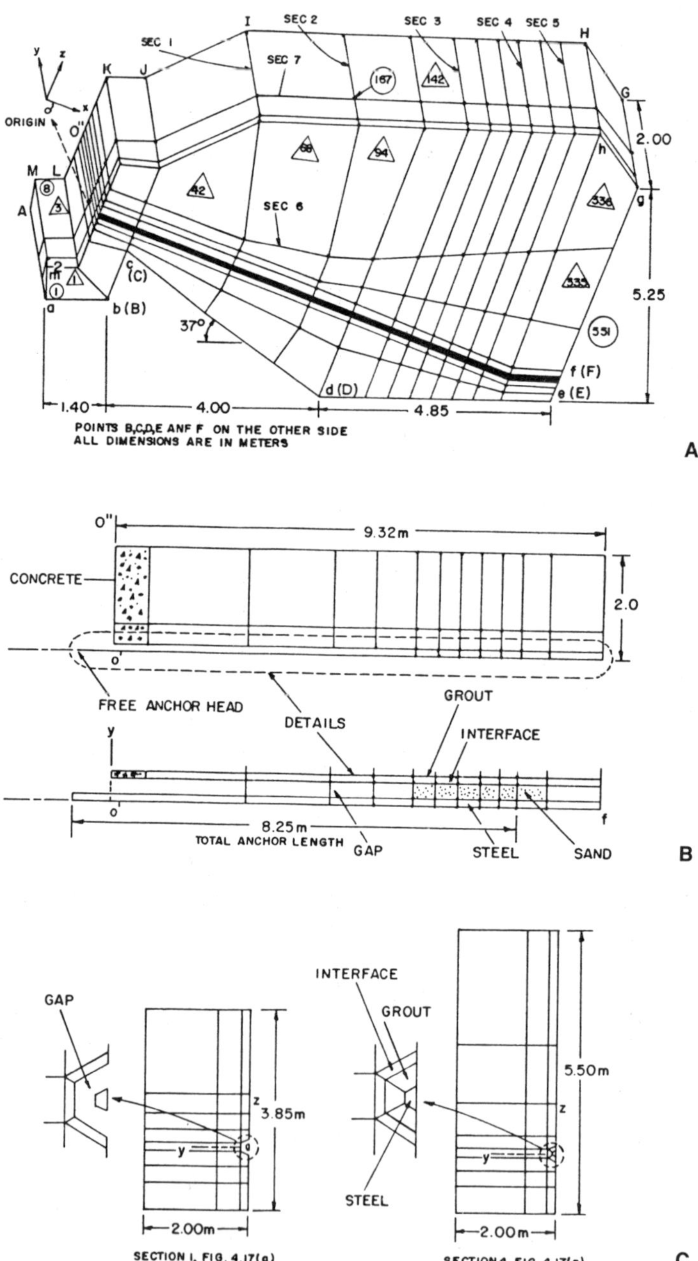

Figure 4-17. Details of finite element mesh: (A) Overall. (B) Along grouted anchor. (C) Across grouted anchor.

Figure 4-17 shows the three-dimensional finite-element mesh for one of the anchors, soil, and interfaces. The load was applied in increments and in the same manner as was done in the field.

Figure 4-18 shows comparisons of observed and computed (linear and nonlinear) load-displacement curves at free and fixed anchor heads and Figure 4-19 shows distribution of normal and shear stresses along the anchor at various load levels. This analysis clearly indicates the effect of nonlinear behavior and relative motions at the interfaces with resulting redistribution of stresses at the ends of the anchor.

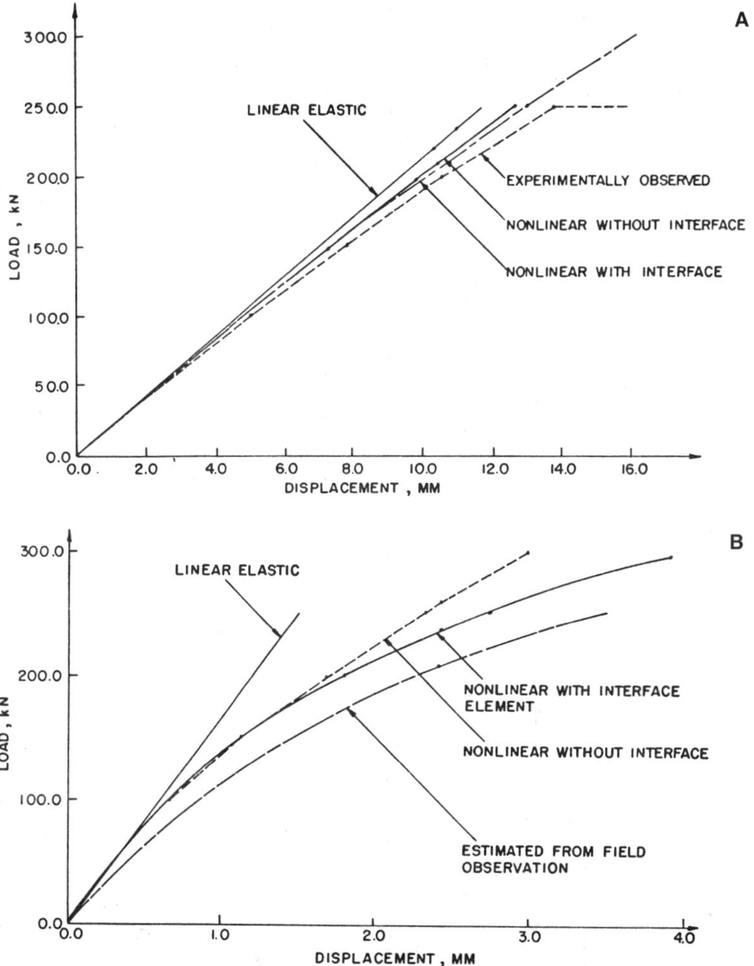

Figure 4-18. Load displacement curves at free and fixed anchor heads: (A) At free anchor head (Point A in Figure 4-16). (B) At fixed anchor head (Point B in Figure 4-16).

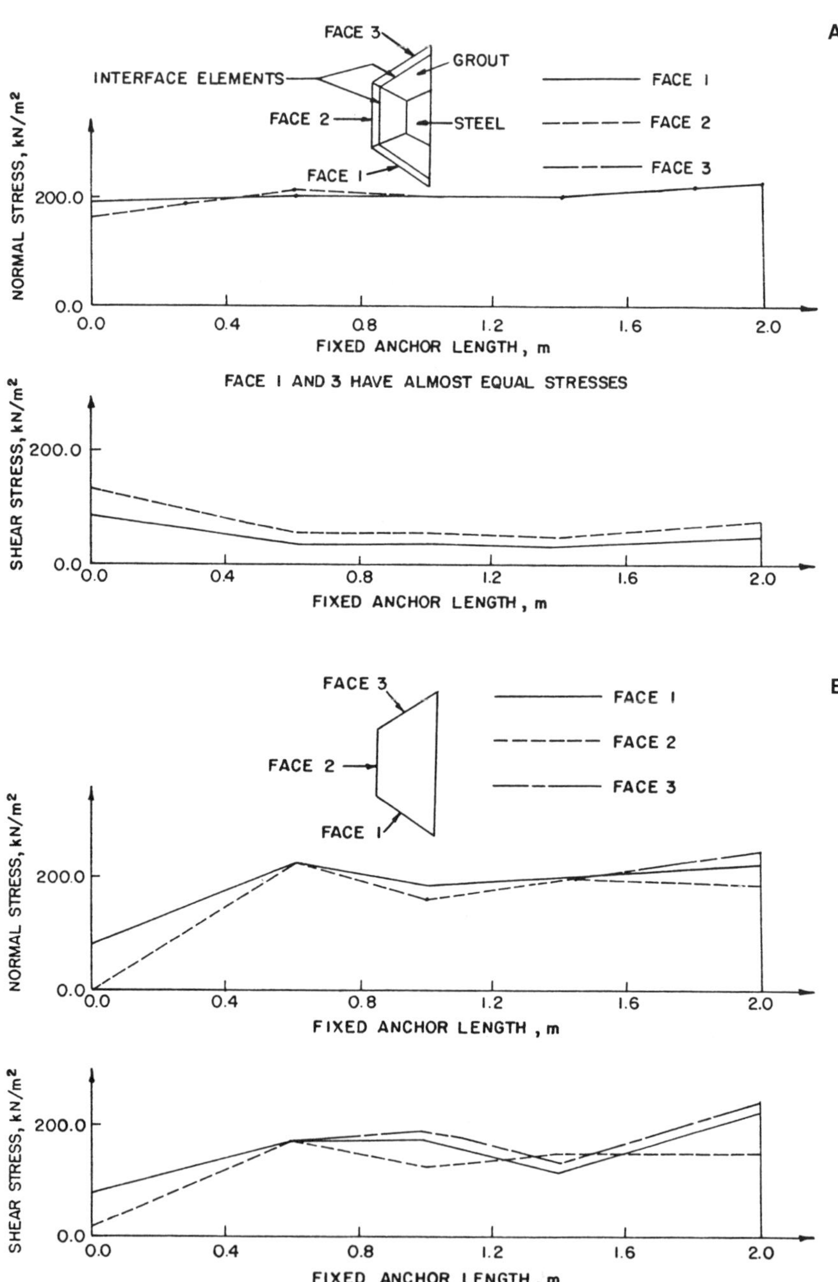

Figure 4-19. Normal and shear stress distributions in interface elements on grouted anchor: (A) At P = 50 kN. (B) At P = 150 kN.

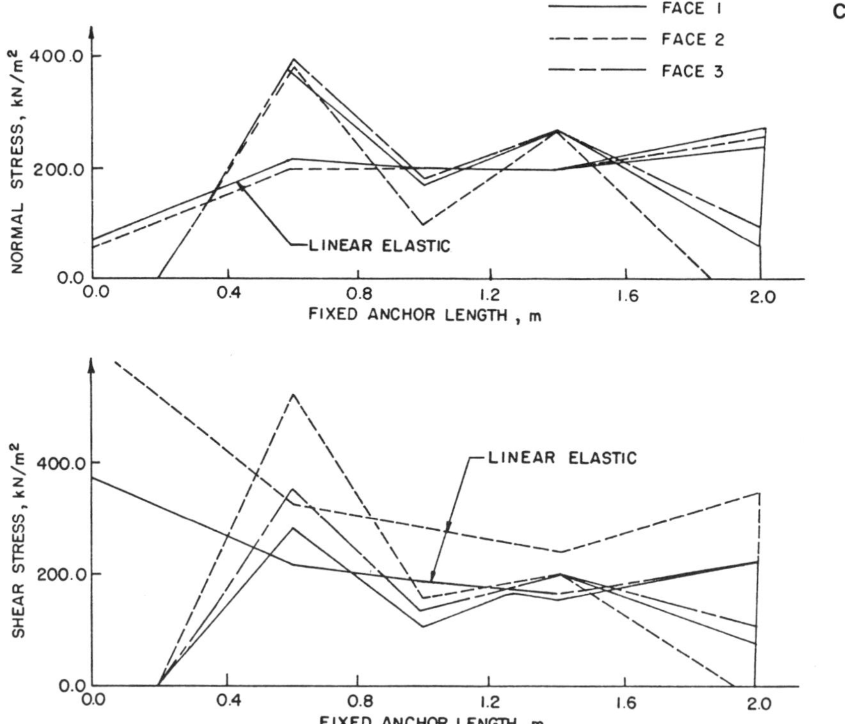

Figure 4-19 (continued). Normal and shear stress distributions in interface elements on grouted anchor: (C) At P = 250 kN.

SIMQUAKE TEST

Figure 4-20B shows the finite element mesh of the one-eighth-scale model of nuclear power plant structure (Figure 4-20A) tested in the field; this test is referred to as SIMQUAKE test [Vaughn and Isenberg, 1983]. The system was subjected to blast loads and velocities were measured at the outside boundaries; these were integrated to obtain the displacement-time history that was introduced in the present finite element analysis. The sand in the foundation was modeled by using the cap model [Vaughn and Isenberg, 1983], whereas the interface behavior was characterized from the laboratory cyclic tests [Zaman et al., 1984] using a similar sand with the thin-layer element.

Figures 4-21A, B, and C show typical comparisons between the finite element predictions and observations for x- and y-velocities at points P and Q, and the contact stresses at the base of the structure. In view of the complexity of the system, the correlation is considered satisfactory.

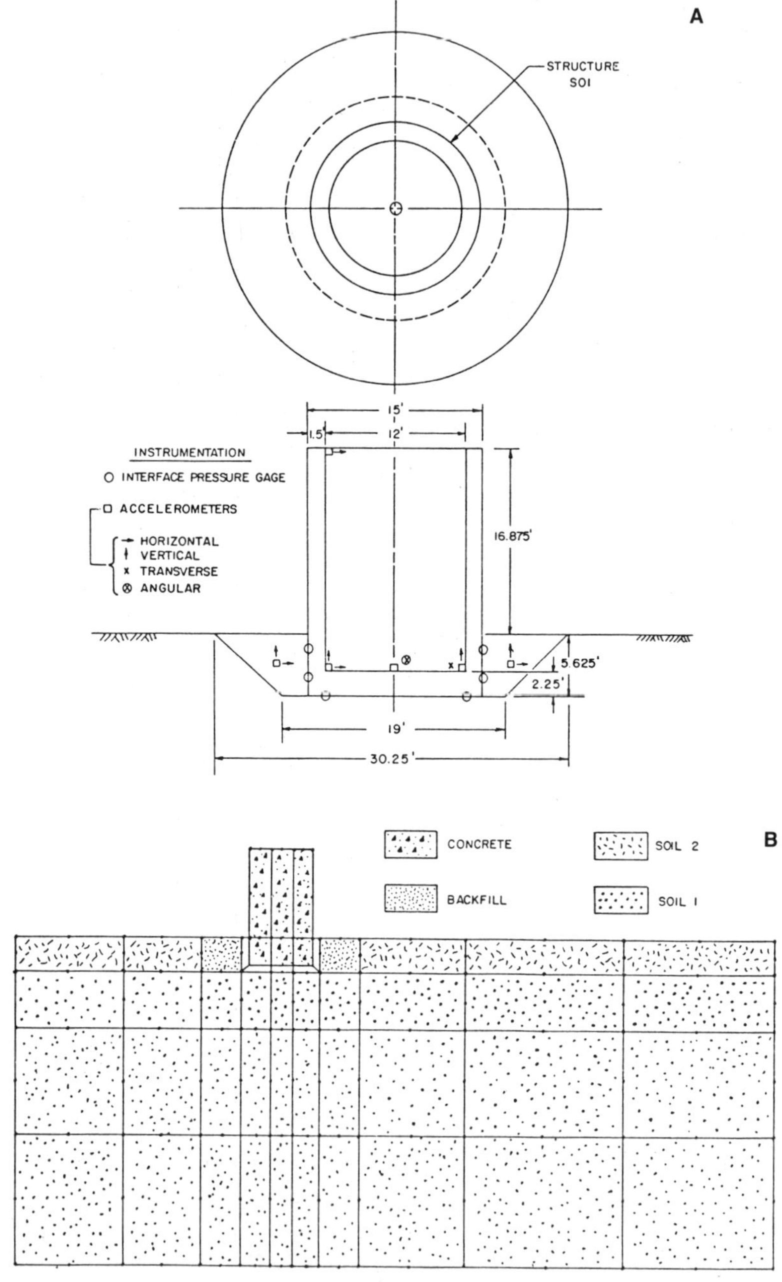

Figure 4-20. One-eighth-scale model SIMQUAKE structure (SO1) including structural and near-field instrumentation and mesh: (A) SIMQUAKE structure [after Vaughn and Isenberg, 1983]. (B) Mesh.

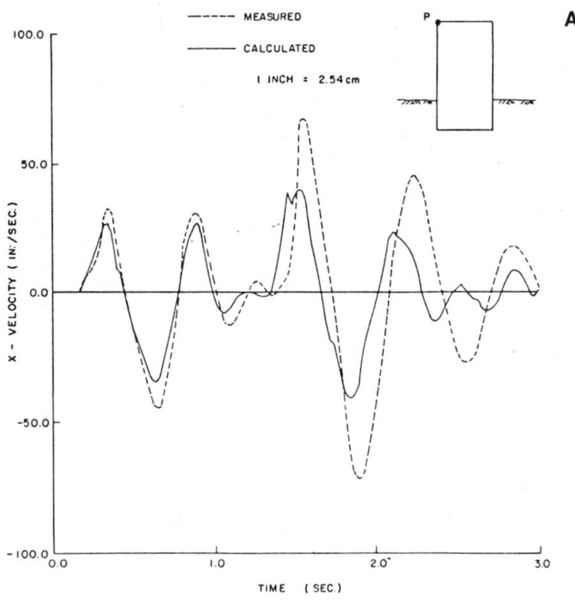

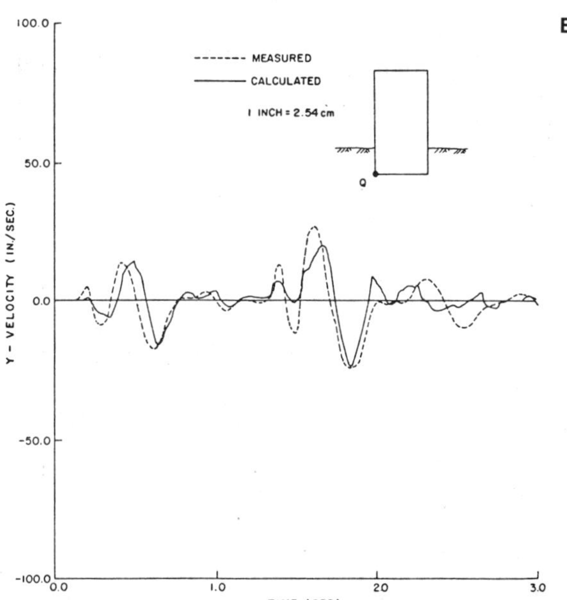

Figure 4-21. Comparison of computed and measured velocities and contact pressures: (A) x-velocity at Point P. (B) y-velocity at Point Q.

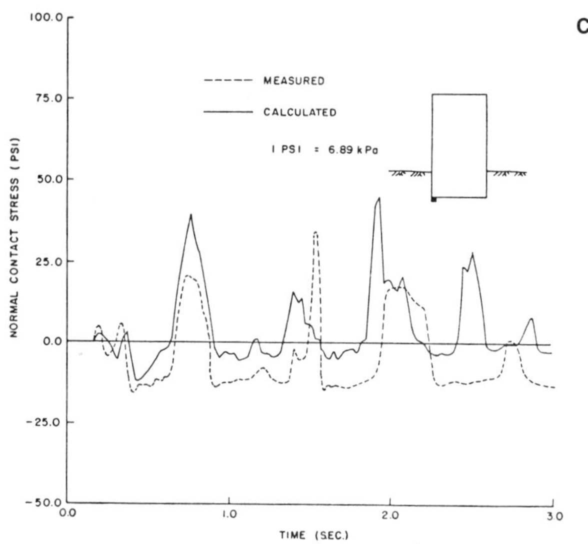

Figure 4-21 (continued). Comparison of computed and measured velocities and contact pressures: (C) contact pressures.

Also, the results indicate that the predictions from the frictional case that allows for relative motions and the bonded case that does not allow for them are significantly different [Zaman et al., 1984]. Overall, the former yields improved and realistic predictions.

CONCLUSIONS

For realistic simulation of soil-structure interaction problems subjected to static and dynamic loadings, it is necessary to characterize appropriately the (nonlinear) behavior of interfaces. Analytical models that allow for the relative motions at the interfaces are proposed for the shear and normal responses. The constants involved are determined based on comprehensive laboratory tests using a new shear device. The models are verified with respect to observed laboratory shear and normal responses under static and cyclic loadings. The models are also implemented in numerical (finite element) procedures, which are used to predict behavior of various practical problems. The predictions are compared with closed-form solutions, and laboratory and field observations for various geotechnical problems involving static and dynamic loadings. Overall, the proposed models are considered to be more realistic in comparison with previous approaches, and provide improved predictive capability for complex soil-structure interaction problems.

Acknowledgments

A number of individuals, Drumm, E. C., Lightner, J. G., Muqtadir, A., Nagaraj, B. K., Siriwardane, H. J., and Zaman, M. M., participated in various aspects of the results presented herein. The research was partly supported by Grant No. 81-CE-94 from the National Science Foundation, Washington, D.C., and No. DOT-05-80013 from Office of University Research, Department of Transportation, Washington, D.C.

REFERENCES

1. Bowden, F. P., and Tabor, D., *The Friction and Lubrication of Solids, Part II*, Clarendon Press, Oxford, England, 1964.
2. Burns, J. W., and Richards, R. M., "Attenuation of Stresses for Buried Cylinders," *Proc., Symp. on Soil-Structure Interaction*, Univ. of Arizona, Tucson, AZ, 1964.
3. Clausen, C. J. F., "Finite Element Analysis of the Strutted Excavation at Vaterland 1," *Report No. 52601-2*, Norwegian Geotech. Inst., Oslo, Norway, 1971.
4. Clough, G. W., and Duncan, J. M., "Finite Element Analyses of Retaining Wall Behavior," *J. Soil Mech. & Found. Div., ASCE*, Vol. 97, SM12, 1971.
5. Crandall, S. H., Lee, S. S., and Williams, J. H., "Accumulated Slip of Friction Controlled Mass Excited by Earthquake Motions," *J. of Appl. Mech., ASME*, 1974, pp. 1094–1098.
6. Desai, C. S., "Numerical Design Analysis of Piles in Sands," *J. Geotech. Eng. Div., ASCE*, Vol. 100, GT6, 1974, pp. 613–625.
7. Desai, C. S., "Finite Element Method for Analysis and Design of Piles," *Misc. Paper S-76-21*, U.S. Army Waterways Expt. Stn., Vicksburg, Miss., October 1976.
8. Desai, C. S., "Soil-Structure Interaction and Simulation Problems," Chap. 7 in Gudehus, G. (Ed), *Finite Elements of Geomechanics*, Wiley, Chichester, 1977.
9. Desai, C. S., "Some Aspects of Constitutive Laws of Geologic Media," *Proc., 3rd Int. Conf. on Num. Methods in Geomech.*, Aachen, W. Germany, 1979.
10. Desai, C. S., "Behavior of Interfaces Between Structural and Geologic Media," *State-of-the-Art Paper, Proc., Int. Conf. on Recent Advances in Geotech. Earthquake Eng. & Soil Dynamics*, St. Louis, Missouri, 1981.
11. Desai, C. S., "Dynamic Multi-Degree-of-Freedom Shear Device," *Report No. 80-36*, Dept. of Civil Eng., Virginia Tech, Blacksburg, VA, 1981.

12. Desai, C. S. and Appel, G. C., "3-D Analysis of Laterally Loaded Structures," *Proc., 2nd Int. Conf. on Num. Meth. in Geomech.*, Blacksburg, VA, ASCE, 1976.

13. Desai, C. S., Drumm, E. C., and Zaman, M. M., "Cyclic Testing and Modelling of Interfaces," *J. of Geotech. Eng. Div., ASCE*, Vol. 111, No. 6, 1985, pp. 793–815.

14. Desai, C. S., Galagoda, H. M., and Wathugala, G. W., "Hierarchical Modelling for Geologic Materials and Discontinuities, Joints, Interfaces," *Proc., 2nd Int. Conf. Const. Laws for Eng. Materials*, Tucson, AZ, Elsevier Publ. Co., N.Y., 1987.

15. Desai, C. S., and Holloway, D. M., "Load-Deformation Analysis of Deep Pile Foundations," *Proc., Symp. on Appl. of Finite Elem. Meth. in Geotech. Eng.*, Vicksburg, Mississippi, 1972.

16. Desai, C. S., Johnson, L. D., and Harget, C. M., "Analysis of Pile Supported Gravity Lock," *J. Geotech. Eng. Div., ASCE*, Vol. 100, GT9, 1974.

17. Desai, C. S., Muqtadir, A., and Scheele, F., "Interaction Analysis of Anchor-Soil Systems," *J. of Geotech. Eng. Div., ASCE*, Vol. 112, GT5, May 1986.

18. Desai, C. S., and Nagaraj, B. K., "Modelling for Cyclic Normal and Shear Behavior of Interfaces," *J. of Eng. Mech. Div., ASCE*, submitted, 1986.

19. Desai, C. S., Phan, H. V., and Perumpral, J. V., "Mechanics of Three-Dimensional Soil-Structure Interaction," *J. of Eng. Mech. Div., ASCE*, Vol. 108, No. EM5, October 1982.

20. Desai, C. S., and Siriwardane, H. J., "Numerical Models for Track Support Structures," *J. of Geotech. Eng. Div., ASCE*, Vol. 108, No. GT3, March 1982.

21. Desai, C. S., Somasundaram, S., and Frantziskonis, G., "Hierarchical Approach for Constitutive Modelling of Geologic Materials," *Int. J. Num. Analyt. Meth. in Geomech.*, Vol. 10, 1986, pp. 225–257.

22. Desai, C. S., and Wu, T. H., "A General Function for Stress-Strain Curves," *Proc. 2nd Int. Conf. Num. Methods in Geomech.*, Blacksburg, VA, published by ASCE, 1976.

23. Desai, C. S., et al., "Thin-Layer Elements for Interfaces and Joints," *Int. J. for Num. & Analyt. Methods in Geomech.*, Vol. 8, 1984, pp. 19–43.

24. Drumm, E. C., "Testing, Modeling, and Application of Interface Behavior in Dynamic Soil-Structure Interaction," *Ph.D. Thesis*, University of Arizona, Tucson, AZ, 1983.

25. Drumm, E. C., and Desai, C. S., "Determination of Parameters for a Model for the Cyclic Behavior of Interfaces," *Earthquake Eng. & Struct. Dyn.*, Vol. 14, 1986, pp. 1–18.

26. Fishman, K. L., and Desai, C. S., "A Constitutive Model for Hardening Behavior of Rock Joints," *Proc., 2nd Int. Conf. Const. Laws for Eng. Materials*, Tucson, AZ, Elsevier Publ. Co., N.Y., 1987.

27. Ghaboussi, J., Wilson, E. L., and Isenberg, J., "Finite Element for Rock Joint and Interfaces," *J. of Soil Mech. & Found. Div.*, ASCE, Vol. 99, SM10, 1973, pp. 833–848.

28. Goodman, R. E., Taylor, R. L., and Brekke, T. L., "A Model for the Mechanics of Jointed Rock," *J. of Soil Mech. & Found. Div.*, ASCE, Vol. 94, SM3, 1968, pp. 637–659.

29. Herrmann, L. R., "Finite Element Analysis of Contact Problems," *J. of Eng. Mech. Div.*, ASCE, Vol. 104, EM5, 1978, pp. 1043–1057.

30. Idriss, I. M., Dobry, R., and Singth, R. D., "Nonlinear Behavior of Soft Clays During Cyclic Loading," *J. Geotech. Eng. Div.*, ASCE, Vol. 104, No. 12, 1978, pp. 1427–1447.

31. Idriss, I. M., et al., "Analysis of Soil-Structure Interaction Effects for Nuclear Power Plants," *Report* by ad hoc Group on Soil-Structure Interaction, ASCE, N.Y., 1979.

32. Isenberg, J., Lee, L., and Agbabian, M. S., "Response of Structures to Combined Blast Effects," *J. of Transp. Eng. Div.*, ASCE, Vol. 99, TE4, 1973, pp. 887–908.

33. Katona, M. G., "A Simple Contact-Friction Interface Element with Applications to Buried Culverts," *Int. J. Num. Analyt. Meth. in Geomech.*, Vol. 7, 1983, pp. 371–384.

34. Kausel, E., et al., "Seismically Induced Sliding of Massive Structures," *J. of Geotech. Eng. Div.*, ASCE, Vol. 15, GT12, 1979, pp. 1471–1488.

35. Lightner, J. G., and Desai, C. S., "Improved Numerical Procedures for Soil-Structure Interaction Including Simulation of Construction Sequences," *Report* No. VPI-E-79.32, Dept. of Civil Eng., Virginia Tech, Blacksburg, VA, 1979.

36. Mana, A. I., "Finite Element Analysis of Deep Excavation Behavior in Soft Clay," *Ph.D. Dissertation*, Stanford University, Stanford, California, 1978.

37. Matlock, H., Bogard, D., and Cheang, L., "A Laboratory Study of Axially Loaded Piles and Pile Groups Including Pore Pressure Measurements," *Proc., Behavior of Offshore Structures, 3rd Int. Conf.*, Washington, D.C., 1983.

38. Muqtadir, A., and Desai, C. S., "Three-Dimensional Analysis of Pile-Group Foundation," *Int. J. Num. Analyt. Meth. in Geomech.*, Vol. 10, 1986, pp. 41–58.

39. Nagaraj, B. K., "Modelling of Normal and Shear Behavior of Interfaces in Dynamic Soil-Structure Interaction," *Ph.D. Dissertation*, Univ. of Arizona, Tucson, AZ, 1986.

40. Newmark, N. M., "Effect of Earthquakes on Dams and Embankments," *Geotechnique*, Vol. XV, No. 2, 1965.

41. Ngo, D., and Scordelis, A. C., "Finite Element Analysis of Reinforced Concrete Beams," *ACI Journal*, Vol. 64, No. 3, 1967.

42. Scheele, F., "Tragfaigekeit von Verpressankern in Nichtbinddigem Boden, Neue Erkenntnisse durch Dehnungsmessungen im Verankernungsbereich," Doctoral Dissertation, Univ. of Munich, W. Germany, Munich, 1981.

43. Toki, T., Sato, T., and Miura, F., "Separation and Sliding Between Soil and Structure During Strong Ground Motion," *Earthquake Eng. & Struct. Dyn.*, Vol. 9, 1981, pp. 263–277.

44. Vaughan, D. K., and Isenberg, J., "Nonlinear Rocking Response to Model Containment Structures," *Earthquake Eng. & Struct. Dyn.*, Vol. 11, 1983, pp. 275–296.

45. Wolf, J. P., "Soil-Structure Interaction with Separation of Base Mat from Soil," *Nuclear Eng. & Design*, Vol. 38, 1976, pp. 357–384.

46. Wolf, J. P., *Dynamic Soil-Structure Interaction*, Prentice-Hall, Inc., Englewood Cliffs, N.J., 1985.

47. Zaman, M. M., Desai, C. S., and Drumm, E. C., "Interface Model for Dynamic Soil-Structure Interaction," *J. Geotech. Eng. Div.*, *ASCE*, Vol. 110, No. 9, 1984, pp. 1257–1273.

48. Zienkiewicz, O. C., et al., "Analysis of Nonlinear Problems With Particular Reference to Jointed Rock Systems," *Proc., 2nd Int. Conf. Soc. of Rock Mech.*, Belgrade, Vol. 3, 1970, pp. 501–509.

5

Stress, Degradation, and Shear Strength of Granular Material

Donald J. Murphy
Vice President
Langan Engineering Associates, Inc.
Elmwood Park, New Jersey USA

The influence of normal stress levels on the shearing resistance of soil was recognized more than two hundred years ago. Until the late 1960s, data used in applying this concept have been limited generally to those obtained from conventional triaxial compression tests conducted at confining pressures of 150 lb per sq in. or less. As shown in Table 5-1, stress conditions in many geotechnical situations are outside this limitation. The solution to related engineering problems has required extensive extrapolation from low-pressure results. As the nonlinear nature of the relationship between normal stress and shearing resistance precluded rational extrapolation, prudence necessitated the employment of high "factors of uncertainty." Elimination of the need for such factors is possible only if strength envelope curvature and other pressure-dependent phenomena can be related to fundamental material characteristics. A substantial part of past research on the effects of high stress levels was directed toward this goal.

As indicated in Table 5-2, many investigators have studied the behavior of materials subjected to high pressures. Detailed reviews of this research have been given elsewhere [Murphy, 1970]. The results of much of this work, particularly the early studies, were essentially qualitative, as most efforts were restricted to confined compression tests on very small samples. Many of the later investigations provided valuable qualitative and quantitative data. Still, most of these were somewhat limited as emphasis was placed on only a single-stress level, or a rather narrow range of stress levels. The few investigators who were able to observe behavior over a reasonably broad range of stress levels did so for quartz sands only.

Table 5-1
Practical Occurrences of High Pressure

Probable Maximum Stress Level (lb per sq in.)	Type of Facility or Situation
1,000	Dam foundations and abutments, tunnel linings, mine shafts.
10,000	Deep wells, mine pillars, residual stresses in near surface evaporate deposits.
50,000	Pile foundations, points of concentration in mine pillars.
150,000 +	Engineering use of conventional and nuclear explosives, relationship between brittle–ductile transition and the occurrence of deep focus earthquakes.
90,000,000	Pressure at the earth's core.
150,000,000	Maximum pressure attainable in research.

The focus of this chapter is a formulation of general hypotheses concerning the behavior of soils and rocks over wide ranges of stress levels and mineralogical compositions. Of particular interest is the curvature of the Mohr strength envelope and the significance (at both low and high stress levels) of the various components of shearing resistance shown in Equation 5-1

$$\phi_d = \phi_u + \phi_r + \phi_\delta + \phi_{deg} \qquad (5\text{-}1)$$

where ϕ_d = the measured angle of shearing resistance
         ϕ_u = the angle of mineral friction
         ϕ_r = the component of shearing resistance caused by particle reorientation
         ϕ_δ = the dilatancy component of shearing resistance
        ϕ_{deg} = the component of shearing resistance attributable to particle degradation

Past research has indicated that each of these components varies. A number of investigators have isolated the dilatancy component. The role of mineral friction has been fairly well established for a variety of minerals by Tschebotarioff and Welch [1948], Horn and Deere [1962], Bromwell [1966] and others. The influence of particle degradation on the shearing resistance of granular materials began to attract formal attention

Table 5-2

Previous Research on the Influence of Stress Levels

Period	Materials Studied	Phenomena of Concern	Principal Investigators
1901–1912	Sandstone, marble	Brittle-ductile transition.	Adams; von Karman; Bridgman
1918	Quartz, feldspar, calcite	Volumetric strain due to particle degradation.	Bridgman
1918–1948	Various rocks and minerals	Brittle-ductile transition, creep, orientation, and re-crystallization of minerals; inter-dependence of time, temperature, stress level and interstitial solution effects.	Bridgman; Griggs; Balsley
1948– present	Various rocks and minerals	Continuation of 1918–1948 efforts, frictional characteristics, folding, failure mechanisms, kinking, shock behavior, field testing.	Griggs; Handin; Heard; Christie; Byerlee; Jaeger; Patterson and Weiss; Turner; Borg; Maxwell; Mann; Ahrens; Lundborg; Youash; Maurer; Ukhov; Krsmanovic;
1948– present	Soils and granular assemblies of minerals	Strength and deformation characteristics under extreme loading conditions, influence of saturation, stress state, end restraints, grain size and mineralogy	Terzaghi and Peck; Roberts and de Souza; DeBeer; Golder and Akroyd; Leslie; Hall and Gordon; Insley and Hillis; Hirschfeld and Poulos; Vesić and Barksdale; Vesić and Clough; Bishop; Bishop, Webb, and Skinner; Hillis and Skermer; Lee and Seed; Lee and Farhoomand; Tai; Murphy; Olson and Parola; Lee and Haley; Mazanti and Holland; Banks and McIver; Sowers, Williams and Wallace; Marsal; Fumagalli; Lee; Leslie; Marachi, Seed, and Chan.

in the late 1960s and early 1970s. Lee and Seed [1967] speculated that the crushing of grains during shear results in the absorption of energy and thus causes the angle of shearing resistance (corrected for dilatancy) to be higher than the angle of mineral friction. Further, they indicated that ϕ_r is not constant but increases with increasing stress level as failure strains (at high pressures) are higher and more work is done by moving particles.

Vesic and Clough [1968] have shown that for Chattahoochee River sand, dilatancy effects vanish at octahedral normal stress levels of about 1,500 lb per sq in. and that in the range of mean normal stresses from 1,500 to about 15,000 lb per sq in., the Mohr strength envelope is linear. Tai [1970] presented similar data for Chattahoochee River sand. In addition, he found that for Ottawa sand dilatancy effects vanish at an octahedral normal stress of about 8,500 lb per sq in. Murphy [1970] has shown that both the Chattahoochee River and Ottawa sands have linear Mohr strength envelopes over a mean normal stress range from 8,500 up to 80,000 lb per sq in.

The contributions of Lee and Seed [1967], Tai [1970] and Murphy [1970] indicate that even small compositional variations between their samples of different quartz sands can result in substantial behavioral differences, particularly at "very high"* stress levels. Vesic and Clough [1968] have indicated that such behavioral differences may be directly attributable to the crushing characteristics of the materials concerned.

PURPOSE AND SCOPE

The primary objectives of this chapter are to isolate the major components of shearing resistance and to study further the influence of degradation on the shear strength of granular materials. To accomplish these goals, it is necessary first to isolate and examine the influence of initial density, degree of saturation, state of stress, initial grain size, and type of end restraint on the behavior of soils subjected to "high" stress levels and to compare these to already well-established low pressure responses. A systematic program of testing over a wide range of stress levels was conducted. To assure broad mineralogical representation, the tests were conducted on samples formed from essentially pure quartz, feldspar, calcite, and chlorite sands. Initially, the intent of this study was to extend some of Koerner's [1968, 1970] work on these minerals at low stress levels (30 lb per sq in.) into the "high" (4,500 lb per sq in.) stress range where parti-

---

* Herein, the terminology given by Vesic and Clough [1968] is used to describe the stress level, i.e., low = 0 to 10 kg per sq cm; elevated = 10 to 100 kg per sq cm; high = 100 to 1,000 kg per sq cm; very high = 1,000 to 10,000 kg per sq cm; ultra high = over 10,000 kg per sq cm.

cle crushing would be expected to be significant. As shown elsewhere herein, the results of tests conducted at 4,500 lb per sq in. implied that mineralogical composition had a significant influence on mechanical behavior. To study the extent of this influence, testing at higher stress levels was necessary. As the effects of minor increases in stress levels could be masked by experimental scatter, an order of magnitude increase to 45,000 lb per sq in. was chosen.

EXPERIMENTAL PROGRAM

The experimental phase of this study consisted of approximately sixty tests. These included drained isotropic and triaxial compression tests at both 4,500 and 45,000 lb per sq in. confining pressure. Tests were performed on samples of sand-sized and silt-sized materials of essentially pure quartz (Mohs hardness $M=7$), feldspar ($M=6$), calcite ($M=3$), and chlorite ($M=1$). Conventional mechanical analyses and electron microscopy techniques were employed to evaluate the extent of particle degradation. Typical results of the 4,500 lb per sq in. tests and of the 45,000 lb per sq in. tests are presented herein. The complete data base, along with details concerning equipment, materials and testing procedures have been presented elsewhere [Murphy, 1970, 1971].

TEST RESULTS AND DISCUSSION

4,500 psi Tests

This part of the program was divided into seven series as described in Table 5-3. The results of these series may be summarized as follows:

Effect of Initial Density. As illustrated in Figure 5-1 for quartz, the strength of each mineral tested essentially was independent of the initial state of density. In all cases, the true axial and true volumetric strains at failure were somewhat greater for the initially loose than for the initially dense samples. The variation of void ratio with mean normal stress during both the consolidation and shear stages is shown in Figure 5-2. The purpose of such a plot is to indicate, qualitatively, the significance of the volumetric strain component attributable to octahedral shear. In cases where it was significant compared to the isotropic volumetric strain component an abrupt downward curvature was noted at the start of the shearing stage of the plot. The extent to which octahedral shear contributed to volumetric strain appeared to be greatest for the hard minerals. As might have been expected, the void ratio at failure was, in all cases, independent of initial density.

Table 5-3
Experimental Series—"High" Stress Level Phase*

Series	Type of Test	Sample Conditions	Objective
I	Triaxial compression.	Initially loose, saturated sand, rough ends, 2.80 in. diameter by 6.50 in. high.	Isolation of the influence of initial density on behavior at "high" stresses.
II	Triaxial compression.	Initially dense, saturated sand, rough ends, 2.80 in. diameter by 6.50 in. high.	Same as I.
III	Triaxial compression.	Loose to intermediate, dry sand, rough ends, 2.00 in. diameter by 4.00 in. high.	Comparison of the behavior of dry and saturated sand at "high" stresses.
IV	Triaxial compression.	Loose to intermediate, dry sand, lubricated ends 4.00 in. diameter by 4.00 in. high.	Determination of the effects attributable to end restraint at "high" stresses.
V	Triaxial compression.	Dry silt, rough ends 2.00 in. diameter by 4.00 in. high.	Investigation of differences between the behavior of sand and silt at "high" stresses.
VI	Isotropic compression.	Same as I.	Segregation of volume changes occurring during consolidation from those occurring during shear.
VII	Isotropic compression.	Same as III.	Same as VI.

* All tests conducted at a confining stress of 4,500 lb per sq in.

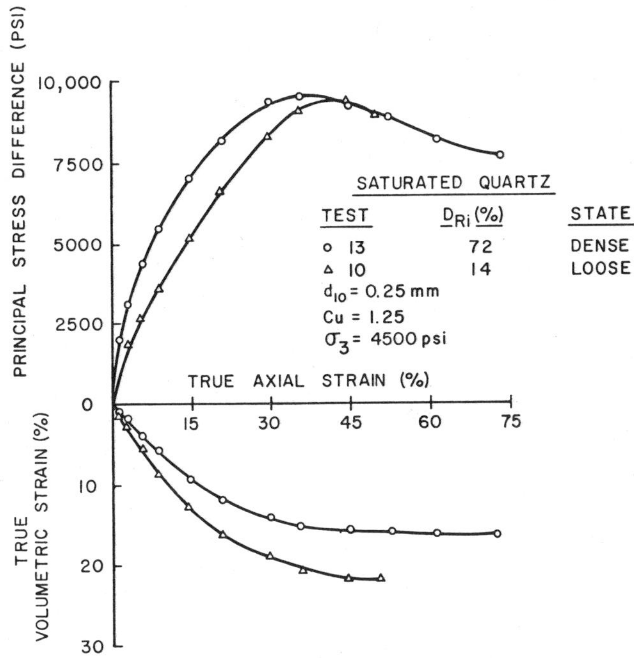

Figure 5-1. Comparison of triaxial compression tests on initially dense and initially loose saturated quartz, under 4,500 psi confinement.

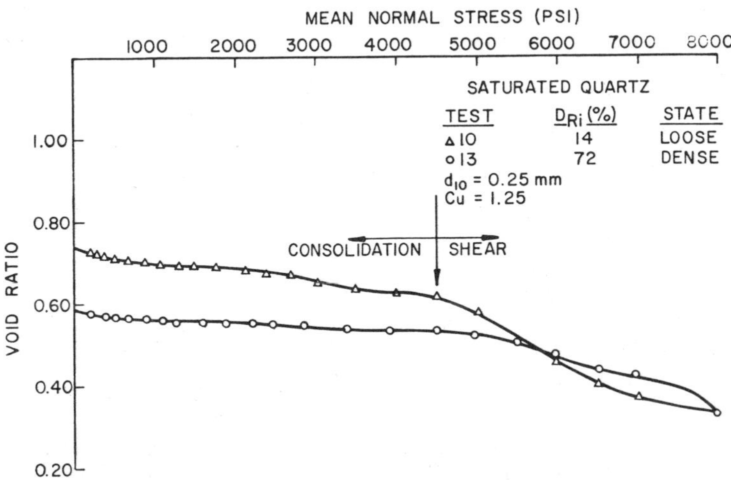

Figure 5-2. Void ratio vs. mean normal stress for initially dense and initially loose saturated quartz.

The results of some of the Series I and II tests are summarized in Table 5-4. In preparing this, and subsequent tables similar to it, it was necessary to compute the dilatancy component of the angle of shearing resistance. In all cases this component was equal or nearly equal to zero regardless of which of the many available separation techniques was used. For this reason. Rowe's method [1962], being the simplest, was employed. The angle of shearing resistance was computed by:

$$\sin\phi_f = \frac{(\sigma_1/\sigma_3) - (1 - k)}{(\sigma_1/\sigma_3) + (1 - k)} \tag{5-2}$$

and

$$k = \frac{dv}{d\epsilon} \ @ \ \text{failure} \tag{5-3}$$

where ϕ_f = the angle of shearing resistance corrected for dilatancy by Rowe's [1962] method

σ_1 = the major principal stress at failure

σ_3 = the minor principal stress at failure

v = the true volumetric strain

ϵ = the major principal strain

In the Series I and II tests all materials, with the exception of quartz, reached the critical state and thus had k values of zero.

The moduli of deformation shown in Table 5-4 were defined using initial tangent concepts and were obtained directly from the stress-strain data. The results checked favorably with those obtained by means of a hyperbolic stress-strain formulation used for defining the initial tangent moduli.

Although it appears that the angle of shearing resistance and the void ratio at failure are at the 4,500 lb per sq in. confining level essentially independent of initial density, some differences did exist between the behavior of initially dense and initially loose samples. Notably, the denser samples failed at lower axial and volumetric strains, underwent less volumetric strain during isotropic compression, and had somewhat higher initial tangent moduli. In addition, the dense samples had slightly higher computed angles of shearing resistance. In part, this may have been attributable to the great differences in lateral strain occurring in the initially loose as compared to the initially dense samples. The shape differences, illustrated in Figure 5-3 for calcite, were typical and necessitated development of parabolic area correction techniques. Though the method

Table 5-4

Comparison of Triaxial Compression Tests Based on Initial Relative Density

Material (in order of decreasing hardness)	Condition	Test	Relative Density D_{r_i}(%)	Void Ratio			Initial Tangent Modulus E_i (psi)	ϵ_f(%)	ϕ_f	v(%)*	
				e_i	e_c	e_f				v_c	v_s
Quartz	Loose	10	14	0.75	0.61	0.32	60,000	45.9	30.7	8.7	21.8
Quartz	Dense	13	72	0.59	0.53	0.33	74,000	36.0	31.0	4.2	14.7
Feldspar	Loose	13	14	1.21	0.52	0.29	50,000	37.8	33.0	45.9	14.7
Feldspar	Dense	6	92	0.83	0.49	0.32	60,000	34.3	35.5	22.8	12.9
Calcite	Loose	12	21	1.00	0.30	0.21	57,000	41.1	32.8	54.3	7.3
Calcite	Dense	7	100	0.66	0.31	0.23	67,000	31.7	35.1	26.4	7.0
Chlorite	Loose	4	26	1.21	0.48	0.41	47,000	29.5	25.3	49.2	4.8
Chlorite	Dense	5	95	0.86	0.30	0.22	55,000	23.9	27.5	42.7	6.3

For all samples $d_{10} \simeq 0.25$, Cu $\simeq 1.25$, S(%) = 100, end conditions—rough,

where d_{10} = effective diameter, Cu = uniformity coefficient; S = degree of saturation.

* v = true volumetric strain and subscripts c, s, and i refer to consolidation, shear, and initial, respectively.

was proven by a limited number of comparisons to the results of lubricated end tests, it still is possible that part of the difference between the computed angles of shearing resistance for initially loose and initially dense samples is attributable to shape differences.

Figure 5-3. Calcite after shear under 4,500 psi confinement: left—rough ends; right—lubricated ends; background—original shape of samples.

Effect of Degree of Saturation. As shown in Figure 5-4 for feldspar (M=6), dry samples were stronger than saturated samples. In terms of the developed secant angles of shearing resistance the strength differences (dry less saturated) were: quartz $+2.6°$; feldspar $+2.5°$; calcite $-0.8°$, chlorite $+0.2°$. In all cases, the true volumetric and axial strains at failure were not significantly influenced by degree of saturation. Interestingly, as illustrated in Figure 5-5 for feldspar (M=6), the void ratio at failure was the same for dry and saturated samples only for the hard minerals. As shown in Figure 5-6 for calcite (M=3), the void ratio at failure for the soft minerals was lower for saturated than for dry samples. This is understandable as it is well known that the crushing strength of certain

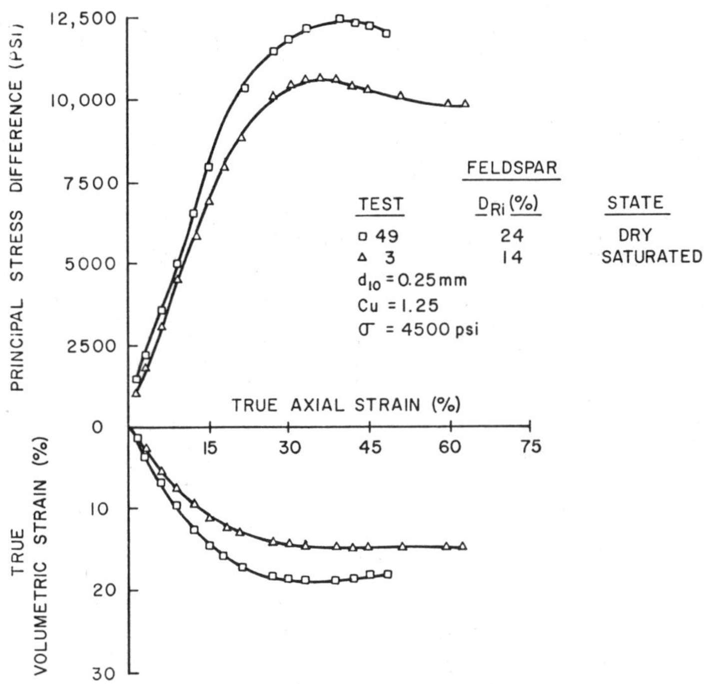

Figure 5-4. Comparison of triaxial compression tests for dry and saturated feldspar at a confining pressure of 4,500 psi.

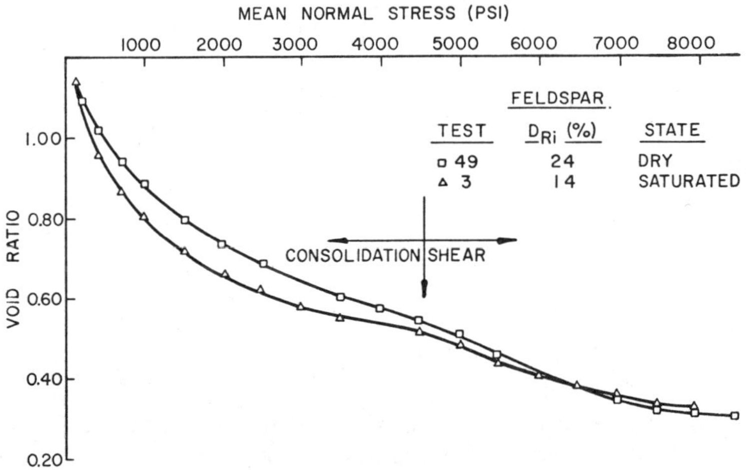

Figure 5-5. Void ratio vs. mean normal stress for dry and saturated feldspar.

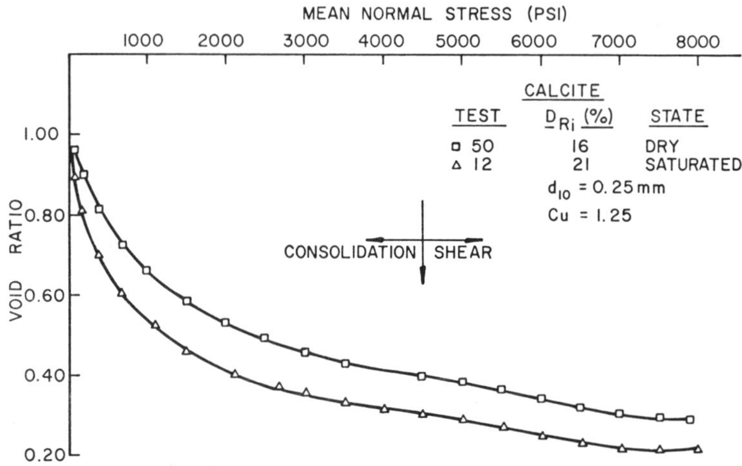

Figure 5-6. Void ratio vs. mean normal stress for dry and saturated calcite.

minerals decreases in the presence of water. The causative mechanism is thought to be the flow of water into microfissures. As such flow requires a pore water pressure differential, the effective stress at different points along a flow path must be different and a shearing stress, in an otherwise isotropic field, must develop. This additional stress induces the crushing attributable to saturation. The nearly parallel nature (at stresses below about 1,500 lb per sq in.) of the two curves shown in Figure 5-6 indicates that these differences in degradation probably were most significant in the low stress range.

The results of the Series III tests are summarized in Table 5-5. As shown, it appears that the difference between the angles of shearing resistance of the dry and of the saturated samples increases with increasing hardness. Most probably, this difference is but one manifestation of the interdependence of hardness, grain microstructure, the angle of mineral friction, and the state of stress. As already indicated, the crushing strength of minerals having significant microstructure is reduced in the presence of water. Horn and Deere [1962] have presented data indicating that for quartz, feldspar, calcite and other "bulky" minerals the angle of mineral friction is higher for saturated than for dry material. Similar results have been shown by Tschebotarioff and Welch [1948]. The angle of mineral friction for chlorite and other latticed minerals is lower in the saturated than in the dry state [Horn and Deere, 1962]. Bromwell [1966] has shown that the influence of saturation on the angle of mineral friction may vanish in the presence of extremely rough surfaces. With these trends established, the differences noted between the strengths of saturated and of dry samples may be explained in the following terms:

Table 5-5

Comparison of Triaxial Compression Tests Based on Degree of Saturation

Material (in order of decreasing hardness)	Condition	Test	Relative Density, D_{r_i}(%)	Void Ratio			Initial Tangent Modulus E_i (psi)	ϵ_f(%)	ϕ_f	v(%)*	
				e_i	e_c	e_f				v_c	v_s
Quartz	Saturated	10	14	0.75	0.61	0.32	60,000	45.9	30.7	8.7	21.8
Quartz	Dry	48	0	0.79	0.52	0.29	45,000	38.3	33.3	17.9	17.8
Feldspar	Saturated	3	14	1.21	0.52	0.29	50,000	37.8	33.0	45.9	14.7
Feldspar	Dry	49	24	1.16	0.54	0.30	55,000	37.9	35.5	40.1	18.6
Calcite	Saturated	12	21	1.00	0.30	0.21	57,000	41.1	32.8	54.3	7.3
Calcite	Dry	50	16	1.03	0.40	0.28	63,000	33.3	32.0	45.4	9.2
Chlorite	Saturated	4	26	1.21	0.48	0.41	47,000	29.5	25.3	49.2	4.8
Chlorite	Dry	51	9	1.29	0.57	0.50	48,000	17.8	25.5	45.9	5.1

For all samples $d_{10} \simeq 0.25$, Cu $\simeq 1.25$, end conditions—rough; density—loose,
where d_{10} = effective diameter, Cu = uniformity coefficient; S = degree of saturation.

* v = true volumetric strain and subscripts c, s, and i refer to consolidation, shear, and initial, respectively.

Quartz. Although the crushing strength of quartz (it being relatively free of microfissures) is not significantly altered by saturation, it appears that particle degradation during shear is more intense in dry than in saturated samples. The cause of this is probably decreased grain mobility resulting from the antilubricating effect of water. In such a situation, the degradation component of the angle of shearing resistance for dry quartz would be greater than that for saturated quartz.

As shown elsewhere herein, the degradation component assumes major significance at high pressures. Most probably, at these pressures the increase in the degradation component, dry over saturated, is more significant than the decrease in mineral friction. The result is that, at high pressures, quartz in the dry state is somewhat stronger than in the saturated state.

At low pressures, the significance of the roles of degradation and mineral friction may be reversed. Koerner's [1968, 1970] finding that a 30-lb-per-sq-in. saturated quartz is slightly stronger than dry quartz is evidence of this possibility.

Since crushing strength is not only a function of hardness, saturation, microstructure and stress state but also of grain size, grain shape, grain size distribution and void ratio, the qualitative explanation offered in the foregoing may require modification in the presence of additional variables.

Feldspar. The crushing strength of feldspar, owing to the presence of microscopic twinning planes, probably is reduced somewhat in the presence of water. The degradation component of the angle of shearing resistance is therefore less for saturated than for dry samples. At high pressures, this difference is more (but not as much more as in the case of quartz) significant than differences in mineral friction attributable to the antilubricating effect of water. The result is that at these pressures feldspar is slightly stronger in the dry than in the saturated state.

Calcite. The presence of three perfect cleavage planes allows the crushing strength of calcite to be significantly decreased by saturation. In the presence of water, the degradation component of the angle of shearing resistance appears to decrease as much as the mineral friction component increases.

As a result the strength of dry and of saturated calcite at 4,500 lb per sq in. is identical.

Chlorite. Owing to the plate-like construction of individual chlorite grains this mineral behaves similarly to calcite.

Effect of End Restraint. As expected, differences between samples tested under rough-end and lubricated-end conditions, were "trendwise" the same at high as at low pressures. As illustrated in Figure 5-7 for feldspar, the measured strength of rough-ended samples was higher than that of samples with lubricated ends. The summary given in Table 5-6 indicates that the magnitude of the difference in strength decreases with decreasing hardness. This, however, may be coincidental. It is more likely that the angle of friction between the steel end platens and the mineral grains is lower for softer materials and that a certain amount of natural lubrication occurred even under rough-end conditions.

The techniques used to ensure end lubrication at high stress levels and the procedures used to make area corrections for samples tested under rough-end conditions are described in detail elsewhere [Murphy, 1970]. The fact that the computed angles of shearing resistance of the two samples shown in Figure 5-3 agree within 1° indicates that the methods developed are promising.

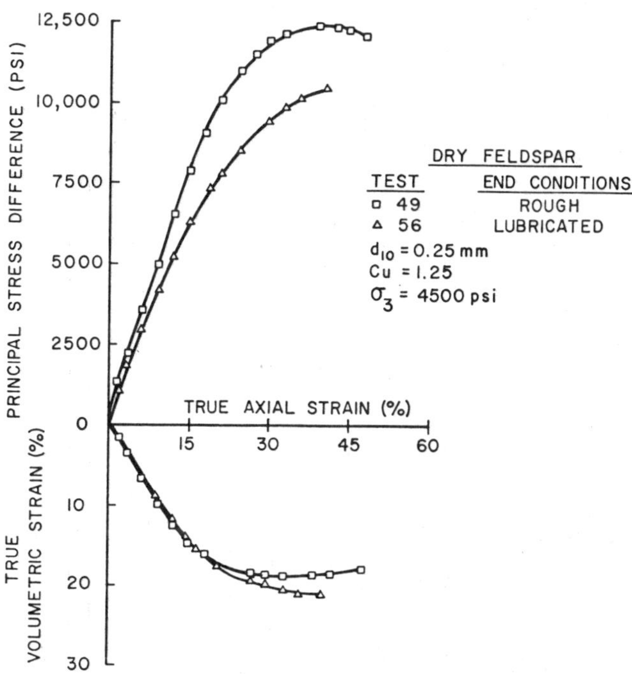

Figure 5-7. Comparison of triaxial compression tests under rough and lubricated end conditions for dry feldspar under 4,500 psi confinement.

Table 5-6

Comparison of Triaxial Compression Tests Based on End Conditions

Material (in order of decreasing hardness)	Condition	Test	Relative Density, D_{r_i}(%)	Void Ratio			Initial Tangent Modulus E_i (psi)	ϵ_f(%)	ϕ_f	v(%)*	
				e_i	e_c	e_f				v_c	v_s
Quartz	Rough	48	0	0.79	0.52	0.29	45,000	38.3	33.2	17.9	17.8
Quartz	Lubricated	58	40	0.68	0.50	0.28	34,000	51.0	30.2	11.5	17.8
Feldspar	Rough	49	24	1.16	0.54	0.30	55,000	37.9	35.5	40.1	18.6
Feldspar	Lubricated	56	48	1.05	0.47	0.21	42,000	43.0	32.6	32.6	21.0
Calcite	Rough	50	16	1.03	0.40	0.28	63,000	33.3	32.0	45.4	9.2
Calcite	Lubricated	22	41	0.92	0.39	–**	48,000	47.0	31.4	38.8**	–**
Chlorite	Rough	51	9	1.29	0.57	0.50	48,000	17.8	25.5	45.9	5.1
Chlorite	Lubricated	54	47	1.10	0.36	0.28	39,000	26.8	25.0	54.9	6.2

For all samples $d_{10} \simeq 0.25$, Cu $\simeq 1.25$, S(%) = 0,

where d_{10} = effective diameter, Cu = uniformity coefficient; S = degree of saturation.

* v = true volumetric strain and subscripts c, s, and i refer to consolidation, shear, and initial, respectively.

** Drainage port blocked by Teflon segments during consolidation.

Effect of Initial Grain Size. As shown in Figure 5-8 for quartz and summarized in Table 5-7 for all the materials tested, the strength of the samples of initially silt size material was, in all cases, greater than or equal to that of the samples of initially sand sized material. This was expected, since in an earlier investigation Koerner [1968, 1970] noted that for quartz tested at 30 lb per sq in. silt size material was substantially stronger than sand size. It may be that this behavior is another manifestation of the influence of the degradation component of the angle of shearing resistance. For example, it is well known that the crushing strength of most materials increases with decreasing size. On this basis, a possible explanation for the behavior noted by Koerner is loss of strength occurring as a result of very minor degradation of the sand grains during consolidation. If, as is likely, the crushing of silt during consolidation at low stresses was even less significant than the minor crushing of sand, at the same stress level, the measured strength of silt would be higher. As is shown elsewhere herein, the difference in strength developed would be the difference in latent strength lost during consolidation. An additional factor contributing to the difference between silt and sand behavior noted by Koerner may be the increase in the mineral friction angle with decreasing size reported by Rowe [1962].

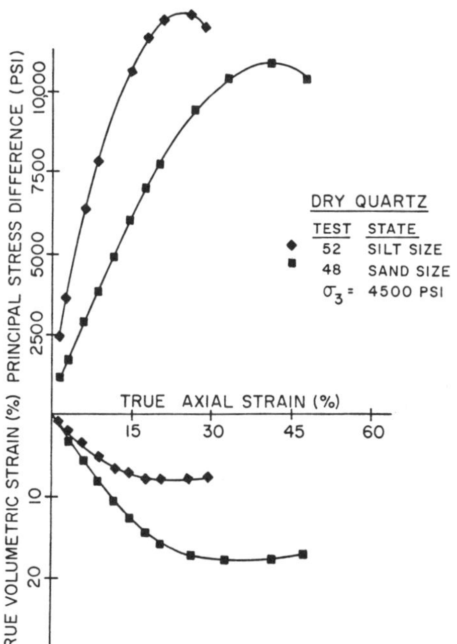

Figure 5-8. Comparison of triaxial compression tests for medium sand size and silt size dry quartz under 4,500 psi confinement.

Table 5-7

Comparison of Triaxial Compression Tests Based on Initial Grain Size

Material	d_{10}	Cu	Test	Relative Density, $D_{r_i}(\%)$	Void Ratio			Initial Tangent Modulus E_i (psi)	$\epsilon_f(\%)$	ϕ_f	v(%)*	
					e_i	e_c	e_f				v_c	v_s
Quartz	0.25	1.25	48	0	0.79	0.52	0.29	45,000	38.3	33.2	17.9	17.8
Quartz	0.074	1.25	14	NA	0.94	0.66	0.44	48,000	42.9	33.7	16.9	15.0
Quartz	0.022	2.65	52	NA	0.93	0.67	0.55	79,000	23.5	35.3	15.5	7.8
Feldspar	0.25	1.25	49	24	1.16	0.54	0.30	55,000	37.9	35.5	40.1	18.6
Feldspar	0.002	20.0	55	NA	1.15	0.89	0.76	89,000	20.4	36.5	13.7	7.5
Calcite	0.25	1.25	50	16	1.03	0.40	0.28	63,000	33.3	32.0	45.4	9.2
Calcite	0.018	3.60	57	NA	0.83	0.53	0.42	88,000	25.3	37.3	19.6	7.8
Chlorite	0.25	1.25	51	9	1.29	0.57	0.50	48,000	17.8	25.5	45.9	5.1
Chlorite	0.018	3.33	53	NA	1.16	0.27	0.17	61,000	17.9	25.2	67.5	8.9

NA—not applicable. For all samples, end conditions—Rough; S(%) = 0 (except test 14 S(%) = 100), where d_{10} = effective diameter, Cu = uniformity coefficient, S = degree of saturation.

* v = true volumetric strain and subscripts c, s, and i refer to consolidation, shear, and initial, respectively.

Influence of Mineralogical Composition. Data from the five series of tests discussed in the foregoing all indicated that behavioral characteristics at high stress levels are dependent strongly on grain mineralogy, particularly the degradation potential or crushing strength of the mineral concerned. Figures 5-9 and 5-10 hint that a relationship exists between mineral hardness and the mechanical behavior of monomineral granular assemblies. Specifically, arrangement of the minerals in decreasing order of true volumetric strain occurring during the isotropic compression stage, or increasing order of true volumetric strain occurring during shear coincides with arrangement in increasing order of Mohs hardness. In addition, trends in the response of the materials to varying conditions of initial density, degree of saturation, type of end restraint, and initial grain size seemed to be related more than casually to mineralogical hardness. The extent to which mineralogical differences influenced degradation is illustrated by the data in Figure 5-11. Electron micrographic indications of this have been presented elsewhere [Murphy, 1970].

It appears that as a consequence of degradation alone, compressibility must increase with decreasing hardness. This is indicated by the increase in the change in mean size (occurring during consolidation) with decreasing hardness as noted in Table 5-8. In addition, the ratio of the change in mean size occurring during shear to that occurring during consolidation, increases with increasing hardness. As is shown in the following section, this last phenomenon implies that at a given stress level, the degradation component of shearing resistance is greatest for hard and least for soft minerals.

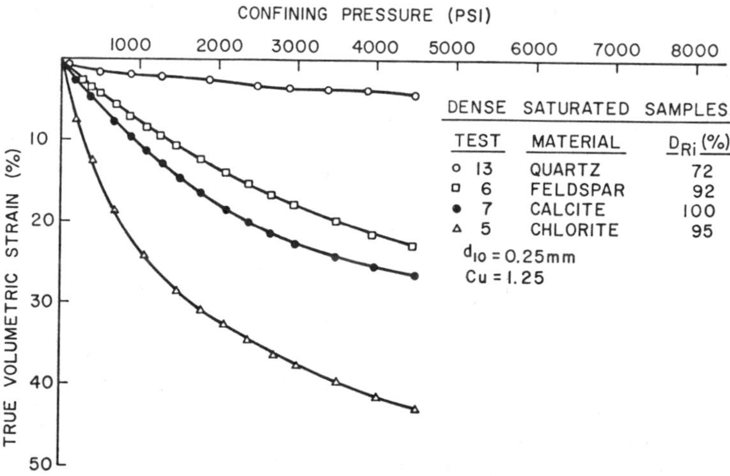

Figure 5-9. Isotropic compression tests for initially dense saturated samples.

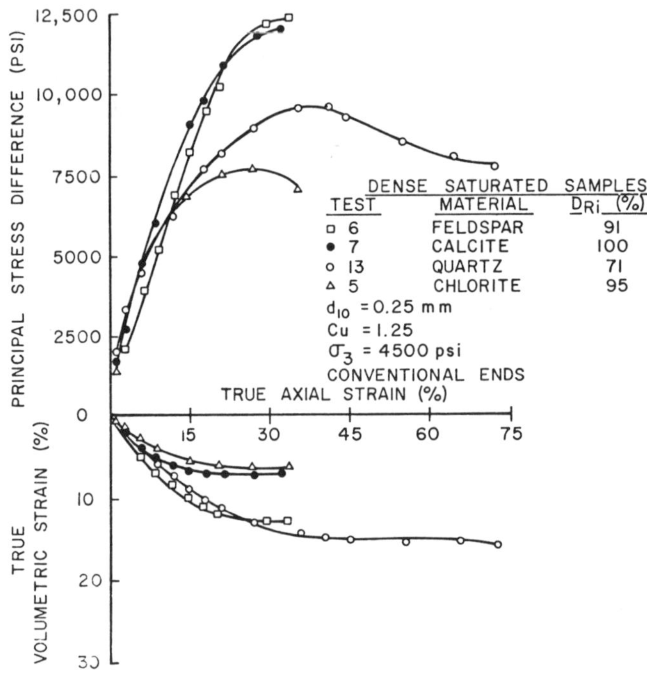

Figure 5-10. Triaxial compression tests on initially dense saturated samples under 4,500 psi confinement.

Table 5-8
Analysis of Degradation

Material (Mohs Hardness Scale)	Condition	Computed Change in Mean Size During	
		Consolidation at 4,500 psi	Shear at 4,500 psi
Quartz	Sat. Sand	less than 0.01	0.11
7	Dry Sand	0.03	0.13
	Dry Silt	–	0.02*
Feldspar	Sat. Sand	0.03	0.17
6	Dry Sand	0.06	0.13
	Dry Silt	–	0.00*
Calcite	Sat. Sand	0.07	0.15
3	Dry Sand	0.06	0.14
	Dry Silt	–	0.00*
Chlorite	Sat. Sand	0.20	0.03
1	Dry Sand	0.10	0.08
	Dry Silt	–	0.03*

* Change resulting from consolidation and shear at 4,500 psi.

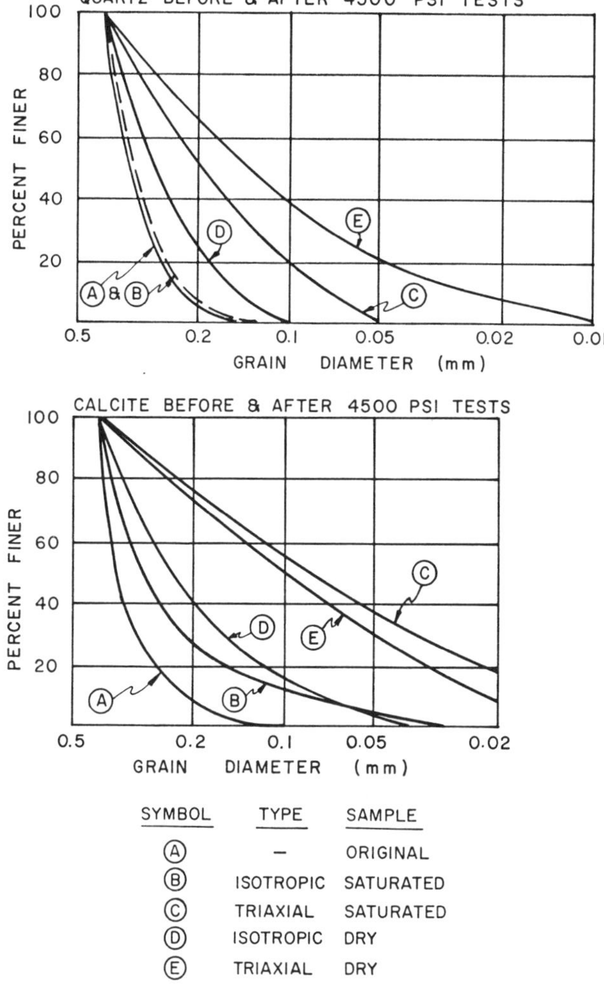

Figure 5-11. Typical gradation curves for quartz and calcite.

45,000 psi Tests

The primary purpose of these tests was to explore further the hypotheses developed from the results of the 4,500 psi tests concerning the existence of a relationship between mineralogical hardness and mechanical behavior. Typical results of the isotropic and triaxial compression tests conducted at confining pressures of 45,000 lb per sq in. are shown in Figures 5-12 through 5-14. Based on these, it is obvious that, the strength and true volumetric strain occurring during shear increase and

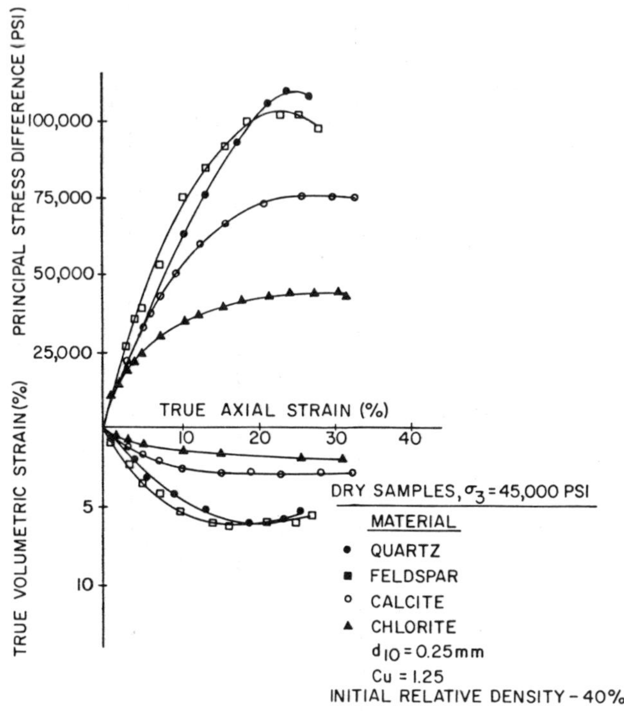

Figure 5-12. Triaxial compression tests at 45,000 psi confinement.

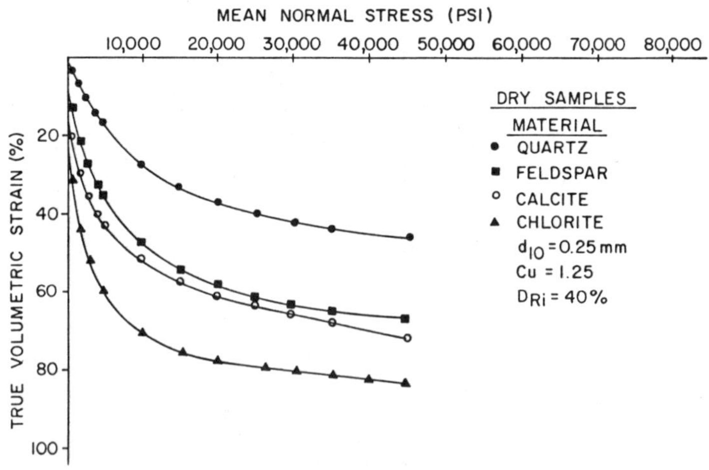

Figure 5-13. Isotropic compression tests.

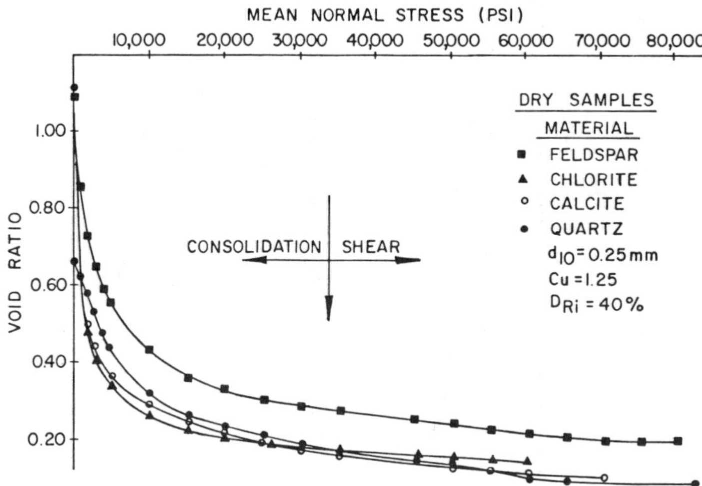

Figure 5-14. Void ratio vs. mean normal stress.

the true volumetric strain occurring during consolidation decreases with increasing hardness. Quantification of the relationship between volumetric strain at high stress levels (where all or most volume change must be a consequence of particle breakdown) and mineralogical hardness is possible since indentation hardness and crushing strength are each related to molecular structure. Tabor has shown that a unit change in Mohs hardness corresponds to a 60% change in indentation hardness. This concept was utilized in plotting true volumetric strain against Mohs hardness so that, in arithmetic terms, each hardness abscissa was 60% greater than the one preceding it. As shown in Figure 5-15, the results are linear for both the strain occurring during consolidation and that occurring during shear. The linearity of these plots prompted speculation concerning a quantitative relationship between strength at high stress levels (where the degradation component of the angle of shearing resistance intuitively is most significant) and mineralogical hardness. As shown in Figure 5-16, the principal stress ratio at failure was found to increase linearly with increasing logarithm of Mohs hardness.

Envelope Curvature. A comparison between the data shown in Figures 5-12 through 5-14 and related information presented elsewhere herein, for tests conducted at confining pressures of 4,500 lb per sq in. is presented in Table 5-9. As shown numerically in this table and schematically in Figure 5-17, the Mohr strength envelope for each of the minerals

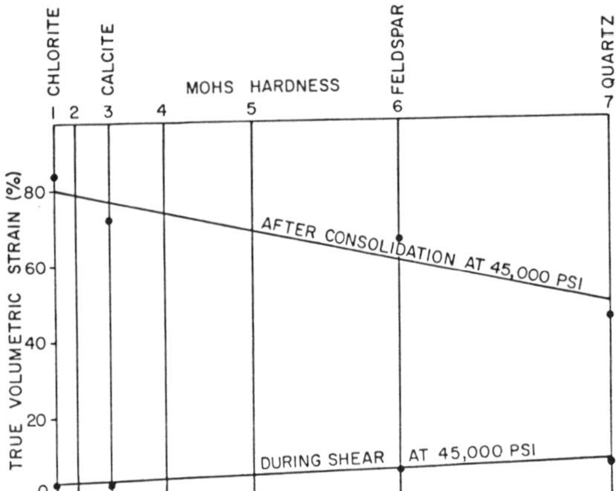

Figure 5-15. Relationship between hardness and compressibility.

Figure 5-16. Relationship between strength and hardness.

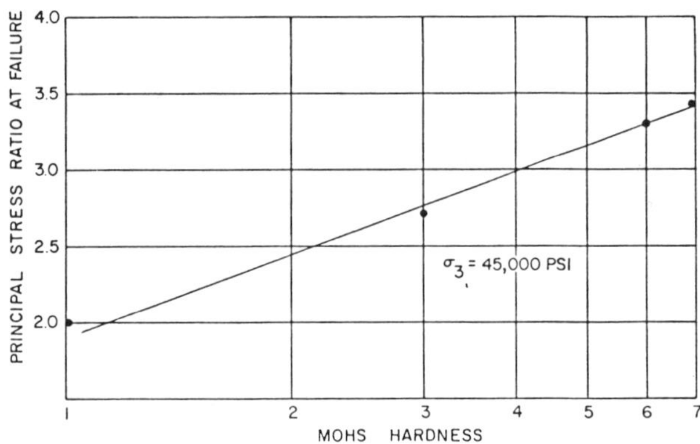

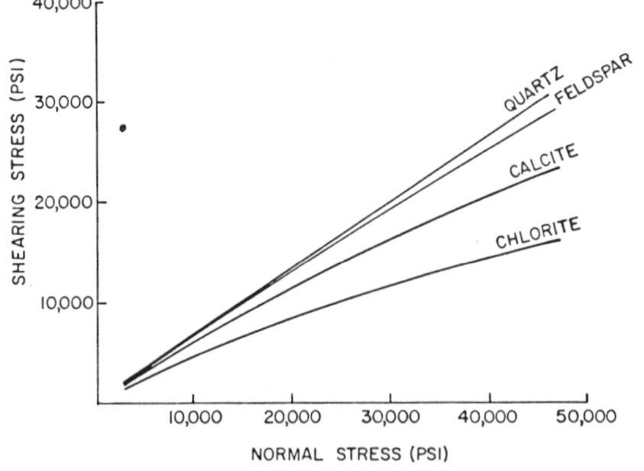

Figure 5-17. Envelope curvature at high stress levels.

Table 5-9
Comparison of Triaxial Compression Tests Based on Confining Pressure

Material	Cell Pressure (psi)	Initial Tangent Modulus (psi)	True Axial Strain at Failure (%)	True Volumetric Strain (%)		Angle of Shearing Resistance
				Consolidation	Shear	
Quartz	30*	1	22	–	–	37
	4,500	45	38	18	18	33
	45,000	630	23	47	6	33
Feldspar	30*	2	26	–	–	40
	4,500	55	38	40	19	36
	45,000	760	24	68	6	33
Calcite	30*	2	25	–	–	40
	4,500	63	33	45	9	32
	45,000	610	29	72	3	27
Chlorite	30*	1	21	–	–	29
	4,500	48	18	46	5	26
	45,000	710	31	84	2	19

* Data from Koerner [1968].

tested (except quartz) is curved over all or part of the stress range from 4,500 to 45,000 lb per sq in. The degree of curvature is illustrated by the differences between the secant angles of shearing resistance at 45,000 and those at 4,500 lb per sq in. shown in Table 5-10. It appears that this manifestation of the effect of stress level on the mechanical behavior of cohesionless soils is related to mineralogical composition. As shown in Figure 5-18, the angle of shearing resistance (for each of the minerals tested) decreases linearly with increasing logarithm of mean normal stress over the entire stress range from 50 lb per sq in. to 80,000 lb per sq in. The relationships illustrated in Figure 5-18 may be stated as:

$$\phi_{f_m} = \phi_{f_i} - C \log \sigma_{o_m}/\sigma_{o_i} \tag{5-4}$$

where ϕ_{f_i} = the angle of shearing resistance at mean normal stress, σ_{o_i}
  ϕ_{f_m} = the angle of shearing resistance at mean normal stress, σ_{o_m}
  C = a material parameter

Table 5-10
Variation of Angle of Shearing
Resistance with Stress Level

Material	Hardness	ϕ_f @ 45,000 psi–ϕ_f @ 4,500 psi
Quartz	7	0
Feldspar	6	2.5
Calcite	3	4.2
Chlorite	1	6.6

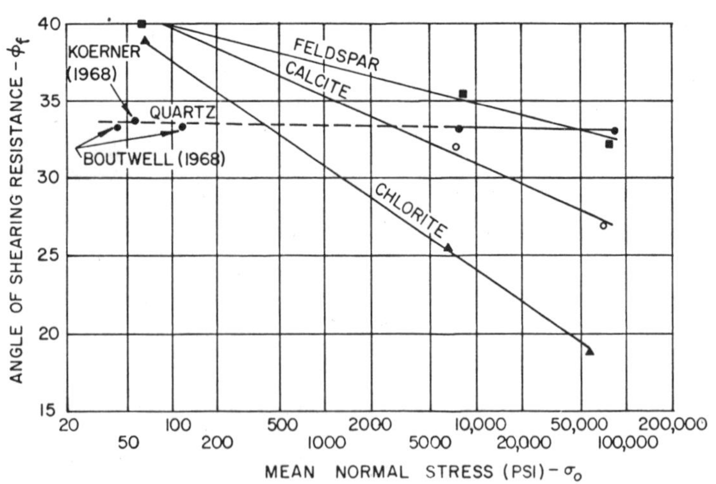

Figure 5-18. Variation of ϕ_f with σ_o.

For the materials tested, the parameter C has values shown in Table 5-11. Examination of these data indicates that decreases in ϕ_{f_i} with increasing σ_{o_i} are proportional to Mohs mineralogical hardness (M). In rough terms,

$$C = 7.7 - 1.1M \tag{5-5}$$

Table 5-11
The Material Constant C

Material	C
Quartz (M = 7)	0
Feldspar (M = 6)	2.5
Calcite (M = 3)	4.2
Chlorite (M = 1)	6.6

DISCUSSION

The test results presented here indicate that degree of saturation, type of end restraint, and initial grain size influence behavioral characteristics in the same manner, at least "trendwise," at high as at low stress levels. At high stress levels, the influence of initial density on mechanical behavior is insignificant since dilatancy effects no longer prevail. It appears that at such stress levels, the deformational and possibly the strength characteristics of a given grain assembly are influenced significantly by the mineralogical properties of the material concerned.

The distinctness of some of the aforementioned relationships between mechanical behavior and mineralogical composition is an indication of the significance of particle degradation at high stress levels. Vesic and Clough [1968] have speculated that beyond some stress level particle crushing may be the sole factor contributing to the curvature of the Mohr envelope. Yet, the mechanism by which this curvature occurs remains undefined. Data presented in Figures 5-15 through 5-18 indicate that the extent of curvature at high stress levels is dependent on the magnitude of the true volumetric strain (primarily particle degradation) occurring both during consolidation and during shear. The greatest curvature was experienced by the minerals that degraded *least* during shear but *most* during isotropic compression. This observation coupled with the intuitive premise that the magnitude of the degradation component of the angle of shearing resistance must depend primarily on the extent of crushing during shear allows for development of a hypothesis concerning the role of particle degradation. The following is proposed:

1. A given granular assembly has a fixed capacity for total volumetric strain. This capacity depends on initial density, assembly structure, and the crushing strength or mineralogical hardness of the material concerned.
2. Beyond some stress level (the magnitude of which varies with mineralogical composition) volumetric strain occurs primarily as a result of particle crushing. Therefore, hardness must be the most significant parameter governing volumetric strain at or beyond this stress level.
3. The degradation component of the angle of shearing resistance could be considered *latent* strength. The magnitude of this component is, as mentioned earlier, intuitively dependent on the extent of degradation occurring during shear. However, the crushing that *can* occur during shear depends on how much of the total volumetric strain (or crushing) capacity *is* spent during consolidation. Thus, the magnitude of the degradation component of the angle of shearing resistance must depend on the degree of crushing that occurs during isotropic compression.
4. The distribution of volumetric strain (or crushing) that occurs during consolidation and shear depends on both mineralogical composition and stress level. At a given level, crushing during isotropic compression is most extensive for the softer minerals. For a given mineral, the percentage of the crushing capacity that *is* spent during the isotropic stage increases with increasing consolidation pressure. Likewise, the percentage remaining *to be* spent during shear decreases. Thus, ϕ_{deg} and therefore ϕ_{f} decrease with increasing σ_{o}.

The last facet of this hypothesis is illustrated in Figure 5-19. Though schematic in nature, the concepts presented in this figure, are supported by a compilation of the data presented herein. Specifically, the relatively high angles of shearing resistance reported by Koerner [1968] for chlorite at 30 lb per sq in., indicate that at very low pressures even the soft minerals can have significant degradation components. In addition, it was shown herein, that at 4,500 lb per sq in. soft minerals crush more during consolidation than during shear and that the angle of shearing resistance of chlorite at high stress levels is much lower than that measured at low stress levels by Koerner. The difference is a manifestation of the latent nature of the shear strength attributable to degradation. The interrelationship of stress level and mineralogical composition is illustrated by the mechanical behavior of the harder minerals. Below some stress level these crush more during shear than during consolidation. The mechanism leading to the increase in crushing in the presence of shear is thought to be intensification of tensile stress, present at grain contacts in

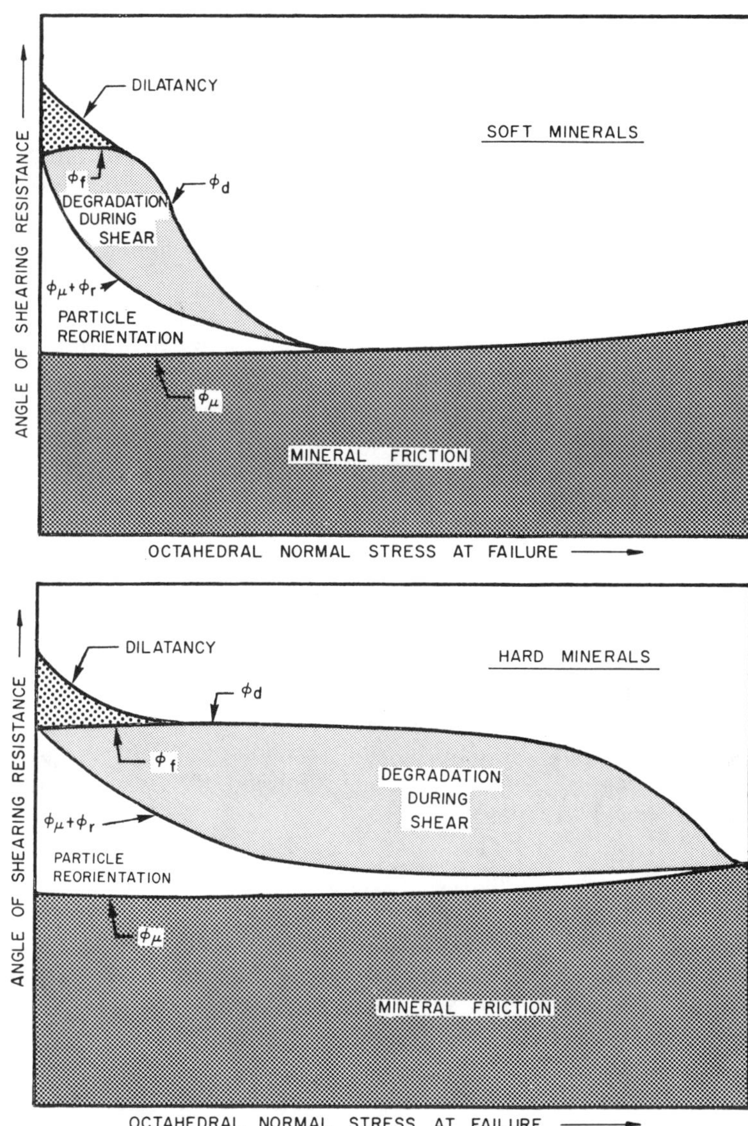

Figure 5-19. The sources of shearing resistance.

an otherwise isotropic stress field. It seems evident that such tensile stresses can assume substantial significance only below some level of compressive stress. Thus, at very high stress levels, the degradation component of the angle of shearing resistance should become insignificant even for hard minerals (as at these levels most crushing should occur during consolidation).

CONCLUSION

The data presented herein lead to the following conclusions:

1. At high stress levels particle degradation is significant and distinct relationships exist between mineralogical hardness (a measure of resistance to such crushing) and true volumetric strain (a measure of the occurrence of such crushing) during consolidation and during shear.
2. Curvature of the Mohr strength envelope is influenced by both stress level and mineralogical composition. For a given mineral, the angle of shearing resistance decreases linearly with the logarithm of mean normal stress. The rate of decrease increases with decreasing mineral hardness.
3. The magnitude of the degradation component of the angle of shearing resistance may be influenced more by the extent of particle crushing during isotropic compression than by the crushing during shear.

REFERENCES

1. Bromwell, L. G., "The Friction of Quartz in High Vacuum," Sc.D. thesis, Massachusetts Institute of Technology, 1966 (as quoted in *Soil Mechanics,* Lambe, T. W. and Whitman, R. V., John Wiley and Sons, New York 1969).
2. Horn, H. M., and Deere, D. U., "Frictional Characteristics of Minerals," *Geotechnique,* Vol. 12, 1962, pp. 319–335.
3. Koerner, R. M., "The Behavior of Cohesionless Soils Formed from Various Minerals," Ph.D. thesis, Duke University, 1968.
4. Koerner, R. M., "Behavior of Single Mineral Soils in Triaxial Shear," *Journal of the Soil Mechanics and Foundations Division,* American Society of Civil Engineers, Vol. 96, No. SM4, 1970, pp. 1373–1390.
5. Lee, K. L., and Seed, H. B., "Drained Strength Characteristics of Sands," *Journal of the Soil Mechanics and Foundations Division,* American Society of Civil Engineers, Vol. 93, No. SM6, 1967, pp. 117–142.
6. Murphy, D. J., "Soils and Rocks: Composition, Confining Level and Strength," Ph.D. dissertation, Duke University, 1970.
7. Murphy, D. J., "The Strength and Deformation Characteristics of Ottawa and Chattahoochee Sands at 45,000 psi Confining Pressures," *unpublished report,* Soil Mechanics Laboratory, Duke University, 1970.

8. Murphy, D. J., "High Pressure Experiments on Soil and Rock," Proceedings, 13th Symposium on Rock Mechanics, University of Illinois, 1971.

9. Rowe, P. W., "The Stress Dilatancy Relation for Static Equilibrium of an Assembly of Particles in Contact," Proceedings, Royal Society of London, Series A, Vol. 269, 1962, pp. 500–527.

10. Tabor, D., "Mohs Hardness Scale—A Physical Interpretation," Proceedings, Physical Society of London, Vol. 67, Section B, 1954, pp. 249–257.

11. Tai, T., "Strength and Deformation Characteristics of Cohesionless Materials at High Pressures," Ph.D. thesis, Duke University, 1970.

12. Tschebotarioff, G. P., and Welch, J. D., "Lateral Earth Pressures and Friction Between Soil Minerals," Proceedings, 2nd International Conference on Soil Mechanics and Foundation Engineering, Rotterdam, 1948, Vol. 7, pp. 135–138.

13. Vesic, A. S., and Clough, G. W., "Behavior of Granular Materials Under High Stresses," *Journal of the Soil Mechanics and Foundations Division,* American Society of Civil Engineers, Vol. 94, No. SM3, 1968, pp. 661–688.

6

Analysis of Shallow Foundations

E. E. De Beer
State University of Ghent
Ghent, Belgium

INTRODUCTION

Foundations should be designed in such a way that they are sufficiently safe with respect to:

1. *The ultimate bearing capacity of the soil supporting the foundation.* The ultimate bearing capacity is by definition the load causing a state of failure of the soil underneath a foundation with given shape, dimensions, and foundation depth. The ultimate bearing capacity is thus not a constant for a given soil but depends also on the mentioned geometrical characteristics of the foundation as well as on the eccentricity and the inclination of the load.
2. *The loading state corresponding to the limit of the tolerance of the deformation.* This state is by definition the load (taking not only into account the load imposed on the soil by the foundation itself, but also those of adjacent foundations and other factors, such as fills, excavations, changes in groundwater pressures, etc.) causing such deformations of the soil that the corresponding movements and differential movements of the structures are at the limit of what is still tolerable for the stability, the function, and the aspect of the structure.
3. *The stresses in the constitutive materials.* The stresses in the foundation may not exceed the allowable values. The foundations are to be made with materials that are sufficiently resistant and that are not attacked by chemical constituents present in the soil layers.
4. *Equilibrium against overturning.*
5. *Equilibrium against sliding.* This equilibrium has to be considered when the foundation transmits horizontal forces to the soil, as well as in the absence of horizontal force components, when the foundation is established in a sloping soil mass.

6. *Scour.* The problem of scour has to be considered when the foundation is exposed to running water. It is evident that scour endangers the foundation, when it extends underneath the level of the foundation. But also smaller depths of scour can endanger the equilibrium. Indeed the ultimate bearing capacity of the soil is increasing with the lateral overburden existing around the foundation. When during flood the river deepens its bed in the vicinity of a bridge pier, the lateral overburden q' decreases ($q' < q$) and consequently also the ultimate bearing capacity. It is possible that for a given value of q', the bearing capacity becomes insufficient to bear the load P.

7. *Upward water pressure.* When a construction is established underneath the phreatic level, its bottom is subjected to upwards water pressures. When the construction is hollow, the bottom slab is submitted to bending moments and shear forces. A sufficient safety must exist against uprising.

8. *Seepage problems.* Such problems are to be considered for water-retaining structures.

9. *Internal erosion—piping.* Groundwater flow in the soil can cause a more or less rapid displacement of solid particles. The dragging of these particles is called internal erosion, which can end in piping. This phenomenon can have very harmful consequences on the equilibrium of the construction. This problem not only has to be considered in the case of water-retaining structures but also in the case of the existence of a natural groundwater flow, especially when it is activated by artificial pumping.

This chapter will discuss only ultimate bearing capacity and soil deformations. It will also restrict discussion to the problem of shallow foundations. Shallow foundations are those for which the part of the bearing capacity due to the shearing strength of the layers of overburden can be neglected with respect to that derived from the shearing strength of the bearing stratum. This is the case when the foundation is established at a small depth below the soil surface. But this is also the case when the foundation is located at shallow depth in, or on the surface of a resisting layer, overlaid by a large thickness of layers with very small shearing strength.

ULTIMATE BEARING CAPACITY

Modes of Failure

When the load on a foundation is increased, a value is finally obtained under which the failure state in the soil is reached, and the penetration of the foundation is no longer under control. In extreme conditions this uncontrolled penetration can be caused by a lateral and upward expulsion of the soil from beneath the foundation, without change in volume, or by a volume decrease in the soil due to a reorganization of the fabric, crushing of the soil particles, etc. In reality uncontrolled penetration occurs mostly under the combination of expulsion and compression, corresponding to the least needed energy. The first extreme case corresponds to the state of general shear failure, the second to that of punching shear failure, and the combination to that of local shear failure [Vesic, 1963].

In other words general shear failure is characterized by a well-defined pattern of failure in which the soil adjacent to the foundation tends to heave and a shear surface develops that extends from one edge of the foundation to the surface (Figure 6-1a).

In punching failure the foundation penetrates the ground immediately beneath the foundation due to the soil's compression, and penetration continues as vertical shear develops about the periphery of the foundation (Figure 6-1c). The soil outside the immediate foundation zone remains unaffected.

In the case of local shear (Figure 6-1b) wedge and slip surfaces develop at the edges of the foundation, and some heave of the adjacent soil occurs. Vertical compression is the more significant factor, and the slip surfaces only reach the ground surface after appreciable foundation penetration has taken place.

The general trend of the load-settlement diagrams corresponding to these three modes of failure are schematized on the right side of Figure 6-1. In the case of general shear, failure may be sudden, as for example, in a rotational failure. In case of local and punching shear during the load test small jerks can occur, but no visible total collapse nor substantial tilting occurs. In order to maintain the footing movement in the vertical direction, an increase of the load is needed.

Since nature always looks for the solution of least energy, punching shear will have an advantage in the case of very compressible or collapsible structures (loose sand), and general shear in the case of nearly incompressible materials with limited shearing strength (for instance, a saturated clay loaded in such a way that during the loading practically no consolidation can occur). However, the mode of failure not only depends on the nature of soil and its relative density, but also on the geometrical conditions of the foundation. Indeed the load needed to induce a possible general shear depends on the depth of the foundation and on its shape and absolute dimensions—and this in a much more pronounced way as with the punching problem. For instance, the general shear value increases considerably with the foundation depth, which is much less the case for punching shear. By increasing foundation depth for a given soil with a

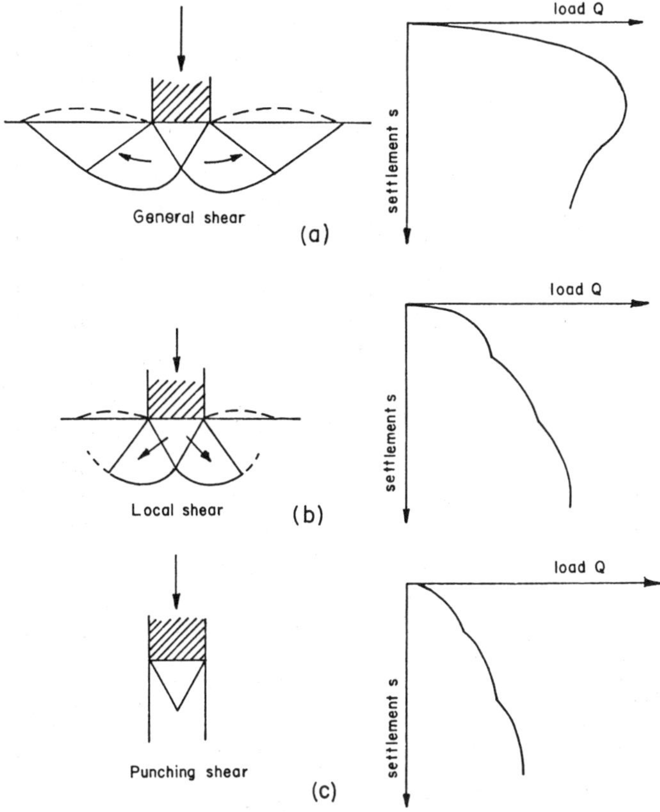

Figure 6-1. Different modes of failure [after Vesic, 1963a].

given relative density, the failure mode will gradually move from general shear failure to local shear failure and to punching shear. This has been clearly explained by Vesic [1963a], and he proposed for sand the relation between the mode of failure, the relative density D_r, and the relative embedment depth of the foundation D/R as shown on Figure 6-2.

The symbol R represents the hydraulic radius of the foundation

$$R = \frac{A}{\chi} \tag{6-1}$$

where A = area of the foundation

χ = perimeter of the foundation.

For a rectangular foundation with width B and length L

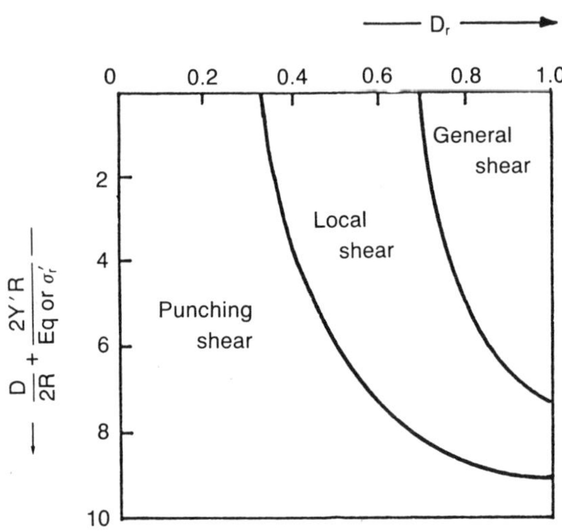

Eq = Young Modulus of quartz

σ_r' = crushing strength of quartz particles

Figure 6-2. Influence on the mode of failure, of the relative density and the relative embedment depth in sand.

$$R = \frac{BL}{2(B+L)} \tag{6-2}$$

for a square $B \times B$

$$R = \frac{B}{4} \tag{6-3}$$

Vesic [1963,1965] proposed to characterize the relative influence of the compressibility related to that of the shearing strength, by a ratio I_r called the rigidity index:

$$I_r = \frac{G}{c' + p_t' \tan\phi'} \tag{6-4}$$

where G = shear modulus of the soil

$$G = \frac{E}{2(1+\nu)} \tag{6-5}$$

where E = Young's modulus of the soil

       ν = Poisson's ratio

       p_t' = effective overburden stress

       c' = cohesion

       ϕ' = angle of internal friction

In this approach the soil is considered as a linear elastic material below the yield state. Equation 6-4 shows that the rigidity index decreases when the effective overburden stress and thus the foundation depth increases. Small values of I_r are related to punching shear, large values of I_r to general shear failure.

In the ratio of Equation 6-4 only the volumetric change in the so-called "elastic" range is taken care of. In order to take also into account the volumetric change that may occur in the plastic range, decreasing the rigidity, Vesic [1965] introduced a reduced rigidity index I_{rr} given by:

$$I_{rr} = \frac{I_r}{1 + I_r \Delta_m} \tag{6-6}$$

where Δ_m = the average volumetric strain in the plastic range

Since, however, it is difficult, except by advanced numerical methods, to have sufficient guidance for the choice of Δ_m, this refinement will not be applied for most current practical problems.

Another factor that can have an influence on the modes of failure is the rate of application of the load. Whereas by slow and gradual application of the load on a foundation with limited extent placed at the surface of a very dense sand, the failure will occur in general shear, in the same conditions. In the case of a transient dynamic loading the failure will occur in punching, as shown by tests of Heller [1964], Vesic, Banks, and Woodward [1965].

In the case of a dense sand layer with limited relative thickness $H_1{:}B$ underlaid by a much less resistant layer (for instance soft clay) the failure will also occur in punching, as shown by Mayer [1953] and Vesic [1975].

The increase of the width of the footing increases the bearing capacity. All other conditions being equal, even in very dense sands, by still increasing width of the footing, the mode of failure will gradually pass from general shear to local shear and punching. To take this influence into account, we propose to define a modified rigidity index:

$$I_r' = \frac{G}{c' + p_t' \tan\phi' + \gamma'R} \tag{6-7}$$

with γ' = effective unit weight of the soil underneath the foundation

According to Vesic a value of I_{rr} larger than 250 implies a phenomenon in which the compressibility plays a negligible role, while values of ten and lower correspond to a punching problem. If there is any doubt concerning the modes of failure, several modes have to be considered, and, according to the law of minimum energy, the solution giving the smallest value for the ultimate bearing capacity has to be retained.

Ultimate Load Criterion

In principle the ultimate bearing capacity is the load Q for which in the load-settlement $Q = f(s)$ diagram the gradient $\Delta s/\Delta Q$ becomes infinite, corresponding to a vertical tangent in the diagram $x = Q$, $y = s$. Such a tangent can be observed in the case of a general shear failure. In case of a loading program with controlled strain rate, the peak value corresponding to this vertical tangent is followed by a decrease of the load (Figure 6-1a).

In the case of local shear and punching shear generally no point with $\Delta s/\Delta Q = \infty$ is observed (Figure 6-1b and c), except perhaps for very large relative settlements s/B.

Consequently a more or less conventional ultimate bearing capacity has to be defined. According to Terzaghi [1943], Vesic [1967] and others, the conventional ultimate bearing capacity is the load causing a relative settlement s/B equal to 10%. The symbol B represents the width of the footing. Another definition could be by considering the load for which s/4R = 10%.

For square and circular footing these two definitions are equivalent. For an infinitely long strip footing, the second definition corresponds to s/B = 20%. As, however, the definition is in a certain way based on a conventional assumption, for shallow foundations one mostly simply adopts s/B = 10%.

Computation of the Ultimate Bearing Capacity

Most formulas of ultimate bearing capacity are based on the fundamental formula of Prandtl [1921]. This fundamental formula is valid for the case of a weightless, rigid-plastic material.

Therefore in the fundamental formula the influence of the compressibility of the soil on the bearing capacity is not taken into consideration. One must be aware of the limitation of the validity of the classical ultimate bearing capacity formulas. They are not applicable for the case of local shear and punching shear; as can be seen in Figure 6-2 this is the case for shallow foundations on very loose and loose-to-medium dense sand.

Centrally Vertical-Loaded Strip Footing. The simplest case to start with is that of a horizontal undeformable centrally vertical-loaded strip footing, supported by a weightless material characterized by a straight intrinsic law (Figure 6-3):

$$\tau = c' + \sigma' \tan \phi' \qquad (6\text{-}8)$$

The shearing strength of the layers above the foundation level can be neglected.

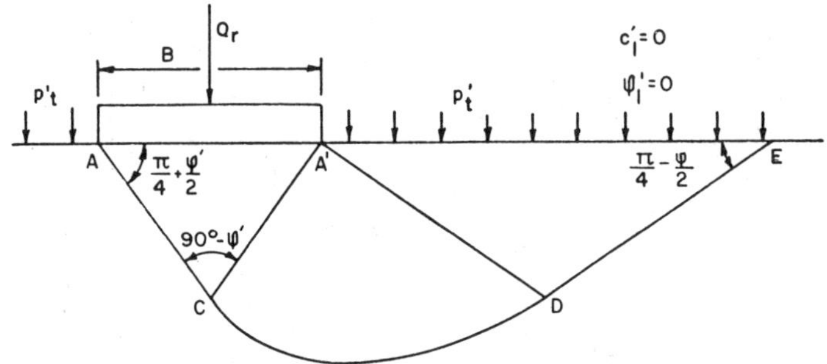

Figure 6-3. Rupture line under a shallow foundation in case of general shear and a weightless material.

The rupture line ACDE consists of a straight line AC with inclination angle $(\pi/4) + (\phi'/2)$ to the horizontal, the logarithmic spiral BC, and the straight line with inclination $(\pi/4) - (\phi'/2)$ to the horizontal.

If in the contact face AA$'$ tangential stresses do not exist, the normal unit actions $q_{r,1}$ are uniformly distributed, and given by

$$q_{r,1} = N_q p'_t + N_c c' \qquad (6\text{-}9)$$

$$\text{with } N_q = e^{\pi \tan \phi'} \tan^2 \left(\frac{\pi}{4} + \frac{\phi'}{2} \right) \qquad (6\text{-}10)$$

$$N_c = (N_q - 1) \cot \phi' \qquad (6\text{-}11a)$$

For $\phi' = 0$

$$N_c = \pi + 2 = 5.14 \qquad (6\text{-}11b)$$

The values of N_q are given versus ϕ' by the curve $\delta = 0°$ on Figure 6-4, and also by the values in the column $\delta = 0$ of Table 6-1.

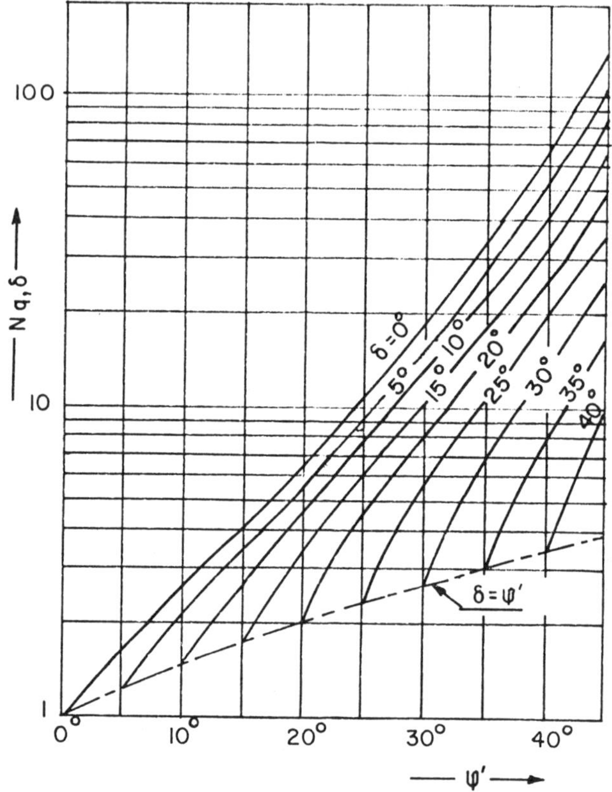

Figure 6-4. Bearing capacity factor $N_{q,\delta}$ vs. ϕ' with inclination angle δ as a parameter.

In order to take into account the weight of the supporting material the case of a foundation resting on the surface of a rigid cohesionless soil, with a straight intrinsic law, ϕ' is considered.

A closed analytical solution of this problem has not yet been found, and approximate numerical solutions have been proposed by several authors: Terzaghi [1943], Terzaghi and Peck [1948], Buisman [1940], Caquot-Kérisel [1956], Meyerhof [1950], Hansen [1961], and several others.

The rupture line now has the shape as ACD′E′ located above the rupture line ACDE valid for a weightless material (Figure 6-5). The reactions in the contact face AA′ are not uniformly distributed, but increase approximately linearly with the distances to the edge. The mean value of the unit reactions can be expressed by

$$q_{r,2} = N_\gamma^* \gamma' B \qquad (6\text{-}12)$$

Table 6-1
**Bearing Capacity Factor $N_{q,\delta}$ for a Continuous Footing Subjected to
an Inclined and Eccentric Load**

ϕ' \ δ	0°	5°	10°	15°	20°	25°	30°	35°	40°	45°
0°	1.000	—	—	—	—	—	—	—	—	—
5°	1.568	1.238	—	—	—	—	—	—	—	—
10°	2.471	2.155	1.502	—	—	—	—	—	—	—
15°	3.94	3.45	2.84	1.79	—	—	—	—	—	—
20°	6.40	5.56	4.65	3.65	2.09	—	—	—	—	—
25°	10.66	9.17	7.65	6.13	4.62	2.41	—	—	—	—
30°	18.40	15.64	12.94	10.37	7.96	5.67	2.75	—	—	—
35°	33.30	27.86	22.77	18.12	13.94	10.23	6.94	3.08	—	—
40°	64.19	52.63	42.37	33.33	25.39	18.52	13.16	8.40	3.43	—
45°	134.87	108.10	85.16	66.00	49.24	35.94	25.25	16.81	10.20	3.74

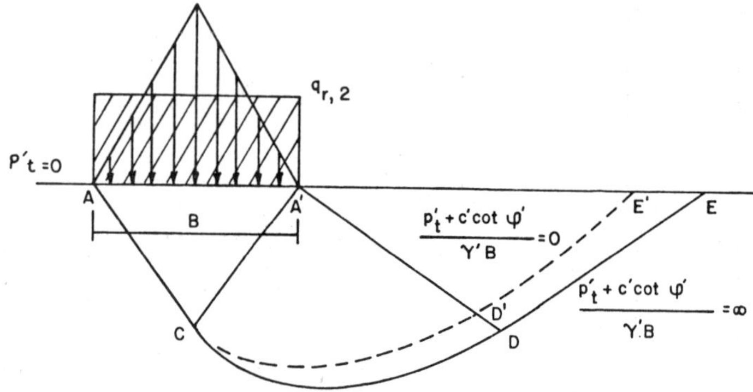

Figure 6-5. Rupture line under a shallow foundation in case of general shear, on a material with effective unit weight γ'.

On the basis of experimental investigations by Muhs [1957], De Beer-Vesic [1958], Van de Perre-Ramelot [1957], among others, it is considered that the values computed by Meyerhof [1950] agree most closely with the actual conditions.

The values of N_γ^* are given versus ϕ' by the curve $\delta = 0$ of Figure 6-6, and in the column $\delta = 0$ of Table 6-2.

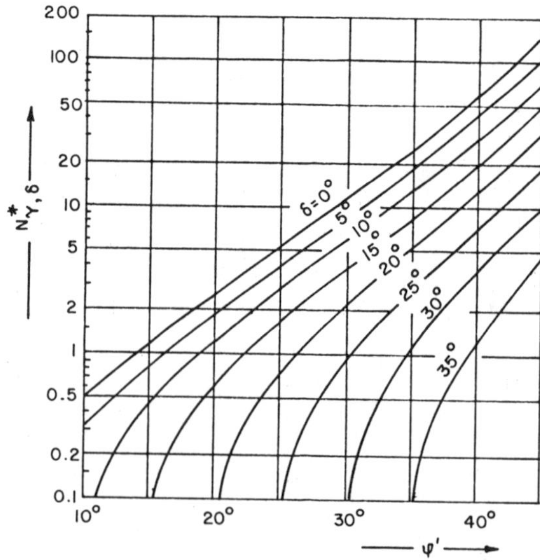

Figure 6-6. Bearing capacity factor $N_{\gamma,\delta}^*$ vs. ϕ' with inclination angle δ as a parameter.

Table 6-2

Bearing Capacity Factor $N^*_{\gamma,\delta}$ for a Continuous Footing Subjected to an Inclined and Eccentric Load

ϕ' \ δ	0°	5°	10°	15°	20°	25°	30°	35°	40°	45°
10°	0.505	0.312	—	—	—	—	—	—	—	—
15°	1.160	0.836	0.444	—	—	—	—	—	—	—
20°	2.485	1.849	1.198	0.588	—	—	—	—	—	—
25°	5.202	3.939	2.671	1.616	0.742	—	—	—	—	—
30°	10.912	8.135	5.764	3.698	2.121	0.915	—	—	—	—
35°	24.481	18.050	12.689	8.389	5.204	2.788	1.092	—	—	—
40°	56.551	40.883	27.872	19.332	12.035	7.045	3.368	1.198	—	—
45°	149.226	106.709	72.196	48.495	31.901	17.595	9.536	4.459	1.508	—
50°	457.192	308.392	188.315	132.576	85.571	49.288	26.921	12.157	5.515	1.845

According to Meyerhof and De Beer [1974].

It is worthwhile to notice that in the literature Equation 6-12 is mostly written

$$q_{r,2} = \frac{N_\gamma}{2} \gamma' B \tag{6-13}$$

We prefer the presentation of Equation 6-12. Consequently

$$N_\gamma^* = \frac{N_\gamma}{2} \tag{6-14}$$

One should not forget this relation when comparing N_γ^* values with those of other authors.

For values of $20° \leq \phi' \leq 45°$ the values of N_γ^* can be obtained with an error less than 10% by the expression of Vesic [1975]:

$$N_\gamma^* = (N_q + 1) \tan \phi' \tag{6-15}$$

For all cases, for which

$$0 < \frac{p_t' + c' \cot\phi'}{\gamma' B} < \infty$$

one simply adds the values of $q_{r,1}$ and $q_{r,2}$ corresponding to the cases

$$\frac{p_t' + c' \cot\phi'}{\gamma' B} = \infty$$

and

$$\frac{p_t' + c' \cot\phi'}{\gamma' B} = 0$$

although $q_{r,1}$ and $q_{r,2}$ correspond to two different rupture lines (see Figure 6-5).

$$q_r = N_q p_t' + N_c c' + N_\gamma^* \gamma' B \tag{6-16}$$

The general shape of Equation 6-16 has been presented, although with different assumptions, independently by Buisman [1940], Terzaghi [1943] and Caquot [1934].

In reality there is but a unique rupture line located between the two extremes ACDE and ACD′E′ (Figure 6-5). The superposition included in Equation 6-16 leads to an error located on the safe side.

Numerical calculations [Lundgren-Mortensen, 1953; Hansen and Christensen, 1969; De Poorter, 1962] have shown that for angles ϕ' from 30° to 40° the safety margin does not exceed 17% to 20%. For frictionless soils ($\phi' = 0°$) the error is of course zero.

Case of a Vertical and Eccentric Load. In the case of a strip footing with a vertical load with an eccentricity e, one defines the complementary eccentricity by

$$u = \frac{B}{2} - e \quad \text{(Figure 6-7)} \tag{6-17}$$

As proposed by De Beer [1949], Meyerhof [1953], with a good approximation the ultimate bearing capacity per unit length can be obtained by

$$Q_{r,1} = 2u(N_q p_t' + N_c c' + N_\gamma^* \gamma' 2u) \tag{6-18}$$

The computation of the actual foundation is therefore based upon an imaginary centrally loaded foundation of a width 2u. This simplification may introduce errors of about 10%; these errors are located on the safe side.

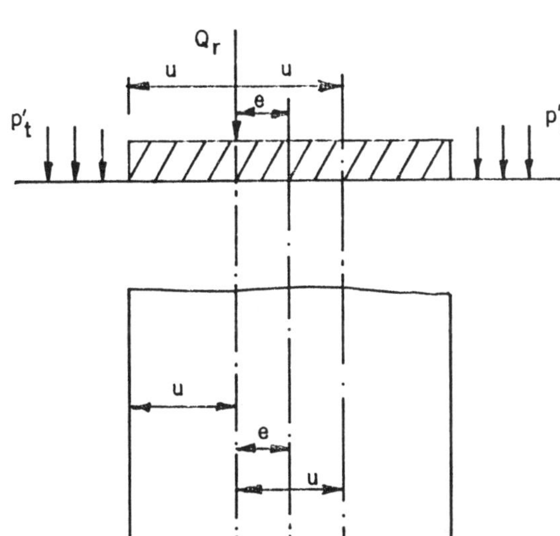

Figure 6-7. Strip footing subjected to a vertical and eccentric load.

Case of a Central, Inclined Loading. In the case of a strip footing loaded by a central load with inclination δ (Figure 6-8), the mean unit vertical ultimate bearing capacity q_r is expressed by

$$q_r = N_{q,\delta}\, p_t' + N_{c,\delta}\, c' + N_{\gamma,\delta}^*\, \gamma' B \qquad (6\text{-}19)$$

where, according to Sokolovski [1960]

$$N_{q,\delta} = \frac{\cos\alpha(\cos\delta + \sqrt{\sin^2\phi' - \sin^2\delta})e^{(\pi-\alpha)\tan\phi'}}{1 - \sin\phi'} \qquad (6\text{-}20)$$

in Equation 6-20 the angle α is defined by

$$\alpha = \delta + \arcsin\frac{\sin\delta}{\sin\phi'} \qquad (6\text{-}21)$$

The values of $N_{q,\delta}$ are given versus ϕ', with the values of δ as parameter, by the curves of Figure 6-4 and are also given in Table 6-1

$$N_{c,\delta} = (N_{q,\delta} - 1)\cot\phi' \qquad (6\text{-}22)$$

The values of $N_{\gamma,\delta}^*$, as computed by Meyerhof [1953], are represented by the curves of Figure 6-6, as a function of ϕ', with δ as a parameter. They are also given in Table 6-2.

There is of course a limitation of the inclination angle δ related to the danger of a sliding along the contact face AA'. A limit P_{max} for the horizontal component P per unit of length of the load is given by

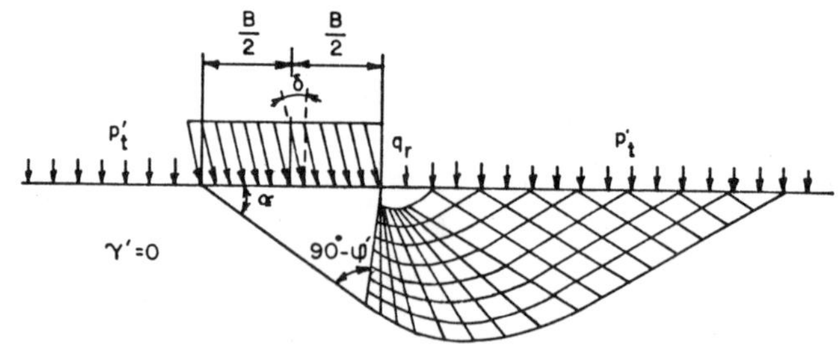

Figure 6-8. Rupture line for a continuous footing subjected to an inclined centric load, according to Brinch Hansen [1961].

$$P_{max} = B \, c_a + Q \tan \phi_c \tag{6-23a}$$

$$\tan \delta_{max} = \frac{B \, c_a}{Q} + \tan\phi_c \tag{6-23b}$$

where c_a = adhesion in the contact face
ϕ_c = angle of friction between the soil and the footing

In the case of a plane contact face, without indentations, one generally puts $c_a = 0$. For concrete made in situ, tamped, or vibrated, one can generally put that the friction angle between the soil and the concrete equals the internal angle of friction, with a maximum value of 35°

$$\phi_c \leqq \phi'$$

$$\phi_c \leqq 35°$$

Instead of using the values of the bearing capacity factors $N_{q,\delta}$, $N_{c,\delta}$, $N^*_{\gamma,\delta}$ many authors introduce so-called inclination factors i_q, i_c, i_γ defined by

$$i_q = N_{q,\delta}/N_q \tag{6-24a}$$

$$i_c = N_{c,\delta}/N_c \tag{6-24b}$$

$$i_\gamma = N^*_{\gamma,\delta}/N^*_\gamma \tag{6-24c}$$

Consequently Equation 6-19 is transformed to

$$q_r = N_q i_q p_t' + N_c i_c c' + N^*_\gamma i_\gamma \gamma' B \tag{6-25}$$

The inclination factor i_q is given versus ϕ' and δ in Table 6-3. The values of i_q are represented in terms of ϕ' with δ as parameter on Figure 6-9. One has

$$i_c = \frac{N_q i_q - 1}{N_q - 1} \tag{6-26}$$

Based on the deduction of Vesic [1975] one has for the case $\phi' = 0$

$$q_r = \frac{p_t' + (\pi + 2)c'}{1 + 2\tan\delta} \tag{6-27}$$

Table 6-3
Inclination Factor i_q for a Continuous Footing Subjected to
an Inclined and Eccentric Load

ϕ' \ δ	0°	5°	10°	15°	20°	25°	30°	35°	40°	45°
0°	1.000	—	—	—	—	—	—	—	—	—
5°	1.000	0.790	—	—	—	—	—	—	—	—
10°	1.000	0.871	0.609	—	—	—	—	—	—	—
15°	1.000	0.875	0.721	0.455	—	—	—	—	—	—
20°	1.000	0.870	0.726	0.570	0.327	—	—	—	—	—
25°	1.000	0.860	0.718	0.575	0.434	0.226	—	—	—	—
30°	1.000	0.850	0.704	0.563	0.433	0.308	0.149	—	—	—
35°	1.000	0.836	0.684	0.545	0.419	0.308	0.208	0.093	—	—
40°	1.000	0.820	0.660	0.520	0.396	0.288	0.205	0.131	0.054	—
45°	1.000	0.803	0.631	0.490	0.365	0.266	0.187	0.125	0.076	0.028

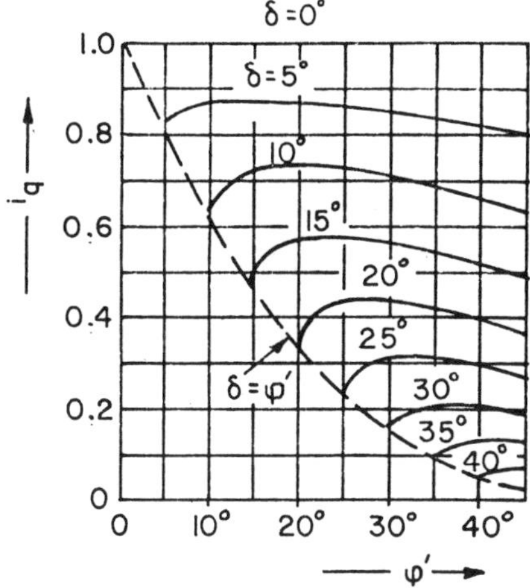

Figure 6-9. Inclination factor i_q vs. ϕ' with inclination angle δ as a parameter.

The values of i_γ are given versus ϕ' and δ in Table 6-4. The vertical component Q_r of the ultimate load is given by

$$Q_r = B \, q_r \tag{6-28}$$

The value q_r can be obtained either by Equation 6-19 or by Equation 6-25.

The horizontal component P_r is given by

$$P_r = Q_r \tan \delta \tag{6-29}$$

The components Q_r and P_r of the ultimate load can be determined for various inclinations δ. The resultant R_r is represented by a vector with an inclination δ, and the curve $R_r = f(\delta)$, which links the end-points of all vectors, can be plotted. This has been done in Figure 6-10 for the case of a cohesionless soil ($c' = 0$) and $\phi' = 35°$. Because the factors $N_{q,\delta}$ and $N_{\gamma,\delta}^*$ always have a value larger than zero for $\delta = \phi'$, the curve $R_r = f(\delta)$ does not end at the origin 0 of the coordinate system for $\delta = \phi'$, but at a point A upon the straight line plotted from the origin under the angle ϕ'. If one starts from the assumption that failure in the bearing layer develops along a curved slide surface, the value obtained for R_r differs from zero even for $\delta \geqq \phi'$. However in most cases such a value has no practical significance since the sliding resistance in the contact face exceeds the internal soil resistance only on rare occasions.

When the condition $\delta = \phi_c = \phi'$ is met, a curved slide surface no longer develops inside the soil, but the foundation slides along the contact face.

Table 6-4

Inclination Factor i_γ for a Continuous Footing Subjected to an Inclined and Eccentric Load

ϕ' \ δ	0°	5°	10°	15°	20°	25°	30°	35°	40°	45°
10°	1.000	0.617	—	—	—	—	—	—	—	—
15°	1.000	0.720	0.383	—	—	—	—	—	—	—
20°	1.000	0.744	0.482	0.237	—	—	—	—	—	—
25°	1.000	0.757	0.514	0.311	0.143	—	—	—	—	—
30°	1.000	0.746	0.528	0.339	0.194	0.084	—	—	—	—
35°	1.000	0.737	0.518	0.343	0.213	0.114	0.045	—	—	—
40°	1.000	0.723	0.493	0.342	0.213	0.125	0.060	0.021	—	—
45°	1.000	0.715	0.484	0.325	0.214	0.118	0.064	0.030	0.010	—
50°	1.000	0.675	0.412	0.290	0.186	0.108	0.059	0.027	0.012	0.004

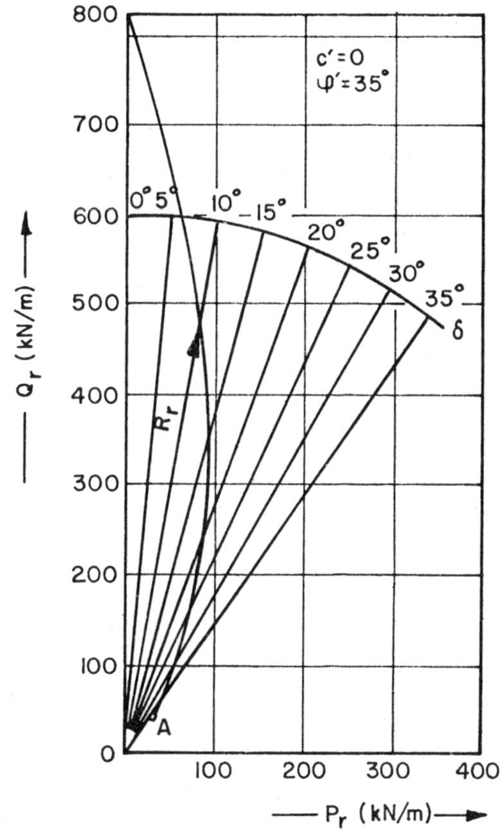

Figure 6-10. Resultant ultimate load R_r in function of the inclination angle δ for a numerical example: c' = 0; φ' = 35°; B = 1 m; p'_t = 10 (kN/m²); γ' = 20 (kN/m³).

Between point A and the origin 0 the slide conditions along the contact face are decisive; they are represented by the straight line 0A.

Case of a Central, Vertical Load on a Rectangular Footing B × L or on a Circular Footing. The influence of the shape can be taken care of by introducing in Equation 6-16, valid for a strip footing, so-called shape factors s_q, s_c, and s_γ.

$$q_r = N_q s_q p'_t + N_c s_c c' + N_\gamma^* s_\gamma \gamma' B \qquad (6\text{-}30)$$

Assuming that the bearing stratum consists of a material characterized by a unique intrinsic straight line (c', φ') independent of the state of strain, from calculations, dimensional analysis, and results of experiments, the following shape factors are proposed

$$s_q = 1 + \frac{B}{L} \sin \phi' \tag{6-31}$$

$$s_c = \frac{s_q N_q - 1}{N_q - 1} \tag{6-32a}$$

for $\phi' = 0$

$$s_c = 1 + 0.2 \, \frac{B}{L} \tag{6-32b}$$

$$s_\gamma = \frac{1 + 0.2 \, \dfrac{B}{L}}{1 + \dfrac{B}{L}} \tag{6-33}$$

In the literature, instead of Equation 6-33, one often finds

$$s_\gamma = 1 - 0.4 \, \frac{B}{L} \tag{6-34}$$

For B/L = 1, both equations give the same value.

Case of a Centrally Vertical-Loaded Footing, Considering Shearing Resistance of the Overburden Layer. A shallow foundation is established at a certain depth D underneath the soil surface. In several cases it is also placed at a relatively small depth D′ underneath the surface of the bearing stratum (Figure 6-11). The overburden soil is usually weaker or cracked, while the foundation is placed by excavation and backfilling. Therefore it is advised only to take the shearing resistance of the overburden into account if measures are taken to reestablish the soil in its previous natural condition. This shearing strength can be accounted for by introducing in Equation 6-16 so-called depth factors d_q, d_c and d_γ.

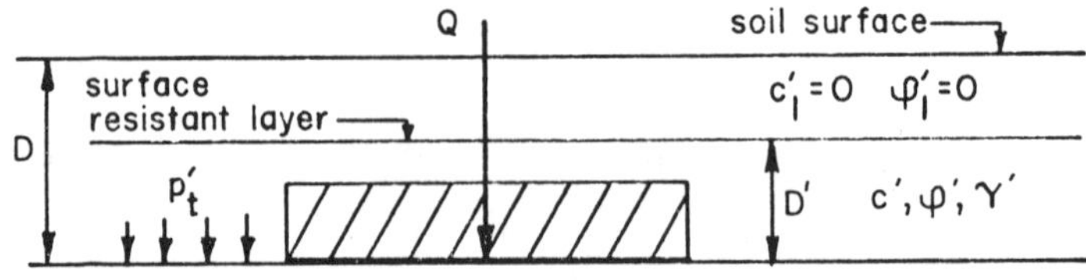

Figure 6-11. Scheme of a strip footing placed at a certain depth in the bearing stratum.

$$q_r = N_q d_q p_t' + N_c d_c c' + N_\gamma^* d_\gamma \gamma' B \tag{6-35}$$

with

$$d_q = 1 + \left[\tan^2 \left(\frac{\pi}{4} - \frac{\phi'}{2} \right) e^{\pi \tan\phi'} - 1 \right] e^{-\pi \frac{2u}{D'} \tan\phi'} \tag{6-36}$$

$$d_c = \frac{N_q d_q - 1}{N_q - 1} \tag{6-37}$$

$$d_\gamma = 1$$

The Equation 6-36 is only valid for $B/D > 0.5$.

Case of a Rectangular Footing with an Eccentric and Inclined Load in the Transversal Cross Section of Symmetry. Figure 6-12 illustrates this case and can be referred to along with the following paragraphs.

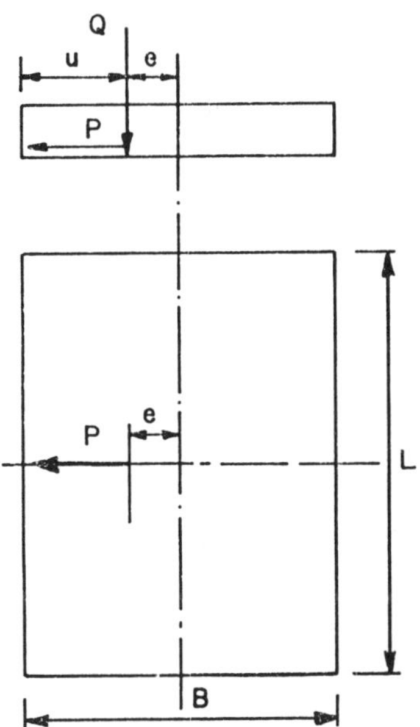

Figure 6-12. Scheme of a rectangular footing with an eccentric and inclined load acting in the transversal cross section of symmetry.

Calculation with Independent Factors. Assuming that the inclination factors i obtained for the case of a central loading (Tables 6-3 and 6-4 and Equation 6-26) the shape factors s obtained for the case of a central vertical loading (Equations 6-31, 6-32, and 6-33) and the depth factors valid for the case of a central and vertical loaded strip footing (Equations 6-36 and 6-37) remain valid for the case of a footing subjected to an inclined and eccentric load and embedded in the load-bearing layer, one gets

$$Q_r = 2uLq_r = 2uL(N_q i_q s_q d_q p_t' + N_c i_c s_c d_c c' + N_\gamma^* i_\gamma s_\gamma d_\gamma \gamma' 2u) \tag{6-38}$$

In all expressions of the parameters in which B appears, B has to be replaced by $B' = 2u$. Even though the i, s, and d values for the special cases should maintain their validity for the compound case as well, the total error for the compound case may yet be considerably greater than in the special cases.

Let each factor computed each time for a special case have an error of, for example, 10% on the safe side, i.e., $\bar{i} = (1 - 0.10)i$, $\bar{s} = (1 - 0.10)s$ and $\bar{d} = (1 - 0.10)d$.

If the factors are also valid for the general case, the final result could present an error (on the safe side) of

$$\bar{i}\,\bar{s}\,\bar{d} = (1 - 0.10)i\,(1 - 0.10)s\,(1 - 0.10)d$$
$$= 0.719\ isd \tag{6-39}$$

Furthermore also in the bearing capacity factor N_γ^* a possible safe error of 10% exists, and also the assumption of a fictive width $B' = 2u$ may include a safe error of 10%. Finally in some compound cases Equation 6-38 may include an error on the safe side, of

$$\overline{Q}_r = (1 - 0.10)^5\ Q_r = 0.59\ Q_r \tag{6-40}$$

Consequently it is worthwhile to improve the calculation procedure when it is possible without increasing the calculation time.

Calculation with Compound Factors. All factors appearing in Equation 6-38 depend on one specific parameter and are independent of the other ones. The application of the formulas for the case of a strip footing with an inclined and eccentric load (when the overburden resistance can be neglected), gives

$$Q_{r,1} = 2u(N_q i_q p_t' + N_c i_c c' + N_\gamma^* i_\gamma \gamma' 2u) \tag{6-41}$$

or

$$Q_{r,1} = 2u(N_{q,\delta} p_t' + N_{c,\delta} c' + N_{\gamma,\delta}^* \gamma' 2u) \tag{6-42}$$

The introduction in the ultimate bearing capacity formulas of factors that depend only on one parameter and are independent of the other parameters can only be a crude approximation.

In the case of a strip footing with an eccentric and inclined load, one thus should more exactly write [De Beer, 1966]

$$Q_{r,1} = 2u(N_{q,\delta,u}p_t' + N_{c,\delta,u}c' + N^*_{\gamma,\delta,u}\gamma'2u) \qquad (6\text{-}43)$$

in which $N_{q,\delta,u}$, $N_{c,\delta,u}$, and $N^*_{\gamma,\delta,u}$ are bearing capacity factors which depend not only on the inclination angle δ, but also on the eccentricity u.

Indeed in the case of an inclined and eccentric load the shape of the sliding surface differs from the shape that applies to an inclined and central load.

Equation 6-43 can also be written

$$Q_{r,1} = 2u(N_q i_{q,u}p_t' + N_c i_{c,u}c' + N^*_\gamma i_{\gamma,u}\gamma'2u) \qquad (6\text{-}44)$$

with $i_{q,u} = \dfrac{N_{q,\delta,u}}{N_q}$

$$i_{c,u} = \frac{N_{c,\delta,u}}{N_c}$$

$$i_{\gamma,u} = \frac{N^*_{\gamma,\delta,u}}{N^*_\gamma} \qquad (6\text{-}45)$$

The subscript u indicates that the inclination factors are no longer independent of the complementary eccentricity u, but vary with this eccentricity.

In the case of an inclined and eccentric load, even in the most simple case $\gamma' = 0$, the exact form of the slide surface is not known. An approximate method is obtained by the slide surfaces introduced by Hansen [1961] for the case of a weightless soil. The Hansen slide surface consists of the arc A''C (Figure 6-13), which meets the boundary conditions at the extremity A''. This arc changes in another arc CD that connects with the straight line DE under an angle $(\pi/4) - (\phi'/2)$ with the horizontal.

If the slide surface does not cover the entire width B of the footing, the boundary condition says that arc A''C must have a horizontal tangent at end point A'' (Figure 6-13a). If the slide surface ends at the edge point A of the footing, the tangent at the end of the arc forms an angle β with the horizontal (Figure 6-13c); this angle is a function of the inclination and of the eccentricity of the load.

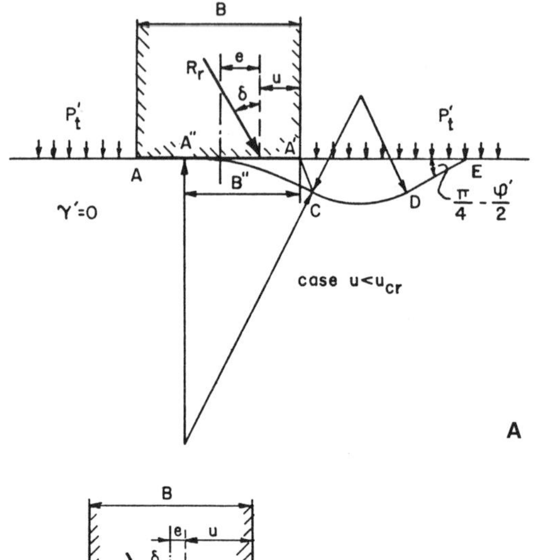

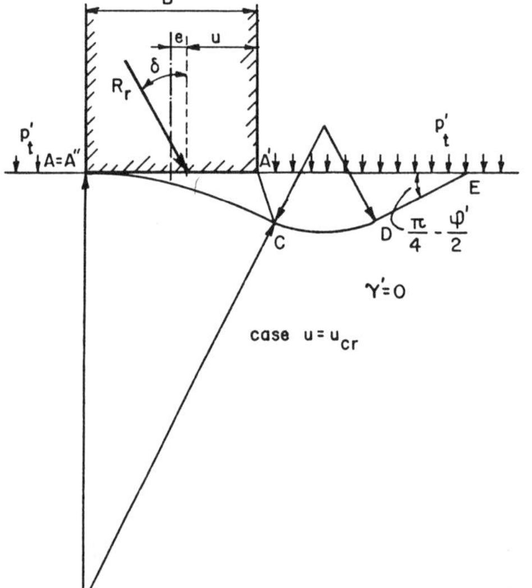

Figure 6-13. Slide surfaces for a continuous footing subjected to an inclined and eccentric load [after Brinch Hansen, 1961] $\gamma' = 0$.

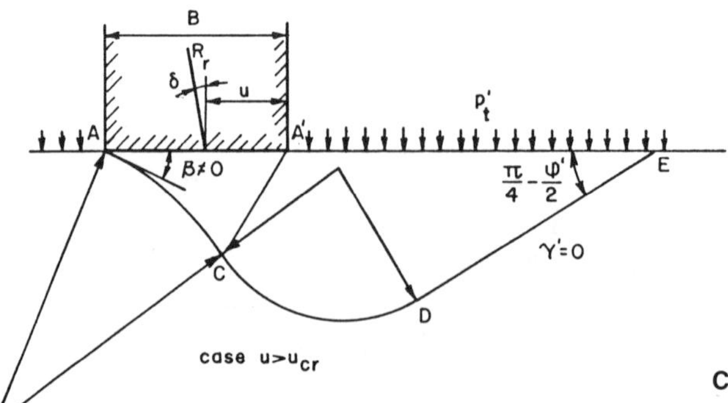

If the distance B″ between the end point A″ of the arc A″C and the edge A′ of the footing is given (Figure 6-13), we obtain as a result of the computation, for each pair of values ϕ', δ and for $\beta = 0$, both the complementary eccentricity u of the load $Q_{r,1}$ and the value of this load. Hence a determinate complementary eccentricity u is associated with given ϕ' and δ values for each point A″. If point A″ coincides with the edge A for $\beta = 0$, a corresponding critical value u_{cr} of the complementary eccentricity is associated with it (Figure 6-13b).

For u values $> u_{cr}$, arc AC does not have a horizontal tangent at A, but a sloping one ($\beta \neq 0$). A specific value of the complementary eccentricity $u > u_{cr}$ and a specific value of the load Q_r then corresponds to each value of the angle β (Figure 6-13). So long as $u < u_{cr}$ and hence A′A″ smaller than A′A, the slide surfaces A″CDE (Figure 6-13a) are similar to those in the case $u = u_{cr}$ (Figure 6-13b).

A strip footing for which $u < u_{cr}$, and $\gamma' = 0$, therefore behaves as if its width were B″.

$$\frac{u_{cr}}{B} = \frac{u}{B''}$$

$$u = \frac{B''}{B} u_{cr} \tag{6-46}$$

Hence for values $u < u_{cr}$, a single computation is sufficient, namely the computation for u_{cr}. When the pertinent load (vertical component) $Q_{r,cr}$ has been determined, we have

$$\frac{Q_r}{Q_{r,cr}} = \frac{B''}{B} = \frac{u}{u_{cr}}; \qquad Q_r = \frac{u}{u_{cr}} Q_{r,cr} \tag{6-47}$$

From the $Q_{r,cr}$ and Q_r values obtained according to this computation, the corresponding $N_{q,\delta,u}$ and the inclination factors $i_{q,u}$ can be derived.

For the case $\gamma' = 0$, $c' = 0$ one has

$$Q_r = 2uN_q i_{q,u} p_t' \tag{6-48}$$

$$Q_{r,cr} = 2u_{cr}N_q i_{q,u_{cr}} p_t' \tag{6-49}$$

and when $u < u_{cr}$, according to Equation 6-47

$$Q_r = \frac{u}{u_{cr}} 2u_{cr}N_q i_{q,u_{cr}} p_t' \tag{6-50}$$

$$Q_r = 2uN_q i_{q,u_{cr}} p_t'$$

(6-51)

Hence when $i_{q,u_{cr}}$ is known, the normal component Q_r of the ultimate load can be determined for any complementary eccentricity $u \leq u_{cr}$. The values of $i_{q,u_{cr}}$ and u_{cr}/B as calculated by Ladanyi are given in Tables 6-5 and 6-6 in terms of the friction angle ϕ' and the inclination angle δ. The inclination factor $i_{q,u_{cr}}$ is also shown in Figure 6-14 versus ϕ' with δ as parameter.

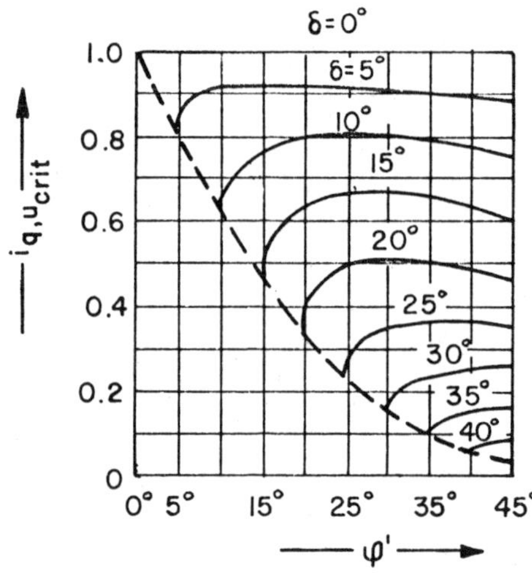

Figure 6-14. Inclination factor $i_{q,u_{cr}}$ vs. ϕ' with δ as a parameter.

For $B/2 > u > u_{cr}$ the computation must be repeated for each value of u because a different angle β (Figure 6-13c) corresponds to each u value; the slide surfaces are no longer similar to each other. However for $u_{cr} < u < B/2$ sufficiently accurate $i_{q,u}$ and $N_{q,\delta,u}$ values can be determined by interpolation.

$$i_{q,u} = i_q + (i_{q,u_{cr}} - i_q) \frac{0.5 - \dfrac{u}{B}}{0.5 - \dfrac{u_{cr}}{B}}$$

(6-52)

To take the influence of the weight γ' of the soil into account, for lack of more precise data, it is simply assumed that the bearing capacity factor $N_{\gamma,\delta}^*$ or the inclination factor $i_{\gamma,u}$ is independent of the eccentricity $i_{\gamma,u} = i_\gamma$.

In order to take the influence of the cohesion into account, the theorem of the corresponding states of Caquot [1956] is used. A cohesive soil

Table 6-5

Bearing Capacity Factor $i_{q,u_{cr}}$ for a Continuous Footing Subjected to an Inclined and Eccentric Load for $\dfrac{u}{B} \leqq \dfrac{u_{cr}}{B}$

ϕ' \ δ	0°	5°	10°	15°	20°	25°	30°	35°	40°	45°
0°	1.000	—	—	—	—	—	—	—	—	—
5°	1.000	0.790	—	—	—	—	—	—	—	—
10°	1.000	0.906	0.609	—	—	—	—	—	—	—
15°	1.000	0.911	0.760	0.455	—	—	—	—	—	—
20°	1.000	0.911	0.795	0.632	0.327	—	—	—	—	—
25°	1.000	0.907	0.801	0.659	0.495	0.226	—	—	—	—
30°	1.000	0.902	0.795	0.661	0.505	0.349	0.149	—	—	—
35°	1.000	0.897	0.782	0.648	0.496	0.360	0.231	0.093	—	—
40°	1.000	0.888	0.765	0.625	0.481	0.359	0.246	0.153	0.054	—
45°	1.000	0.880	0.745	0.595	0.462	0.350	0.246	0.161	0.081	0.028

Table 6-6

Ratio $\dfrac{u_{cr}}{B}$ vs. the Angle of Internal Friction ϕ' and the Inclination Angle δ

ϕ' \ δ	0°	5°	10°	15°	20°	25°	30°	35°	40°	45°
0°	0.500	—	—	—	—	—	—	—	—	—
5°	0.500	0.500	—	—	—	—	—	—	—	—
10°	0.500	0.481	0.500	—	—	—	—	—	—	—
15°	0.500	0.465	0.457	0.500	—	—	—	—	—	—
20°	0.500	0.450	0.425	0.431	0.500	—	—	—	—	—
25°	0.500	0.440	0.405	0.402	0.430	0.500	—	—	—	—
30°	0.500	0.430	0.386	0.373	0.383	0.415	0.500	—	—	—
35°	0.500	0.420	0.366	0.345	0.347	0.362	0.400	0.500	—	—
40°	0.500	0.407	0.347	0.321	0.316	0.320	0.338	0.382	0.500	—
45°	0.500	0.394	0.323	0.295	0.284	0.283	0.290	0.311	0.368	0.500

characterized by c' and ϕ' is in a limit state of equilibrium under a given external load when a cohesionless soil of the same form and with the same internal friction is at the limit of equilibrium under the same load and an overall constant external pressure $\sigma_c' = c' \cot \phi'$.

In the case of an inclined load however the application of the theorem includes a difficulty concerning the inclination angle δ. Indeed by introducing the overall pressure $c' \cot \phi'$ one changes the inclination angle δ to a fictive value δ_f.

One has (Figure 6-15)

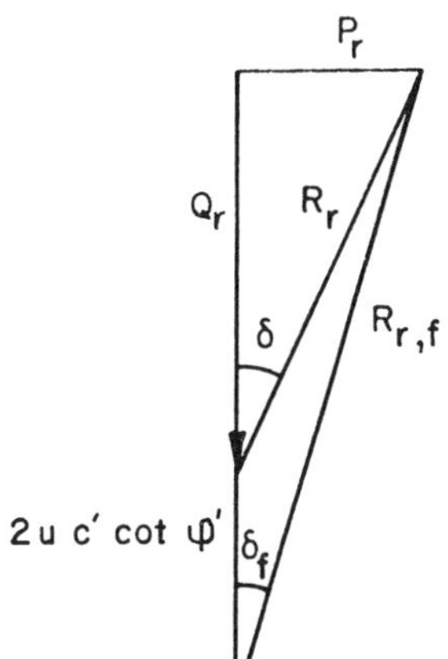

Figure 6-15. Fictive inclination angle δ_f in case of a cohesive soil.

$$P_r = Q_r \tan \delta = (Q_r + 2u\, c' \cot \phi') \tan \delta_f$$

$$\tan \delta_f = \frac{\tan\delta}{1 + \dfrac{2uc'\cot\phi'}{Q_r}} \tag{6-53}$$

The value of δ_f depends on the unknown value Q_r. Consequently the problem has to be solved in steps. A first value Q_r' of Q_r is calculated neglecting the differences between δ_f and δ.

$$Q_r' = 2u(N_q i_{q,u,\delta} p_t' + N_c i_{c,u,\delta} c' + N_\gamma^* i_\gamma \gamma' 2u) \tag{6-54}$$

with $i_{q,u,\delta}$ the value of $i_{q,u}$ for the given inclination δ.

$$i_{c,u,\delta} = \frac{N_q i_{q,u,\delta} - 1}{N_q - 1} \tag{6-55}$$

$i_{q,u,\delta}$ is given by Table 6-5, with δ when $u \leq u_{cr}$, and by Equation 6-52 when $u \geq u_{cr}$.

The value Q_r' is introduced in Equation 6-53 and a first value δ_f' of the fictive angle δ_f is found.

$$\tan \delta_f' = \frac{\tan\delta}{1 + \dfrac{2uc'\cot\phi'}{Q_r'}} \tag{6-56}$$

With this value δ_f' new values for i_{q,u,δ_f} and i_{c,u,δ_f} are calculated, giving a second value Q_r''. With this value Q_r'', a second value of δ_f'' of the angle δ_f is calculated

$$\tan \delta_f'' = \frac{\tan\delta}{1 + \dfrac{2uc'\cot\phi'}{Q_r''}}$$

and these calculations are continued until $\delta_f^{(j)} = \delta_f^{(j-1)}$. The value $Q_r^{(j-1)}$ gives the solution of the problem.

Numerical Examples

A strip footing of width $B = 2m$ rests upon a cohesive soil with $\phi' = 30°$, $c' = 50$ kPa at a depth of 1 m under the soil surface. The water table is at the foundation level.

$\gamma_n = 20$ kN/m^3

$p_t' = 20$ kPa

$\gamma' = 10$ kN/m^3

The strip footing is subjected at a distance $u = 0.70$ m from its outer edge to an inclined load R under an inclination angle δ to the vertical, where $\tan \delta = 1/5 = 0.20$; $\delta = 11°20'$

- One has $u/B = 0.70/2.00 = 0.35$; $c'\cot \phi' = 50 \times 1.732 = 86.6$ kPa

- From Tables 6-1 and 6-2 for $\phi' = 30°$, $N_q = 18.4$ and $N_\gamma = 10.91$.

- From Equation 6-11, $N_c = (18.4 - 1)\, 1.732 = 30.1$.

Calculation with Independent Factors

The inclination factors are considered to be independent of the complementary eccentricity u.

- From Table 6-3 for $\phi' = 30°$ and $\delta = 11°20'$

$$i_q = 0.56 \times (0.70 - 0.56) \, \frac{3.67}{5} = 0.664$$

- From Equation 6-26

$$i_c = \frac{18.4 \times 0.664 - 1}{18.4 - 1} = 0.645$$

- From Table 6-4 for $\phi' = 30°$ and $\delta = 11°20'$

$$i_\gamma = 0.34 \times (0.53 - 0.34) \, \frac{3.67}{5} = 0.481$$

- From Equation 6-41

$$Q_r' = 2 \times 0.70 \, [18.4 \times 0.664 \times 20 + 30.1 \times 0.645$$
$$\times 50 + 10.91 \times 0.481 \times 10 \times 2 \times 0.70]$$

$$= 1.40 \times 1,288.5 = 1,804 \, \frac{kN}{m}$$

- From Equation 6-56

$$\tan\delta_f' = \frac{0.20}{1 + \dfrac{2 \times 0.70 \times 86.6}{1,804}} = 0.188 = \tan 10°40'$$

- From Table 6-3 for $\phi' = 0°$ and $\delta = 10°40'$

$$i_q = 0.56 \times (0.70 - 0.56) \, \frac{4.337}{5} = 0.681$$

- From Equation 6-26

$$i_c = \frac{18.4 \times 0.681 - 1}{18.4 - 1} = 0.66$$

- From Table 6-4 for $\phi' = 30°$ and $\delta = 10°40'$

$$i_\gamma = 0.34 \times (0.53 - 0.34) \frac{4.337}{5} = 0.505$$

$$Q_r^* = 1.40 \, [18.4 \times 0.681 \times 20 + 30.1 \times 0.66 \\ \times 50 + 10.91 \times 0.505 \times 10 \times 1.40]$$

$$= 1.40 \times 1{,}325.5 = 1{,}856 \, \frac{kN}{m}$$

- From Equation 6-56

$$\tan\delta_f' = \frac{0.20}{1 + \dfrac{1.40 \times 86.6}{1{,}856}} = 0.187 = \tan 10°38' \approx \tan \delta_f'$$

Consequently the first method gives

$$Q_r = 1{,}856 \, \frac{kN}{m}$$

$$P_r = 1{,}856 \, \tan\delta = 1{,}856 \times 0.20 = 371 \, \frac{kN}{m}$$

$$R_r = \sqrt{1{,}856^2 + 371^2} = 1{,}893 \, \frac{kN}{m}$$

Calculation with Compound Factors

The inclination factor $i_{q,u}$ is a function of u.

- From Table 6-6 for $\phi' = 30°$ and $\delta = 11°20'$

$$\frac{u_{cr}}{B} = 0.37 + (0.39 - 0.37) \frac{3.67}{5} = 0.385$$

$$\frac{u}{B} = 0.35 < \frac{u_{cr}}{B} = 0.385$$

$$\frac{u}{u_{cr}} = \frac{0.35}{0.385} = 0.909$$

- From Table 6-5 for $\phi' = 30°$; $\delta = 11°20'$

$$i_{q,u_{cr}} = 0.66 \times (0.79 - 0.66)\,\frac{3.67}{5} = 0.756$$

- From Equation 6-55

$$i_{c,u_{cr}} = \frac{18.4 \times 0.756 - 1}{18.4 - 1} = 0.742$$

- From Table 6-4

$$i_{\gamma,u} = i_\gamma = 0.481$$

- From Equation 6-41 and 6-51

$$Q_r' = 2 \times 0.70\,[18.4 \times 0.756 \times 20 + 30.1 \times 0.742$$
$$\times 50 + 10.91 \times 0.481 \times 10 \times 2 \times 0.70]$$

$$= 1.40 \times 1{,}468.5 = 2{,}055\,\frac{kN}{m}$$

- From Equation 6-56

$$\tan\delta_f' = \frac{0.20}{1 + \dfrac{2 \times 0.70 \times 86.6}{2{,}055}} = 0.189 = \tan 10°42'$$

- From Table 6-6 for $\phi' = 30°$

$$\frac{u_{cr}}{B} = 0.37 \times (0.39 - 0.37)\,\frac{4.30}{5} = 0.387$$

$$\frac{u}{B} = 0.35 < \frac{u_{cr}}{B} = 0.387$$

- From Table 6-5 for $\phi' = 30°$ and $\delta_f' = 10°42'$

$$i_{q,u_{cr}} = 0.66 \times (0.79 - 0.66)\, \frac{4.30}{5} = 0.771$$

- From Equation 6-55

$$i_{c,u_{cr}} = \frac{18.4 \times 0.771 - 1}{18.4 - 1} = 0.757$$

$$i_{\gamma} = 0.481$$

$$Q_r'' = 1.4\,[18.4 \times 0.77 \times 20 + 30.1 \times 0.757$$
$$\times 50 + 10.91 \times 0.481 \times 10 \times 1.4]$$

$$= 1.4 \times 1{,}496 = 2{,}095\ \frac{kN}{m} \simeq Q_r'$$

$$\frac{Q_r' + Q_r''}{2} = \frac{2{,}055 + 2{,}095}{2} = 2{,}075\ \frac{kN}{m}$$

with the second method, which takes into account the influence of the eccentricity on the inclination factors, one obtains

$$Q_r = 2{,}075\ \frac{kN}{m}$$

$$P = 2{,}075 \times 0.20 = 415\ \frac{kN}{m}$$

$$R_r = \sqrt{2{,}075^2 + 415^2} = 2{,}116\ kN$$

This is a gain of about 10% in comparison with the first method.

Case of a Rectangular Footing with an Inclined and Eccentric Loading in the Transversal Section of Symmetry. Although the values of $N_{q,\delta,u}$ and $i_{q,\delta,u}$ have only been calculated for the case of strip footings, and negligible overburden shearing strength, the method outlined for the simple case of the strip footing can be extended to the more intricate case.

With such extension, the following is obtained:

$$Q_r = 2uL(N_q i_{q,u} s_q d_q p_t' + N_c i_{c,u} s_c d_c c' + N_\gamma^* i_\gamma s_\gamma d_\gamma \gamma' 2u) \qquad (6\text{-}57)$$

Case of a Rectangular Footing with Central Inclined Load in an Arbitrary Direction. So far inclined loads located in the transversal symmetry plane have been considered (Figure 6-16a). For such a case, one has

$$Q_r = BL(N_q s_q i_q p_t' + N_c s_c i_c c' + N_\gamma^* s_\gamma i_\gamma \gamma' B) \qquad (6\text{-}58)$$

It is however possible that the inclined load is located in a plane making an angle θ_n with the longitudinal symmetry plane (Figure 6-16c).

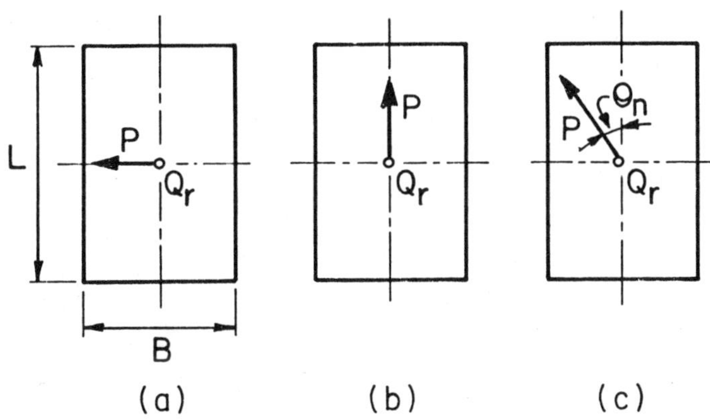

Figure 6-16. Rectangular footing with central inclined loads in an arbitrary direction.

Based on the proposals of Vesic [1975] one could write

$$Q_{r,\theta} = BL[N_q s_q i_q F_q(\theta_n) p_t' + N_c s_c i_c F_c(\theta_n) c' + N_\gamma^* s_\gamma i_\gamma F_\gamma(\theta_n) B] \qquad (6\text{-}59)$$

with

$$F_q(\theta_n) = [1 - \tan\delta_f]^{[(B-L)/(B+L)]\cos^2\theta_n} \qquad (6\text{-}60)$$

$$\tan\delta_f = \frac{\tan\delta}{1 + \dfrac{BLc'\cot\phi'}{Q_r}} \qquad (6\text{-}61)$$

As Q_r is unknown, when $c' \neq 0$, there must be a stepwise solution of the problem.

By applying the theorem of Caquot one gets

$$F_c(\theta_n) = \frac{N_q s_q i_q F_q(\theta_n) - 1}{N_q s_q i_q - 1} \tag{6-62}$$

According to Vesic [1975] one has

$$F_\gamma(\theta_n) = F_q(\theta_n) = (1 - \tan \delta_f)^{[(B-L)/(B+L)]\cos^2\theta_n}$$

- *Strip footing* $B/L = 0$

$$F_q(\theta_n) = F_\gamma(\theta_n) = (1 - \tan \delta_f)^{-\cos^2\theta_n} \tag{6-64}$$

Transversal inclination $\theta_n = 90°$, $\cos \theta_n = 0$

$$F_\gamma(\theta_n) = F_q(\theta_n) = 1 \tag{6-65}$$

Longitudinal inclination $\theta_n = 0$, $\cos \theta_n = 1$

$$F_\gamma(\theta_n) = F_q(\theta_n) = (1 - \tan \delta_f)^{-1} \tag{6-66}$$

$$F_c(\theta_n) = \frac{N_q i_q (1 - \tan \delta_f)^{-\cos^2\theta_n} - 1}{N_q i_q - 1} \tag{6-67}$$

Transversal inclination

$$F_c(\theta_n) = 1 \tag{6-68}$$

Longitudinal inclination

$$F_c(\theta_n) = \frac{N_q i_q (1 - \tan \delta_f)^{-1} - 1}{N_q i_q - 1} \tag{6-69}$$

- *Square footing* $B = L$

$$F(\theta_n) = F_q(\theta_n) = F_c(\theta_n) = 1 \tag{6-70}$$

Case of Inclined and Eccentric Loading on Footings Other than a Rectangle. For such cases Hansen [1961] proposes to consider an equivalent rectangle whose geometric center coincides with the point of application of the load and that follows as closely as possible the adjacent contour of the actual base area.

A few examples taken from Hansen are shown in Figure 6-17. In Equations 6-58 through 6-61 the lengths B and L are to be replaced by the lengths B′ and L′ of the equivalent rectangle as defined in Figure 6-17.

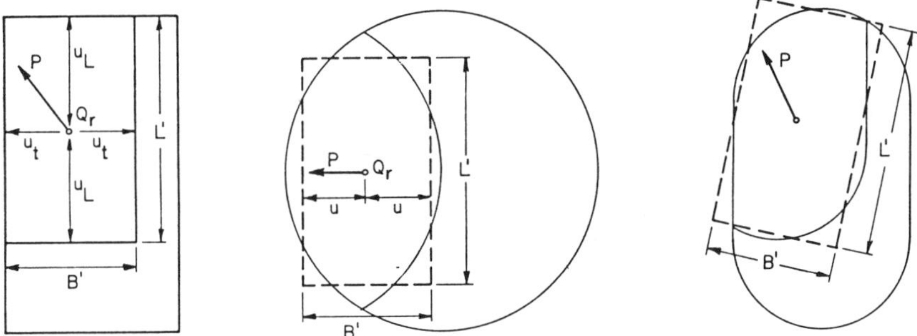

Figure 6-17. Equivalent foundation areas [after Brinch Hansen, 1961].

Case of an Inclined Footing. In order to increase the safety against sliding in the contact face AA′ (Figure 6-18) in some situations this face is given a certain angle α to the horizontal. In the case of a weightless and cohesionless soil and a load Q_r normal to the face AA′, one gets (Figure 6-19)

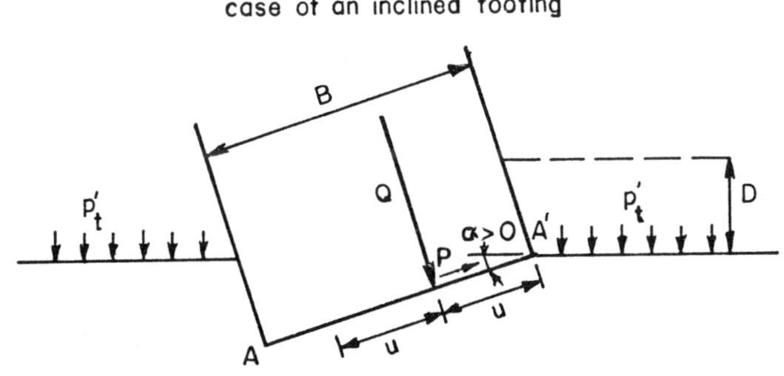

Figure 6-18. Case of an inclined footing.

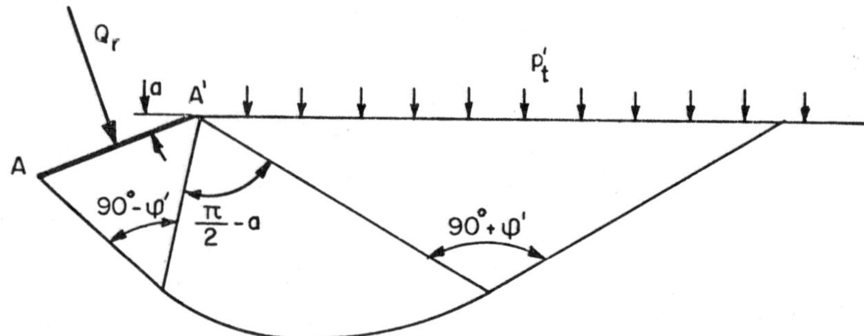

Figure 6-19. Rupture line under an inclined footing, when $\gamma' = 0$ and $c' = 0$.

$$N_{q,\alpha} = \tan^2\left(45° + \frac{\phi'}{2}\right) e^{(\pi - 2\alpha)\,\tan\phi'} \tag{6-71}$$

$$N_{q,\alpha} = N_q e^{-2\alpha\tan\phi'} \tag{6-72}$$

$$N_{q,\alpha} = N_q a_q \tag{6-73}$$

$$a_q = e^{-2\alpha\tan\phi'} \tag{6-74}$$

$$a_q = e^{-2\alpha\,\tan\phi'} = 1 - 2\alpha\,\tan\phi' + \frac{(2\alpha\,\tan\phi')^2}{2} - \frac{(2\alpha\,\tan\phi')^3}{6} + \dots$$

$$= (1 - \alpha\,\tan\phi')^2 + \alpha^2\,\tan\phi'^2 - \frac{(2\alpha\,\tan\phi')^3}{6} + \dots \tag{6-75}$$

Neglecting the higher exponents, one obtains the expression proposed by Vesic [1975]

$$a_q = (1 - \alpha\,\tan\phi')^2 \tag{6-76}$$

For the case of a cohesive soil, and taking into account the weight of the soil, one gets

$$q_r = N_q a_q p_t' + N_c a_c c' + N_\gamma^* a_\gamma \gamma' B \tag{6-77}$$

For the case of a cohesive soil one gets

$$a_c = \frac{N_q a_q - 1}{N_q - 1} \tag{6-78}$$

for $\phi' = 0$

$$a_c = 1 - \frac{2\alpha}{\pi + 2} \tag{6-79}$$

Finally for the weight term Hansen [1961] and Vesic [1975] propose

$$a_\gamma = a_q = (1 - \alpha \tan \phi')^2 \tag{6-80}$$

For the general case of an eccentric load on an inclined rectangular footing one gets

$$Q_r = B'L'(N_q s_q i_q d_q a_q p_t' + N_c s_c i_c d_c a_c c' + N_\gamma^* s_\gamma i_\gamma a_\gamma \gamma' B') \tag{6-81}$$

with

$$a_c = \frac{N_q s_q i_q d_q a_q - 1}{N_q s_q i_q d_q - 1} \tag{6-82}$$

Strip Footing Near a Slope. Let us first consider the case of a horizontal strip footing vertically and centrally loaded and established at a depth D on a cohesionless and weightless soil, limited by a slope with angle ω (Figure 6-20).

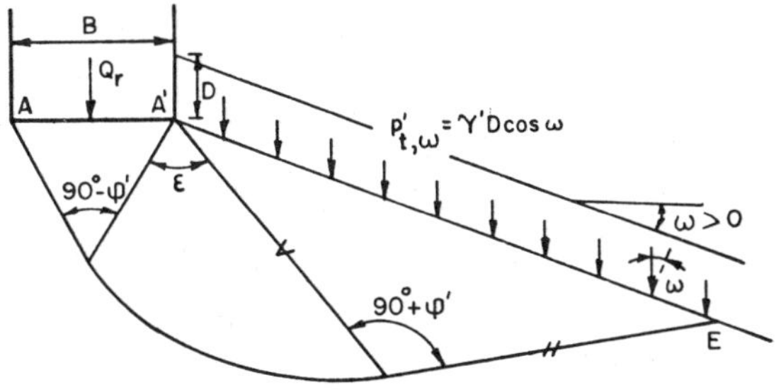

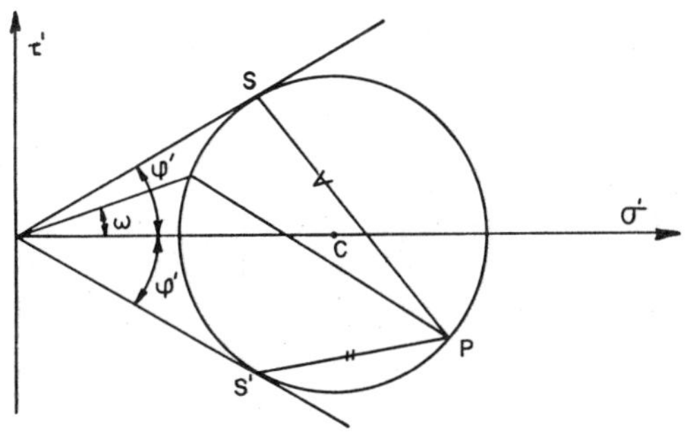

Figure 6-20. Case of a strip footing near a slope.

On the plane $A'E$ parallel to the soil surface exists an overburden stress $p'_{t,\omega} = \gamma'$, D cos ω making an angle ω with the normal. Consequently $A'E$ is not a principal surface. The center angle ϵ of the logarithmic spiral decreases accordingly. The effect of the slope angle ω of the soil surface on the bearing capacity is analogous to that of the inclination angle δ of the load.

Introducing soil slope factors $F_q(\omega)$, $F_c(\omega)$, $F_\gamma(\omega)$ one writes

$$q_r = N_q F_q(\omega) p'_{t,\omega} + N_c F_c(\omega) c' + N_\gamma^* F_\gamma(\omega) \gamma' B \tag{6-83}$$

Hansen [1970] and Vesic [1975] propose the following approximate formulas for $F_q(\omega)$ and $F_\gamma(\omega)$

$$F_q(\omega) = F_\gamma(\omega) = (1 - \tan \omega)^2 \tag{6-84}$$

$$F_c(\omega) = \frac{N_q F_q(\omega) - 1}{N_q - 1} \tag{6-85}$$

for $\phi' = 0$

$$F_c(\omega) = 1 - \frac{2\omega}{\pi + 2} \tag{6-86}$$

and as shown by Vesic [1975] for the case $\phi' = 0$ a negative value has to be introduced for the absence of weight due to the slope. One has $N_\gamma^* = -\sin \omega$.
Finally for $\phi' = 0$

$$q_r = (\pi + 2) \left(1 - \frac{2\omega}{\pi + 2}\right) c' - \sin\omega(1 - \tan\omega)^2 \gamma' B$$

$$= (\pi + 2 - 2\omega)c' - \sin\omega(1 - \tan\omega)^2 \gamma' B \tag{6-87}$$

The formulas established for a strip footing are generally not of great help for practical problems, as the footing has a limited dimension L parallel to the slope.

One could of course introduce the same shape factors, inclination factors, depth factors, orientation factors, as for the case of a horizontal soil surface. However there is no theoretical nor experimental evidence concerning the error involved in such extrapolation. Such calculations can thus only be considered as a rough guess. In each case such calculations should be completed by a classical slope analysis, taking into account the three-dimensional character of the problem.

Numerical Example

Consider the case of Figure 6-21.

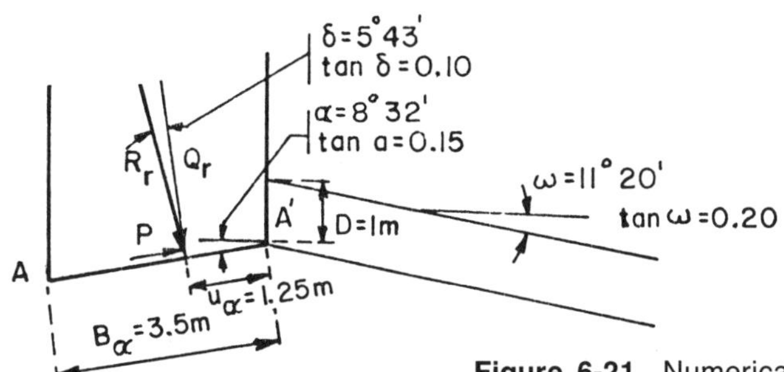

Figure 6-21. Numerical example of an in-clined footing near a slope.

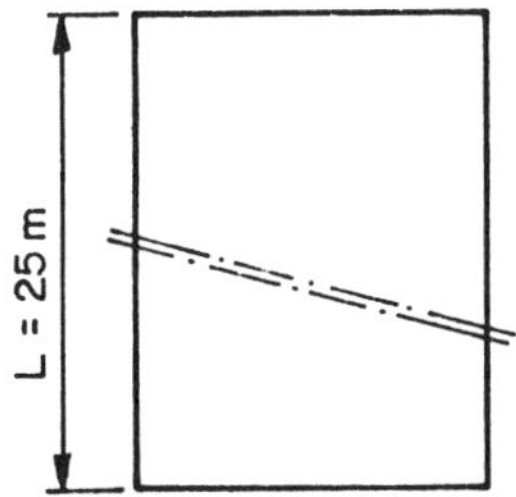

Footing with an inclination: $\alpha = 8°32'$; $\tan \alpha = 0.15$
Width parallel to contact surface: $B_\alpha = 3.50$ m
Slope of the soil surface: $\omega = 11°20'$; $\tan \omega = 0.20$
Complementary eccentricity: $u_\alpha = 1.25$ m
Inclination angle δ of the load: $\delta = 5°43'$; $\tan \delta = 0.10$
Length of the footing: $L = 25$ m; $B_\alpha/L = 3.50/25 = 0.14$
Foundation depth: $D = 1$ m

Groundwater is at great depth.

$$\gamma_n = 16 \frac{kN}{m^3}$$

$$\phi' = 30°$$

$$c' = 20 \frac{kN}{m^2}$$

Find the component Q_r of the ultimate load.

The shearing resistance of the overburden layer is neglected. The general formula is

$$Q_r = 2\ uL\ (N_q s_q i_q a_q F_q(\omega) p'_{t,\omega} + N_c s_c i_c a_c F_c(\omega) c' + N_\gamma^* s_\gamma i_\gamma a_\gamma F_\gamma(\omega) \gamma' 2u)$$

- From Table 6-1 $\phi' = 30°$ and $N_q = 18.4$ and from Table 6-2 $N_\gamma^* = 10.91$.

- From Equation 6-11

$$N_c = \frac{18.4 - 1}{0.577} = 30.1$$

- From Equation 6-31

$$s_q = 1 + 0.14 \sin 30° = 1.07$$

- From Equation 6-32

$$s_c = \frac{1.07 \times 18.4 - 1}{18.4 - 1} = 1.074$$

- From Equation 6-33

$$s_\gamma = \frac{1 + 0.2 \times 0.14}{1 + 0.14} = 0.902$$

with $\phi' = 30°$; $\delta = 5°43'$; $\tan \delta = 0.10$

- From Table 6-3 for $\phi' = 30°$; $\delta = 5°43'$; $i_q = 0.704$
  $+ (0.850 - 0.704)\ 4.28/5 = 0.829$

- From Table 6-4 for $\phi' = 30°$; $\delta = 5°43'$; $i_\gamma = 0.518$
  $+ (0.740 - 0.518) \times 0.856 = 0.708$

$$i_c = \frac{N_q s_q i_q - 1}{N_q s_q - 1} = \frac{18.40 \times 1.07 \times 0.829 - 1}{18.4 \times 1.07 - 1} = 0.821$$

with $\phi' = 30°$; $\alpha = 8°32'$; $\tan \alpha = 0.15$; $\alpha = 0.149$ rad

- From Equation 6-76, $a_q = (1 - 0.149 \times 0.577)^2 = 0.835$

- From Equation 6-80, $a_\gamma = 0.835$

$$a_c = \frac{N_q s_q i_q a_q - 1}{N_q s_q i_q - 1} = \frac{16.34 \times 0.835 - 1}{16.34 - 1} = 0.824$$

with $\phi' = 30°$; $\omega = 11°20'$; $\tan \omega = 0.20$

- From Equation 6-84

$$F_q(\omega) = (1 - 0.20)^2 = 0.64$$

$$F_\gamma(\omega) = (1 - 0.20)^2 = 0.64$$

$$F_c(\omega) = \frac{N_q s_q i_q a_q F_q(\omega) - 1}{N_q s_q i_q a_q - 1}$$

$$= \frac{13.64 \times 0.64 - 1}{13.64 - 1} = 0.612$$

$$p'_{t,\omega} = \gamma' \, D \cos \omega = 16 \times 1 \times \cos 11°20' = 15.69 \text{ kN/m}^2$$

$$\begin{aligned}
Q_r = 2 \times 1.25 \times 25 \, [&18.4 \times 1.07 \times 0.829 \times 0.835 \times 0.64 \times 15.69 \\
&+ 30.1 \times 1.074 \times 0.821 \times 0.824 \times 0.612 \times 20 \\
&+ 10.91 \times 0.902 \times 0.708 \times 0.835 \times 0.64 \times 16 \times 2 \times 1.25] \\
= {} & 62.5 \times 553.5 = 34{,}594 \text{ kN}
\end{aligned}$$

Shearing Strength Parameters to be Introduced in the Formulas. All formulas are established for the case of a rigid material, characterized by a unique and straight intrinsic law (c', ϕ'). However, not only is the soil not rigid, but its shearing strength is also much more complicated than that corresponding to a unique straight line.

First the intrinsic law depends on the state of strain. For plane strain conditions the intrinsic angle ϕ'_p should be about 10% larger than for triaxial strain conditions ϕ'_{tr} [Bishop, 1961; Bjerrum, 1961]:

$$\phi'_p = 1.10\phi'_{tr} \tag{6-88}$$

For strip footings (L = ∞) the plane strain condition exists. Some advocate that a state of triaxial strain exists under square and circular footings. Thus different angles ϕ' should prevail for strip and circular footings. However, all formulas have been established with the assumption that but one value of ϕ' has to be considered. All values of the shape factors are based on that assumption.

If for a strip footing and a circular footing different values ϕ'_p and ϕ'_{tr} should be introduced, the expressions of all the factors should be modified. However there are still other difficulties. Indeed, for a given relative density and a given structure, the intrinsic law of a soil, under a given state of strain, is not a straight line, but a curve turning its concavity towards the σ' axis [De Beer, 1965; Ladanyi, 1961]. By increasing stress state σ' the angle ϕ' (either ϕ'_{sec} or ϕ'_{tan}) decreases (Figure 6-22). Consequently the value to be introduced will depend on the state of stress.

How does one proceed in practice? The routine tests are triaxial tests, performed under a spherical pressure of the order of 50-100-200 kN/m². Plane strain tests are rather exceptional. Therefore, for a first calculation, preference should be given to the shearing strength characteristics c', ϕ', obtained in routine triaxial tests, and these values entered in the formulas given in the previous sections. The value of the obtained results will be discussed in the next section.

Scale Effect. When the constancy of ϕ' is accepted, the bearing capacity factors N_q, N_c, N_γ^* and all other factors should be independent of the size of the footing, and the embedment depth. However, the reality is completely different. For instance, as shown by Figure 6-23, the experimen-

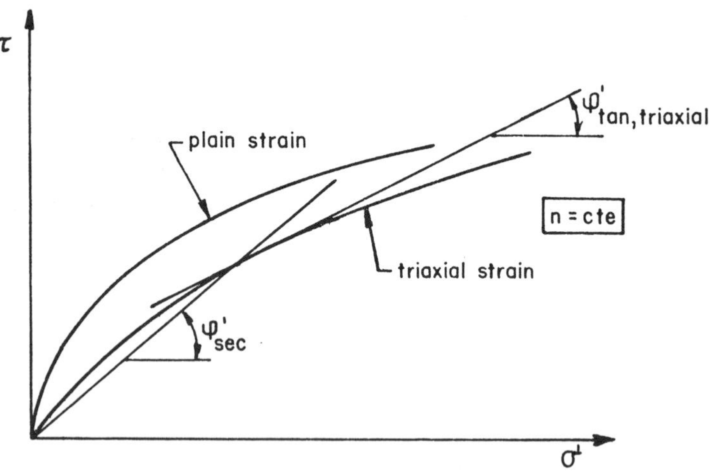

Figure 6-22. Curved intrinsic law of sand.

tal values of N_γ^* decrease considerably when the dimensions of the footing increase. According to the general bearing formula with constant ϕ' the unit ultimate bearing capacity increases linearly with B, and should become infinite when B tends to infinity.

The unit bearing capacity for a footing on the surface cannot become infinite, but it cannot be greater than the resistance of deep footings on the same soil [Vesic, 1975]. Very large footings should fail exclusively in punching shear, as also do deep foundations. We could improve the scheme of Figure 6-2 by putting as ordinate instead of D/2R the values

$$\frac{D}{2R} + 2\frac{\gamma' R}{E_q \text{ or } \sigma_r'}$$

with E_q = Young's modulus of quartz
$\quad\ \sigma_r'$ = crushing strength of the quartz particles.

The zones of general shear, local shear, and punching shear, are given versus the relative density D_r, and the relative depth and width parameters

$$\frac{D}{2R} + \frac{2\gamma' R}{E_q \text{ or } \sigma_r'}$$

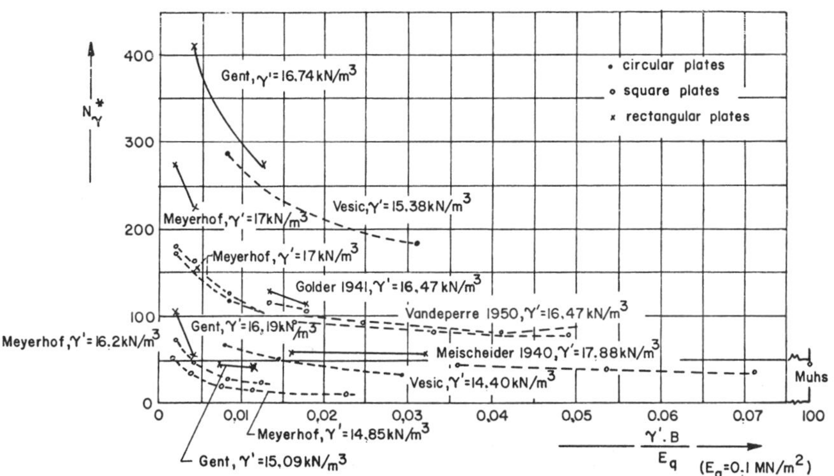

Figure 6-23. Variation of bearing capacity factor N_γ^* with foundation size.

From this scheme it follows that even for a footing at the surface $(D = 0)$, for very dense sand $(D_r = 1)$, punching shear will occur when $2\gamma'\,R/E_q$ or $2\gamma'\,R/\sigma_r'$ becomes sufficiently large. This is thus a first reason why the ultimate bearing capacity depends on the size of the footing, and on its embedment depth. A second obvious reason is that the intrinsic law is not a straight line for a soil with a given density (Figure 6-22). By increasing mean stress level, the values of ϕ_{sec}' and ϕ_{tan}' become smaller. The mean stress level increases by increasing embedment depth and by increasing dimensions of the footing. As an average mean normal stress $(\sigma_g')_m$ along the potential gliding surface, the following value is proposed [De Beer, 1970]:

$$(\sigma_g')_m = \frac{1}{4}\,(q_r + 3p_t')(1 - \sin\phi') \tag{6-89}$$

The value $(\sigma_g')_m$ depends on the unknown value q_r. Consequently the problem has to be solved stepwise. A first value q_r' is calculated with the ϕ_1' value corresponding to a classical triaxial test $(\sigma_3' = 100 \text{ kN/m}^2)$.

With the values q_r' and ϕ_1' a first value of $(\sigma_g')_m$ is calculated

$$(\sigma_g')_m = \frac{1}{4}\,(q_r' + 3p_t')(1 - \sin\phi_1') \tag{6-90}$$

With this value in the curved intrinsic law, the value $\phi_{sec,2}'$ is deduced; with this value, a second value q_r'' is calculated.

$$(\sigma_g')_m'' = \frac{1}{4}\,(q_r'' + 3p_t')(1 - \sin\phi_{sec,2}') \tag{6-91}$$

and the process is continued, until

$$(\sigma_g')_m^i = (\sigma_g')_m^{i-1}$$

A third reason is the progressivity of the rupture phenomenon. Due to the progressivity of the rupture, in the region AC, (Figure 6-3) the peak shear strength has already been surpassed, when in the region ED it is not yet attained. The influence of this progressivity increases with the absolute dimensions of the footing [De Beer, 1965].

Influence of the Compressibility. The formulas established starting from the hypothesis of a rigid-plastic material are certainly not directly applicable for the case of a compressible material. Terzaghi [1948] proposed for the case of loose sands, to still accept the basic formulas valid for a rigid material, but introducing in the formulas reduced or notional values of the shearing strength characteristics given by

$$\tan\phi_n' = \frac{2}{3}\tan\phi' \tag{6-92}$$

$$c_n' = \frac{2}{3}c' \tag{6-93}$$

This approach can only be a first aid, as in reality the reduction should have to be a function of the relative density. Between the loose sands and the dense sands there should be a jump. Furthermore for saturated sands below the critical density, the equilibrium is precarious, as a small vibration can already cause the liquefaction of the sand.

A better approach for taking into account the compressibility of the soil is presented by Vesic [1963]. He considers the case of an elastic-plastic material, characterized by a Young modulus E, a Poisson ratio ν, and a unique strength intrinsic law (c', ϕ').

His deductions are based on the hypothesis that a wedge AA'C (Figure 6-3) is pushed into the soil under high pressure, the normal components of the stresses on the faces AC and A'C being equal to the pressure needed to expand a cylindrical cavity (strip footing) in an elastic-plastic material (E, μ, c', ϕ') or with rigidity index I_r (Equation 6-4). For taking into account the influence of the compressibility, Vesic proposes to introduce in the basic formulas rigidity factors $F(I_r)$.

For a centrally and vertically loaded strip footing

$$q_r = N_q F_q(I_r)p_t' + N_c F_c(I_r)c' + N_\gamma^* F_\gamma(I_r)\gamma'B \tag{6-94}$$

For the case of a strip footing, Vesic gets

$$F_q(I_r) = e^{-4.4\tan\phi' + (3.07\sin\phi'\log 2I_r)/(1+\sin\phi')} \tag{6-95}$$

$$F_c(I_r) = \frac{N_q F_q(I_r) - 1}{N_q - 1} \tag{6-96}$$

which for $\phi' = 0$ gives

$$F_c(I_r) = 0.32 + 0.60 \log_{10}I_r \tag{6-97}$$

Vesic puts:

$$F_\gamma(I_r) = F_q(I_r)$$

Vesic has also considered the case of shapes other than that of strip footing. He then obtains compressibility factors $F(I_r)$, which are no longer only a function of I_r but also of the shape ratio B/L. If these more intricate formulas of $F(I_r)$ are applied, the shape factors s_q, s_c, s_γ should no longer be introduced. But by doing so some contradictions appear; Vesic himself stresses that the information produced by his approximate analysis must be treated as tentative and qualitative in nature. Therefore we prefer only to retain values of the rigidity factors $F(I_r)$, obtained for the case of strip footings, introducing for other shapes the value of the shape factors as before. Furthermore, instead of using the rigidity index I_r, for safety we prefer to introduce the rigidity index I_r' of Equation 6-7.

In the case of a strip footing the values of $F_q(I_r)$ are plotted versus ϕ', with the rigidity index I_r' as parameter in Figure 6-24.

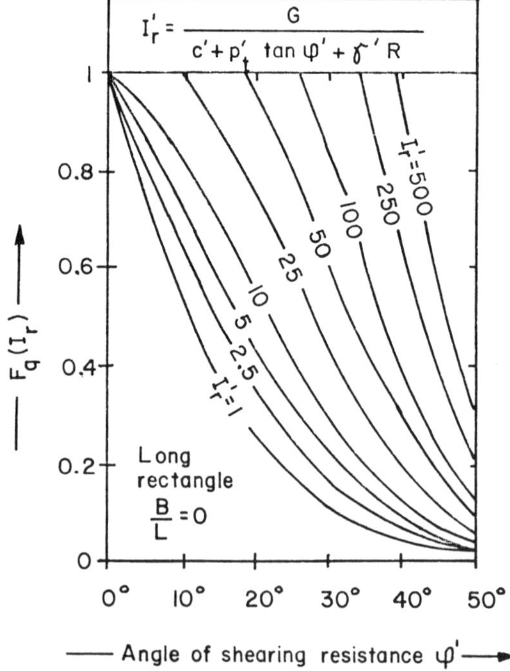

Figure 6-24. Theoretical compressibility factors [after Vesic, 1970]; B/L = 0.

The critical value of I_r, for which $F_q(I_r) = 1$, is given by

$$I_{r,cr} = \frac{1}{2}\, e^{3.30\cot\left(45° - \frac{\phi'}{2}\right)}$$
(6-98)

The values of $I_{r,cr}$ are given in Table 6-7.

For values of I_r larger than the values of Table 6-7 the material could be considered as rigid. The method proposed by Vesic respects the continuity. However, the real soil properties are more intricate than those of a material with constant E and ν, and with constant c' and ϕ'. The method does not fully take into account this variability.

Table 6-7
Values of $I_{r,cr}$

ϕ'	$I_{r,cr}$
0°	13.6
5°	18.3
10°	25.5
15°	36.9
20°	55.7
25°	88.8
30°	152
35°	283
40°	592
45°	1,442

Strip footing $\dfrac{B}{L} = 0$

Numerical Example

A footing B = 2m, L = 6m is established at a depth of 1m in a soft clay

$\gamma_n = 16\ \text{kN/m}^3$
$c_u = 50\ \text{kN/m}^2$
$\phi_u = 0$
$c' = 10\ \text{kN/m}^2$
$\phi' = 20°$

An odometer test gives an odometer modulus M = 500 kN/m²

The watertable corresponds with the foundation level. The footing is vertically loaded with an eccentricity e = 0.25 m. Investigate the effect of compressibility on ultimate bearing capacity. Assimilating the clay to a purely elastic material, one has Poisson's ratio of the solid skeleton

$$\nu_s = \frac{1 - \sin\phi'}{2 - \sin\phi'} \tag{A}$$

$$\nu_s = \frac{1 - \sin 20°}{2 - \sin 20°} = 0.397$$

The Young modulus E_s of the skeleton is

$$E_s = \frac{1 - \nu_s - 2\,\nu_s^2}{1 - \nu_s} M \tag{B}$$

$$E_s = \frac{1 - 0.397 - 2(0.397)^2}{1 - 0.397} M = 0.477\,M$$

$$E_s = 0.477 \times 500 = 238.5 \text{ kN/m}^2$$

Thus

$$G_d = G_u$$

$$\frac{E_s}{2(1 + \nu_s)} = \frac{E_u}{2(1 + \nu_u)} \tag{C}$$

and with

$$\nu_u = 0.5$$

$$E_u = \frac{3}{2}\frac{E_s}{1 + \nu_s} \tag{D}$$

$$E_u = \frac{1.5}{1 + 0.397} E_s$$

$$= 1.074\,E_s$$

$$E_u = 1.074 \times 238.5 = 256 \text{ kN/m}^2$$

However the soil skeleton is not an elastic body, and the modulus E_u obtained in an unconfined compression test can be quite different from the value given by the Equation C. Such a test gives for the considered clay for instance $E_u = 1,000$ kN/m^2

Short Term Stability

In case of rapid application of the load, undrained conditions are considered: $\phi_u = 0$; $Q_r = 2uLq_r = 2uLN_cF_c(I_r)c_u$.

- From Equation 6-98

$$I'_{r,cr} = \frac{1}{2}\, e^{3.30} = \tfrac{1}{2}\, 27.1 = 13.6$$

- The rigidity index is (Equation 6-7):

$$I'_r = \frac{E_u}{2(1 + \nu_u)(c_u + \gamma' R)}$$

$$\gamma' = 16 - 10 = 6 \text{ kN/m}^3$$

$$R = \frac{2uL}{4u + 2L} = \frac{2 \times 0.75 \times 6.00}{4 \times 0.75 + 2 \times 6.00} = 0.60 \text{ m}$$

$$I'_r = \frac{1,000}{2\left(1 + \dfrac{1}{2}\right)(50 + 0.6 \times 0.6)} = 6.62 < 13.6$$

- Consequently the compressibility is to be taken into account. From Equation 6-97

$$F_c(I_r) = 0.32 + 0.60 \log 6.62 = 0.81$$

$$Q_r = 2uLq_r = 2uLN_cF_c(I_r)s_c c_u$$

$$s_c = 1 + 0.2\,\frac{B'}{L} = 1 + 0.2\,\frac{1.5}{6} = 1.050$$

$$Q_r = 2 \times 0.75 \times 6.00 \times 5.14 \times 0.81 \times 1.050 \times 50 = 9 \times 218.6$$
$$= 1,967 \text{ kN}$$

Long Term Stability

- From Table 6-7, $\phi' = 20°$; $I_{r,cr} = 55.7$

  $p_t' = \gamma_n D = 16 \times 1.00 = 16$ kN/m^2

- From Equation 6-7

$$I_r' = \frac{240}{2(1 + 0.397)(10 + 16 \tan 20° + 0.60 \times 0.60)} = 5.3 < 55.7$$

$$Q_r = 2\ uLq_r$$
$$= 2\ uL\ [N_q F_q(I_r)s_q p_t' + N_c F_c(I_r)s_c c' + N_\gamma^* F_\gamma(I_r)s_\gamma \gamma'\ 2u]$$

- From Table 6-1, $\phi' = 20°$; $N_q = 6.40$.

- From Table 6-2, $\phi' = 20°$; $N_\gamma^* = 2.48$.

- From Equation 6-11

$$N_c = \frac{N_q - 1}{\tan\phi'} = \frac{6.40 - 1}{\tan 20°} = \frac{5.40}{0.365} = 14.84$$

- From Equation 6-31

$$s_q = 1 + \frac{1.50}{6.00} \sin 20° = 1.0855$$

- From Equation 6-32

$$s_c = \frac{s_q N_q - 1}{N_q - 1} = \frac{1.0855 \times 6.40 - 1}{6.40 - 1} = 1.1079$$

- From Equation 6-33

$$s_\gamma = \frac{1 + 0.20 \times 0.250}{1 + 0.250} = 0.840$$

- From Figure 6-24 and Equation 6-95, $I_r' = 5.3$, $\phi' = 20°$

  $F_q(I_r') = 0.448$

$$F_c(I'_r) = \frac{N_q s_q F_q(I_r) - 1}{N_q s_q - 1} = \frac{6.40 \times 1.0855 \times 0.448 - 1}{6.40 \times 1.0855 - 1} = 0.36$$

$$Q_r = 2 \times 0.75 \times 6.00 \, [6.40 \times 0.448 \times 1.0855 \times 16 + 14.84$$
$$\times 0.36 \times 1.10 \times 10 + 2.480 \times 0.448 \times 0.84 \times 6 \times 2 \times 0.75]$$

$$Q_r = 9.00 \times 117 = 1{,}053 \text{ kN} < 1{,}967 \text{ kN}$$

Because of the influence of the compressibility, the ultimate bearing capacity after full consolidation can be smaller than the ultimate bearing capacity existing immediately after applying the load. This should not be the case when considering a rigid-plastic material. In such material the ultimate bearing capacity after dissipation of the excess-water pressures should be always higher than in the nonconsolidated state.

Influence of the Roughness of the Foundation Base. In the literature much controversy exists concerning the rupture pattern beneath a centrally loaded, infinitely smooth foundation [Vesic, 1975; De Beer and Vesic, 1958; Meyerhof, 1955; Kishida, 1977]. In practice foundations have a sufficiently rough base as to behave as rough foundations. Consequently the case of an infinitely smooth foundation has mostly a purely academic importance.

Prediction of the Ultimate Bearing Capacity from Field Tests

The ultimate bearing capacity of the soil surrounding a foundation depends not only on the shearing strength of the soil, but also on its compressibility characteristics, the relative depth of embedment, the initial stress tensor in the soil and also on the stress tensor produced by installing the foundation. It is necessary to know all these factors and it is not an easy task to simulate them all in laboratory tests performed on more or less undisturbed samples. On the other hand some in situ tests provide an in situ value which is influenced by several of these factors. Among these in situ tests are plate bearing tests, cone penetration tests (CPT), and pressuremeter tests.

Plate Bearing Tests. With these tests it is possible to obtain deformability characteristics as well as the ultimate bearing capacity of the soil around the plate. These data together with those from other field and laboratory tests afford the possibility of making a better choice of the parameters to be used for predicting the ultimate bearing capacity for the considered foundation. It is difficult, especially in the case of stiff clays, to correctly determine the value of c_u to be used in the formulas based only on results of laboratory tests. Indeed in the case of stiff fissured clays the laboratory value of c_u depends on the size of the samples tested. Furthermore it is by no means easy to obtain undisturbed samples of fissured clay. Accordingly, especially in stiff clays, it is worth carrying out plate-loading tests using plates with large diameters, 200 mm and larger [Marsland, 1973]. Moreover for very large foundations, because of the very large deformations at rupture and the resulting progressive nature of the rupture phenomenon, it may be necessary to choose a value between the peak shear strength and the residual shear strength remaining after large distortions.

Cone Penetration Tests (CPT). In the cone penetration test a small cone $B_c = 3.6$ cm is pushed into the soil with a rather limited speed and the corresponding cone resistance q_c is measured at different depths. Since the relative depths D/B_c in the bearing stratum are mostly large, the stress field around the cone corresponds to that of a deep foundation, and is very different from that around a shallow footing. Furthermore the stress tensor around the cone is much higher than that around the shallow footing, with the consequence that this is also the case for the influence of the deformability. The real phenomenon causing the measured value q_c is the combination of expulsion and compression corresponding to the least energy.

Thus $q_{c,\text{gen.shear}}$ = the ultimate bearing capacity corresponding to a rigid material, with the c' and ϕ' values of the real soil corresponding to the considered stress level, and $q_{c,\text{punching}}$ = the ultimate bearing capacity in punching shear for a material having the deformability characteristics of the real soil, corresponding to the considered stress level.

Then one has

$$q_c \leq q_{c,\text{general shear}}$$

$$q_c \leq q_{c,\text{punching}}$$

The value q_c in general shear in the case of a cohesionless soil is given by the expression

$$q_{c,\text{gen.shear}} = p_t' \, F(\phi') \tag{6-99}$$

$$F(\phi') = 1.3 \; e^{2\pi\tan\phi'} \; \frac{1 + \sin\phi'}{1 - \sin\phi'} \tag{6-100}$$

$$q_c \leqq q_{c,\text{gen.shear}} = p_t' \, F(\phi') \tag{6-101}$$

$$F(\phi') \geqq \frac{q_c}{p_t'} \tag{6-102}$$

Equation 6-102 thus gives a notional value ϕ_n' smaller than the real value ϕ'; $\phi_n' < \phi'$. By introducing the measured value q_c, the influence of the compressibility can be accounted for, in a more continuous way, as done by Terzaghi. However, the notional value ϕ_n' corresponds to a much higher stress level than will exist around the shallow footing. Since the intrinsic law of a soil is not a straight line, but a curve, the characteristic angle for the shallow footing is larger than for the deepseated cone. For shallow footings one can introduce an adapted angle ϕ_n'' given by

$$\phi_n'' = \phi_n' + 5° \, \log \frac{q_c}{q_r} \tag{6-103}$$

If for instance $q_c = 10$ MPa
$$\phi_n' = 22°$$
$$q_r = 1 \text{ MPa}$$

then $\phi_n'' = 22 + 5 \, \log \dfrac{10}{1}$

$$\phi_n'' = 27°$$

Since, however, q_r is an unknown quality, an iterative calculation has to be performed.

Numerical Example

A site is characterized by the results of CPT shown in Figure 6-25. The phreatic level is at a depth of 2.41 m underneath the soil surface. At a depth of 2.60 m underneath the soil surface a dense soil layer is encountered. The problem is the foundation of a bridge pier to be established on the sand layer. The dimensions of the footing of the pier are B = 3.00 m; L = 15.00 m; B/L = 0.2. The pier is to be considered as vertically and centrally loaded. The unit weights are

$$\gamma_n = 16 \, \frac{\text{kN}}{\text{m}^3} \quad \text{above}$$

and

$$\gamma_n = 20 \, \frac{\text{kN}}{\text{m}^3} \quad \text{below the phreatic level}$$

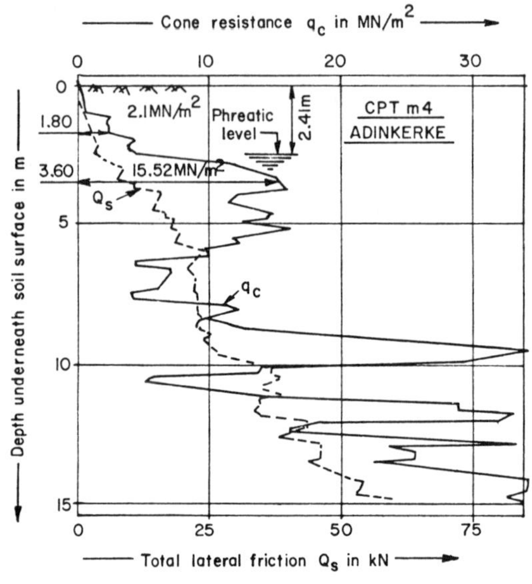

Figure 6-25. CPT diagram in a sandy formation.

Find the ultimate bearing capacity.

- For $z < 2.41$ m

$$p_t' = \gamma_n\, z$$

- For $z > 2.41$ m

$$p_t' = 2.41 \times 16 + (z - 2.41)(20 - 10) = 14.46 + 10\, z$$

$$F(\phi_n') = \frac{q_c}{p_t'}$$

- For instance, at a depth of 1.80 m

$$q_c = 2.1\,\frac{MN}{m^2} = 2{,}100\,\frac{kN}{m^2}$$

$$p_t' = 1.80 \times 16 = 28.8\,\frac{kN}{m^2}$$

$$F(\phi_n') = \frac{q_c}{p_t'} = \frac{2{,}100}{28.8} = 73$$

- The values of $F(\phi_n')$ are given versus ϕ_n' in Figure 6-26.

- For $F(\phi_n') = 73$, one gets $\phi_n' = 26°$.

- From Figure 6-25, at a depth of 3.60 m, one has

  $q_c = 15.52$ MPa

  and

$$p_t' = 14.46 + 10 \times 3.60 = 50.46 \ \frac{kN}{m^2}$$

$$F(\varphi_n') = \frac{15{,}520}{50.46} = 308$$

- Figure 6-26 gives $\phi_n' = 34°$.

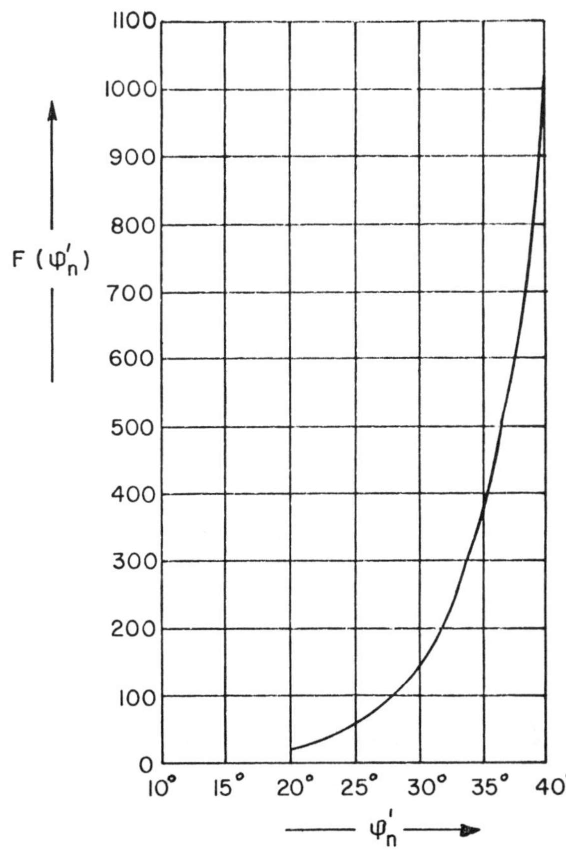

Figure 6-26. Values of F (ϕ_n') vs. ϕ_n'.

The values of ϕ_n' versus depth are given in Figure 6-27. Between 2.60 m and 5.62 m depth the ϕ_n' values are located between 32° and 34°.

Let us make a first step calculation with $\phi_n' = 32°$; $c' = 0$. One gets

$$q_r = N_q s_q p_t' + N_\gamma^* s_\gamma \gamma' \ B$$

- From Table 6-1, for $\phi_n' = 32°$

$$N_q = 18.40 + (33.30 - 18.40)\ 2/5 = 24.36$$

- From Equation 6-31

$$s_q = 1 + 0.2 \sin 32° = 1.106$$

$$p_t' = 14.46 + 10 \times 2.60 = 40.46 \ \frac{kN}{m^2}$$

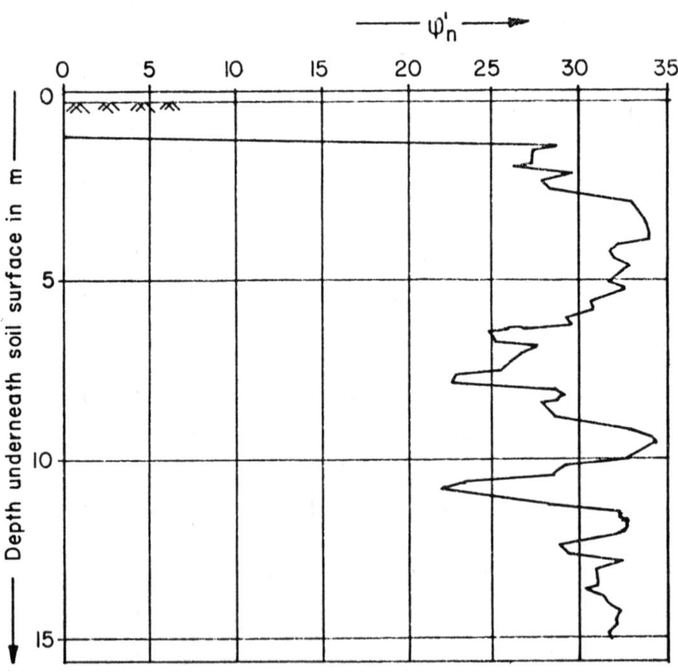

Figure 6-27. Values of φ_n' vs. depth z beneath the soil surface.

- From Table 6-2

$$\phi_n' = 32°; \ N^*_\gamma = 10.91 + (24.48 - 10.91) \ 2/5 = 16.34$$

- From Equation 6-33

$$s_\gamma = \frac{1 + 0.2 \times 0.2}{1 + 0.2} = 0.867$$

$$q_r = 24.36 \times 1.106 \times 40.46 + 16.340 \times 0.867 \times 10 \times 3$$

$$= 1{,}515 \ \frac{kN}{m^2}$$

$$= 1.534 \ MPa$$

Let us guess

$$q_r' = 2 \ MPa$$

One gets (Equation 6-103)

$$\phi''_n = \phi'_n + 5° \ \log \frac{q_c}{q_r}$$

$$q_c \cong 14 \ MPa$$

$$\phi''_n = 32° + 5° \ \log \frac{14}{2} = 36°13'$$

Because the thickness of the dense sand is only 3.00 m (Figures 6-25 and 6-27) (thus only equal to the width of the footing B = 3.00 m) and beneath the dense sand is a looser inclusion with only $\phi_n' = 26°$, ϕ_n'' will be limited to $\phi' = 35°$. One gets

- From Table 6-1, $\phi' = 35°$; $N_q = 33.30$.

- From Equation 6-31

$$s_q = 1 + 0.20 \sin 35° = 1.114.$$

- From Table 6-2

$$\phi' = 35°; \; N^*_\gamma = 24.48$$

$$q_r = 33.30 \times 1.114 \times 40.46 + 24.48 \times 0.867 \times 10 \times 3$$

$$= 1{,}500 + 637 = 2{,}137 \; kN/m^2 = 2.173 \; MPa$$

The guess for the second step of $q_r' = 2$ MPa was thus sufficiently correct.

This approximate method based on CPT tests is very useful in the case of sand layers, where it is very difficult or even impossible to take undisturbed samples.

Pressuremeter Test. The method of predicting the ultimate bearing capacity of a foundation from the results of pressuremeter tests is described by Ménard [1963]. The unit ultimate bearing capacity q_r is given by

$$q_r - p_{v,o} = k(p_\ell - p_{h,o}) \tag{6-104}$$

where p_ℓ = the total limit pressure measured in the test (irrespective if it is taken by the pore water or by the grain skeleton)

$p_{h,o}$ = the total initial natural horizontal stress in the soil at the considered depth.

$p_{v,o}$ = the total initial natural vertical stress in the soil at the considered depth.

The bearing capacity factor k depends on the nature of the soil layers involved, the shape of the foundation and on the depth of embedment. As far as the type of soil is considered, Ménard considers four categories, which are given in Table 6-8. When the foundation rests on a layer, the resistance of which varies with depth, an equivalent limit pressure $p_{\ell,e}$ is used and defined as the geometrical mean of the limit values found in the vicinity of the foundation level

$$p_{\ell,e} = (p_{\ell,1} \times p_{\ell,2} \times p_{\ell,3})^{1/3} \tag{6-105}$$

where $p_{\ell,1}$ = geometrical mean of the limit pressures measured between 1.5 B and 0.5 B above the foundation level.

$p_{\ell,2}$ = geometrical mean of the limit pressures measured 0.5 B above and 0.5 B beneath the foundation level.

$p_{\ell,3}$ = geometrical mean of the limit pressure measured between 0.5 B and 1.5 B beneath the foundation level.

Table 6-8
Soil Categories in the Pressuremeter Method*

Range of the limit pressure p kPa	Nature of the Soil	Category of the Soil
0 – 1,200	Clay	
0 – 700	Loam	I
1,800 – 4,000	Stiff clay and marl	
1,200 – 3,000	Dense loam	II
400 – 800	Loose sand	
1,000 – 3,000	Soft and weathered rock	
1,000 – 2,000	Sand and gravel	III
4,000 – 10,000	Rock	
3,000 – 6,000	Very dense sand and Gravel	III bis

* According to Ménard.

In the case of heterogeneous layers, a weighted depth of embedment D_e is introduced and is defined by

$$D_e = \frac{1}{p_{\ell,e}} \int_o^D p_\ell(z)dz \qquad (6\text{-}106)$$

in which D is the physical depth of the foundation beneath the soil surface.

In Figure 6-28 the bearing capacity factor k for shallow foundations is given versus the ratio D_e/B for the different soil categories as a parameter, for a strip footing $L/B = \infty$ and for a square footing ($L/B = 1$). In the case $1 < L/B < \infty$ the value of k is obtained by graphical interpolation between the values valid for square and strip footings (see Figure 6-28). Other diagrams of k values are given by Baguelin et al. [1978]. It must also be remarked that the semi-empirical method of the pressuremeter test does not always respect the relative influences of all parameters as appearing in the classical formulas of bearing capacity. The advantage of the semi-empirical approach of the pressuremeter test, is that the calculation starts from experimental results, which are not only influenced by the shearing strength parameters, but also by the deformability characteristics of the involved soil layers.

Numerical Example

A vertical and centrally loaded raft B = 3.0 m and L = 6.0 m is established at a depth of 2.0 m beneath the soil surface in a clay layer belonging to category I (Table 6-8). Pressuremeter tests have been performed every meter, and the limit pressures are given in Figure 6-29. The unit weight of the clay is $\gamma_n = 17 \text{kN/m}^3$, and the water table is 1.0 m beneath the surface of the soil. Find the ultimate bearing capacity.

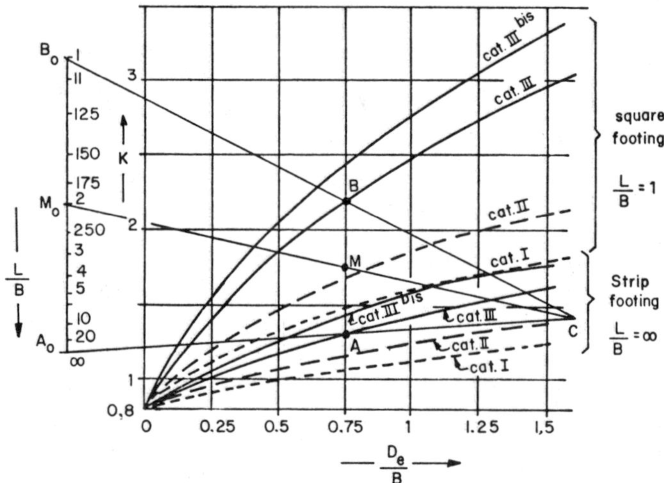

Figure 6-28. Bearing capacity factor k for shallow foundations vs. the relative embedment depth D_e/B, with the soil category as a parameter [after Ménard 1963].

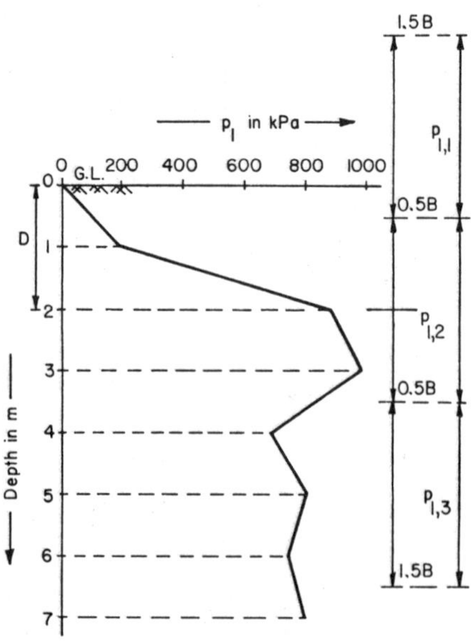

Figure 6-29. Variation of the limit pressure with depth for the case of the numerical example.

Between 0.5 B = 1.5 m and 1.5 B = 1.5 × 3.00 = 4.50 m above the foundation level, no readings exist. Therefore the geometrical mean $p_{\ell,1}$ in Equation 6-104 is not considered, and this expression is replaced by

$$p_{\ell,e} = (p_{\ell,2} \times p_{\ell,3})^{1/2}$$

$$p_{\ell,2} = (200 \times 900 \times 1{,}000)^{1/3} = 565 \text{ kPa}$$

$$p_{\ell,3} = (700 \times 820 \times 760)^{1/3} = 758 \text{ kPa}$$

$$p_{\ell,e} = (565 \times 758)^{1/2} = 654 \text{ kPa}$$

- From Equation 6-106

$$D_e = \frac{1}{654}\left(\frac{200+0}{2} \times 1.0 + \frac{900+200}{2} \times 1.0\right)$$

$$D_e = 0.993 \text{ m}$$

$$\frac{L}{B} = \frac{6.00}{3.00} = 2$$

$$\frac{D_e}{B} = \frac{0.993}{3.00} = 0.331$$

Figure 6-28 gives for $\dfrac{D_e}{B} = 0.331$

$$\frac{L}{B} = \infty; \qquad k = 0.94$$

$$\frac{L}{B} = 1; \qquad k = 1.18$$

by interpolation $\dfrac{L}{B} = 2; \qquad k = 1.06$

Hence $p_{v,o} = 17 \times 2 = 34 \text{ kPa}$
$p_{h,o} = \sigma_{\omega,o} + K_o p'_{v,o}$
$p_{h,o} = 10 + 0.5\,[17 \times 1.0 + (17 - 10) \times 1.0]$
$p_{h,o} = 22 \text{ kPa}$

- From Equation 6-104

$$q_r - 34 = 1.06 \, (654 - 22)$$

$$q_r = 34 + 670 = 704 \text{ kPa}$$

Influence of the Nonhomogeneity of the Soil

Cases of homogeneous soil conditions over the site and over a sufficient thickness are rather rare. Only for these rare cases can the results of laboratory tests on an always limited number of samples be sufficient to solve the problem.

In order to disclose heterogeneities it is necessary to perform a sufficient number of CPT tests, in which some parameters can be measured in a continuous manner with depth. One must be aware that nature loves the law of least energy, and that consequently the rupture proceeds along surfaces of weakness.

If, thanks to the data of a sufficient number of investigations, a definite pattern in the nonhomogeneity can be disclosed, then one can try to take this nonhomogeneity into account, by using one of the following procedures.

Statistical Approach. For important structures often the number of borings and test samples is increased. By way of statistics one can find which samples belong to the same family or collection. For this collection the average value and the standard deviation can be determined.

By no means is it justified to base the calculation of the bearing capacity on the average value, but on a much lower value corresponding to a given probability that no lower values will be found (compare the calculation of concrete). However, for soils the possibility of such an approach will be rather exceptional.

Case of Two Layers with Similar Shearing Strength Characteristics. When the shearing parameters of stratified layers which can influence the ultimate bearing capacity do not differ very much, the German specifications [DIN 4017, 1975] propose to make a calculation for each of the characteristic shear strength parameters, adopting the mean of the calculated values.

Case of an Intermediate Watertable. If in a homogeneous layer the water table is located beneath the foundation level but in the potential rupture zone, for the effective unit weight γ' to be used in the expression of ultimate bearing capacity, a weighted value between the unit weight γ_n valid above the water table and the effective unit weight $\gamma' = \gamma_n - \gamma_w$ beneath the watertable can be introduced.

Case of Stratified Layers with Different Strengths. When strip footings (L $\geq$ 5 B) are founded on stratified layers with different shearing strength parameters, a first limit of the ultimate bearing capacity is determined by trial and error, considering several possible rupture lines in a plane strain case and determining graphically the pressure line respecting all statical conditions. As an example, in Figure 6-30 [DIN 4017, 1975] the case is considered of a strip footing established on a layer of sand resting on a layer of soft clay. Assuming that the rupture line consists of a circle linked to straight lines with inclinations θ_1 and θ_2, by trial and error the most unfavorable rupture line with the assumed general shape has been found.

For rafts (L $\leq$ 5 B) founded on stratified layers of different strengths graphical calculations analogous to those described above for strip footings, are made, but separately for two extreme cases:

1. Calculations are made neglecting the self weight of the soil beneath the foundation level

$$\frac{p_t' + c'\cot\phi'}{\gamma'B} = \infty$$

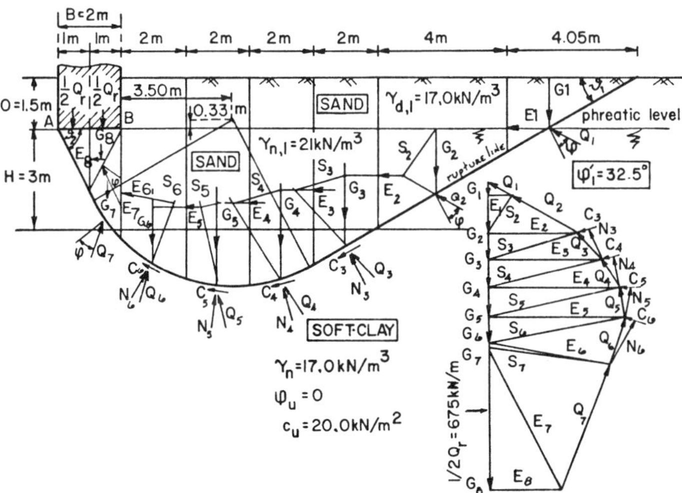

Figure 6-30. Graphical determination of ultimate bearing capacity Q$_r$ in case of stratified layers.

2. Calculations are made neglecting the influence of the overburden pressure p_t' and of the cohesion c'.

The most unfavorable rupture line is defined for each of these cases by trial and error, and the corresponding rupture loads are multiplied by the appropriate shape factors. The sum of both cases represents the load at failure.

Case of a Soft Layer Between the Footing and a Stronger Layer. In case of a soft layer with limited thickness H resting on a stronger layer, the problem of the squeezing should also be considered.

In the case of a frictionless soil ($\phi_u = 0$) the ultimate unit load is given by

$$q_{r,sq} = \left(\pi + 2 + \frac{B}{2H} - \frac{\sqrt{2}}{2} \right) c_u + p_t' \tag{6-107}$$

valid for $\dfrac{B}{2H} - \dfrac{\sqrt{2}}{2} \geqq 0$

$$q_{r,sq} = (\pi + 2) \left(1 + \frac{\dfrac{B}{2H} - \dfrac{\sqrt{2}}{2}}{\pi + 2} \right) c_u + p_t'$$

$$F_{sq} \left(\frac{B}{H} \right) = 1 + \frac{\dfrac{B}{2H} - \dfrac{\sqrt{2}}{2}}{\pi + 2} \tag{6-108}$$

Calculations have also been performed by Mandel and Salençon [1969]. The results for the case of a strip footing and $\phi_u = 0$ are expressed by

$$q_r = N_c F \left(\frac{B}{H} \right) c_u + \gamma' D \tag{6-109}$$

The values of F (B/H) and F_{sq} (B/H) are given on the second and third line of Table 6-9.

The values F_{sq} (B/H) do not differ very much from the values F (B/H). Consequently Equation 6-107 can be accepted for practical applications.

In the case of a soft layer (c', ϕ') Buisman [1940] proposed the following expression for $q_{r,sq}$ of a strip footing (L > 5B)

Table 6-9
Values of the Thickness Factors for Frictionless Soils
vs. the Width-to-Thickness Ratio

$\dfrac{B}{H}$	2	3	4	5	6	8	10
$F\left(\dfrac{B}{H}\right)^*$	1.02	1.11	1.21	1.30	1.40	1.59	1.76
$F_{sq}\left(\dfrac{B}{H}\right)^\dagger$	1.057	1.154	1.252	1.35	1.446	1.641	1.835

* Buisman
† Mandel-Salençon

$$q_{r,sq} = \frac{H}{B} K_a \frac{1}{\tan \phi'} \left(\frac{c'}{\tan \phi'} + q_{r,o}\right) \left(e^{\frac{B}{H} \frac{\tan \phi'}{K_a}} - 1\right)$$

$$- \frac{c'}{\tan \phi'} \tag{6-110}$$

$$q_{r,o} = \frac{1}{K_a}\left[p_t' + 2\,c'\,\tan\left(\frac{\pi}{4} + \frac{\phi'}{2}\right)\right] + 2\,c'\,\tan\left(\frac{\pi}{4} + \frac{\phi'}{2}\right) \tag{6-111}$$

$$K_a = \frac{1 - \sin \phi'}{1 + \sin \phi'} \tag{6-112}$$

Mandel and Salençon [1969] treated the problem of the bearing capacity of a strip footing with the method of the characteristics. They expressed the results as

$$q_r = N_q\,F_q\left(\frac{B}{H}\right) p_t' + N_c\,F_c\left(\frac{B}{H}\right) c' + N_\gamma^*\,F_\gamma\left(\frac{B}{H}\right) \gamma'\,B \tag{6-113}$$

The values of the thickness factors F (B/H) are given in Table 6-10.

Numerical Example

Consider: B = 6.00 m; L = 30.00 m; D = 1.00 m and the water table at the foundation level. Clay layer H = 2.00 m; γ_n = 18 kN/m³; c' = 10 kN/m²; ϕ' = 20°.

Table 6-10
Thickness Factor for Soil with Friction in Function of B/H and the Friction Angle*

ϕ'	$\dfrac{B}{H}$	1	2	3	4	5	6
0°	$F_q\left(\dfrac{B}{H}\right)$	1	1	1	1	1	1
	$F_c\left(\dfrac{B}{H}\right)$	1	1.02	1.11	1.21	1.30	1.40
	$F_\gamma\left(\dfrac{B}{H}\right)$	1	1	1	1	1	1
10°	$F_q\left(\dfrac{B}{H}\right)$	1	1.07	1.21	1.37	1.56	1.79
	$F_c\left(\dfrac{B}{H}\right)$	1	1.11	1.35	1.62	1.95	2.33
	$F_\gamma\left(\dfrac{B}{H}\right)$	1	1	1	1	1.01	1.04
20°	$F_q\left(\dfrac{B}{H}\right)$	1	1.33	1.95	2.93	4.52	7.14
	$F_c\left(\dfrac{B}{H}\right)$	1.01	1.39	2.12	3.29	5.17	8.29
	$F_\gamma\left(\dfrac{B}{H}\right)$	1	1	1.07	1.28	1.63	2.20

* According to Mandel-Salençon.

The clay layer rests on a limestone formation. Find the ultimate bearing capacity.

$$p_t' = \gamma_n D = 18\ \frac{kN}{m^2}$$

$$\frac{L}{B} = \frac{30}{6} = 5$$

- Formula of Buisman:

$$K_a = \frac{1 - \sin 20°}{1 + \sin 20°} = \frac{1 - 0.342}{1 - 0.342} = 0.490$$

$$\tan\left(45° + \frac{20°}{2}\right) = 1.428$$

From Equation 6-111:

$$q_{r,o} = \frac{1}{0.490}\,(18 + 2\times 10\times 1.428) + 2\times 10\times 1.428 = \times 123.6\,\frac{kN}{m^2}$$

$$q_{r,sq} = \frac{2.0}{6.0}\,0.490\,\frac{1}{\tan 20°}\left(\frac{10}{\tan 20°} + 123.6\right)\left(e^{\frac{6.0}{2.0}\,\frac{\tan 20°}{0.49}} - 1\right)$$

$$- \frac{10}{\tan 20°}$$

$$= 534.2\,\frac{kN}{m^2}$$

- Formula of Mandel:

$$\phi' = 20°$$

From Table 6-1; $N_q = 6.40$

From Equation 6-11; $N_c = \dfrac{6.40 - 1}{\tan 20°} = 14.8$

From Table 6-2; $N_\gamma^* = 2.4S$

From Table 6-10; $\dfrac{B}{H} = 3$

$$F_q\left(\frac{B}{H}\right) = 1.95$$

$$F_c\left(\frac{B}{H}\right) = 2.12$$

$$F_\gamma\left(\frac{B}{H}\right) = 1.07$$

From Equation 6-113

$$q_r = 6.40 \times 1.95 \times 18 + 14.84 \times 2.12 \times 10$$

$$+\ 2.48 \times 1.07 \times (18 - 10) \times 6.00 = 667\ \frac{kN}{m^2}$$

Considering that in Equation 6-110 the weight of the squeezed soil is not taken into account, the numerical example shows that Equations 6-110 of Buisman and Equation 6-113 of Mandel give the same order of magnitude for q_r.

Punching Through a Strong Layer Resting on a Soft Layer. In the case of a strong layer resting on a soft layer, the possibility of the punching phenomenon is to be considered, as often this rupture pattern will necessitate less energy than the phenomenon of lateral expulsion.

In the case of a strong cohesionless layer resting on cohesive soil without friction, the ultimate bearing capacity, q_r, corresponding to the punching hypothesis, is according to Meyerhof [1974], given by

$$q_r = \left(1 + 0.2\ \frac{B}{L}\right) N_c\ c_u + \left(1 + \frac{B}{L}\right) \gamma_1'\ H^2 \left(1 + 2\ \frac{D}{H}\right) K_s\ \frac{\tan \phi_1'}{B}$$

$$+\ \gamma_1' D \tag{6-114}$$

where ϕ_1' = angle of shearing strength of the cohesionless layer
$\quad\quad c_u$ = undrained shearing strength of the frictionless layer
$\quad\quad H$ = thickness of the cohesionless layer under the foundation level
$\quad\quad D$ = embedment depth of the foundation
$\quad\quad K_s$ = punching shear coefficient. This punching shear coefficient is a function of the angle ϕ_1' and is given in Figure 6-31.

Numerical Example

A rectangular footing $B = 3.00$ m, $L = 6.00$ m, vertically and axially loaded is established at a depth $D = 1.00$ m beneath the soil surface in a dense sand layer

$$\phi_1' = 35°; \quad \gamma_{d,1} = 16\ \frac{kN}{m^3}; \quad \gamma_{n,1} = 20\ \frac{kN}{m^3}$$

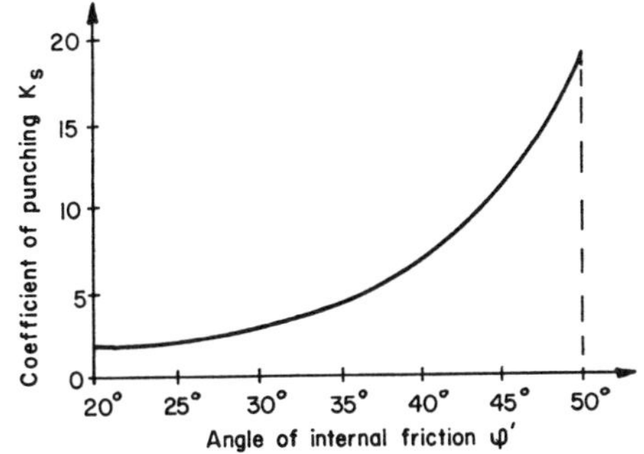

Figure 6-31. Coefficient of punching K_s vs. the angle ϕ_1' of the punched layer [after Meyerhof, 1974].

The thickness of the sand layer beneath the foundation level is $H = 2.50$ m. The sand layer is underlaid by a thick layer of soft clay, $c_u = 30\text{kPa}$, $\phi_u = 0$. The water table is found at foundation level. Determine the load at which the soil will fail in punching shear.

$$1 + 0.2\,\frac{B}{L} = 1 + 0.2\,\frac{3.0}{6.0} = 1.1$$

$$1 + \frac{B}{L} = 1 + \frac{3.0}{6.0} = 1.5$$

$$1 + 2\,\frac{D}{H} = 1 + 2\,\frac{1.0}{2.5} = 1.8$$

- From Figure 6-31, with $\phi_1' = 35°$

$$K_s = 4.4$$

- From Equation 6-114

$$q_r = 1.1 \times 5.14 \times 30 + 1.5(20 - 10)\overline{2.5}^2$$

$$\times 1.8 \times 4.4 \times \frac{\tan 35°}{3.00} + 16 \times 1.0$$

$$q_r = 169.6 + 173.3 + 16 = 358.9 \text{ kPa}$$

$$Q_r = 0.3589 \times 3.0 \times 6.0 = 6.46 \text{ MN}$$

Case of a Soft Frictionless Layer with Limited Thickness Resting on a Frictionless Layer with a Somewhat Larger Shearing Strength. The Equation 6-107 is based on the assumption that the layer supporting the soft layer has a much larger shearing strength than the squeezed soft layer and that the foundation resting on the soft layer is very rigid.

The case of a soft frictionless layer with cohesion c_1, resting on a stronger frictionless layer with cohesion c_2 has been treated by Vesic [1975]. Based on the results of Vesic, we propose the following formula, valid for rectangular footings

$$q_r = \left(1 + 0.2\,\frac{B}{L}\right)\left[\pi + 2 + \left(1 - \frac{c_1}{c_2}\right)\frac{\dfrac{B}{H} - \sqrt{2}}{2\left(\dfrac{B}{L} + 1\right)}\right]c_1 + p_t' \tag{6-115}$$

For the case $c_2 = \infty$, and a strip footing $B/L = 0$, Equation 6-107 is retrieved. For the case $c_2 = \infty$ and a square footing $B/L = 1$, Equation 6-11 becomes

$$q_{r,sq} = 1.2\left[\pi + 2 + \frac{1}{4}\left(\frac{B}{H} - \sqrt{2}\right)\right]c_1 + p_t' \tag{6-116}$$

The unit squeezing resistance of a square footing will be larger than that of a strip footing with the same width, when

$$1.2\left[\pi + 2 + \frac{1}{4}\left(\frac{B}{H} - \sqrt{2}\right)\right] > \pi + 2 + \frac{1}{2}\left(\frac{B}{H} - \sqrt{2}\right) \tag{6-117}$$

when $B/H < 6.56$

$$q_{r,sq,square} > q_{r,sq,strip} \tag{6-118}$$

The proposed formulas are a simplification of the proposals of Vesic [1975]. They show that for a given relative thickness B/H the squeezing resistance is a function of the shape ratio B/L. In case of a soft layer resting on an infinitely resistant layer ($c_2 = \infty$), the formula shows that as long as the thickness H of the soft layer is larger than $H > 0.152\,B$, the squeezing resistance for a square footing is larger than for a strip footing. For $H = 0.152\,B$ both resistances are equal. For $H < 0.152\,B$ the unit squeezing resistance of the strip footing becomes larger than that of the square footing.

Case of a Soft Layer (c_1, $\phi_1' = 0$) Resting on a Stronger Layer with Friction and Cohesion (c_2', ϕ_2'). To that case a tentative formula is

$$
q_{r,sq,rectangle} = p_t' + c_1 \left(1 + 0.2\,\frac{B}{L}\right) \left\{ \pi + 2 \right.
$$

$$
\left. + \frac{\dfrac{B}{H} - \sqrt{2}}{2\left(\dfrac{B}{L}+1\right)} \left[1 - \frac{c_1}{(c_2' + p_{t,2}' \tan \phi_2')\,\tan^2\left(\dfrac{\pi}{4}+\dfrac{\phi'}{2}\right)}\right] \right\}
$$

$$(6\text{-}119)$$

Combination of Punching and Squeezing. In the case of a strong cohesionless layer with limited thickness, H_1, resting on a thin soft frictionless layer with thickness H_2, supported by a thick strong layer (Figure 6-32) a combination of punching and squeezing can occur.

The combination of the Equation 6-119 and 6-114 gives

$$
q_r = \left(1 + \frac{B}{L}\right)\gamma_1'\,H_1^2\left(1 + 2\,\frac{D}{H_1}\right)K_s\,\frac{\tan \phi_1'}{B} + \gamma_1'\,D
$$

$$
+ \left(1 + 0.2\,\frac{B}{L}\right)\left\{\pi + 2 + \left[1 - \frac{c_{u,2}}{(c_3' + p_{t,3}'\tan \phi_3')\,\tan^2\left(\dfrac{\pi}{4}+\dfrac{\phi_3'}{2}\right)}\right]\right.
$$

$$
\left. \frac{\dfrac{B}{H_2} - \sqrt{2}}{2\left(\dfrac{B}{L}+1\right)}\right\}\,c_{u,2}
$$

$$(6\text{-}120)$$

Numerical Example

Consider:

$$
c_{u,2} = 36.4\,\frac{kN}{m^2} << (c_3' + p_{t,3}'\tan\phi_3')\,\tan^2\left(\frac{\pi}{4}+\frac{\phi_3'}{2}\right)
$$

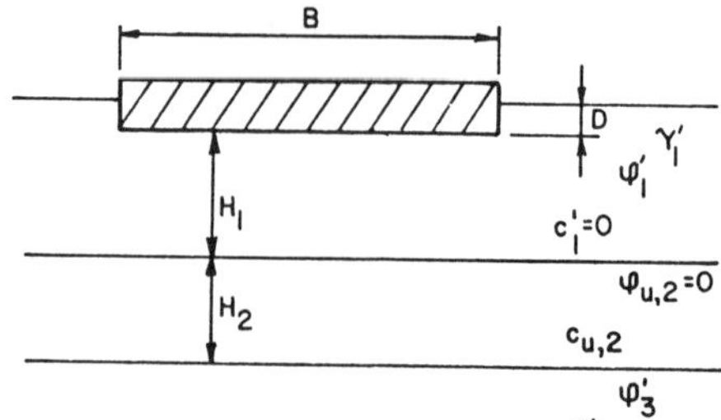

Figure 6-32. Combination of punching and squeezing.

Circular raft $B/L = 1$; $D = 0$; $\phi_1' = 30°$; $\tan \phi_1' = 0.577$; $B = 45.50$ m; $H_1 = 15.00$ m; $H_2 = 7.20$ m; $\gamma_1' = 10$ kN/m³.

Find the ultimate bearing capacity.

- From Figure 6-31 with $\phi_1' = 30°$; $K_s = 2.8$.
- From Equation 6-120:

$$q_r = 1.2 \left[5.14 + \frac{1}{4} \left(\frac{45.5}{7.20} - 1.41 \right) \right] 36.4$$

$$+ 2 \times 10 \times 15^2 \, \frac{2.8 \times 0.577}{45.5}$$

$$q_r = 438 \, \frac{\text{kN}}{\text{m}^2}$$

Dynamic Loading

The influence of the rate of loading has been clearly presented and analyzed by Vesic [1965–1975]. In the case of dynamic loading inertia forces and viscous resistances get into action. Consequently the resistance to impact loads of very short duration in dense sands and stiff clays is much larger than under static loading conditions. Furthermore the rupture pattern is completely different. In impact loadings the soil fails in punching shear, the inertia forces playing the same role as an overburden pressure. Consequently the presented formulas are not directly applicable to such impact conditions.

Problems with vibrating loads require a dynamic analysis, and cannot be completely solved by static formulas.

Vesic has shown by model tests that the ultimate bearing capacity of dense sands first decreases slightly with increasing, but still moderate, loading velocity, goes through a minimum, and then increases with increasing loading velocity.

With model tests with footings B = 100 mm placed on dense sand Vesic has found that the ultimate bearing capacity reaches a minimum when the penetration velocity is about 10^{-1} (mm/sec) (Figure 6-33). However, one ignores how this finding has to be transposed to the case of footings with larger dimensions.

For the case of stiff clays no direct experimental results concerning the variation of the ultimate bearing capacity with the rate of loading exist. There exist however data concerning the variation of the shearing strength with the rate of deformation in laboratory tests. The shearing strength increases with increasing rate of deformation. However, for deformation rates not exceeding 1 mm/sec the increase in shearing strength is small, and is not larger than 10%. For deformation rates larger than 1 mm/sec, the shearing strength starts to increase much more rapidly reaching by impact loading values twice as large as those reached by very slow static loading. Again one ignores how these data are to be transposed to real problems with much larger dimensions.

Safety Factors

Having determined the ultimate bearing capacity, a first limitation of the allowable load is obtained, by imposing a certain safety margin with respect to the ultimate bearing capacity.

This safety margin is needed for the following reasons:

1. The acting forces are not exactly known, i.e., nor their maximum value, nor their point of application, nor their inclination.

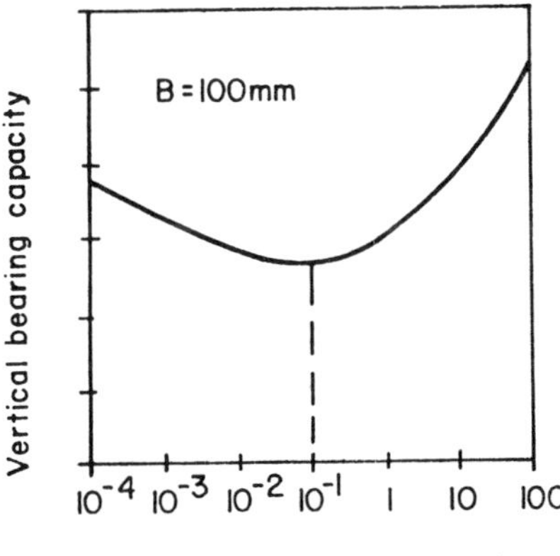

Figure 6-33. Effect of loading velocity on bearing capacity of dense sand beneath a surface footing.

2. The calculation methods at disposal are based on simplified models, which do not exactly correspond to reality.
3. The shearing and deformability characteristics of the involved soil layers are not exactly known.

Two different ways are used to solve the problem.

1. The introduction of a total safety factor, which is more or less a traditional approach.
2. The introduction of partial safety factors, introducing the principles of probability.

Total Safety Factors.

Definition. The traditional approach consists in dividing the calculated ultimate bearing capacity by a safety factor, in order to obtain a first limitation against the ultimate bearing capacity. However, let us analyze this way of doing so.

In order to perform a calculation one must dispose of the thickness of the different layers involved, their shearing and deformability characteristics, and know the eccentricity and the inclination of the applied load.

Concerning the characteristics of the soil in most cases the laws of probability do not work. Furthermore for rupture phenomena, weak inclusions play a very important role (law of minimum energy). Disposing of all data, geology, field tests, and laboratory tests, over the whole building site and at a sufficient depth, taking into account the scale effect and the progressivity of the rupture phenomena, values of the shearing strength and deformability characteristics are to be chosen, which are located on the safe side. These values generally cannot be obtained only by a statistical analysis. Here also the judgment of an experienced engineer is needed. On the chosen values no other factors of safety are to be applied. The safety is included in the choice of the soil characteristics and this choice depends on the complexity of the soil structure, its variability, the available amount of field and laboratory data and the difficulties of their interpretation and use.

The imprecision concerning the acting forces and the imperfection of the calculation methods are covered by the introduction of adequate safety factors.

The acting forces are known by the values of their resultant R, in which the static water pressures on the foundation AA$'$, its eccentricity e, and inclination angle δ are included (Figure 6-34). The values of R, e (or u), and δ are only known between certain limits.

For given δ and u the vertical component Q_r of the ultimate bearing capacity is given by a formula

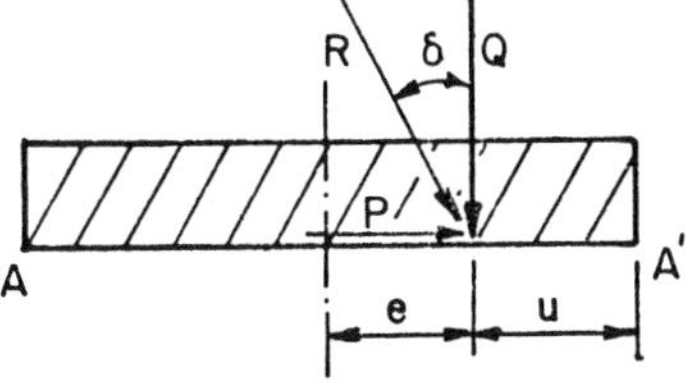

Figure 6-34. Footing loaded by an eccentric and inclined load.

$$Q_r = 2uLF(\delta, u, \ldots) \qquad (6\text{-}121)$$

and a first safety margin is

$$F_Q = \frac{Q_r}{Q} \qquad (6\text{-}122)$$

However, one can also determine for which eccentricity e_{max} (or u_{min}) the given value Q by given δ corresponds to the ultimate bearing capacity $Q_{r,u_{min}}$. The value u_{min} is given by

$$Q = 2u_{min}LF(\delta, u_{min}, \ldots) \qquad (6\text{-}123)$$

a second safety margin is

$$F_u = \frac{u}{u_{min}} \qquad (6\text{-}124)$$

Finally, one can determine by given e (or u), for which inclination δ_{max} the given value Q corresponds to the ultimate bearing capacity $Q_{r,\delta_{max}}$. The value δ_{max} is given by

$$Q = 2uLF(\delta_{max}, u, \ldots) \qquad (6\text{-}125)$$

The difference $\delta_{max} - \delta$ must be larger than the imposed safety value ϵ. One can define a safety factor F_δ:

$$F_\delta = \frac{\tan(\delta_{max} - \delta)}{\tan \epsilon} > 1 \qquad (6\text{-}126)$$

This safety factor must be larger than 1.

Numerical Example

Consider the bridge abutment of Figure 6-35: $c' = 0$; $\phi' = 30°$; watertable at great depth; $\gamma_n = 17$ kN/m^3; $D = 1$ m. Determine the factor of safety against bearing capacity failure.

$$p_t' = 1 \times 17 = 17 \ \frac{kN}{m^2}$$

$$B = 5.00 \text{ m}; \ L = 15.00 \text{ m}; \ e = 1.00 \text{ m}; \ u = 1.50 \text{ m}; \ \delta = 10°; \ \frac{B}{L} = \frac{1}{3}.$$

- From Table 6-6: $\delta = 10°$; $\phi' = 30°$;

$$\frac{u_{cr}}{B} = 0.39.$$

$$u_{cr} = 0.39 \times 5.00 = 1.95 \text{ m}$$

$$u = 1.50 \text{ m} < u_{cr} = 1.95 \text{ m}$$

- From Table 6-5: $\delta = 10°$; $\phi' = 30°$; $i_{q,u_{cr}} = 0.79$.

- From Table 6-1: $\phi' = 30°$; $N_q = 18.4$

- From Equation 6-31:

$$s_q = 1 + \frac{2u}{L} \sin \phi' = 1 + \frac{2 \times 1.5}{15} \sin 30° = 1.10$$

- From Table 6-2: $\phi' = 30°$; $\delta = 10°$.

$$N^*_{\gamma,\delta} = 5.76$$

- From Equation 6-33:

$$s_\gamma = \frac{1 + 0.2 \dfrac{2u}{L}}{1 + \dfrac{2u}{L}} = \frac{1 + 0.2 \dfrac{3}{15}}{1 + \dfrac{3}{15}} = 0.87$$

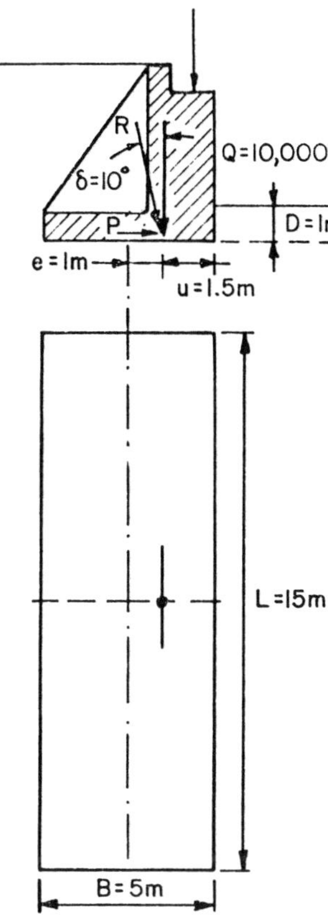

Figure 6-35. Bridge abutment.

$$Q_r = 2\,u\,L\,(N_q s_q i_{qu_{cr}}\,p_t' + N^*_{\gamma,\delta}\,s_\gamma\,\gamma'\,2\,u)$$

$$Q_r = 2 \times 1.50 \times 15.00\,(18.4 \times 1.10 \times 0.79 \times 17$$
$$+\, 5.76 \times 0.87 \times 17 \times 2 \times 1.50)$$
$$= 45 \times 527 = 23{,}732 \text{ kN}$$

$$F_Q = \frac{Q_r}{Q} = \frac{23{,}732}{10{,}000} = 2.37$$

$$Q = 2\,u_{min}\,L(N_q s_q i_{q,u_{cr}} p_t' + N^*_{\gamma,\delta} s_\gamma \gamma'\,2u_{min})$$

At first we neglect the influence of u on s_q and s_γ

$$(2\,u_{min})^2\,L\,N^*_{\gamma,\delta} s_\gamma \gamma' + 2\,u_{min}\,L\,N_q\,i_{q,u_{cr}}\,p_t'\,s_q - Q = 0$$

$$A = L\ N^*_{\gamma,\delta} s_\gamma \gamma' = 15.00 \times 5.76 \times 0.87 \times 17 = 1{,}278$$

$$B = L\ N_q\ i_{q,u_{cr}}\ p'_t\ s_q = 15.00 \times 18.4 \times 0.79 \times 17 \times 1.10 = 4{,}077$$

$$2\ u_{min} = \frac{-4{,}077 + \sqrt{4.077^2 \times 10^6 + 4 \times 1{,}278 \times 10 \times 10^6}}{2 \times 1{,}278}$$

$$2\ u_{min} = 1.64\ \text{m}\ ;\ u_{min} = 0.825\ \text{m}$$

$$s'_q = 1 + \frac{2\ u_{min}}{L}\ \sin\phi' = 1 + \frac{0.81}{15} \times 0.5 = 1.05$$

$$s'_\gamma = \frac{1 + 0.2\,\dfrac{1.62}{15}}{1 + \dfrac{1.62}{15}} = 0.922$$

$$A' = 1{,}278\,\frac{0.922}{0.87} = 1{,}354$$

$$B' = 4{,}077\,\frac{1.05}{1.1} = 3{,}892$$

$$2\ u_{min} = \frac{-3{,}892 + \sqrt{3{,}892^2 + 4 \times 1{,}354 \times 10{,}000}}{2 \times 1{,}354}$$

$$2\ u_{min} = 1.64\ \text{m} \hspace{4cm} u_{min} = 0.825\ \text{m}$$

Let us take the mean

$$u_{min} = \frac{0.81 + 0.82}{2} = 0.815\ \text{m}$$

$$F_u = \frac{u}{u_{min}} = \frac{1.50}{0.815} = 1.84$$

The third safety margin has to be determined by trial and error.

- For $u = 1.50$ m: $\delta = 10°$, one has $Q_r = 23{,}732$ kN

- For $u = 1.50$ m: $\delta = \phi' = 30°$, one has $Q_r = 0$

- For $u = 1.50$ m: $\delta = 20°$, one has from Table 6-6

$$\frac{u_{cr}}{B} = 0.38$$

$u_{cr} = 0.38 \times 5.00 = 1.90$ m

$u = 1.50$ m $< u_{cr} = 1.90$ m

- From Table 6-5: $i_{q,u_{cr}} = 0.50$

- From Table 6-2: $N^*_{\gamma,\delta} = 2.12$

$$Q_r = 2 \times 1.50 \times 15.00 \,(18.4 \times 0.50 \times 17 \times 1.10$$
$$+ \, 2.12 \times 0.87 \times 17 \times 2 \times 1.50)$$
$$= 45 \times 266.1 = 11{,}975 \text{ kN}$$

- For $u = 1.50$ m; $\delta = 25°$, one has from Table 6-6:

$$\frac{u_{cr}}{B} = 0.41$$

$u_{cr} = 0.41 \times 5.00 = 2.05$ m

$u = 1.50$ m $< u_{cr} = 2.05$ m

- From Table 6-5: $i_{q,u_{cr}} = 0.35$

- From Table 6-2: $N^*_{\gamma,\delta} = 0.91$

$$Q_r = 45 \,(18.4 \times 0.35 \times 17 \times 1.10 + 0.91 \times 0.87 \times 17$$
$$\times \, 2 \times 1.50)$$

$$Q_r = 45 \times 160.8 = 7{,}236 \text{ kN}$$

- By interpolation (Figure 6-36) one gets for $Q_r = 10{,}000$ kN $\delta_{max} = 21.6°$

$$\delta_{max} - \delta = 21.6° - 10 = 11.6°$$

$$\tan(\delta_{max} - \delta) = \tan 11.6° = 0.205$$

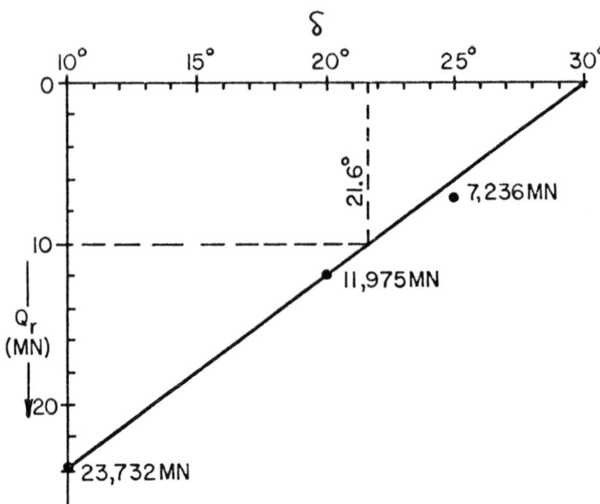

Figure 6-36. Interpolation curve for the determination of δ_{max} in the numerical example.

$$F = \frac{\tan (\delta_{max} - \delta)}{\tan \epsilon}$$

if for instance $\epsilon = 9°$; $\tan \epsilon = 0.1584$

$$F_\delta = \frac{0.205}{0.1581} = 1.29$$

Required Safety Factors. The safety factors which should be required, depend on several parameters:

1. The loading conditions: normally occurring loads, exceptional loading conditions, catastrophic loading conditions.
2. The more or less thoroughness of the soil exploration.
3. The more or less perfection of the used calculation method.

Loading conditions.

L_1: Static loads and frequently occurring transient loads (including winds).
L_2: Same loads as under L_1, but also includes irregularly occurring heavy mobile loads and loads only occurring during the building period.
L_3: In addition to the loads under L_2, includes exceptional loads that only occur from deficiency of safety measures and catastrophic loads. Dynamic loading conditions (earthquakes), however, also necessitate a dynamic analysis of the problem.

Thoroughness of soil exploration.

E_1: Exploration at distances not larger than 20 m, and to sufficient depth underneath the foundation level (larger than 3 times the width of the footing and 1.5 times the width of the building) by means of field tests (SPT, CPT, pressuremeter) and laboratory tests.

E_2: Investigation not fulfilling all conditions corresponding to E_1.

Calculation methods.

C_1: Calculation method based on the results of laboratory tests:

$C_{1,a}$: Without taking into account the influence of compressibility and progressivity.

$C_{1,b}$: Taking into account the influence of compressibility and progressivity.

C_2: Calculation methods based on the results of CPT tests.

C_3: Calculation methods based on the results of pressuremeter tests.

The proposed safety factors are given in Table 6-11.

The values obtained when applying the proposed safety factors are only a first limitation of the allowable load. A second limitation concerns the tolerable soil deformations, and will be considered later.

Partial Safety Factors. Lundgren and Hansen [1965] proposed the partial factors of safety mentioned in Table 6-12.

Case L_2 covers the cases of exceptional loading conditions during the building period, temporary structures, and exceptional loading combinations.

The strength factors are introduced for taking into account the imperfection of the soil exploration and of the calculation method. These partial factors are only the equivalent for the factor F_q of Table 6-11. Furthermore they must be considered to correspond to the case E_1 of the thorough soil exploration and the case $C_{1,b}$ of calculation taking into account the influence of compressibility and progressivity.

For the case of limited exploration E_2 the strength factors of Table 6-12 are to be multiplied by 1.10. However, when the influence of the compressibility is not taken into account, the factors are multiplied by 1.2 (Table 6-13).

Especially in layered soils the introduction of lower, fictive values of $\tan \phi'$, has an influence on the shape of the potential rupture lines, and therefore can give misleading results. This is one of the reasons we are reluctant to use the partial safety factors. Except for the rare cases of soils that are sufficiently homogeneous to allow the use of a statistical

Table 6-11
Safety Factors

Loading Conditions	Soil Exploration	Calculation Methods	F_q and F_δ	ϵ	$F_g \epsilon$
L_1 Normal	E_1 Thorough	$C_{1,a}$ and C_3 Rigid; pressuremeter	2.75	9°	0.16
		$C_{1,b}$ and C_2 Compression; CPT	2.25	7°	0.12
	E_2 Limited	$C_{1,a}$ and C_3 Rigid; pressuremeter	3.00	10°	0.18
		$C_{1,b}$ and C_4 Compression; CPT	2.50	7°	0.12
L_2 Including exceptional loads	E_1 Thorough	$C_{1,a}$ and C_3 Rigid; pressuremeter	2.50	7°	0.12
		$C_{1,b}$ and C_2 Compression; CPT	2.00	5°	0.09
	E_2 Limited	$C_{1,a}$ and C_3 Rigid; pressuremeter	2.75	9°	0.16
		$C_{1,b}$ and C_2 Compression; CPT	2.25	7°	0.12
L_3 Including catastrophic situations	E_1 Thorough	$C_{1,a}$ and C_3 Rigid; pressuremeter	2.25	7°	0.12
		$C_{1,b}$ and C_2 Compression; CPT	1.75	5°	0.09
	E_2 Limited	$C_{1,a}$ and C_2 Rigid; pressuremeter	2.50	8°	0.14
		$C_{1,b}$ and C_2 Compression; CPT	2.00	6°	0.11

approach, the values of c' and $\tan \phi'$ have to be more or less chosen by judgment. In such a case the partial strength factors have only the aim to cover the imprecision of the calculation methods, should it not be preferable to include them in the load factors. In that case one comes back to the first method.

PROBLEM OF DEFORMATIONS

General Considerations

The second problem to be solved for a foundation design consists of determining if the soil deformations to be expected are tolerable for the stability, the function, and the aspect of the superstructure.

It must first be stressed that the ultimate bearing capacities, as considered earlier are generally only reached by high values of the relative settlements s_r/B.

Tests performed by Muhs [1957] and Leussink et al. [1966] on footings of 1 m width show that depending on the relative density, the relative settlements needed to reach the rupture state represent easily between 5% and 10%. For footings with a width of 2.00 m, this leads to settlements of 100 to 200 mm. The tolerable settlements are generally much less than these values. Consequently it is not sufficient to have a margin of safety against the calculated ultimate bearing capacity for the reasons given in the section on safety factors, but one must also be able to ensure that deformations are still acceptable under the expected load conditions.

Table 6-12
Load and Strength Factors

	Case L_1 (Normal)	Case L_2 (Exceptional)
Load factors		
Dead load	1	
Static water pressure	1	
Fluctuating water pressure	1.20	1.10
Live loads (general)	1.50	1.25
Wind loads	1.50	1.25
Earth pressure	1.20	1.10
Strength factors		
Cohesive C'	2.00	1.80
Tan ϕ'	1.20	1.10

Table 6-13
Partial Strength Factors

Loading Conditions	Soil Exploration	Calculation Methods	F_c	$F_{\tan \phi'}$
L_1 Normal	E_1 Thorough	C_1 and C_3 Rigid; pressuremeter	2.40	1.45
		$C_{1,b}$ and C_2 Compression; CPT	2.00	1.20
	E_2 Limited	$C_{1,a}$ and C_3 Rigid; pressuremeter	2.65	1.60
		$C_{1,b}$ and C_2 Compression; CPT	2.20	1.32
L_2 Exceptional	E_1 Thorough	$C_{1,a}$ and C_3 Rigid; pressuremeter	2.15	1.32
		$C_{1,b}$ and C_2 Compression; CPT	1.80	1.10
	E_2 Limited	$C_{1,a}$ and C_3 Rigid; pressuremeter	2.37	1.45
		$C_{1,b}$ and C_2 Compression; CPT	1.98	1.20

In order to be able to predict the soil deformations one has to know the distribution of stresses with depth and the deformation parameters of the soil.

The deformability properties of a soil skeleton are very complex. This is also the case for the exact stress distribution in such a skeleton. A soil skeleton is composed of independent particles, which cannot transmit tensile stresses. The real stresses in the contacts are very different from the so-called "effective stresses," which are in fact fictive values. In soils the statistical values of the contact stresses can be 20 times larger than the "effective stress."

The simplest way to make a mathematical approach should of course be to assimilate the soil skeleton to a perfectly linear elastic body, characterized by the elastic parameters E_s and ν_s. If this should really be the case, one could use not only the formulas established by Boussinesq for the stresses but also those established for the strains. However, experience has clearly shown that the soil does not behave as predicted by all of Boussinesq's formulas. The same experience shows that the distribution of the vertical normal stresses do not differ too much from the values predicted by Boussinesq but that this is not the case for the stresses in the horizontal directions, and for the shearing stresses. A very large difference exists between the deformations predicted by the formulas given by Boussinesq for the strains and actual ones. Numerical methods based on perfectly linear elastic models can only be of limited help for solving that problem. They must inevitably give Boussinesq's answer. Of course one could always "adapt" the values introduced for E_s and ν_s for obtaining correspondence between computation and reality for a certain point, but one has to accept divergences in other points.

Only numerical methods based on nonlinear and inelastic behavior of the soil, taking into account the relationship between the incremental stress field and deformation field as obtained in laboratory tests on representative perfectly undisturbed samples, could give a nearly correct answer. As, however, this second condition is never fully satisfied, the numerical method can give only a trend, and a very good means to determine the influence of each of the parameters. Furthermore numerical methods based on the real laws between stresses and deformations, especially in three-dimensional cases, require the use of very powerful computers.

Lambe [1967] advocates that the deformations could be exactly predicted by subjecting in a triaxial test the undisturbed soil sample to the stress-path it will experience in the field, and by measuring the deformations. However this method, presupposes that the incremental stress field should be known at each point of the soil.

A further difficulty is the following: a soil layer is never homogeneous neither vertically nor in a horizontal direction. The presence of less resistant spots is very detrimental for the ultimate bearing capacity, and this detrimental effect increases with the dimensions of the footing, and thus the volume of soil involved in the bearing capacity problem. This is the reason the ultimate bearing capacity is not governed by the mean of the values of "undisturbed" samples taken at random in the soil layer, but by a much lower value. This value becomes lower, the larger the dimensions of the footing become.

The contrary is true for the deformation problem. Indeed if more deformable spots are present, which are distributed at random, the movements of the footing will not be governed by the mean value of the deformation characteristics determined on "undisturbed" samples taken at random in the layer, but by the values of the samples taken in the less deformable spots. Since the tendency exists to calculate the settlement with the mean of the deformability characteristics, or even, for safety, with a value lower than the mean, settlements larger than the real ones would be computed.

The general routine is to perform oedometer tests on the soil samples. Already Terzaghi [1943] has shown that for normally consolidated clays and virgin loading, the final strain $\epsilon_{z,f}$ by prevented lateral deformation varies with the log of the stress increase. In a general way one can write

$$\epsilon_{z,f} = \frac{\Delta h}{h} = \frac{1}{C} \ln \frac{p_t + \Delta\sigma_z + p_c}{p_t + p_c} \qquad (6\text{-}127)$$

where C = constant of compressibility (dimensionless)
   p_c = a constant with the dimensions of a stress

Between the constant of compressibility C and the oedometer or bulk modulus M for virgin loading exists the relationship

$$M = C(p_t + p_c) \qquad (6\text{-}128)$$

For normally consolidated clays, the value of the stress constant p_c is very small and can generally be neglected. On the contrary for sands the value of p_c cannot be neglected (order of magnitude, for instance, 20 kN/m^2).

On the other hand, Janbu [1963] gives for the case of sands, for the oedometer or bulk modulus, the expression

$$M = \frac{\sigma_1}{a\,(1 - w)} \left(\frac{\sigma}{\sigma_1}\right)^w \qquad (6\text{-}129)$$

For sands, most experiments give

$$w = 0.5$$

$$0.32 < w < 0.794$$

By reloading one should always have

$$w = 0.5$$

The symbol "a" represents the strain when the sample is loaded from zero to the arbitrarily chosen reference stress σ_1.

Restricted Meaning of the Oedometer Tests

Concerning the meaning of the deformation parameters obtained from oedometer tests for solving the real deformation problems, some very valuable tests performed by Kérisel [1963] can be mentioned. A sand with a given initial relative density is subjected in a triaxial apparatus to a spherical stress tensor σ_m. Subsequently the vertical stress σ_1 is increased, and the lateral stresses are decreased, in such a way that the mean stress remains constant.

The vertical strains $\epsilon_{z,f}$ corresponding to the stress increases $\Delta\sigma_1$ are measured, and a modulus of compressibility K is determined by the relationship

$$K = \frac{\Delta\sigma_1}{\epsilon_{z,f}} \tag{6-130}$$

If the soil skeleton should be perfectly linearly elastic and characterized by a Young modulus E_s and a Poisson's ratio ν_s between the moduli M, K, and E_s the following relationships should exist.

$$M = \frac{1 - \nu_s}{(1 + \nu_s)(1 - 2\,\nu_s)}\, E_s \tag{6-131}$$

$$K = \frac{E_s}{1 + \nu_s} \tag{6-132}$$

$$M = \frac{1 - \nu_s}{1 - 2\,\nu_s}\, K \tag{6-133}$$

The tests are repeated for different values of σ_m. The Figure 6-37 gives the variation of the modulus K versus the ratio of the deviatoric stress $\sigma_1 - \sigma_3$ to the rupture deviatoric stress $(\sigma_1 - \sigma_3)_r$ with the mean normal stress σ_m as a parameter. It clearly shows that the defined deformation modulus K is a function of the mean normal stress σ_m and the deviatoric stress ratio.

For a given deviatoric stress ratio the defined modulus K increases markedly with increasing mean normal stress. For a given mean normal stress the modulus K remains approximately constant, as long as the deviatoric stress ratio is smaller than 0.25, thus, as long as one is far away from the plastic state; but, when the deviatoric stress ratio becomes larger than 0.25, the compressibility modulus K rapidly decreases by increasing deviatoric stress ratio. Of course the defined modulus becomes zero, when the deviatoric stress ratio is equal to one. For the sand tested by Kérisel the value of the modulus K varies between 280 MPa and zero.

In the oedometer test, because of the prevented lateral deformation, the value of the deviatoric stress ratio $(\sigma_1 - \sigma_3)/(\sigma_1 - \sigma_3)_r$ is located between narrow limits. On the contrary for a footing the values of σ_m and of the deviatoric stress-ratio have different values from one point to another. This is thus also the case for the K modulus. For this reason an oedometer test, which on Figure 6-37 corresponds to a unique well defined curve

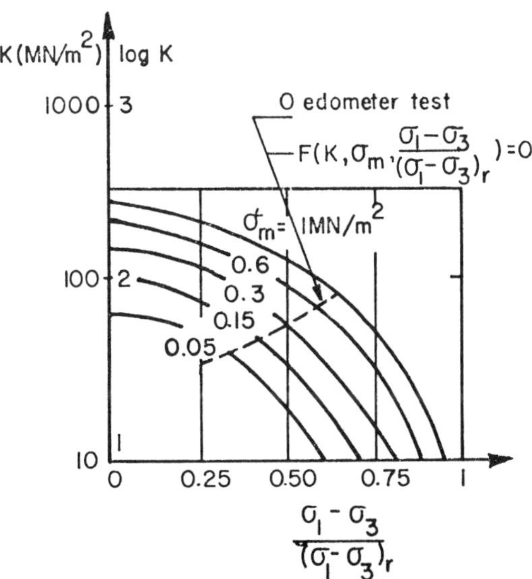

Figure 6-37. Variation of modulus of deformation K with mean normal stress, and deviatoric stress ratio.

$$F = \left(K, \; \sigma_m, \; \frac{\sigma_1 - \sigma_3}{(\sigma_1 - \sigma_3)_r} \right) = 0$$

cannot give a complete picture of the problem. However the test performed by Kérisel shows that as long as a loading case is considered for which in each point $(\sigma_1 - \sigma_3)/(\sigma_1 - \sigma_3)_r$ is smaller than 0.25, one should be not too far from real values, and the estimates of deformation will be on the safe side by using the parameters found in an oedometer test.

Theoretically a more exact solution could be obtained by performing a triaxial test in which the sample should be first reconsolidated under the stress tensor to which it was submitted in the soil. Subsequently a stress tensor is added, corresponding to that which will be generated in the considered point by the load of the footing. The deformations caused by this stress increase are measured. This corresponds to the method of the stress path by Lambe [1967].

However, the horizontal stresses in a given point of a soil mass caused by the load cannot be obtained by linear elastic theory. Only the vertical normal stresses can approximately equal those given by the theory of Boussinesq or of Buisman [1940].

If more exact answers are needed, only numerical methods taking into account the influence of the mean normal stress and the deviatoric stress ratio on K and on the Poisson's ratio ν_s could be tried.

A very complete review about the problem of settlements is to be found in the contribution of W. H. Perloff [1975]. However, much of the data and diagrams are based on the simplified assumption of a linear elastic material.

Deformation in Sands

Laboratory Tests. Sand is a very conservative material; its compressibility is influenced in an important way by its stress history. Now, by taking a sand sample, the stress-history is disturbed. Even when reproducing the in situ stress field, existing in the soil before sampling, in the laboratory, the stress-history cannot be exactly duplicated.

This is one of the reasons field tests for sands are preferred.

Pressuremeter Tests after Ménard. Such tests [Ménard, 1975] give a very good approximation for the compressibility modulus in a horizontal direction under an applied stress field. They allow the influence of the mean normal stress and the deviatoric stress ratio to be traced. Because the soil is disturbed little, at least when the test is correctly performed, the influence of the stress history is also taken into account. Besides, by introducing several experimental factors, Ménard has introduced several refinements.

Plate Bearing Tests. This method has already been described by Terzaghi-Peck [1948]. The difficulty with this method is extrapolating the settlements measured with small plates to the case of the much larger footings. An interpolation formula has been given by Terzaghi [1943]. Bjerrum [1963] has shown that the results given by this formula can largely differ from reality.

CPT Tests. A semi-empirical method for predicting the settlements is based on the result of CPT tests. In this method a distinction is made between normally consolidated sands, which during their history have not been subjected to loads heavier than those corresponding to the actual soil and water levels, and those which in an earlier stage have been subjected to heavier loads than those to be transmitted by the construction.

Based on the deduction of Buisman [1940] and settlement measurements (De Beer-Martens [1957], for the first group of sand layers (mostly Holocene) the following relationship between the constant of compressibility C and the cone resistance q_c is used.

$$C = 3 \frac{q_c}{p_t' + p_c} \qquad (6\text{-}134)$$

Generally there is no major objection to making $p_c = 0$

$$M = C(p_t' + p_c) = 3q_c \qquad (6\text{-}135)$$

On Figure 6-38 the results of a CPT are shown, giving the variation of the cone resistance q_c versus depth. On the same figure the values of the bulk modulus M are also given.

The values of the deduced compressibility constant C are given in Figure 6-39.

The settlements are calculated with the formula of Terzaghi:

$$s = \int_0^\infty \frac{1}{C} 2.3 \log_{10} \frac{p_t' + \sigma_z + p_c}{p_t' + p_c} \, dz \qquad (6\text{-}136)$$

In the formula σ_z represents the vertical normal stress caused in the considered point by the load transmitted by the footing.

The application of Equation 6-136 is only valid for loads for which the deviatoric stress ratios are sufficiently small ($< 1/3$ till $1/4$); this means that the applied load must present a sufficiently large safety margin with respect to the ultimate bearing capacity.

The method based on the results of CPT tests, has some analogy with the method, first given by Terzaghi and Peck [1948] in which the allowa-

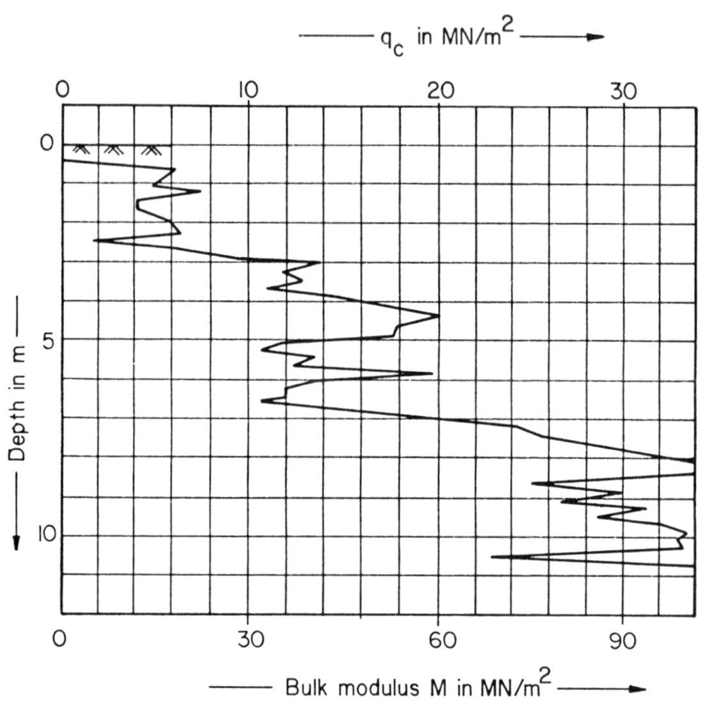

Figure 6-38. Relationship between cone resistance and bulk modulus M.

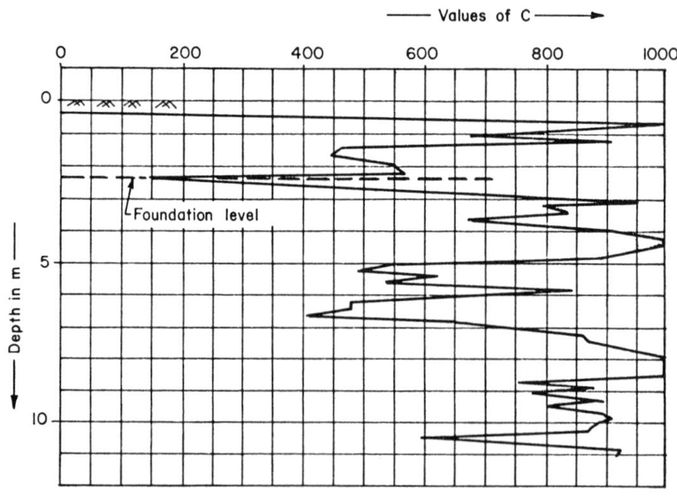

Figure 6-39. Variation of the compressibility constant C vs. depth.

ble unit loads related to the limits of the allowable settlements are deduced from the results of SPT tests. The method based on static tests (CPT) is however more reliable than the purely empirical method of the dynamic SPT tests.

When the soil has already been subjected to larger stresses than those caused by the construction, the following method is applied. The constants of compressibility C are deduced from the q_c values as indicated before. But now also samples are taken from the sand layer, and after these samples have been brought to about the same compactness as in the field, oedometer tests are performed on them. The samples are subjected to a cycle of loading and unloading (Figure 6-40).

From the loading branch one gets:

$$\frac{\Delta h}{h} = \frac{1}{C_{oed}} \ln \frac{p_1 + \Delta p + p_c}{p_1 + p_c} \tag{6-137}$$

A semi-log law with a swelling constant A_{oed} corresponds to the unloading branch:

$$\frac{\Delta h}{h} = \frac{1}{A_{oed}} \ln \frac{p_1 - \Delta p + p_t'}{p_1 + p_c'} \tag{6-138}$$

From such an oedometer test the ratio A_{oed}/C_{oed} is obtained.

The values of C deduced from the CPT tests are now multiplied by this ratio; in this way an A value is obtained, given by

$$A = C \frac{A_{oed}}{C_{oed}} \tag{6-139}$$

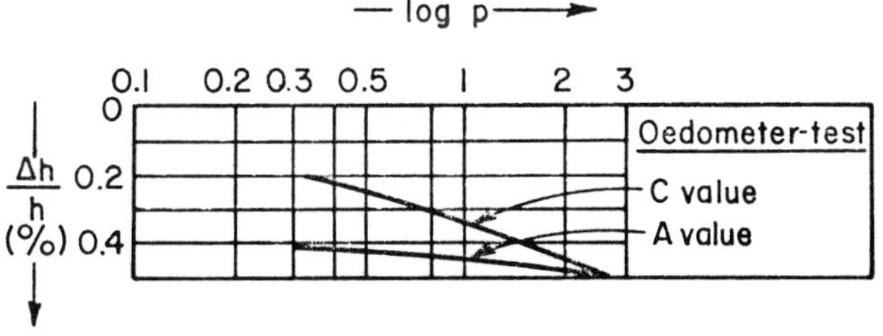

Figure 6-40. Cycle of loading and unloading in an oedometer test on sand.

The settlements are now calculated with the formula

$$s = \int_0^\infty \frac{1}{A} \, 2.3 \, \log_{10} \frac{p_t' + \sigma_z}{p_t'} \tag{6-140}$$

The fact that the sand has previously been subjected to larger loads than will be imposed by the foundation is obtained from geological evidence or from the special features of the works (for instance, foundations in a big permanent excavation).

For sands the ratio A/C has mostly values between 5 and 10. Neglecting the question of preloading can lead to settlements that are much too large.

The method of estimation of the settlements based on the results of CPT's further has the advantage that the inevitable lack of homogeneity of the sand layers vertically and horizontally can be taken into consideration. With only the results of laboratory tests on an always restricted number of samples, it is very difficult, or even impossible to take care of this lack of homogeneity.

The consequence of this lack of homogeneity is that two identical footings loaded with the same load can show rather large settlement differences. Terzaghi and Peck [1948], their conclusions based on their experience, admitted that the settlement differences could amount to 50% of the total settlement, when considering identical footings.

Even if a correct theory for the evaluation of the settlements of a perfectly homogeneous sand layer should be available, the settlement differences due to the lack of homogeneity ought still to be estimated. If a sufficiently large number of CPT's are performed, it is however possible to get a closer idea of the settlement differences related to the lack of homogeneity.

Thus if the method based on CPT's is more crude than those based on laboratory tests on samples, it will have the advantage of giving a better picture of the influence of heterogeneity on the settlement differences.

Deformations in Saturated Cohesive Soils

General Considerations. In saturated cohesive soils a distinction has to be made between the immediate settlements occurring by constant volume, the hydrodynamic settlements occurring during the period of excess pore water pressure, and the secondary or creep settlements.

In the following cases, a large safety margin with respect to the ultimate bearing capacity is assumed to prevail.

Normally Consolidated Saturated Cohesive Soils. From oedometer tests the values of the constant of compressibility C or the bulk modulus M are obtained. The total settlement at the end of the hydrodynamic period is given by

$$s = \int_0^\infty \frac{1}{C} \, 2.3 \, \log_{10} \frac{p_t' + \sigma_z}{p_t'} \, dz \tag{6-141}$$

Although not directly apparent, it can also be admitted that the immediate settlement is included in Equation 6-141. The only way to get an order of magnitude of the immediate settlement (except by complicated numerical methods) is to make the incorrect assumption that the soil behaves as a perfectly linear elastic material.

The immediate settlement should be given by

$$s_i = q \, B \, \frac{1 - \nu_u^2}{E_u} \, I_\rho \tag{6-142}$$

$I_\rho = F$ (B/L, x/B, y/B, distribution of the soil reactions underneath the footing) $\tag{6-143}$

The factor I_ρ is obtained with the stress distribution of Boussinesq. Applications are to be found in Perloff [1975].

where E_u = undrained modulus obtained in undrained unconsolidated triaxial tests or in unconfined compression tests.

$$\nu_u = \frac{1}{2}$$

The hydrodynamic settlement s_H is obtained by $s_H = s - s_i$, s being given by Equation 6-141.

The secondary settlements due to creep phenomena in the soil increase with the log of time. Data about the governing parameters can be obtained from long-lasting oedometer or triaxial tests. It must however be stressed that these parameters markedly depend on the magnitude and the rate of the load increments. By larger load increments the structure of the soil skeleton is more thoroughly broken, with the consequence that the creep phenomena become much smaller. By very small load increments, or by a very slow continuous load increase, the creep phenomena under the final load become much more important. Since the load increase proceeds much slower in reality than in the classical laboratory tests, an arithmetic series (3 kPa, 6 kPa, 9 kPa) instead of the geometric series for the load steps (10 kPa, 20 kPa, 40 kPa, 80 kPa, etc.) should be used to determine the parameters of the secondary settlements for soft soils. With an arithmetic series larger creep deformations appear, but lower deformations corresponding to the hydrodynamic period are obtained.

For determining the increase of the deformations with time during the hydrodynamic period, the classical solutions of Fröhlich-Terzaghi, et al. can be used. These solutions require the knowledge of the coefficient of permeability. Here also the lack of homogeneity of the involved soil layers plays an important role. Indeed, the water expulsion phenomena will not be governed by the mean value of the coefficient of permeability but by that of the more pervious inclusions. Hydrodynamic settlements often proceed more quickly than predicted because these more pervious parts are not always detected. The piezocone is a very good method for detecting such inclusions [Jamiolkowski,1985].

Overconsolidated Clays. Two different methods can be used to solve the problem of the settlements of overconsolidated clays.

Method of Skempton-Bjerrum [1957]. In this method the immediate settlement is calculated with Equation 6-142 as it is for the normally consolidated clays. In the case of stiff fissured clays it is, however, not an easy matter to deduce the value of E_u from laboratory tests.

In the case of an increase in the spherical tensor $\Delta\sigma_3$ and of the deviatoric tensor $(\Delta\sigma_1 - \Delta\sigma_3)$ the immediate hydrodynamic pressure increase Δu_o is given by

$$\Delta u_o = \Delta\sigma_3 + A_o \, (\Delta\sigma_1 - \Delta\sigma_3) \tag{6-144}$$

where A_o = coefficient of Skempton [1954].

Consequently Δu_o is generally different from $\Delta\sigma_1$ (see also [Henkel, 1956]).

To take this difference into account, the hydrodynamic settlement is given by :

$$s_H = \mu_{sk} \int_0^Z \frac{\sigma_z}{M} \, dz \tag{6-145}$$

The value of μ_{sk} as calculated by Skempton and Bjerrum is given as a function of A_o and the ratio of the thickness Z of the layer to the width B of the footing in Figure 6-41.

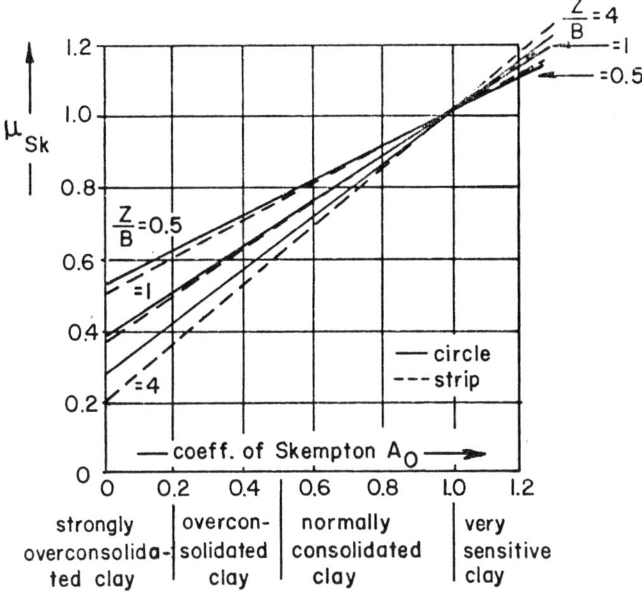

Figure 6-41. Variation of the reduction factor μ_{Sk} in function of the coefficient of Skempton A_o, with the relative thickness Z/B as a parameter.

In overconsolidated clays for current problems the influence of the secondary settlements can generally be neglected.

Method of Hansen [1964]. In the case of overconsolidated clays the usual oedometer tests give values much too low for the compressibility parameters, with the consequence that with conventional formulas (such as Equation 6-141) settlements that are much too large are obtained. In order to get more reliable data, the sample is subjected to a geologic oedometer test. First the sample is loaded to the maximum stress σ'_{max}, to which it has been subjected during its geological history (Figure 6-42). After that the sample is unloaded to the minimum stress σ'_{min} to which the layer has been unloaded after the time of existence of the maximum load. Then the sample is loaded to the actual overburden pressure p'_t (point D).

From this stress state on, the load on the sample is increased (point E) in the case of a loading problem or decreased in the case of an unloading problem (E'). From the curves DE or DE' the compression or the swelling parameter can be deduced. The curvature of the curves DE and DE' indicates that small stress-changes cause relatively smaller movements than larger stress changes.

The predicted settlement by the method of Hansen is the sum of the immediate and hydrodynamic settlement.

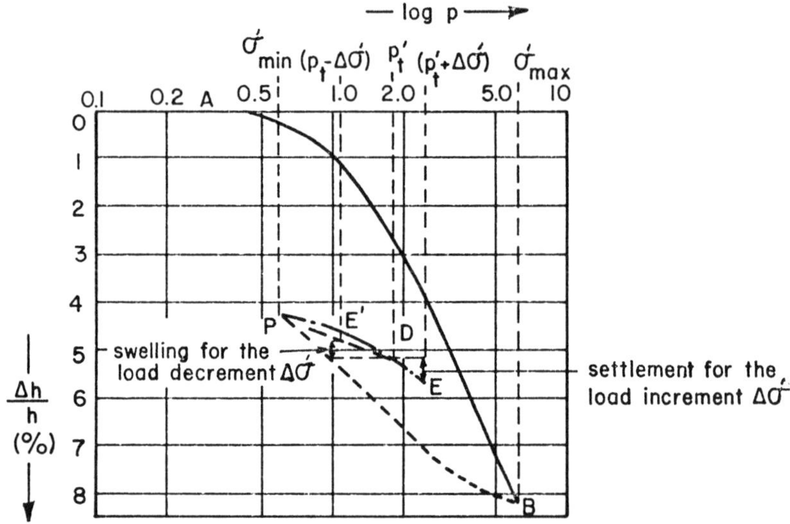

Figure 6-42. Oedometer test on an overconsolidated clay [after Brinch Hansen, 1961].

The methods of Skempton-Bjerrum and of Hansen do not give the same results. However practical application shows that the results usually do not differ too much.

Settlement of a Centrally Loaded Footing

In general the settlement of a point of a footing depends on the distribution of the soil reactions underneath the footing. There are however certain points, called singular points, for which the stress distribution along the vertical is independent of the distribution of the reactions. For the case of rectangular footings and for the law of Buisman, the location of these singular points is shown and the stress distribution influence factor $i = \sigma_z/q$ is given versus the relative depths z/B in Figure 6-43.

The values of σ_z can be introduced in the formulas given before, in order to get the mean settlement of the footing or its settlement if it is infinitely rigid.

Bearable Deformations

From experience Terzaghi-Peck [1948] concluded that in the case of sands settlement differences of 20 mm for normal offices and buildings founded on isolated footings and strips are not detrimental. Eurocode [1981, 1982] endorses the limits proposed by Bjerrum [1963] for the bearable angular distortions $\Delta s/L$, depending on the kind of building (Figure 6-44), where Δs represents the settlement difference between adjacent supports at distance L.

A more rational criterion is the radius of curvature, and further programs in this sense may be expected in the near future [Burland, 1974].

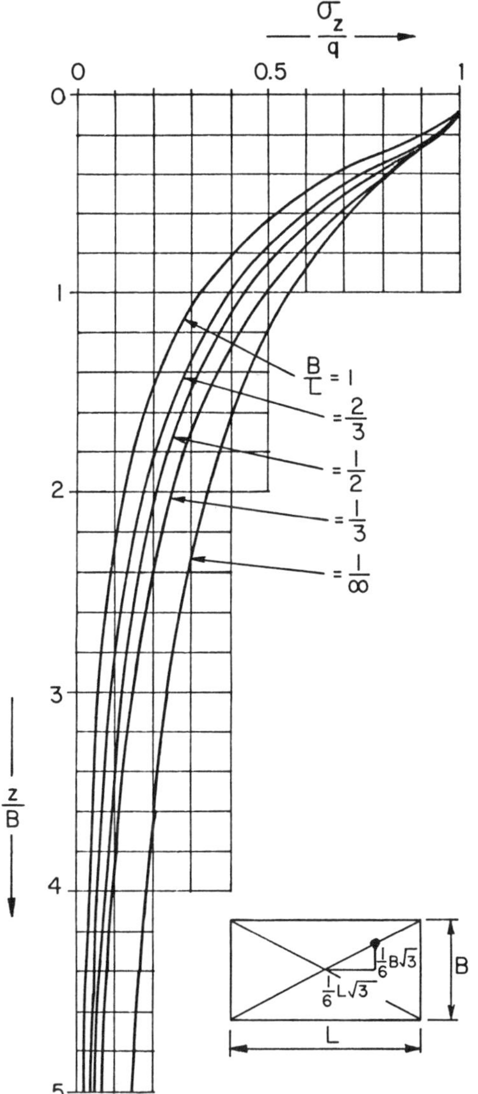

Figure 6-43. Location of the singular points and stress distribution σ_z/q vs. relative depth z:B.

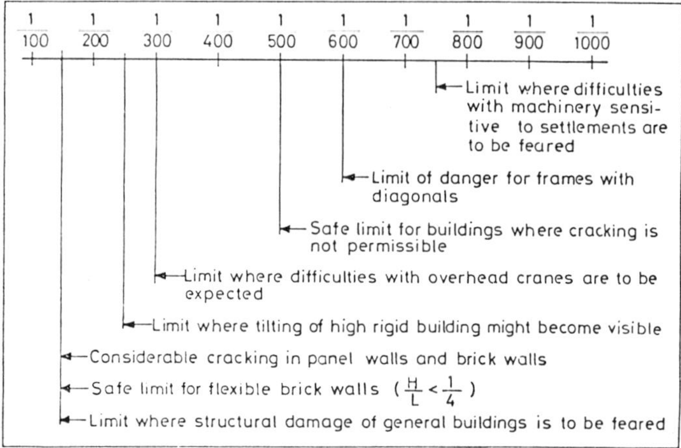

Figure 6-44. Bearable angular distortions depending on the kind of building [after Bjerrum, 1963].

PROBLEM OF THE DISTRIBUTION OF THE SOIL REACTIONS UNDERNEATH A FOUNDATION RAFT

The distribution of the soil reactions under a foundation raft depends strongly on the rigidity of the foundation raft and the superstructure relative to that of the soil.

There are two extreme ways to characterize the rigidity of the soil:

1. By introducing a modulus of soil reaction k_s (Winkler hypothesis).
2. By assimilating the soil to a linearly elastic material (E_s, ν_s) (law of Boussinesq).

As in general the compressibility decreases with depth z, it should be more appropriate to suppose

$$E_s = E_{s,o} + k'z \tag{6-146}$$

Schultze [1969] defines a coefficient α as:

$$\alpha = \frac{E_{s,o}}{R} \Bigg/ \frac{\Delta E}{\Delta z} \tag{6-147}$$

where R = hydraulic radius of the loading surface.

If $\alpha < 0.04$ the Winkler hypothesis should be sufficiently correct.

If $\alpha > 100$ the method based on the law of Boussinesq should be acceptable. For intermediate values of α the model of Repnikov [1967] could be used. In that model the soil is replaced by a medium with constant Young's modulus $E_{s,o}$ in which are embedded springs with a spring constant k_s. Both systems take a part of the total load such that their respective deformations are identical.

In such a model the settlement s is given by

$$s = \frac{B\ q_m}{B\ k_s + \dfrac{E_{s,o}}{2(1 - v_{s,o}^2)F\left(\dfrac{B}{L}\right)}} \tag{6-148}$$

where F (B/L) = function of the shape ratio B/L.

For a given footing, and a given loading q_m the values of the mean settlement s can be calculated for two different values of the dimension B, according to the method outlined earlier for computing the deformations.

$$s_1 = \frac{B_1\ q_m}{B_1\ k_s + \dfrac{E_{s,o}}{2(1 - v_{s,o}^2)F\left(\dfrac{B_1}{L}\right)}} \tag{6-149}$$

$$s_2 = \frac{B_2\ q_m}{B_2\ k_s + \dfrac{E_{s,o}}{2(1 - v_{s,o}^2)F\left(\dfrac{B_2}{L}\right)}} \tag{6-150}$$

From these two formulas the values of k_s and $E_{s,o}$ of the Repnikov model can be deduced, and for the model the distribution of the reactions underneath the footing can be calculated.

The method of Repnikov, applied by Schultze, can give more realistic answers when

$$0.04 < \alpha < 100 \tag{6-151}$$

NUMERICAL METHODS

As clearly stated by Christian [1977] one must be able to represent the nonlinear stress-strain characteristics of the soil in the numerical method. Further the program must permit an incremental analysis.

Actually there exist already numerous computer programs allowing several of the intricate properties of soil (for instance anisotropy) to be taken into account [Desai, 1984; Smith, 1981; Garber, 1977].

For shallow foundations the aim is to find the ultimate bearing capacity, and to predict the load-settlement curve and also the distribution of the soil reactions under the footings and rafts. One must be aware that in general it is very difficult to introduce in the calculations all parameters of the problem in a completely exact way. There is a danger that in modeling some of the parameters exactly but neglecting or distorting the influence of others, an answer that differs more from reality than the results given by much more rough methods will be obtained. Therefore numerical calculations should always be made by varying the parameters in order to detect the sensitivity of the results to their variation.

The advantage of the numerical methods is precisely their allowing an easy way to make all kinds of combinations of the values of the parameters.

A good example of what can be expected by using finite elements for the computation of bearing capacity factors is given by the contribution of B. V. Griffith [1982]. If a plain strain case using the method of finite elements and elasto-plastic material are assumed the values of the bearing capacity factors N_q and N_c as given in the section concerning the ultimate bearing capacity for the case of general shear failure are retrieved. It appears that these values are independent of the degree of rugosity of the footing. Conversely, the value of N_γ^* appears to differ markedly for smooth and rough foundations. For rough foundations the values given for N_γ^* can serve as a guide. However, the values obtained for N_γ^* by the finite element calculation depend on the width of the foundation, and decrease with increasing width.

By the finite element method instead of calculating the ultimate bearing capacity for separated simplified cases (N_q, N_c, N_γ^*), it is also possible to consider directly the combined case, and to find the safety margin included in the separate calculations. Therefore with the numerical methods it becomes also possible to calculate directly more complicated cases of inclined and eccentric loadings. Nevertheless, as is also the case in the contribution of Griffith, the very complex properties of the soil are still replaced by a somewhat less complicated model.

In most cases it will not often be necessary to use the numerical methods and their programs. The ultimate bearing capacity can be predicted as outlined earlier. A relative settlement s_r/B which in a first approximation can be related only to the relative density of the sand corresponds to this calculated ultimate bearing capacity.

$$\frac{s_r}{B} = 0.04 + 0.15 \, (1 - D_r) \qquad\qquad (6\text{-}152)$$

For the load $Q = Q_r/3$ the settlement s can be calculated as already outlined. Still a rough estimate of the load settlement curve can be made (Figure 6-45).

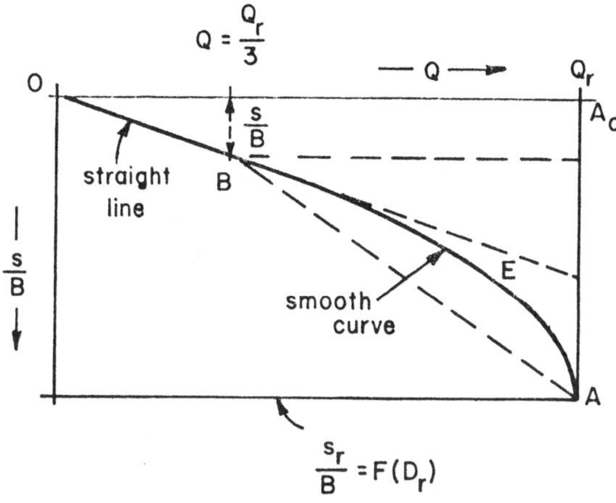

Figure 6-45. Rough estimate of the load-settlement diagram of a footing.

Knowing the points A and B, one draws the straight line OB. Through the points A and B, and with the tangents A_oA and OB one draws a smooth curve A E B. If with this rough method the solution appears to be economically acceptable, no further research is generally needed. When it does not, more intricate research can be worthwhile, in order to obtain a more exact answer. For very important and exceptional cases the application of numerical methods becomes necessary, and use can be made of the already numerous programs available in big computer centers.

Dr. Ing. Coyette of the University of Louvain, who is quite experienced with numerical methods gives this advice for better defining the possibility of numerical analysis in the case of shallow foundations: "The most utilized methods are those of finite elements, and boundary integrals. As the method of finite elements is better adapted for nonlinear materials, it is often utilized in soil mechanics."

Several constitutive laws for modeling the problems related to shallow foundations (prediction of the load settlement curve) have been considered recently.

The numerical evaluation of the ultimate bearing capacity under a foundation is a difficult problem, as it is related to a limit state. The determination of such a limit state requires the adoption of a model based on a particular rupture criterion, and an incremental increase of the load in order to determine the largest multiplier of the load that still fulfills the constitutive laws and the conditions of equilibrium.

The method of the finite elements does not appear as an easy mean for evaluating the ultimate bearing capacity. However, several references show there is interest in such an approach for checking the validity of certain theories utilized for calculating the ultimate bearing capacity [I. M. Smith, 1981]. The application of the variational method for finding the ultimate bearing capacity underneath a foundation seems to be more attractive [M. Garber, 1977]. A method based on kinematical elements (incompressible) [P. Gussman, 1982] for the prediction of the mechanism of rupture and the value of the ultimate bearing capacity is also in use. Consequently, the contribution of the numerical methods to the analysis of shallow foundations is necessarily limited; the prediction of the ultimate bearing capacity requires a costly elasto-plastic analysis in an incremental mode to the level of the loads involving nonconvergency, indicating the rupture. Except for certain exceptional cases, this technique is difficult to apply. Conversely, a prediction of the settlement under service load is more easily obtainable. Because the diversity of proposed laws (nonlinear elastic, elasto-plastic, visco-elastic, visco-plastic, endochronic, etc.), it is difficult to choose a constitutive model.

Finally, one must mention the necessarily limited character of the geometric domain discretized in finite elements. This limitation can be a source of difficulties but these difficulties can be overcome by utilizing infinite elements and coupling finite elements with boundary elements.

REFERENCES

1. Baguelin, F., Jézéquel, J. F., and Schields, R. H., *The Pressuremeter and Foundation Engineering,* Trans Tech Publications, Aedermannsdorf, Switzerland, 1978, p. 617.
2. Bishop, A. W., Discussion on "Soil Properties and Their Measurement," Proceedings of the 5th Int. Conf. Soil Mech. Found. Eng., Vol. 3, 1961, p. 97.
3. Bjerrum, L., "Contribution to the discussion of section VI—Interaction between Structure and Soils Concerning the Problem of Allow-

able Settlements of Structures (Damages Criteria)," Proceedings of the Eur. Conf. Soil Mech. Found. Eng., Wiesbaden, Vol. 2, 1963, pp. 135–137.

4. Bjerrum, L., and Kummeneje, "Shearing Resistance of Sand Samples with Circular and Rectangular Cross Sections," Norwegian Geotechnical Institute, publ. No. 44-1, 1961.

5. Buisman, A. S. K., *Grondmechanica,* Waltman, Delft, 1940, pp. 182–185, 234–236, 241–249.

6. Burland, J. B., and Wroth, C. P., "Allowable and Differential Settlement of Structures, Including Damages and Soil-Structures Interaction," Proc. Brit. Geotech. Soc. Conf., *Settlement of Structures,* London, Pent. Press., 1974, pp. 611–657.

7. Caquot and Kérisel, *Traité de mécanique des sols,* Gauthier Villars, Paris, 3e édition, 1956, pp. 383–394.

8. Christian, J. T., *Numerical Methods in Geotechnical Engineering,* Shallow Foundation, Ch. 6, McGraw Hill Book Co., 1977, p. 211.

9. De Beer, E., "Bearing capacity," *Ground engineer's references book,* F. C. Bell (Ed.), Butterworths Technical Publishers, Borough Green, Sevenoaks, Kent, 1985.

10. De Beer, E., "Computation of the Failure in the Bearing Capacity of Shallow Foundations with Inclined and Eccentric Loads," U.S. Army Engineer Waterways Experiment Station Corps of Engineers, Vicksburg, Mississippi, translation No. 66–13, 1966.

11. De Beer, E., "Experimental Determination of the Shape Factors and the Bearing Capacity Factors of Sand," *Géotechnique,* Vol. 20, No. 4, Dec. 1970, pp. 387–411.

12. De Beer, E., *Grondmechanica, Part 2,* funderingen, Standaard boek handel, Antwerpen, 1949, pp. 41–46.

13. De Beer, E., "The Scale Effect on the Phenomenon of Progressive Rupture in Cohesionless Soils," Proceedings of the 6th Int. Conf. Soil Mech. Found. Eng., Montréal, Vol. 2, 1965, pp. 13–17.

14. De Beer, E., and Martens, A., "Method of Computation of an Upper Limit for the Influence of the Heterogeneity of Sand Layers on the Settlements of Bridges," Proceedings of the 4th Int. Conf. Soil Mech. Found. Eng., London, Vol. 1, 1957, pp. 275–282.

15. De Beer, E., and Vesic, A., "Etude expérimentale de la capacité portante du sable sous des foundations directes établies en surface," Annales des travaux publics de Belgique 58, No. 3, 1958, pp. 5–58.

16. De Poorter, "Kritisch statisch evenwicht van een grondmedium onder een fundering op staal," Thesis Catholic University Leuven (Belgium), 1962.

17. Desai, C. S., and Siriwardane, H. J., *Constitutive Laws for Engineering Materials,* Prentice Hall, 1984.

18. DIN 4017, "Calculation of Ultimate Bearing Capacity of Soil Underneath Shallow Foundations" (in German), Fachnormen Ausschuss Bauwesen (FN Bau), Deutsches Institut für Normung, Beuth Verlag, Berlin, 1975.

19. Eurocode E7, "Design and Construction of Foundations, Retaining Structures and Earth Works," 1981.

20. Eurocode E7, "Verification Procedures for Shallow Foundations," 1982.

21. Garber, M., and Baker, R., "Bearing Capacity by Variational Method," *ASCE J. Geot. Div.*, Vol. 103, 1977, pp. 1209–1235.

22. Griffith, D. U., "Computation of Bearing Capacity Factors Using Finite Elements," *Géotechnique*, Vol. 32, No. 3, 1982, pp. 195–202.

23. Gussmann, P., "Kinematical Elements for Soils and Rocks," 4th Int. Conf. on Num. Meth. in Geomech. Edmonton, Vol. 1, 1982, pp. 47–52.

24. Hansen, J. Brinch., "A General Formula for Bearing Capacity," Danish Geotechnical Institute, Bulletin No. 11, 1961.

25. Hansen, B., and Christensen, N. H., "Discussion on Theoretical Bearing Capacity of Very Shallow Footings," Proceedings ASCE, *Journal of the Soil Mechanics and Foundations,* Division 95, No. SM-6, 1969, pp. 1568–72.

26. Hansen, J. Brinch, and Misa Tadashi, "An Empirical Evaluation of Consolidation Tests with Little Belt Clay," The Danish Geotechnical Institute, Copenhagen, Bulletin No. 16, 1964.

27. Heller, L. W., "Failure Modes of Impact-Loaded Footings on Dense Sand," Technical report R 281 U.S. Naval Civil Engineering Laboratory, Port Hueneme, California, 1964, pp. 1–31.

28. Henkel, D. J., "The Effect of Overconsolidation on the Behavior of Clays During Shear," *Géotechnique,* Vol. 6, 1956, pp. 139–150.

29. Jamiolkowski, M., "New Developments in Field and Laboratory Testing of Soils," Proceedings of the 11th Int. Conf. Soil Mech. Found. Eng., Vol. 1, 1985, pp. 57–153.

30. Janbu, N., "Soil Compressibility as Determined by Oedometer and Triaxial Tests," Proc. of the European Conference of the Int. Soc. Soil Mech. Found. Eng., Wiesbaden, Vol. 1, 1963, pp. 19–25.

31. Kérisel, "Discussion à propos "Théorie générale, consolidation, distribution des contraintes, processus de tassement en fonction du temps," Congrès Européen de mécanique des sols et des travaux fondation, Wiesbaden, Proceedings, Vol. 2, 1963, pp. 33–35.

32. Ladanyi, B., "Etude des relations entre les contraintes et les déformations lors du cisaillement des sols pulvérulents," Annales du travaux, publics de Belgique, No. 3, 1961.

33. Lambe, T. W., "Shallow Foundations on Clay," *Bearing Capacity and Settlement of Foundations,* Duke University, 1967, pp. 35–43.

34. Leussink, H., Blinde, A., and Abel, P. G., "Versuche über die Sohldruck-verteilung unter Starren Gründungskörpern auf Kohäsionslosem Sand," Veröffentlichungen des Institutes für Bodenmechanik und Felsmechanik der Technischen Hochschule Fridericiana in Karlsruhe, Heft 22, 1966.

35. Lundgren, H., and Hansen, Brinch J., *Geotechnik,* Mollers Printing Co., Copenhagen, 2nd edition, 1965, pp. 178–180.

36. Mandel, J., and Salençon, J., "Force portante d'un sol sur une assise rigide," Proc. 7th Int. Conf. Soil Mech. Found. Eng., Mexico, Vol. 2, 1969, pp. 157–164.

37. Marsland, A., "Large In Situ Tests to Measure the Properties of Stiff Fissured Clays," Current papers CP 1/73, Building Research Establishment, Garston Watford, Herts, England, 1973.

38. Mayer, A., "Discussion," Proceedings of the 3rd Int. Conf. Soil Mech. Found. Eng., Vol. 2, 1953, pp. 185–187.

39. Ménard, L., "Calcul de la force portante des fondations sur la base des résultats des essais pressiométriques," Sols-soils, No. 5-9-24, No. 6-9-27, 1963.

40. Ménard, L., "The Interpretation of Pressuremeter Test Results," Sols-soils, No. 26-7-43, 1975.

41. Meyerhof, G. G., "The Bearing Capacity of Foundations under Eccentric and Inclined Loads," Proceedings of the 3rd. Int. Conf. Soil Mech. Found. Eng., Zürich, Vol. 1, 1953, pp. 440–445.

42. Meyerhof, G. G., "The Bearing Capacity of Sand," Ph.D. Thesis, University of London, 1950.

43. Meyerhof, G. G., "Influence of Roughness of Base and Groundwater Conditions on the Ultimate Bearing Capacity of Foundations," *Géotechnique,* Vol. 5, 1955, pp. 227–242.

44. Meyerhof, G. G., "Ultimate Bearing Capacity of Footings on Sand Layer Overlying Clay," *Canadian Geotechnical Journal,* Vol. 11, 1974, p. 223–229.

45. Meyerhof, G. G., "The Ultimate Bearing Capacity of Foundations," *Géotechnique II,* 1950–1952.

46. Mortensen, K., and Lundgren, H., "Determination by the Theory of Plasticity of the Bearing Capacity of Continuous Footings on Sand," Proceedings of the 3rd Int. Conf. Soil Mech. Found. Eng., Zürich, Vol. 1, 1953.

47. Muhs, H., and Kahl, H., "Ergebnisse von Probebelastungen auf grossen Lastflächen zur Ermittlung der Bruchlast im Sand (3. Bericht)," Frank Verlag, Stuttgart, 1957.

48. Perloff, W., "Pressure Distribution and Settlement," *Foundation Engineering Handbook,* Van Nostrand Reinhold, New York, 1975, pp. 148-196.

49. Prandtl, "Uber die Eindringungsfestigkeit plastischer Baustoffe und die Festigkeit von Schneiden," Zeitschrift für angewandte Mathematik und Mechanik, Vol. 1, No. 1, 1921, pp. 15-20.

50. Repnikov, "Calculation of Beams on an Elastic Base Combining the Deformative Properties of a Winkler Base and an Elastic Mass," *Soil Mechanics and Foundation Engineering,* No. 6, 1967, pp. 348, (translation from Osnovaniya, Fundamenty i Mekhanika Grutnov, Moscow 6, pp. 4-6).

51. Schultze, E., "The Combination of Modulus of Subgrade Reaction and Modulus of Compressibility Methods, " International Symposium on Civil Engineering Structures Resting on Soil and Rocks, Academy of Sciences and Arts of Bosnia, Hercegovina, Sarajevo, Vol. 2, No. 15, 1969.

52. Skempton, A. W., "The Pore Pressures Coefficients A and B," *Géotechnique,* Vol. 4, No. 4, 1954.

53. Skempton, A. W., and Bjerrum, L., "A Contribution to the Settlement Analysis of Foundations on Clay, *Géotechnique,* Vol. 7, No. 3, 1957.

54. Smith, I. M., "Numerical Analysis of Shallow Foundations," Nato Advanced Study Institute, Num. Meth. in Geomech., Vimeiro, Portugal, 1981.

55. Sokolovski, V. V., "Statics of Soil Media," Butterworths Scientific Publications, London, 1960.

56. Terzaghi, K., *Theoretical Soil Mechanics,* J. Wiley and Sons, New York, 1943, pp. 120-133.

57. Terzaghi, K., and Peck, R. B., "Soil Mechanics in Engineering Practice," John Wiley and Sons, New York, 1948, pp. 130, 169-172, 421-423.

58. Van de Perre, S. J., and Ramelot, C., "Travaux de la commission d'étude des foundations de pylones," Research report No. 2 of the IRSIA, Brussels, 1957.

59. Vesic, A. S., "Bearing Capacity of Deep Foundations in Sand," National Research Council, Highway Research Record 39, 1963, pp. 112-153.

60. Vesic, A. S., "Bearing Capacity of Shallow Foundations," *Foundation Engineering Handbook,* H. F. Winterkorn and H. Y. Fang (Ed.) Van Nostrand Reinhold Company, 1975, pp. 121-145.

61. Vesic, A. S., "A Study of Bearing Capacity of Deep Foundations," Final Report Project B 189, Georgia Institute of Technology, Atlanta, Georgia, 1967.

62. Vesic, A. S., "Ultimate Loads and Settlements of Deep Foundations in Sand," *Bearing Capacity and Settlement of Foundations,* Proceedings of a Symposium held at Duke University, 1965, pp. 53–68.
63. Vesic, A. S., Banks, D C., and Woodard, J. M., "An Experimental Study of Dynamic Bearing Capacity of Footings on Sand," Proceedings of the 6th Int. Conf. Soil Mech. Found. Eng., Vol. 2, 1965, pp. 209–213.

Static and Dynamic Analyses of Axially and Laterally Loaded Piles and Pile Groups

P. K. Banerjee
Professor of Civil Engineering
State University of New York at Buffalo
Buffalo, New York USA
R. Sen
Assistant Professor of Civil Engineering and Engineering Mechanics
University of South Florida
Tampa, Florida USA
T. G. Davies
Lecturer in Civil Engineering
University of Glasgow
Glasgow, Scotland UK

INTRODUCTION

The response of piles and pile groups to static and dynamic excitation has been the subject of much investigation over the past decade. A review of the research undertaken to date indicates that although some outstanding efforts exist in the published literature, the scope of some of the solution procedures developed is restrictive and cannot always be extended to the generalized analysis of pile groups subjected to axial and lateral loading.

The underlying problem in analyzing pile groups is, of course, in the representation of the soil surrounding the pile and that of the far field, i.e., boundaries at infinity. The simplest model used represents the soil as a Winkler foundation with distributed springs and dashpots that are constant or frequency dependent or with lumped springs concentrated at a finite number of nodes. The spring constants are obtained from analytical considerations or from experimental data. The major advantage of this approach lies in its ability to simulate nonlinearity, inhomogeneity, and hysteretic degradation of the soil surrounding the pile by simply changing the spring and dashpot constants. Such analyses have been proposed by Matlock and Foo [1979], Penzien and others [1964, 1970], and indeed the results of these studies form the basis of the p-y and t-z curves extensively used in the design of piles and pile groups, primarily due to the absence of any reasonable alternative. It is apparent, however, that as this analysis ignores continuity between individual piles through the surrounding medium, it cannot adequately describe the behavior of pile groups.

A number of research workers, notably Novak and others (References 27–34) have attempted to eliminate some of the limitations of the discrete spring and dashpot models by considering approximate wave propagation modes along horizontal layers and also at the pile tip. These outstanding papers were, in fact, the first attempts to include the effects of radiation damping in the analysis in a consistent manner and they do provide a very good insight into the behavior of piles under dynamic loading. Although both Novak and Nogami (References 27–29) describe extensions of their analysis to pile groups, their analysis is difficult to apply to general pile groups under both lateral and axial loading.

Finite element analyses of piles have been carried out by Kuhlemeyer [1979a,b] and Krishnan et al. [1983] in which the piles are represented by axisymmetric elements and energy-transmitting boundaries are used to represent the far field. Extensions of the finite element procedures for the analysis of pile groups have been made by Wolf and Von Arx [1978] and Waas and Hartman [1981]. Considering the obvious limitations of the finite element method for modeling boundaries at infinity, these analyses represent considerable achievements in characterizing the dynamic behavior of pile.

A boundary element formulation of the pile problem was recently reported by Kaynia and Kausel [1982a,b] who obtained (numerically) dynamic buried ring load (axial and lateral) solutions from the work of Apsel [1979]. The major problem here is, of course, the accuracy of the numerically constructed dynamic solution since the convergence of the semi-infinite integrals in Apsel's work is heavily dependent on the frequency parameter.

Almost all of the work just described specifically relates to piles and pile groups embedded in homogeneous soils. Although uniform deposits of stiff clays can be adequately represented by a homogeneous stratum, for most natural soil deposits a more appropriate model is one in which the soil's stiffness increases linearly with depth [Gibson, 1967].

Very few studies of the response of piles embedded in inhomogeneous soils have been conducted. Even for the static case, only Banerjee [1978], Banerjee and Davies [1977] and Davies [1979] address the general pile group problem—the remaining analyses [Nogami, 1980; Novak and Aboul-Ella, 1978; Randolph and Wroth, 1980] being only concerned with single pile behavior.

In this study, a boundary element formulation has been used to determine the steady-state dynamic response of axially and laterally loaded piles and pile groups embedded in an inhomogeneous soil medium. The boundary element method offers considerable advantages over the other numerical methods for this problem primarily because of its ability to take into account the three-dimensional effects of soil continuity and boundaries at infinity. In applications to static (References 2,4–10) and dynamic analyses (References 11,18,19,41–43) of pile groups, it has proved to be the most rigorous, versatile, and economical method of analysis.

STATEMENT OF PROBLEM

Figure 7-1 sketches the system examined. A pile group consisting of vertical or raked piles of diameter d and length L is embedded in an infinite inhomogeneous soil stratum in which the soil is characterized by the following:

1. Linearly varying modulus with depth according to the equation $E_s(z) = E(0) + fz$.
2. Two-layer soil having moduli E_1 and E_2.
3. Homogeneous soil having Young's modulus E_s.

The piles are considered to be linear elastic beams columns with constant Young's modulus E_p, cross-sectional area A_p, second moment of area I_p, and of mass density ρ_p. The soil is assumed to be a linear hy-

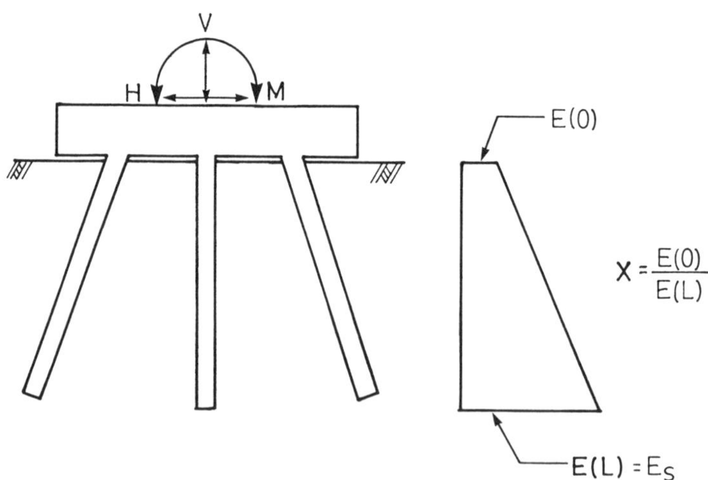

Figure 7-1. Axially and laterally loaded pile group.

steretic medium with a constant Poisson's ratio ν_s, a constant mass density ρ_s, a constant hysteretic damping β.

The soil modulus E_s is defined as the modulus at depth $z = L$ and the definition of a nondimensional frequency parameter a_o is also with respect to the shear wave velocity for E_s, i.e., $a_o = \omega d/C_s(L)$ where ω is the applied circular frequency and $C_s(L)$ is the shear wave velocity at depth $z = L$.

In order to model the effects of internal energy dissipation within the soil (i.e., material damping) complex moduli are introduced into the governing equations for the soil domain. For example, the Young's modulus of elasticity in the equation is replaced by its complex counterpart:

$$E_s^* = E_s[1 + 2i\beta]$$

where β = damping ratio,
$\quad\quad\ i = \sqrt{-1}$

For most soils, the damping ratio is largely insensitive to the frequency of excitation (hysteretic damping). It was found, however, that material damping plays only a minor role in the dynamic response of pile groups.

METHODS OF ANALYSIS

If the piles are represented by compressible columns and flexible beams, the governing equations of motion for these elements can be expressed as:

$$m\frac{\partial^2 u_z}{\partial t^2} - E_p A_p \frac{\partial^2 u_z}{\partial z^2} = \pi d\phi_z \tag{7-1}$$

for the axial dynamic response and

$$E_p I_p \frac{\partial^4 u_x}{\partial z^4} + m\frac{\partial^2 u_x}{\partial t^2} = -d\phi_x \tag{7-2}$$

for the lateral response,

where $\quad u_x$ and u_z = lateral and axial displacements
$\quad\quad\quad\quad\quad m$ = mass per unit length of the pile
$\quad\quad\quad \phi_x$ and ϕ_z = lateral and axial tractions of the pile
$\quad\quad\quad\quad\quad d$ = diameter of the pile

Equations 7-1 and 7-2 are essentially the dynamic counterparts of those derived by Banerjee [1978] for the static behavior of piles and include only one additional term representing the inertia forces mü in the equation of motion. For a completely general transient dynamic problem Equations 7-1 and 7-2 can be solved by coupling with them the equations of soil motion and satisfying compatibility and equilibrium at the pile-soil interface.

If the response is time harmonic, the preceding representations simplify to:

$$m\omega^2 u_z + E_p A_p \frac{d^2 u_z}{dz^2} = -\pi d\phi_z \tag{7-3}$$

$$m\omega^2 u_x - E_p I_p \frac{d^4 u_x}{dz^4} = d\phi_x \tag{7-4}$$

The equation describing the motion of the soil can be written as:

$$u_j(\xi,\omega) = \int_S G_{ij}(x,\xi,\omega)\phi_i(x,\omega)ds \tag{7-5}$$

where u_j = displacements of the soil
   G_{ij} = Green's functions (discussed later)
   ϕ_i = pile-soil interface tractions

Equations 7-3 to 7-5 all reduce to the relevant static case by simply letting $\omega \to 0$. The entire analysis is, of course, dependent on the availability of the fundamental solution for a point force, i.e., G_{ij} in Equation 7-5. For the static analysis in homogeneous soils Mindlin's solution has often been used. Davies and Banerjee [1978] developed a point force solution for a two-layer soil, which they subsequently modified to obtain an approximate displacement field in a soil medium whose modulus increases linearly with depth.

Recently, Kausel [1981, 1982] has developed an explicit solution for the Greens' function corresponding to dynamic loads in the interior of a layered stratum, using a layer stiffness matrix approach. Variations in the soil modulus with depth may be easily accommodated by representing the medium as a layered stratum, each layer having a different elastic modulus.

The formalism to determine the response of a layered soil deposit to dynamic loading was first proposed by Thompson [1950] and Haskel [1950] over a generation ago and is based on the use of transfer matrices in the frequency wave number domain. The solution technique for arbitrary loadings necessitates resolving the loads in terms of their temporal and spatial Fourier transforms, assuming them to be harmonic in time and space. Closed-form solutions are then found for simple cases by contour integration while numerical solutions are needed for arbitrary layered soils.

The transfer matrix approach was modified by Kausel and Roesset [1981] to obtain a computationally more versatile layer stiffness approach, that could be applied and understood very much in the sense used in structural analysis. The external loads applied at the layer interfaces are related to the displacements at the interfaces through layer stiffness matrices, that are functions of both frequency and wave number.

Although the stiffness matrices presented by Kausel and Roesset [1981] are valid for arbitrary layer thicknesses, frequency excitation, and wave number, their application is not surprisingly restricted to the class of problems that could be solved by Haskel and Thompson's original formulation, i.e., closed-form solution of problems involving simple geometries or numerical solutions for multi-layered problems. The basic weakness of the formulation is the dependence of the terms of the stiffness matrix on transcendental functions that make closed-form evaluation of the integral transformation intractable in the general case. However, if the layer thickness is small compared to the wave length of interest, it is possible to linearize these transcendental functions [Lysmer et al., 1969, 1970, 1972] to obtain algebraic expressions that can be handled more easily. This is the approach used by Kausel.

When the soil is homogeneous, the dynamic Green's function G_{ij} can be obtained by using the superposition of the infinite space solution (see [Sen, Davies, and Banerjee, 1985]).

By dividing the pile shaft-soil interface into n cylindrical segments, the base of the piles by circular discs we can use Equation 7-5 to determine the displacements at pile-soil interfaces as

$$u_j(\xi_m,\omega) = \sum_{p=1}^{n} \left[\int_{\Delta S_p} G_{ij}(x,\xi_m,\omega)dS_p \right] \phi_i^p(x,\omega) \qquad (7\text{-}6)$$

where surface tractions ϕ_i^p have been assumed to be constant over each segment of the shaft-soil and pile base-soil interfaces $p(p = 1, \ldots, n)$.

By taking ξ_m successively as the centroidal points of all elements on the boundary we can obtain the following matrix equations for the interfaces:

$$\{u^s\} = [G]\{\phi^s\} \qquad (7\text{-}7)$$

where the superscript s indicates that the displacements and tractions at the pile-soil interface are obtained from a consideration of the soil domain alone. The coefficients of the system Equation 7-7 have all been evaluated by using Gaussian quadrature and a suitable mapping [Banerjee and Butterfield, 1981]. It should be noted that for inclined piles it is convenient to define vectors u and ϕ as the axial and transverse components $\{u_*^s\}$ and $\{\phi_*^s\}$, respectively, with respect to the system of local axes [Banerjee, 1978]. Thus if λ_{ij} are the direction cosines of the angles between these local axes and the global axes, Equation 7-7 then takes the form:

$$\{u_*^s\} = [G_*]\{\phi_*^s\} \tag{7-8}$$

where $[G_*] = [\lambda_a]^T[G][\lambda_b]$, where $[\lambda_a]$ is the direction cosine of the local axis of the pile at which displacements are calculated and $[\lambda_b]$ is that for the loaded pile.

Following the strategy used in the static analysis as outlined by Banerjee, Butterfield [1971a,b,1978], and Poulos [1968, 1979, 1980], we need to solve the governing equations for the pile domain for axial and lateral vibrations (Equations 7-3 and 7-4, respectively) and cast the solution in the form [Sen, Kausel, and Banerjee, 1985]:

$$\{u^p\} = [D]\{\phi^p\} + \{b^p\} \tag{7-9}$$

where $[D]$ = a coefficient matrix
        $\{b\}$ = boundary condition matrix for pile head displacements and rotation

and the superscript p indicates that the pile-soil interface displacements and tractions are obtained from a consideration of the pile domain alone.

In order to solve Equations 7-3 and 7-4 for the prescribed pile head displacements it is necessary to examine the behavior of individual piles when subjected to axial and lateral dynamic loads. This can be best done by considering the effects of applied pile head displacements and rotations as the algebraic sum of the motion of an unsupported pile (i.e., no soil reaction) under the prescribed pile head boundary conditions and those of a fixed head pile (as described below), subjected to soil reaction.

Finally, by satisfying the interface equilibrium and compatibility conditions we obtain from Equations 7-8 and 7-9

$$- [[D] + [G]]\{\phi^p\} = \{b^p\} \tag{7-10}$$

The solution of Equation 7-10 yields the soil reactions. The displacements may then be obtained from 7-8. The total load (axial load, shear force, bending moments) at a depth z in the pile is determined from (for the static case $\omega = 0$):

$$P_z(z) = \int_L^z [\phi_z \pi d + m\omega^2 u_z] dz + A_b \phi_b$$

$$P_x(z) = \int_L^z [\phi_x d + m\omega^2 u_x] dz \qquad (7\text{-}11)$$

$$M(z) = \int_L^z [\phi_x d + m\omega^2 u_x] z dz$$

where A_b and ϕ_b are base area and base reaction, respectively.

The pile group stiffness can be obtained by successively applying the unit vertical, horizontal, and rotational displacement and calculating the resulting system of forces necessary to equilibriate the stresses developed at the pile-soil interface.

Thus, an external loading system V, H, and M acting on the cap can be related to the displacements u_z, u_x, and θ of the cap via

$$\begin{Bmatrix} V \\ H \\ M \end{Bmatrix} = \begin{bmatrix} S_{11} & S_{12} & S_{13} \\ S_{21} & S_{22} & S_{23} \\ S_{31} & S_{32} & S_{33} \end{bmatrix} \begin{Bmatrix} u_z \\ u_x \\ \theta \end{Bmatrix} \qquad (7\text{-}12)$$

The 3×3 matrix in Equation 7-12 which in the dynamic case becomes complex, can be described as the global foundation stiffness matrix, the imaginary part of which represents the effects of geometric (radiation) as well as material damping that can be included in the usual manner by adopting complex elastic moduli.

Thus,

$$S_{ij} = K_{ij} + iC_{ij} \qquad (7\text{-}13)$$

where K_{ij} = the real part of the stiffness

C_{ij} = the imaginary part

By inversion of Equation 7-12, the system flexibility, which is also complex, can be similarly obtained as

$$F_{ij} = I_{ij} + iJ_{ij} \tag{7-14}$$

where I_{ij} = the real part of the flexibility
$\quad\quad J_{ij}$ = the imaginary part

SOLUTION FOR PILES

In order to form Equation 7-9 it is necessary to solve the governing equations of pile motion, Equations 7-3 and 7-4. In all previous work [Poulos and Davis, 1980; Banerjee and Driscoll, 1976; Banerjee, 1978] these were solved approximately for the static case of $\omega = 0$. Kaynia [1982] described an analytical treatment of these equations by assuming that the pile-soil interface tractions could be lumped as point forces. In what follows we have used a similar analytical treatment of Equations 7-3 and 7-4 for distributed pile-soil interface tractions. The exact analytical expressions are described for the dynamic case only. The corresponding static ones can be obtained from these by simply taking the frequency of vibration ω as zero in the limit.

Lateral Vibration of Piles

The general solution of Equation 7-4 for a cantilever subjected to a periodic point force $P_x e^{i\omega t}$ can be represented over the segments I and II (above and below the applied load, respectively) of the pile by:

$$u_x^I = A_1 \cos(\eta z) + A_2 \sin(\eta z) + A_3 \cosh(\eta z) + A_4 \sinh(\eta z) \tag{7-15a}$$

$$u_x^{II} = B_1 \cos(\eta z) + B_2 \sin(\eta z) + B_3 \cosh(\eta z) + B_4 \sinh(\eta z) \tag{7-15b}$$

where $\eta = (m\omega^2/E_p I_p)^{1/4}$ and the constants A_i and B_i are determined by imposing kinematic and force boundary conditions at the two ends along with displacement compatibility and equilibrium condition at the point of application of the load.

The solution for the lateral displacement of the cantilever subjected to a distributed periodic loading intensity of $p_x = d\phi_x$, can be obtained by integrating Equation 7-15a,b over the loaded segment (Figure 7-2). The result can be expressed as:

(i) $z \leq a$
$$u_x = u_x^I = C_1 \cos(\eta z) + C_2 \sin(\eta z) + C_3 \cosh(\eta z)$$
$$+ C_4 \sinh(\eta z) \tag{7-16a}$$

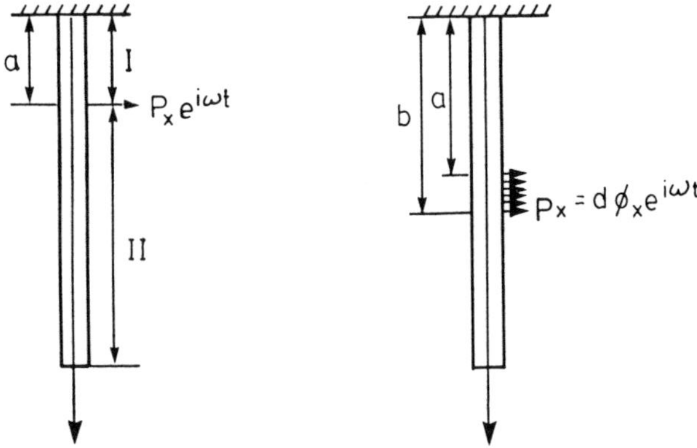

Figure 7-2. Lateral vibration of a cantilever.

(ii) $z \geq b$
$$u_x = u_x^{II} = D_1\cos(\eta z) + D_2\sin(\eta z) + D_3\cosh(\eta z)$$
$$+ D_4\sinh(\eta z) \tag{7-16b}$$

(iii) $a \leq z \leq b$
$$u_x = E_1\cos(\eta z) + E_2\sin(\eta z) + E_3\cosh(\eta z) + E_4\sinh(\eta z)$$
$$+ F_1\cos(\eta z) + F_2\sin(\eta z) + F_3\cosh(\eta z)$$
$$+ F_4\sinh(\eta z) \tag{7-16c}$$

where the constants C_i, D_i, E_i, and F_i are given in the appendix at the end of this chapter. Equations 7-16a,b,c represent the dynamic motion of a fixed head pile (zero displacement and rotation) to which we must add the motion of the unsupported pile subjected to pile head motions.

For the lateral dynamic motion of the unsupported pile (soil reaction $P_x = 0$) the solution for the governing equation can be expressed as:

$$u_x = G_1\cos(\eta z) + G_2\sin(\eta z) + G_3\sinh(\eta z) + G_4\cosh(\eta z) \tag{7-17}$$

where the constants G_i can be determined for the two prescribed pile head motions (lateral displacement and rotation), together with zero force conditions at the free end. The constants for the two cases considered are then given by

(i) $G_1 = \dfrac{T_2}{2T_o}, \quad G_2 = \dfrac{1 - T_4}{2T_o}, \quad G_3 = -G_1, \quad G_4 = 1 - G_2$

(ii) $G_1 = \dfrac{-(1 + T_1)}{2T_o\eta}, \quad G_2 = \dfrac{T_3}{2T_o\eta}, \quad G_3 = -\left(G_1 + \dfrac{1}{\eta}\right), \quad G_4 = -G_2$

where T_i is defined in the appendix. Equations 7-16 and 7-17 can now be added to calculate the system matrix for piles subjected to lateral loads.

Axial Vibration of Piles

The general solution for an axial point force $P_z e^{i\omega t}$ acting at a point along the pile can be expressed as (see Figure 7-3):

$$u_z^I = M_1 \sin\gamma z + M_2 \cos\gamma z \tag{7-18a}$$

$$u_z^{II} = N_1 \sin\gamma z + N_2 \cos\gamma z \tag{7-18b}$$

where $\gamma = \omega \sqrt{\rho_p / E_p A_p}$

The constants M_i and N_i in the preceding equations can be determined following the procedure described for lateral vibration as:

$$N_1 = \frac{P_z}{E_p A_p \gamma} \frac{\sin\gamma a}{\cot\gamma L}, \quad N_2 = \frac{P_z}{E_p A_p \gamma} \sin\gamma a$$

$$M_1 = N_1(1 + \cot\gamma L \cos\gamma a), \quad M_2 = 0$$

The solution for the axial displacement of the pile subjected to a distributed periodic intensity of $p_z = \phi_z \pi d$, can then be obtained by integrating Equations 7-18a,b over the segment (Figure 7-3). The result can be expressed as:

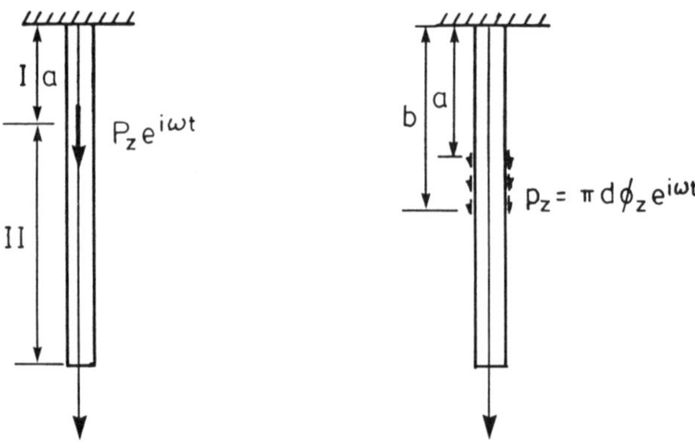

Figure 7-3. Axial vibration of cantilever.

For $0 \leq z \leq a$

$$u_z = u_z^I = \frac{p_z \sin\gamma z}{E_p A_p \gamma^2} \left[\frac{\cos\gamma a - \cos\gamma b}{\cot\gamma L} + \sin\gamma b - \sin\gamma a \right] + u^b(z) \qquad (7\text{-}19a)$$

For $b \leq z \leq L$

$$u_z = u_z^{II} = \frac{p_z}{E_p A_p \gamma^2} \frac{\cos\gamma a - \cos\gamma b}{\cot\gamma L} [\sin\gamma z + \cot\gamma L \, \cos\gamma z] + u^b(z) \quad (7\text{-}19b)$$

For $a \leq z \leq b$

$$u_z = u_z^I \text{ (with } a = z) + u_z^{II} \text{ (with } b = z) + u^b(z) \qquad (7\text{-}19c)$$

where $u^b(z)$ represents the contribution of the base load which can be obtained from Equation 7-19a by substituting $a = L$ and $p_z = A_b\phi_b$ (A_b being the area and ϕ_b being the base load intensity). Thus,

$$u^b(z) = \frac{A_b\phi_b}{E_p A_p \gamma} \frac{\sin\gamma z}{\cos\gamma L} \qquad (7\text{-}19d)$$

For axial dynamic motion of the unsupported pile the general solution of the governing equation is given by:

$$u_z = Q_1 \sin\gamma z + Q_2 \cos\gamma z \qquad (7\text{-}20)$$

where the constant Q_1 is determined for a specified unit pile head axial displacement and zero force at the free end to give the resulting equation of free motion as:

$$u_z = [\tan\gamma L \, \sin z + \cos\gamma z] \qquad (7\text{-}21)$$

Equations 7-19a–d, which are only valid for zero pile head displacement, can now be added to the equation for free motion (Equation 7-21) to provide displacements along the pile for a specified axial head displacement.

RESULTS OF THE ANALYSIS FOR STATIC LOADING

The most widely available results for piles and pile groups are for homogeneous soils. Only Banerjee [1978] and Davies [1979] described the behavior of pile groups embedded in Gibson soil ($E_s = fz$). Some representative results for inhomogeneous soils [Davies, 1979] are described later. The Poisson's ratio has been assumed to be 0.5 and the relative pile flexibility ratio is defined as $K_r = E_p I_p / E_s L^4$ and $K_a = E_p / E_s(L)$.

The influence of a rigid underlying layer on the settlements of some typical axially loaded 2×2 and 3×3 pile groups is shown in Figures 7-6 and 7-7, respectively. It can be seen that the settlement of the larger groups is reduced proportionally more than the smaller groups in such cases. This observation is in accord with intuitive reasoning based on the

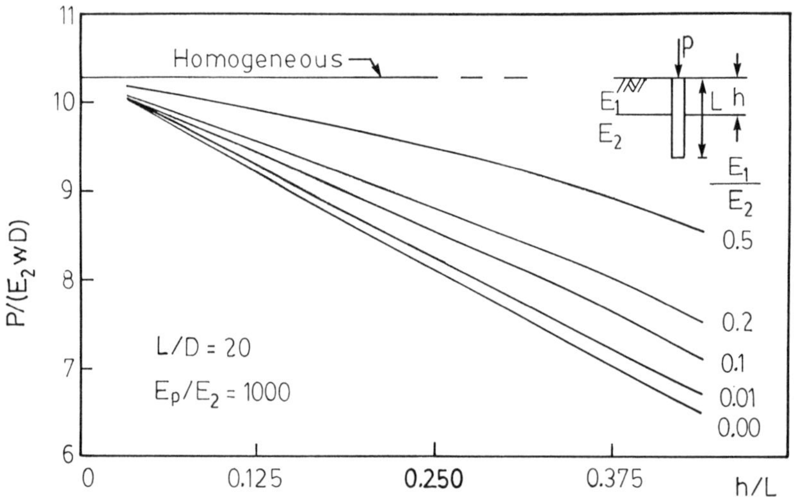

Figure 7-4. Influence of flexible surface layer on axial settlements.

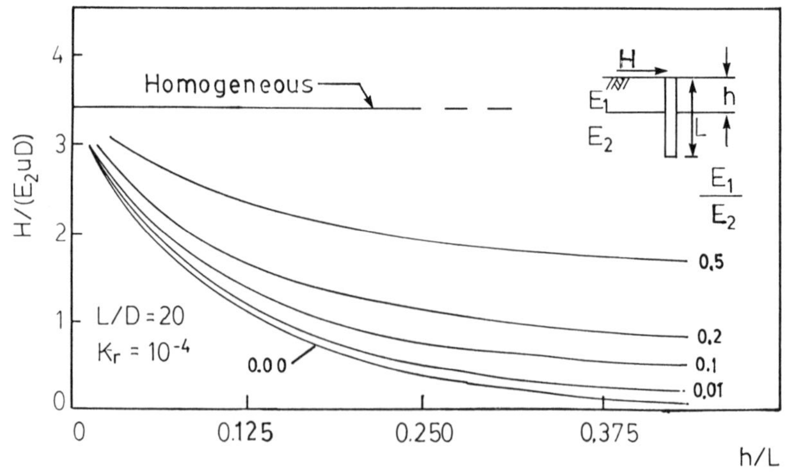

Figure 7-5. Influence of flexible surface layer on lateral displacements.

Figures 7-4 and 7-5 show the effects of a stiffer lower layer on the axial and lateral response of a vertical pile embedded in a two-layer soil. It is easy to see that the lateral response is strongly influenced by the stiffness of the top layer while that for the axial case is influenced more by the lower layer.

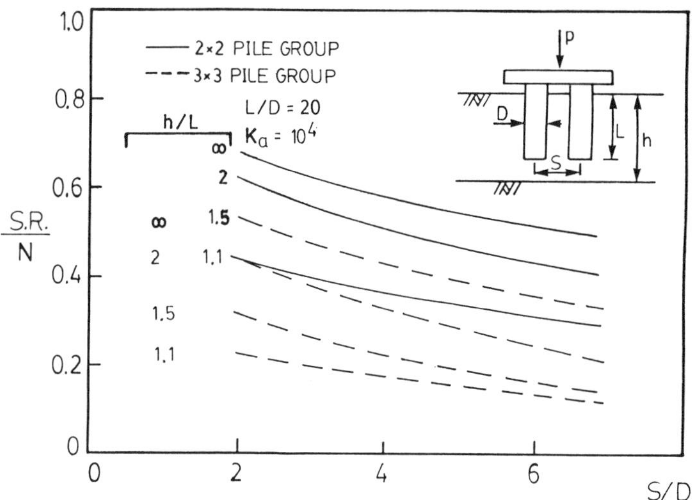

Figure 7-6. Settlement ratios for 2 × 2 and 3 × 3 pile groups. (N = number of piles in group. S.R. = settlement ratio.)

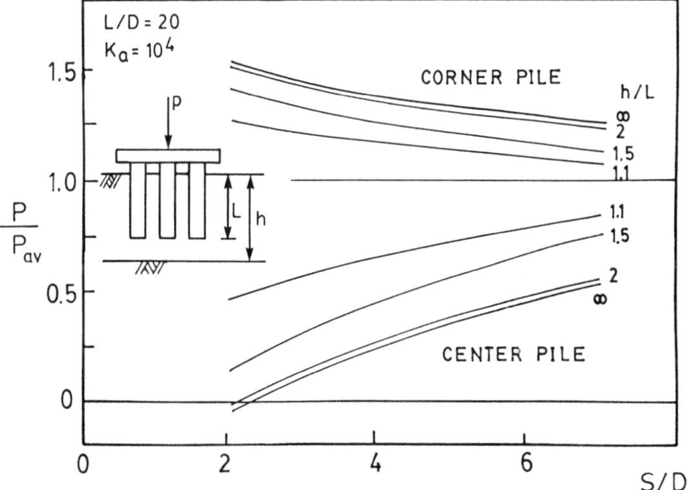

Figure 7-7. Load distribution in 3 × 3 pile groups.

superposition of the "bulb of pressures." The load distribution profiles are much more uniform than those of the corresponding homogeneous cases.

The response of laterally loaded end-bearing pile groups is complicated by the "push-pull" effect due to rotation of the group. This is manifest in Figure 7-8 in which the lateral responses of 2×2 pile groups are much more influenced by rigid basal layers than were the corresponding single piles in Figure 7-5. Figure 7-9 is the corresponding plot for mo-

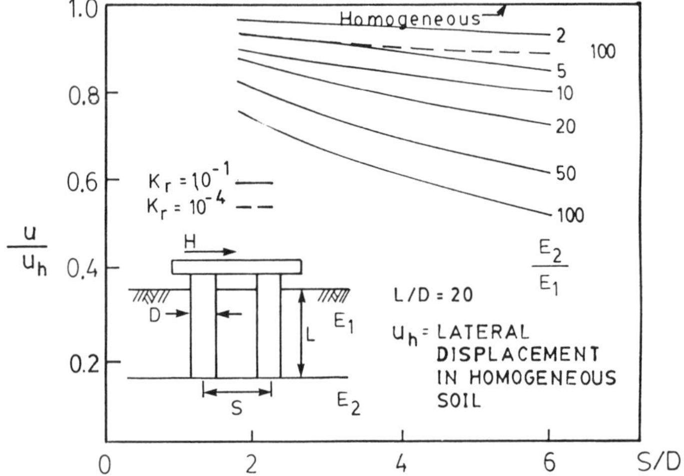

Figure 7-8. Lateral displacement of 2 × 2 group.

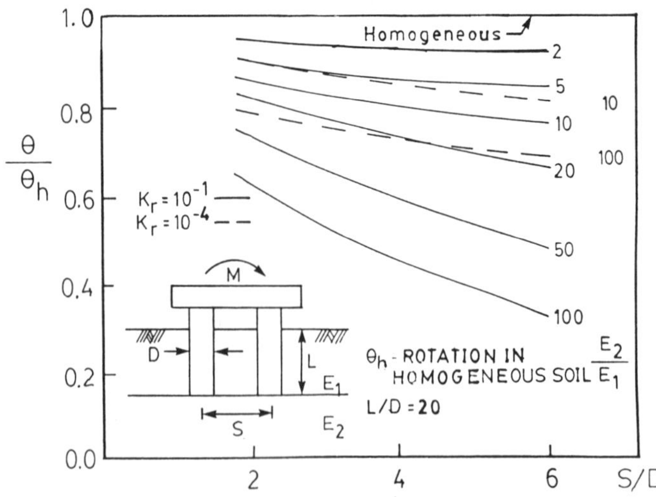

Figure 7-9. Rotation of 2 × 2 group.

ment loading in which the beneficial effects of "push-pull" are clearly in evidence in transferring the load down to the lower layer.

Feagin in 1953 carried out a series of full-scale tests on groups of battered wooden piles fixed in concrete test monoliths subjected to lateral loads. Various configurations of the piles were constructed in order to investigate the effects of group geometry in lateral stiffness.

Details of the tests are as follows:

Embedded length of piles (L) = 30 ft
Mean diameter of the pile head = 13 in.
Mean diameter of the pile base =   9 in.
Angle of batter = 20°

Over the length of the piles the borehole log showed fine to coarse sand containing occasional gravel. Practically the same conditions existed below the pile tips with a slight increase in coarseness up to a depth of 75 ft where the bedrock was encountered.

The computer program developed cannot at present handle tapered piles, although under-reamed piles are permissible. In consequence an average pile diameter of 12 in. was adopted as being representative over the critical region of the pile. The pile cap was assumed to be rigid and free standing.

The sand is characterized by a pseudo-elastic modulus whose magnitude is proportional to the in situ effective stresses. Accordingly, a zero surface modulus was adopted and the parameter f in the nonhomogeneous ($E = fz$) soil model was adjusted until reasonable agreement with the experimental results for the first test monolith was obtained. For $f = 40$ lb-f/in.^{2}/in., the overall theoretical results shown in Figure 7-10 are in substantial agreement with the experimental data and reflect reasonably well the variations associated with changes in the group configurations. It is noteworthy for the first test monolith with $f = 30$ lb-f/in.^{2}/in. and $f = 50$ lb-f/in.^{2}/in. the lateral loads required to produce 0.25 in lateral displacement are 3.7 and 5.3 tons per pile, respectively.

These tests were carried out at relatively close pile spacings ($S/D = 3$) and in consequence, the favorable effects produced by increasing the number of battered piles in the groups are magnified by the decreased interaction of the piles. In addition, an increase in the lateral stiffness occurs in raked pile due to partial transfer of the lateral load by the development of the axial shear stress along the pile shaft.

The lateral deflections due to vertical loads were not specified accurately for each group configuration by Feagin; therefore, some of these test results are in very good agreement with the experimental data.

TEST No.	Schematic Diagram[a]		Lateral load for 1/4 in. deflection, Tons per pile	THEORETICAL results $(M = 40\,lbf\ in^3)$	Lateral deflection due to vertical load of 20 Tons per pile (in inches)	THEORETICAL results $(M = 40\,lbf\ in^3)$
	PLAN	ELEVATION				
1	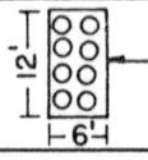	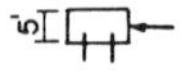	4.8	4.8	0.0	0.0
2	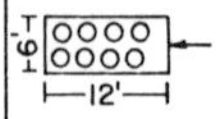	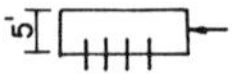	5.8	5.3	0.0	0.0
3	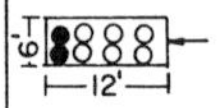	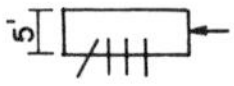	7.0	7.3	0.04	0.06
4	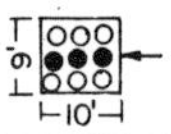	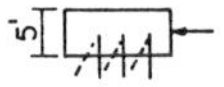	7.1	6.8	0.06	0.08
5	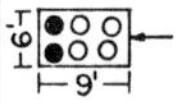	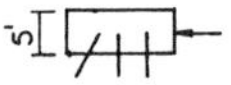	7.3	8.1	0.05	0.07
6	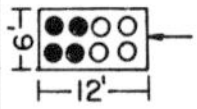	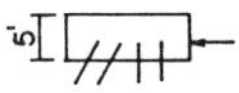	9.0	8.4	0.07	0.11
7	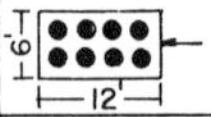	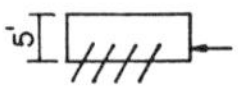	9.0	8.2	0.21	0.27
8	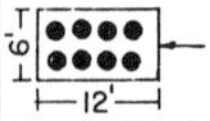	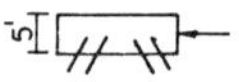	15.8	11.7	0.0	0.0

a o = vertical pile   ● = battered pile

Figure 7-10. Comparisons with full-scale tests by Feagin [1953].

RESULTS OF DYNAMIC ANALYSIS

The results of the present analysis yields in the limiting case of $a_o = 0$, the static solutions described extensively in the published literature by Banerjee, Butterfield, and others as well as by Poulos. The maximum difference being of the order of 5% when compared with the most accurate static solutions. In what follows results have been normalized with respect to the static solutions to demonstrate the effects of dynamic loading on the predicted static behavior. In order to demonstrate the effects of inhomogeneity a parameter $\chi = E(O)/E(L)$ has been introduced.

Figures 7-11 and 7-12 show the effect of the variation of frequency on the stiffness and damping (i.e., the imaginary part of the pile stiffness) of compressible and rigid piles embedded in inhomogeneous soils with $\chi = 0$ and $\chi = 0.5$ normalized with respect to the respective static stiffness. The same graphs also show the variation in the axial stiffness for the case of piles embedded in a homogeneous soil characterized by $\chi = 1$. It may be seen that the variation in the stiffness and damping with frequency is similar for each case.

Figures 7-13–7-15 show the variation of the real and imaginary parts of the lateral (K_{22}, C_{22}), cross-coupled (K_{23}, C_{23}), and rocking (K_{33}, C_{33}) stiffness for flexible ($K_R = 10^{-4}$) piles for three different values of the inhomogeneity index corresponding to $\chi = 0$, 0.5, and 1.0. The most interesting feature in these figures is the similarity in the variation in the pile stiffness in homogeneous ($\chi = 1$) and inhomogeneous medium ($\chi = 0$, $\chi = 0.5$).

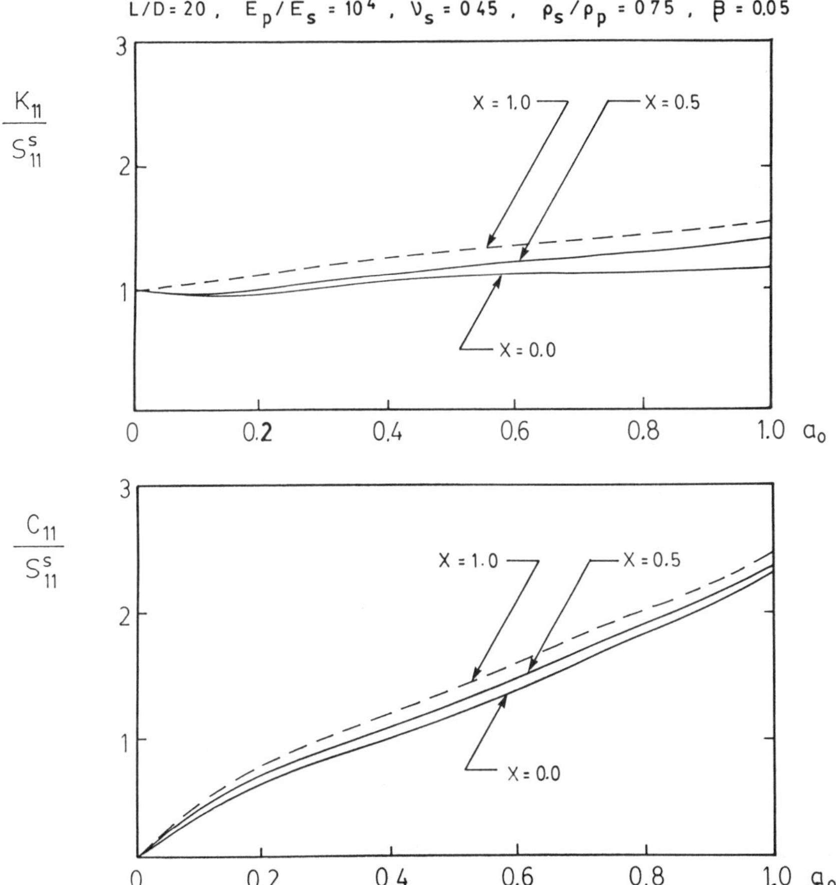

Figure 7-11. Effect of frequency variation on the dynamic vertical stiffness of axially rigid piles embedded in an inhomogeneous soil with $\chi = 0$, 0.5, and comparison with results for homogeneous case ($\chi = 1$).

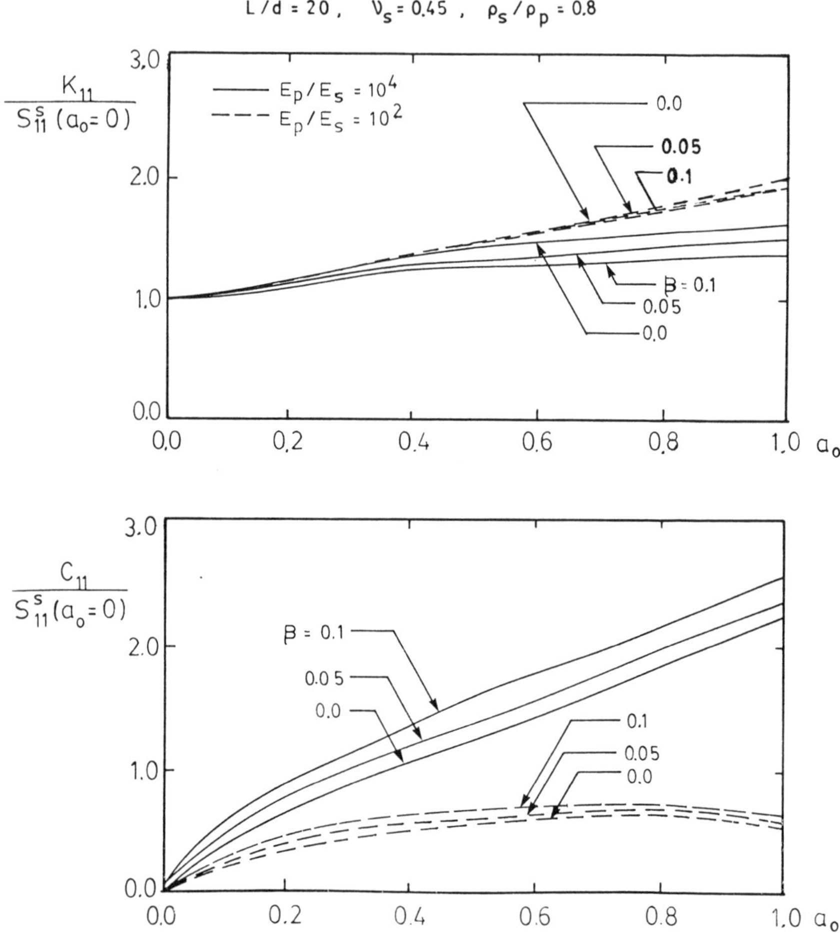

Figure 7-12. Effect of variation of frequency and hysteretic damping on the dynamic stiffness of compressible and axially rigid piles.

For flexible piles ($K_R = 10^{-4}$), the proportionate increase in the stiffness with frequency tends to be larger for piles embedded in inhomogeneous soils than it is for its homogeneous counterparts, especially at higher frequency. This is probably because of larger displacements in the piles embedded in the inhomogeneous soil so that the proportionate contribution of the pile inertia to the overall stiffness is greater.

The behavior of pile groups can be obtained by superposition of the interactions between the piles, conveniently characterized by the so-called "interaction factors." Static interaction factors were originally

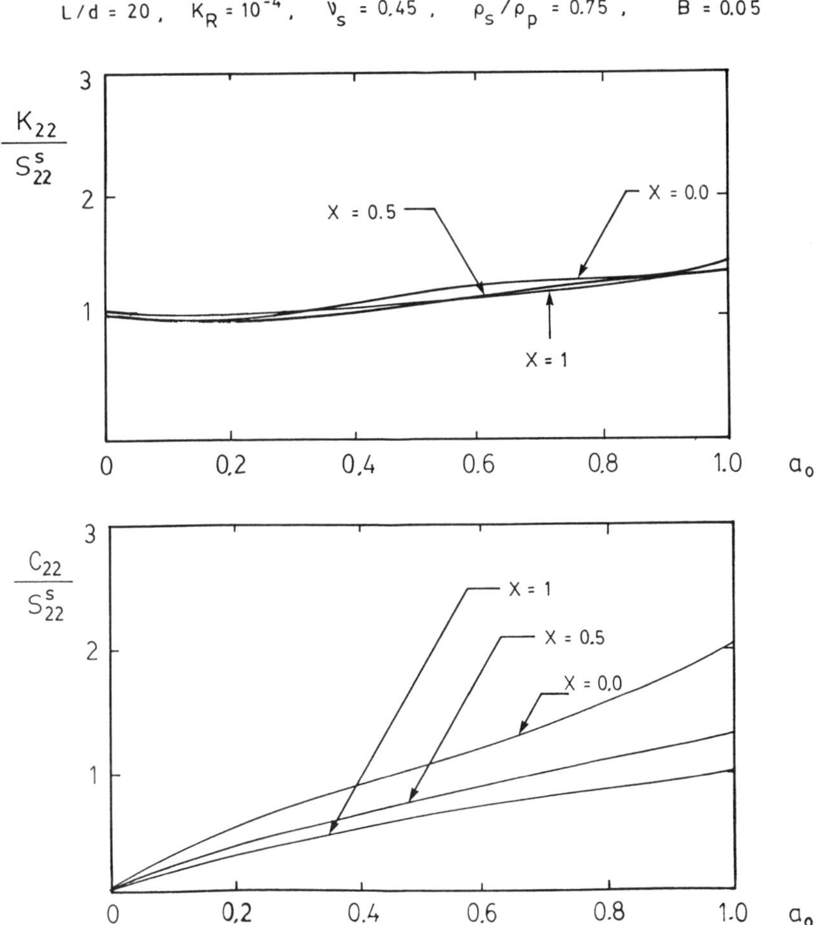

Figure 7-13. Effect of frequency variation on the lateral stiffness of flexible piles embedded in an inhomogeneous medium with χ = 0, 0.5, and comparison with corresponding homogeneous case (χ = 1).

proposed by Poulos [1968] and the concept has been extended to dynamic loading by Kaynia and Kausel [1982]. Formally, the dynamic interaction factor α_{ij} between two piles i and j is defined as:

$$\alpha_{ij} = \frac{\text{(dynamic displacement of pile i}}{\text{(static displacement of pile j}} \tag{7-22}$$

$$\alpha_{ij} = \frac{\begin{array}{c}\text{(dynamic displacement of pile i}\\\text{due to load applied to pile j)}\end{array}}{\begin{array}{c}\text{(static displacement of pile j}\\\text{due to load applied to pile j)}\end{array}} \tag{7-22}$$

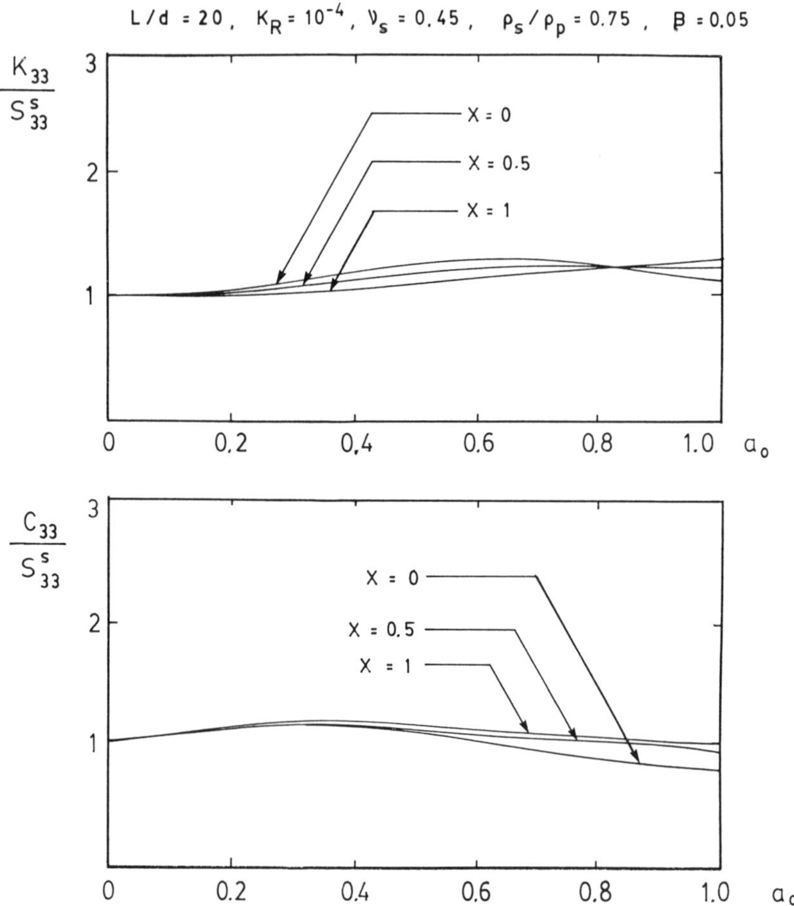

Figure 7-14. Effect of frequency variation on the rocking stiffness of flexible piles embedded in an inhomogeneous medium with χ = 0, 0.5, and comparison with corresponding homogeneous case (χ = 1).

The displacements of all piles in a group of N piles can be obtained from the equation:

$$u_i = u_o \sum_{j=1}^{N} \alpha_{ij} P_j \tag{7-23}$$

where u_o = static displacement of a single pile subjected to unit static load

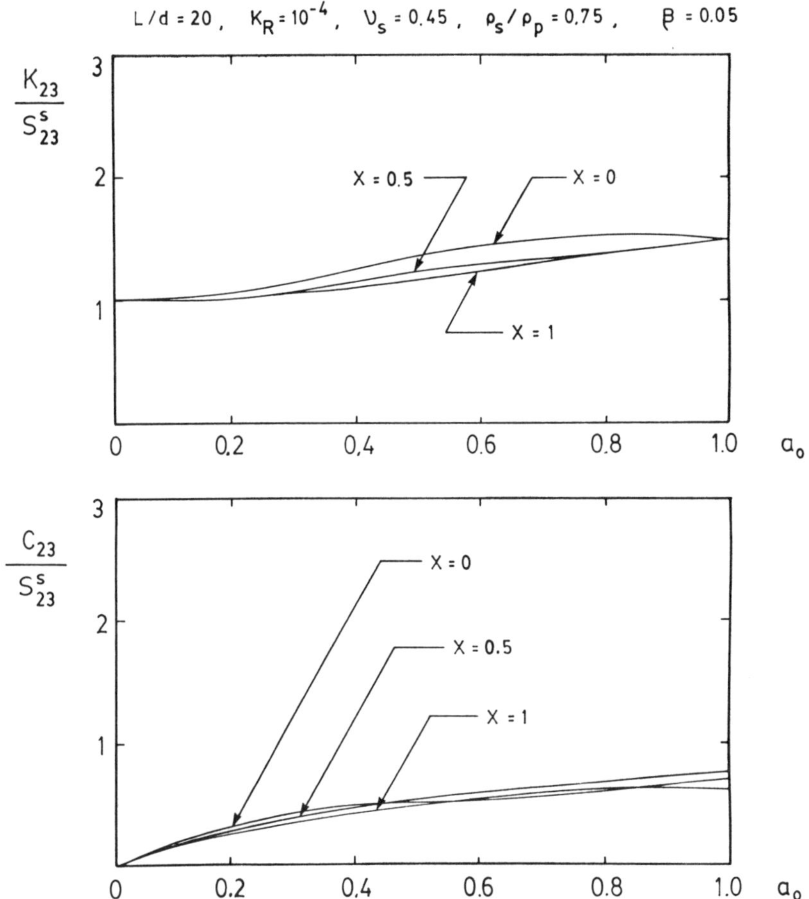

Figure 7-15. Effect of frequency variation on the cross-couple stiffness of flexible piles embedded in an inhomogeneous medium with $\chi = 0$, 0.5, and comparison with corresponding homogeneous case ($\chi = 1$).

u_i = dynamic displacement of the ith pile and

P_j = dynamic load on the jth pile

The real and imaginary parts of the dynamic interaction factors for vertical loading calculated on the basis of Equation 7-22 have been plotted for homogeneous soils in Figures 7-16 and 7-17. The same figures also show comparisons with the dynamic interaction factors presented by Kausel and Kaynia [1982] where it can be seen that the agreement is fairly close except for larger a_o values. This is probably due to the differ-

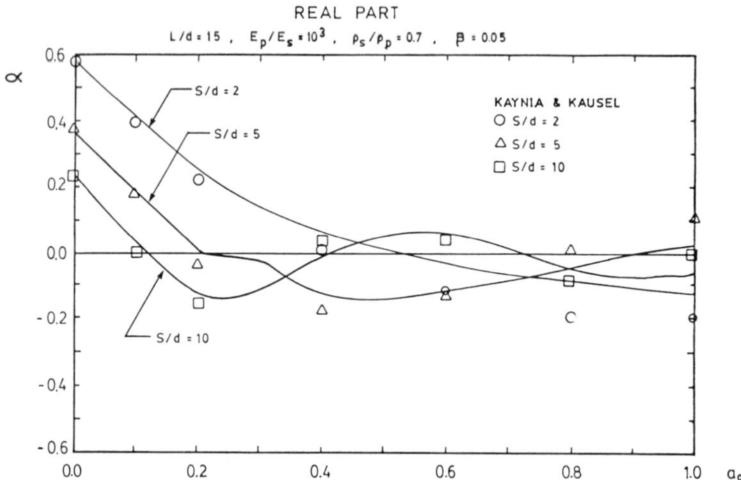

Figure 7-16. Interaction curves for the real vertical displacement of Pile 2 due to a vertical force on Pile 1 and comparison with Kaynia and Kausel's results.

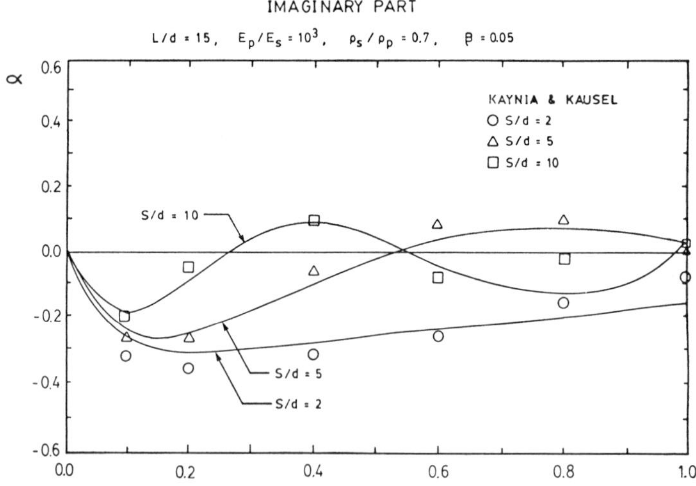

Figure 7-17. Interaction curves for the out-of-phase vertical displacement of Pile 2 due to a vertical force on Pile 1 and comparison with Kaynia and Kausel's results.

ences in the inertia term (see [Sen, Davies, and Banerjee, 1985]) of Equation 7-11.

The interaction factors show certain unusual characteristics in comparison to the static ones. Whereas in the static case, the interaction factors are always positive and tend to decay exponentially to zero with increasing S/d ratios, the dynamic ones show a sinusoidal variation with frequency. In other words, at certain frequencies, the α magnitudes are negative depending on the S/d ratio: the higher the S/d ratio, the lower the frequency at which α begins to be negative.

It is evident that if the interaction factors are negative then the displacement of a 2×1 group is less than that for a single pile subject to the same loading per pile. This can only happen if the displacement of each pile in a group is completely out of phase with respect to the other. In other words, when one pile is displacing downwards, the other pile would be displacing upwards.

Figures 7-18 and 7-19 show the variation in the real and imaginary parts of the dynamic stiffness for 3×3 groups, normalized with respect to the static stiffnesses. It may be seen that both the real and imaginary parts display sharp peaks at certain frequencies and the stiffnesses are

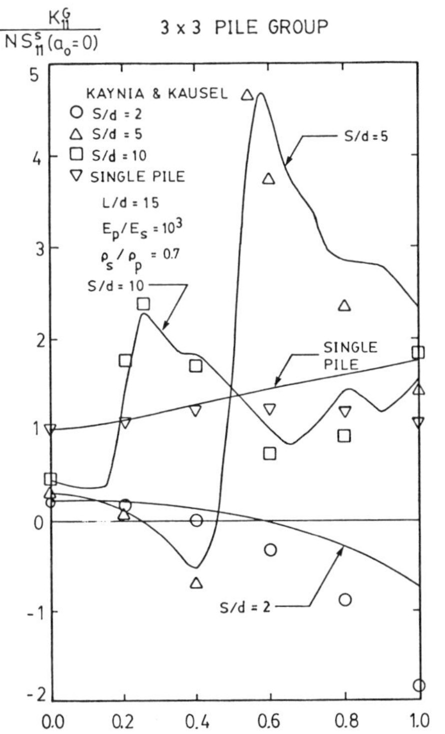

Figure 7-18. Variation in the real vertical dynamic stiffness of a 3 x 3 group with frequency for spacing ratios of 2, 5, and 10 and comparison with Kaynia and Kausel's solution ($\beta = 0.05$).

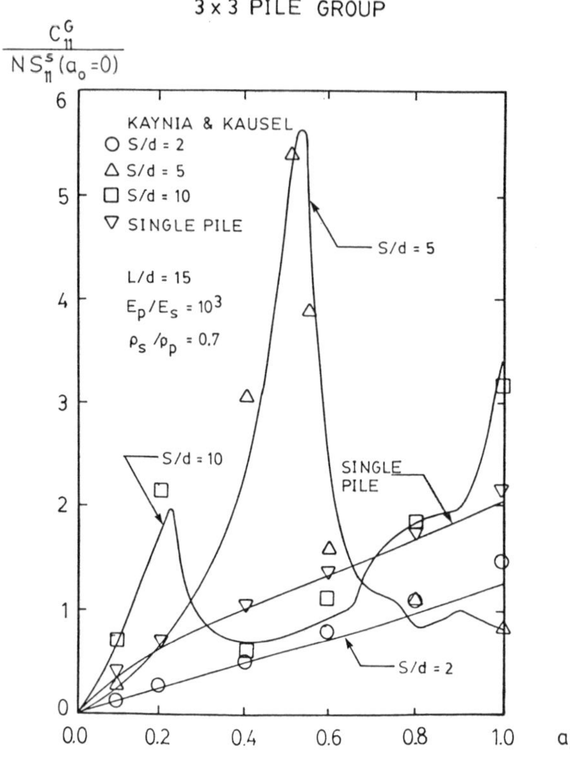

Figure 7-19. Variation in the imaginary vertical dynamic stiffness with frequency for the 3 × 3 group shown in Figure 7-18.

several times the static groups stiffness (which cannot exceed unity for the normalization used). In view of the observations made in the previous section, this variation is to be expected. Indeed, by reference to interaction factors, it may be confirmed that the peaks coincide with negative or zero α values and the troughs with positive α values.

Figures 7-18 and 7-19 also show comparisons with Kausel and Kaynia's [Kaynia, 1982] analysis. It may be seen that the agreement is very good at low frequencies and for frequencies that result in peaks in the stiffnesses. The agreement is poor at higher frequencies where no peaks are present, where the results from the present analysis show higher group stiffness. Although the agreement with Kausel and Kaynia for the peak values of stiffness might seem surprising at first, a little reflection will indicate that this is to be expected. The high peaks are caused by out-of-phase motions of piles when the contribution of the pile inertia from in-phase and out-of-phase piles cancel out.

Figures 7-20–7-23 show the variations in the real and imaginary parts of the horizontal and rocking stiffnesses for the same 3×3 group that was the subject of discussion regarding axial behavior. Only two values of S/ d, S/d = 2 and 5, have been shown. On the basis of the discussion on interaction factors in the previous section, the stiffness varies exactly as anticipated. There is an initial decrease, as piles are in phase, followed by a sharp increase indicating out-of-phase piles. The interaction is much weaker as evidenced by shallower and wider peaks.

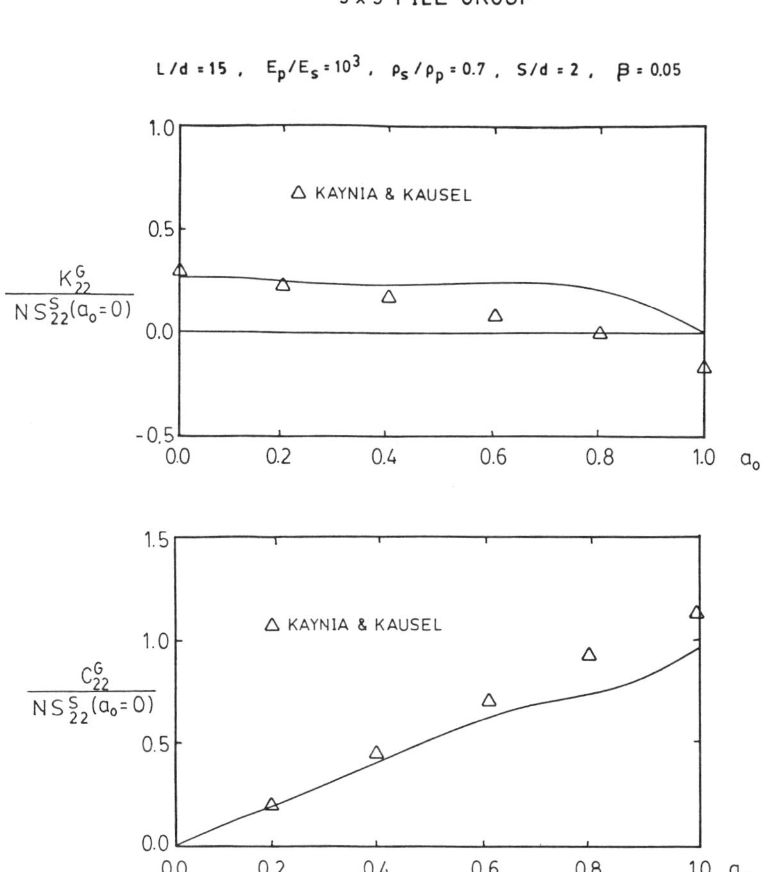

Figure 7-20. Variation in the dynamic horizontal stiffness with frequency for the 3 x 3 group considered in Figure 7-18 for S/d = 2 and comparison with Kaynia and Kausel's results.

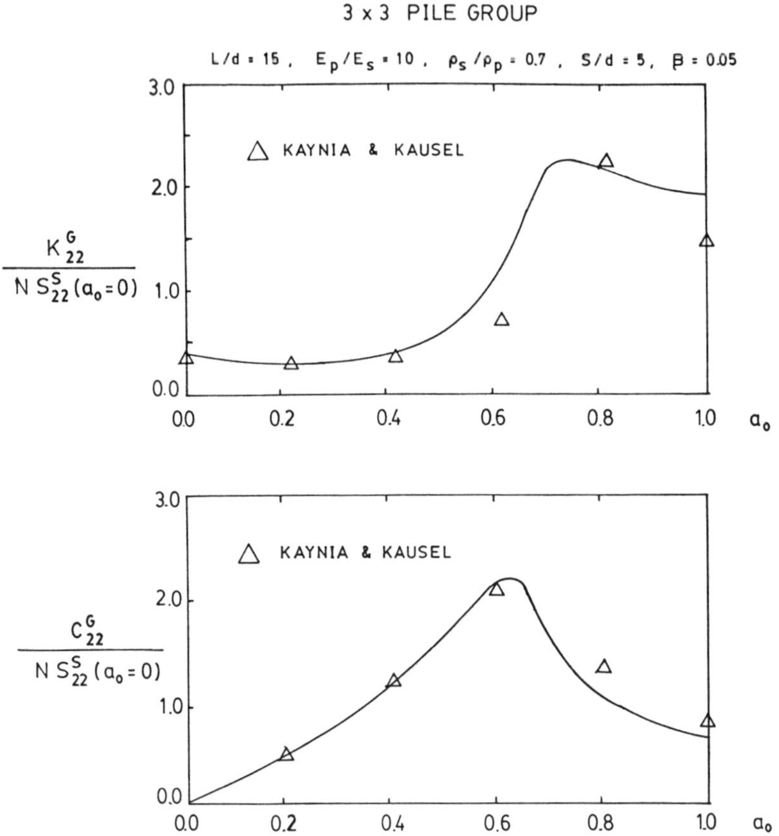

Figure 7-21. Variation in the dynamic horizontal stiffness with frequency for the 3 x 3 group considered in Figure 7-18 for S/d = 5 and comparison with Kaynia and Kausel's results.

CONCLUSIONS

A hybrid boundary element formulation has been presented for the three-dimensional steady-state analysis of piles and pile groups embedded in a nonhomogeneous medium in which the soil modulus varies linearly with depth. The dynamic Green's function developed earlier by Kausel as well as an explicit dynamic point force solution in a half space have been implemented and the equations of motion for the piles have been solved. Comparisons with available static and dynamic solutions confirm the accuracy of the proposed formulation. This formulation

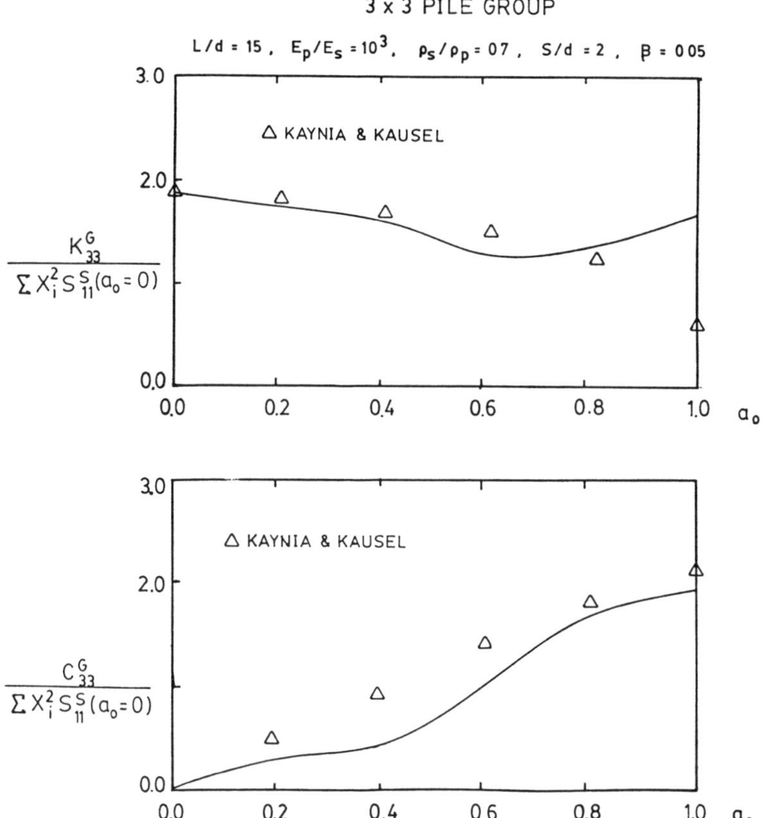

Figure 7-22. Variation in the dynamic rocking stiffness with frequency for the 3 × 3 group considered in Figure 7-18 for S/d = 2 and comparison with Kaynia and Kausel's results.

has been used to obtain the response of single piles and pile groups. Results presented show that in certain instances the normalized stiffness variation of both single piles and pile groups embedded in a nonhomogeneous medium show similarities to their homogeneous counterpart.

APPENDIX—LATERAL VIBRATION: CONSTANTS OF INTEGRATION

$$C_1 = C'[T_3\{\sin(\eta b) - \sin(\eta a)\} - T_3\{\sinh(\eta b) - \sinh(\eta a)\} + (T_4 - 1)\{\cosh(\eta b) - \cosh(\eta a)\} + (T_1 + 1)\{\cos(\eta b) - \cos(\eta a)\}]$$

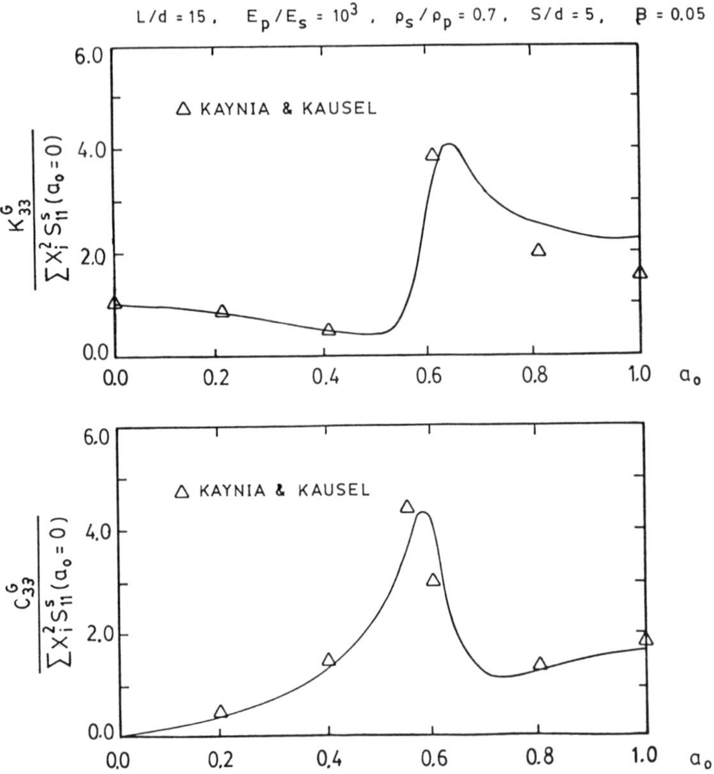

Figure 7-23. Variation in the dynamic rocking stiffness with frequency for the 3 x 3 group considered in Figure 7-18 for S/d = 5 and comparison with Kaynia and Kausel's results.

$$C_2 = -C'[(T_4 - 1)\{\sin(\eta b) - \sin(\eta a)\} - (T_1 + 1)\{\sinh(\eta b) - \sinh(\eta a)\} + T_2\{\cosh(\eta b) - \cosh(\eta a)\} + T_2\{\cos(\eta b) - \cos(\eta a)\}]$$

$$C_3 = -C_1, \; C_4 = -C_2$$

$$D_1 = C'[T_3\{\sin(\eta b) - \sin(\eta a)\} - T_3\{\sinh(\eta b) - \sinh(\eta a)\} + (T_4 - 1)\{\cosh(\eta b) - \cosh(\eta a)\} + (T_4 - 1)\{\cos(\eta b) - \cos(\eta a)\}]$$

$$D_2 = -C'[(T_1 + 1)\{\sin(\eta b) - \sin(\eta a)\} - (T_1 + 1)\{\sinh(\eta b) - \sinh(\eta a)\} + T_2\{\cosh(\eta b) - \cosh(\eta a)\} + T_2\{\cos(\eta b) - \cos(\eta a)\}]$$

$D_3 = (D_1 T_1 + D_2 T_3), D_4 = -(D_1 T_2 + D_2 T_4)$

E_i is identical to C_i with a replaced by z

F_i is identical to D_i with b replaced by z

$T_o = 1 + \cos(\eta L)\cosh(\eta L)$

$T_1 = \cos(\eta L)\cosh(\eta L) + \sin(\eta L)\sinh(\eta L)$

$T_2 = \sin(\eta L)\cosh(\eta L) + \cos(\eta L)\sinh(\eta L)$

$T_3 = \sin(\eta L)\cosh(\eta L) - \cos(\eta L)\sinh(\eta L)$

$T_4 = \sin(\eta L)\sinh(\eta L) - \cos(\eta L)\cosh(\eta L)$

$C' = p_x/(4E1\eta^4 T_o)$

REFERENCES

1. Apsel, R., "Dynamic Green's Functions for Layered Media and Applications to Boundary-value Problems," Ph.D. Thesis submitted to the Department of Applied Mechanics and Engineering Sciences, University of California, San Diego, 1979.
2. Banerjee, P. K., "Analysis of Axially and Laterally Loaded Pile Groups," *Developments in Soil Mechanics,* Scott, C. R. (Ed.), Applied Science Publishers, London, 1978, pp. 317–346.
3. Banerjee, P. K., and Butterfield, R., *Boundary Element Methods in Engineering Science,* McGraw-Hill, London, 1981.
4. Banerjee, P. K., and Driscoll, R. M., "Three-dimensional Analysis of Raked Pile Groups," Proc. of Inst. Civil Engineers, Research and Theory, Vol. 61, 1976, pp. 653–671.
5. Banerjee, P. K., and Davies, T. G., "Analysis of Pile Groups Embedded in Gibson Soil," Proc. 9th International Conference on Soil Mechanics and Foundation Engineering, Tokyo, 1977, pp. 381–386.
6. Banerjee, P. K., and Davies, T. G., "The Behavior of Axially and Laterally Loaded Single Piles Embedded in Nonhomogeneous Soils," *Geotechnique,* Vol. 28, No. 3, 1978, pp. 309–326.
7. Banerjee, P. K., and Davies, T. G., "Analysis of Some Case Histories of Axially and Laterally Loaded Pile Groups," Numerical Methods in Offshore Piling, Institution of Civil Engineers, London, May 1979, pp. 101–108.

8. Butterfield, R., and Banerjee, P. K., "Analysis of Axially Loaded Compressible Piles and Pile Groups," *Geotechnique,* Vol. 21, No. 1, 1971, pp. 43–60.

9. Butterfield, R., and Banerjee, P. K., "Analysis of Axially Loaded Compressible Piles and Pile Groups," *Geotechnique,* Vol. 21, No. 1, 1971, pp. 135–142.

10. Davies, T. G., "Linear and Non-linear Analyses of Pile Groups," Ph.D. Thesis, University College, Cardiff, UK, 1979.

11. Davies, T. G., Sen, R., and Banerjee, P. K., "Dynamic Behavior of Pile Groups in Inhomogeneous Soils," *Journal of Geotechnical Engineering Division,* ASCE, Vol. 111, No. 12, 1985, pp. 1365–1379.

12. Davies, T. G., and Banerjee, P. K., "Displacement Field due to a Point Force at the Interface of a Two-layer Elastic Halfspace," *Geotechnique,* Vol. 28, No. 1, pp. 43–56.

13. Gibson, R. E., "Some Results Concerning Displacements and Stresses in a Non-homogeneous Elastic Halfspace," *Geotechnique,* Vol. 17, 1967, pp. 58–67.

14. Haskel, N. A., "The Dispersion of Surface Waves on Multilayered Media," *Bull. Seis. Soc. Am.,* Vol. 43, pp. 17–34, 1950.

15. Kausel, E., "An Explicit Solution for the Green's Function for Dynamic Loads in Layered Media," MIT Research Report R81-13, Order No. 699, Department of Civil Engineering, MIT, Cambridge, Mass., 1981.

16. Kausel, E., and Roesset, J. M., "Stiffness Matrices for Layered Soils," *Bull. Seis. Soc. Am.,* Vol. 71, 1981, pp. 89–93.

17. Kausel, E., and Peek, R., "Dynamic Loads in the Interior of a Layered Stratum: An Explicit Solution," *Bulletin of the Seismological Society of America,* Vol. 72, No. 5, October, 1982, pp. 1459–1481.

18. Kaynia, A. M., "Dynamic Stiffnesses and Seismic Response of Pile Groups," Thesis presented to the Massachusetts Institute of Technology, at Cambridge, Mass., in 1982 in partial fulfillment of the requirements for the degree of Doctor of Philosophy.

19. Kaynia, A. M., and Kausel, E., "Dynamic Behavior of Pile Groups," 2nd International Conference on Numerical Methods in Offshore Piling, 1982, Austin, TX.

20. Krishnan, R., Gazetas, G., and Velez, A., "Static and Dynamic Lateral Deflection of Piles in Nonhomogeneous Soil Stratum," *Geotechnique,* Vol. 33, No. 3, 1983, pp. 307–326.

21. Kuhlemeyer, R. L., "Vertical Vibration of Piles," *Journal of the Geotechnical Engineering Division,* ASCE, Vol. 105, No. GT2, 1979, pp. 273–287.

22. Kuhlemeyer, R. L., "Static and Dynamic Laterally Loaded Floating Piles," *J. of the Geotech. Engng. Div.,* ASCE, Vol. 105, No. GT2, 1979, pp. 289–304.

23. Lysmer, J., "Lumped Mass Method for Rayleigh Waves," *Bull. Seis. Soc. Am.,* Vol. 60, No. 1, 1970, pp. 89–104.
24. Lysmer, J., and Waas, G., "Shear Waves in Plane Infinite Structures," *J. of the Engng. Mech. Div.,* ASCE, Vol. 98, No. EM1, Proc. Paper 8716, 1972, pp. 85–105.
25. Lysmer, J., and Kuhlemeyer, R. L., "Finite Dynamic Model for Infinite Media," *J. of the Engng. Mech. Div.,* ASCE, Vol. 95, No. EM4, Proc. Paper 6719, 1969, pp. 859–877.
26. Matlock, H., and Foo, S. H. C., "Axial Analysis of Piles Using a Hysteretic and Degrading Soil Model," *Numerical Methods in Offshore Piling,* London, Institution of Civil Engineers, May, 1979, pp. 127–134.
27. Nogami, T., "Dynamic Group Effect of Multiple Piles Under Vertical Vibration," Proceedings of the ASCE Engineering Mechanics Specialty Conference, Austin, Texas, 1979, pp. 750–754.
28. Nogami, T., "Dynamic Stiffness and Damping of Pile Groups in Inhomogeneous Soil," Proc. of Session on Dynamic Response of Pile Foundations: Analytical Aspects, ASCE, Oct. 1980, pp. 31–52.
29. Nogami, T., and Novak, M., "Soil-pile Interaction in Vertical Vibration," *Journal of Earthquake Engineering and Structural Dynamics,* Vol. 4, Jan.–Mar. 1976, p. 277–293.
30. Novak, M., "Vertical Vibration of Floating Piles," *J. of the Engng. Mech. Div.,* ASCE, Vol. 103, No. EM1, Feb. 1977, pp. 153–168.
31. Novak, M., "Dynamic Stiffness and Damping of Piles," *Canadian Geotechnical Journal,* Vol. 11, 1974, pp. 574–598.
32. Novak, M., "Soil Pile Interaction Under Dynamic Loads," Proc. of the Int. Symp. on Num. Methods in Offshore Piling, London, England, 1979, pp. 41–50.
33. Novak, M., and Aboul-Ella, F., "Impedance Functions of Piles in Layered Media," *Journal of the Engineering Mechanics Division,* ASCE, Vol. 104, No. EM3, Proc. Paper 1384, June 1978, pp. 643–661.
34. Novak, M., and Aboul-Ella, F., "Stiffness and Damping of Piles in Layered Media," Proc. Earthquake Engineering and Soil Dynamics, ASCE Specialty Conference, Pasadena, Calif., June 19–21, 1978, pp. 704–719.
35. Penzien, J., "Soil-pile Foundation Interaction," in R. L. Wiegel (Ed.), Earthquake Engineering, Prentice-Hall, Inc., Englewood Cliffs, NJ, 1970.
36. Penzien, J., Scheffley, C. F., and Parmalee, F. A., "Seismic Analysis of Bridges on Long Piles," *J. of the Engng. Mech. Div.,* ASCE, EM3, 1964.
37. Poulos, H. G., "Analysis of the Settlement of Pile Groups," *Geotechnique,* Vol. 18, 1968, pp. 449–471.

38. Poulos, H. G., "Settlement of Single Piles in Non-homogeneous Soil," *J. of the Geotech. Eng. Div.*, ASCE, Vol. 105, No. GT5, 1979, pp. 627–641.
39. Poulos, H. G., and Davis, E. H., *Pile Foundation Analysis and Design*, John Wiley and Sons, New York, 1980.
40. Randolph, M. F., and Wroth, C., "Analysis of Deformation of Vertically Loaded Piles," *Journal of the Geotechnical Engineering Division*, ASCE, Vol. 104, No. GT12, pp. 1465–1488.
41. Sen, R., "Dynamic Analysis of Buried Structures," Ph.D. Dissertation, State Univ. of N.Y. at Buffalo, 1984.
42. Sen, R., Davies, T. G., and Banerjee, P. K., "Dynamic Behavior of Axially and Laterally Loaded Piles and Pile Groups Embedded in Inhomogeneous Soil," *Earthquake Engineering and Structural Dynamics*, Vol. 13, 1985, pp. 53–65.
43. Sen, R., Kausel, E., and Banerjee, P. K., "Dynamic Behavior of Axially and Laterally Loaded Piles and Pile Groups Embedded in Inhomogeneous Soil," *Int. Jour. Numerical and Analytical Methods in Geomechanics*, Vol. 9, No. 6, pp. 507–524.
44. Sheta, M., and Novak, M., "Vertical Vibration of Pile Groups," *Journal of the Geotechnical Engineering Division*, ASCE, Vol. 108, No. GT4, April 1982, pp. 570–590.
45. Thompson, W. T., "Transmission of Elastic Waves through a Stratified Soil Stratum," *J. Appl. Phys.*, Vol. 21, 1950, pp. 89–93.
46. Waas, G., and Hartman, H. G., "Pile Foundations Subjected to Dynamic Horizontal Loads," *Modelling and Simulations of Large Scale Structural Systems*, Capri, Italy, 1981, p. 17.
47. Wolf, J. P., and Von Arx, G. A., "Impedance Functions for a Group of Vertical Piles," Proc. ASCE Specialty Conference on Earthquake Engineering and Soil Dynamics, Pasadena, Calif., II, 1978, pp. 1024–1041.

8

Geotextiles and Related Products

J. P. Giroud
GeoServices Inc. Consulting Engineers
Boynton Beach, Florida USA

Geotextiles have transformed geotechnical engineering to the point that it is no longer possible to do geotechnical engineering without geotextiles. In ten years they have progressively pervaded all branches of geotechnical engineering in what may be one of the most important revolutions in the history of geotechnical engineering.

Geotextiles are used as a practical means to solve construction problems and, at the same time, have opened up new opportunities for creativity for the geotechnical engineer. Geotextiles are used for drainage against basement walls, but also in large dams. They are used for temporary erosion control, but also in sophisticated reinforced structures with a 120-year design life. They are used in such traditional geotechnical applications as landslide rehabilitation, but also in new applications such as hazardous waste containment facilities.

Once simply considered a convenient technique, geotextiles are now a full-fledged discipline. There is now an International Geotextile Society and three international conferences on geotextiles have been held. Considerable knowledge has been gained and an impressive amount of information is now available: more than 2,000 papers on geotextiles have been compiled by Scott and Richards [1985]. This chapter will present the basic principles of geotextile engineering.

DEFINITIONS AND TYPES

Definition of Geotextiles

Geotextiles are textiles (fabrics) used in geotechnical engineering. The following definition has been adopted by the American Society for Testing and Materials (ASTM): "A geotextile is any permeable textile material used with foundation, soil, rock, earth, or any other geotechnical engineering-related material, as an integral part of a manmade project, structure, or system."

The types of geotextiles that strictly comply with the ASTM definition are the knitted, woven, and nonwoven fabrics, which are typical products of the textile industry. Other products such as webbings, geomats, geonets, geogrids, and formed plastic sheets have been developed for use in combination with, or in place of, geotextiles. These products do not fit the ASTM definition of geotextiles because they are not textile materials. However, because of similarities in appearance, function, or application with geotextiles, these products are referred to as "geotextile-related products."

It is important not to confuse geotextiles with geomembranes. The term geomembrane refers to low permeability sheets (membranes) used in geotechnical engineering. The ASTM definition of geomembrane is as follows: "A geomembrane is any impermeable membrane used with foundation, soil, rock, earth, or any other geotechnical engineering-related material, as an integral part of a man-made project, structure, or system."

In fact, nothing is "impermeable" and it is more correct to state that geomembranes, which are intended to be used as fluid barriers, are constructed to be as impermeable as possible. Permeability is the most obvious difference between geotextiles and geomembranes: geotextiles are permeable by construction, and geomembranes are as impermeable as possible by design. In other words, geotextiles allow or conduct fluid flow, while geomembranes restrict fluid flow. The permeability of a material is evaluated by its hydraulic conductivity, also called coefficient of permeability (see the section "Permeability Normal to the Plane"). Typical values of hydraulic conductivity are: 10^{-5} m/s to 1 m/s (10^{-3} to 10^2 cm/s) for geotextiles (or even more in the case of some geotextile-related products such as geomats, geonets, and geogrids) and 10^{-14} to 10^{-13} m/s (10^{-12} to 10^{-11} cm/s) for geomembranes. The hydraulic conductivity of geotextiles is of the same order as the hydraulic conductivity of highly permeable soils such as sand and gravel. The hydraulic conductivity of geomembranes is much smaller than the hydraulic conductivity of clay, which is the least permeable soil.

The term "geosynthetics" is increasingly used to encompass geotextiles, geotextile-related products, and geomembranes. In this chapter only geotextiles and geotextile-related products are discussed, i.e., all geosynthetics except geomembranes.

Types of Geotextiles and Related Products

Twenty types of geotextiles and related products are presented in Figure 8-1. They are classified according to the manufacturing process. Two

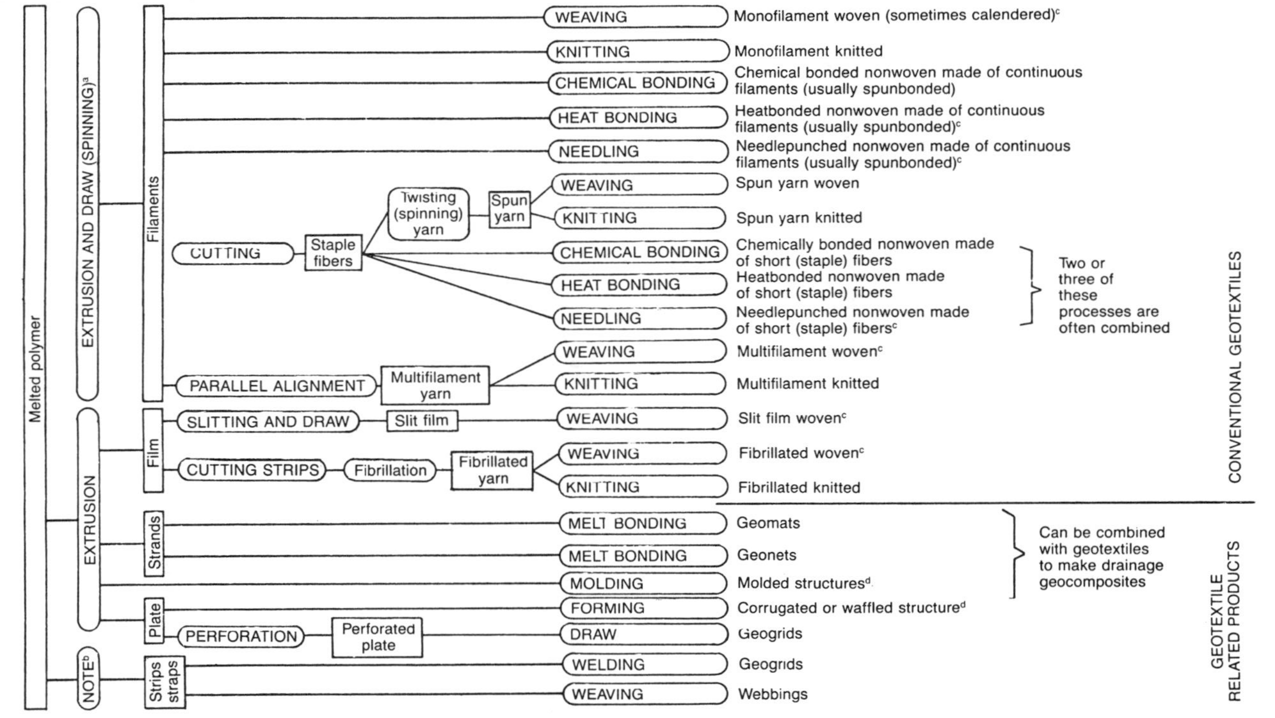

Figure 8-1. Production of geotextiles and related products: products are in lower case letters and processes are in capital letters. Notes: (a) "spinning" has two meanings—extrusion through a spinneret to make a filament and fabrication of yarns from staple fibers; (b) strips or straps can be made using any appropriate process such as extrusion, calendering, weaving, fabric coating, etc.; (c) those conventional geotextiles that are used more than others; (d) corrugated, waffled, or molded structures are generally not used alone—they are used to make geocomposites [Giroud, 1984a].

broad classes of geotextiles can be considered: the "conventional geotextiles," typically products of the textile industry, including woven, knitted, and nonwoven fabrics; and the "geotextile related products," which are products recently developed to be used in combination with, or in place of, conventional geotextiles.

Conventional Geotextiles. The manufacturing process of a conventional geotextile includes two steps.

First Step. The first step consists of making linear elements such as filaments, fibers, slit films (tapes), or yarns.

Filaments are produced by extruding melted polymer through dies or spinnerets. Since this process is continuous, filaments are sometimes called "continuous filaments." After extrusion, a filament is usually drawn (i.e., pulled along its longitudinal axis) to orient its molecules in the same direction. As a result of the draw, the modulus of the filament is increased (i.e., when subjected to a given force, a drawn filament exhibits a smaller elongation than a nondrawn filament).

Short fibers (also called "staple fibers" or simply "fibers") are obtained by cutting filaments to a short length, typically 2 to 10 cm (1 to 4 in.). Almost always, the filaments are crimped before cutting.

Slit films are small tapes, typically 1 to 3 mm (40 to 120 mils) wide, produced by slitting an extruded plastic film with blades. After the slitting of the film, the tapes are drawn. As a result of the draw, molecules become oriented in the same direction and the modulus of the tapes increases.

Fibrillated yarns are film strips which have been nicked and broken up into fibrous strands (in other words, a fibrillated yarn is a bundle of tape-like fibers that can still be partially attached to each other).

Nonwoven fabrics are made with only two types of linear elements: filaments and fibers. Woven and knitted fabrics are made with a variety of linear elements generically called yarns. There are two types of yarns:

- Yarns obtained directly as already indicated, which include: monofilament yarns (i.e., yarns comprised of a single filament), slit films (tapes), and fibrillated yarns.
- Yarns obtained by combining several of the simple linear elements already described, such as: multifilament yarns made from filaments aligned together, and spun yarns made from staple fibers interlaced or twisted together.

Second Step. The second step in the manufacturing process of a conventional geotextile consists of combining linear elements produced in the first step to make a planar permeable structure called a fabric. The three types of fabrics are wovens, knitted, and nonwovens.

Woven geotextiles are composed of two sets of parallel yarns systematically interlaced to form a planar structure. The manner in which the two sets of yarns are interlaced determines the weave pattern. By using various combinations of the three basic weave patterns, i.e., plain, twill, and satin, it is possible to produce an almost unlimited variety of fabric constructions. Generally, the two sets of yarns are perpendicular, but some special looms weave at a skew angle. Yarns in the machine direction are called warp yarns. Yarns in the cross-machine direction are called filling (or weft) yarns. Some woven geotextiles are calendered, which means they are pressed between heated rolls. Yarns are thus softened and deformed at their crossover points and, as a result, are somewhat mechanically interlocked, which gives stability to the fabric. Monofilament and slit film woven geotextiles are generally thin, typically of the order of 0.5 mm (20 mils), while multifilament, spun, and fibrillated woven geotextiles are thicker, typically 3 to 5 mm ($1/8$ to $3/16$ in.) and, in some special cases, up to 10 mm ($3/8$ in.).

Knitted geotextiles include two types: the classical knits and the weft insertion knits.

Classical knits are formed by interlocking a series of loops of one or more yarns to form a planar structure. The way the loops are interlocked identifies the type of knit, such as jersey.

Weft insertion knits are produced by inserting parallel yarns, in the weft (cross-machine) direction, into a knit that is being made in the warp (machine) direction. Using this manufacturing process, it is possible to make geotextiles that have a high modulus in the weft direction because the yarns in that direction can be perfectly straight. (In typical woven fabrics, the yarns cannot be straight because they must interlace with, and, therefore, go over and under, the yarns in the perpendicular direction.) It is also possible to insert warp and weft yarns into a knit that is being made in the warp direction, which produces a fabric with a high modulus in both directions. Although they combine properties of wovens and knitted fabrics, fabrics produced by weft insertion knitting or warp/weft insertion knitting are clearly classified as knitted fabrics because the cohesion of the planar structure results from the knitting process. (Even when yarns are inserted in both weft and warp directions, perpendicular yarns are not interlaced.)

Nonwoven geotextiles are formed from filaments or short fibers, arranged in all directions (but not necessarily at random) and bonded together into a planar structure. The filaments or the short fibers are first arranged into a loose web, then bonded together using one, or a combination, of the following processes:

- Chemical bonding: a cementing medium such as glue, rubber, latex, cellulose derivative or, more frequently, synthetic resin, is added to fix filaments or short fibers together. One thus obtains *chemically bonded nonwoven geotextiles.*
- Thermal bonding: heat causes partial melting of filaments or short fibers which makes them adhere at their crossover points. The *heat-bonded nonwoven geotextiles* created by this process are usually relatively thin, typically 0.5 to 1 mm (20 to 40 mils).
- Mechanical bonding by needlepunching: thousands of small barbed needles, which are set into a board, are punched through the loose web and withdrawn, leaving filaments or fibers entangled. The *needlepunched nonwoven geotextiles* created by this process are relatively thick, typically 1 to 5 mm (40 to 200 mils) or more.

Nonwoven geotextiles made in a continuous line process, in which filaments are extruded, drawn, formed into a loose web, and bonded, are called *spunbonded.* The term spunbonded implies that spinning (extruding and drawing) and bonding of the filaments occurs without interruption (i.e., without any intermediate step, such as cutting filaments to make staple fibers) regardless of the bonding process. Some people tend to restrict the use of the term spunbonded to the heatbonded nonwovens; this is not justified.

Geotextile-Related Products. Geotextile-related products have a coarser structure than conventional geotextiles. They include webbings, geomats, geonets, and geogrids.

Webbings are a kind of very coarse woven fabric made of strips several centimeters (a few inches) wide.

Geomats are thick, open, compressible, and extendible structures that can be made by a variety of processes. They can be made of coarse filaments which have tortuous shapes and are bonded at their intersections. They can also be made by spot-welding several wavy or cuspated layers of lightweight polymer nets stacked on top of each other. Geomats typically have a thickness of 1 or 2 cm ($\frac{1}{2}$ to 1 in.) and opening sizes of the order of 5mm ($\frac{1}{4}$ in.).

Geonets consist of two sets of coarse parallel strands intersecting at a constant angle (generally between 60° and 90°). The two sets of strands are simultaneously extruded, and the strands of one set are melt-bonded to the strands of the other set. Typically, the size of strands is 1 to 5 mm ($\frac{3}{64}$ to $\frac{3}{16}$ in.) and the size of openings is from a few millimeters ($\frac{1}{4}$ in.) to several centimeters (1 in. or more).

Geogrids can be manufactured by heating and stretching a perforated sheet of extruded polymer in one or two perpendicular directions; in the process, the small perforations become large quasirectangular openings between narrow strands, usually 1 to 10 cm ($^1/_2$ to 4 in.). Geogrids can also be manufactured by welding or melt-bonding perpendicular straps made in a first, independent process. The strands, in the first manufacturing process, and the straps, in the second manufacturing process, usually have a high modulus as a result of draw during the manufacturing process.

Geocomposites. Geocomposites are made of several layers of geotextiles and/or geotextile-related products. Examples are:

- Combinations of multiple layers of knitted, woven, and/or non-woven fabrics bonded together by needlepunching, heatbonding, or stitchbonding. Needlepunching and heatbonding have been defined previously as part of the manufacturing of nonwovens. Stitchbonding consists of sewing together two or more fabric layers by means of closely spaced parallel seams (i.e., rows of stitches).
- Combination of one or two geotextiles, acting as filters, with an open synthetic core material, where fluids can flow easily, to form prefabricated drainage structures. Materials used to form the core are typically geomats, geonets, waffled or corrugated plastic plates, and molded plastic structures.

Lately there has been a tendency to restrict the use of the term "geocomposite" to the prefabricated drainage structures just described. It would be more appropriate to call these structures "drainage geocomposites."

Some synthetic materials are never used alone in geotechnical applications but are typically used to form geocomposites. Examples include the waffled and corrugated plastic plates already discussed and the scrims. Scrims are open-weave woven fabrics, which are too loose to be used alone. They have a high modulus (i.e., little deformability) in the direction of their yarns (i.e., in two perpendicular directions), and are typically combined with nonwovens to form geocomposites that are less deformable than nonwovens (at least in two perpendicular directions), and have the desirable properties of nonwovens such as high puncture resistance and retention of fine soil particles (for separation of two different soils or filtration).

Basic Description

A short description of a geotextile or a geotextile-related product should include its composition and a brief summary of its manufacturing process as follows.

Composition. The vast majority of geotextiles (conventional geotextiles as well as geotextile-related products) are usually made from synthetic polymers such as polypropylene, polyester, polyethylene, and polyamide (nylon). These polymers have a very high resistance to chemical and biological degradation.

Some geotextiles are made from glass fibers. Steel wires and cables have also been used in conjunction with synthetic polymers to create geotextiles for special applications.

Natural fibers (i.e., wool, jute, cotton, or acetate) are seldom used to make geotextiles because they are biodegradable. Geotextiles made from natural fibers and even paper may, however, serve temporary functions where biodegradation is desirable (e.g., temporary erosion control).

Identification. A geotextile or a geotextile-related product should be identified by listing:

- The kind of polymer(s)
- The type of fiber or yarn, if appropriate
- The type of geotextile or geotextile-related product
- The mass per unit area
- Additional information if necessary.

The mass per unit area will be defined in the next section. Some engineers (in particular those familiar with geomembranes, whose thickness is easy to measure) tend to identify geotextiles by their thickness. While it is perfectly legitimate to specify minimum thickness values in applications where thickness governs the performance (see the section "Design"), it is inappropriate to use the thickness to identify a geotextile; the thickness of most geotextiles is very difficult to measure because of their irregular surface, whereas the mass per unit area is easy to measure as will be shown later. Examples of descriptions of geotextiles and related products follow: polyester multifilament plain weave woven 270 g/m^2 (8 oz/sq yd); polypropylene monofilament woven 230 g/m^2 (7 oz/sq yd); polypropylene slit film woven 135 g/m^2 (4 oz/sq yd); polypropylene spunbonded needlepunched nonwoven 370 g/m^2 (11 oz/sq yd); polypropylene staple fiber needlepunched nonwoven 340 g/m^2 (10 oz/sq yd); stitchbonded geocomposite made of polypropylene slit film woven between two polyamide heatbonded nonwovens 580 g/m^2 (17 oz/sq yd); polyethylene geonet 440 g/m^2 (13 oz/sq yd) with 8 mm ($^5/_{16}$ in.) openings (in this last example, the opening size, which is easily measured, is a characteristic that should distinguish one geonet from another).

The foregoing identification process provides the minimum information required to identify a geotextile or a geotextile-related product. To determine suitability of a product for a specific application, a more de-

tailed description is required, which includes a list of pertinent characteristics and properties.

Note: The preceding section is an updated version of descriptions of
types of geotextiles and related products already published by the author
[Giroud and Carroll, 1983; Giroud, 1984a; Giroud et al., 1985].

PROPERTIES

Geotextiles and geotextile-related products are similar to soils: they
have physical, hydraulic, and mechanical properties, and their properties
can be altered by the effects of the environment in which they are used.

Physical Properties

Soils are characterized by physical properties which evaluate the solids
in the soil (such as dry density and density index, also called relative density) and properties which evaluate the voids in the soil (such as porosity
and void ratio). Geotextiles also have these two types of physical properties.

Evaluation of Solids. Two physical characteristics are used to evaluate
the solids in a geotextile or a geotextile-related product: the mass per unit
area and the thickness.

Mass per Unit Area. The basic physical characteristic used to identify a
geotextile is its mass per unit area, which is easily determined by weighing a piece of geotextile and dividing the measured mass by the area of
the piece. Typical mass per unit area values range from 100 g/m^2 (3 oz/sq
yd) to 600 g/m^2 (18 oz/sq yd) for the majority of conventional geotextiles, and from 500 g/m^2 (15 oz/sq yd) to 1,500 g/m^2 (44 oz/sq yd) for
geotextile-related products. Larger mass per unit area values are possible
for special products. (Note: Several inappropriate terms such as
"weight," "basis weight," "weight per square yard," "mass per square
meter," are sometimes used instead of mass per unit area.)

Thickness. The thickness of a geotextile or geotextile-related product is
obtained by measuring the distance between two plates placed on each
side of a specimen and subjected to a very small standardized pressure.
The same procedure can be used with incrementally higher pressures to
evaluate the compressibility of a geotextile.

With respect to thickness, geotextiles and related products can be classified in two general categories:

1. Relatively thin and incompressible products which include most wovens, heatbonded nonwovens, and geogrids.
2. Relatively thick products which include needlepunched nonwovens (1 to 5 mm), geonets (5 mm), geomats (10 to 20 mm), and some drainage geocomposites (5 to 25 mm).

Products of the second category are more or less compressible (as discussed later in the section "Compressive Behavior"). Some functions of geotextiles can be performed only by relatively thick products (as discussed later in the section "Relationships Between Functions and Properties").

Evaluation of Voids. Two physical characteristics are used to evaluate the voids in a geotextile or a geotextile-related product: the size of openings and the amount of openings.

Size of Openings. The size of openings in woven and nonwoven geotextiles is on the order of a few dozen to a few hundred micrometers (two to twenty mils), which approximates the range of silt to coarse sand particle sizes. Geomats, geonets, and geogrids have large openings, typically several millimeters to several centimeters ($^1/_4$ in. to a few inches).

The openings of a given geonet or geogrid are usually uniform; therefore, it is sufficient to measure the size of one of the openings to characterize the product. In a nonwoven geotextile, the openings are of different sizes and a gross approximation of the largest opening size of a given nonwoven geotextile may be obtained by sieving calibrated glass beads or sand particles through this geotextile in a standardized manner. The opening size value thus obtained is called the "apparent opening size" (AOS) of the geotextile. This method is also applicable to woven geotextiles because the openings of a given woven geotextile are not uniform due to manufacturing irregularities or deformations.

Ideally, every geotextile would be characterized by an opening size distribution curve (Figure 8-2) similar to the particle size distribution curve of a soil. It is often assumed that the AOS just defined is equivalent to the "O_{95}" of the opening size distribution curve. The O_{95} is defined in a manner similar to the d_{95} definition for soils, where O_{95} is that size which is larger than 95% of the geotextile's openings (in other words, 95% of the openings of the geotextile are smaller than O_{95}). Typical values of O_{95} for geotextiles used as filters are: 100 to 400 micrometers (140 to 40 U.S. Sieve Number) for nonwovens, and 300 to 500 micrometers (50 to 35 U.S. Sieve Number) for monofilament wovens.

Amount of Openings. For thick products such as needlepunched nonwovens, geomats, and geonets, the amount of openings is characterized by the porosity, which is defined, as it is for soils, as the ratio between volume of voids and total volume of a sample. Due to their open structure with linear elements (e.g., fibers, strands) geotextiles, geonets, and geomats have porosities that typically range between 50% and 90%. Typical porosities for granular materials such as soils are between 20% and 50%.

For thin woven geotextiles, the amount of openings is characterized by the percent open area which is the ratio of the open area to the total area of a sample, expressed as a percentage. Typical values of percent open area for woven geotextiles are from less than 1% to 20%.

Hydraulic Properties

All geotextiles and geotextile-related products are permeable, i.e., liquids and gases can easily pass through them. In addition, some thick geotextiles and geotextile-related products can convey liquids and gases within their plane.

Permeability Normal to the Plane. Mechanisms by which liquids and gases seep through thick nonwoven geotextiles and soils are similar. Consequently, the hydraulic conductivity (also known as the "coefficient of permeability"), originally defined for water seeping through soils, can also be used to evaluate the permeability of thick nonwoven geotextiles. The hydraulic conductivity in the direction normal to the plane of a geotextile is defined, using Darcy's equation, as the rate of water flow divided by the unit area and the hydraulic gradient (the hydraulic gradient being the hydraulic head divided by thickness of soil or geotextile). Geotextile permeability can also be evaluated through the use of the hy-

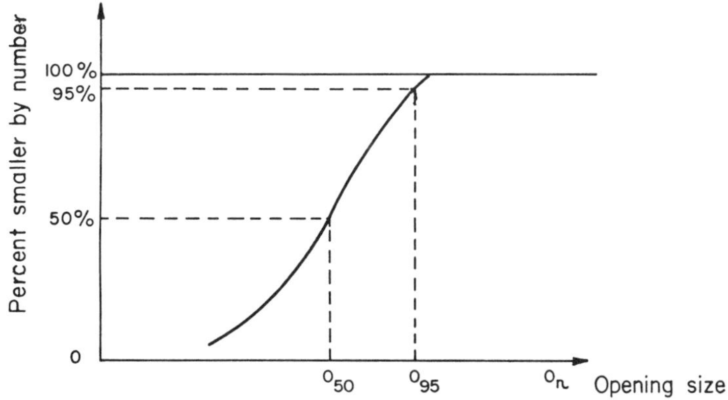

Figure 8-2. Opening size distribution curve of a geotextile.

draulic permittivity, which is defined as the rate of water divided by the unit area and the hydraulic head. The permittivity is measured directly in a flow test without having to measure geotextile thickness. If the flow is laminar, the permittivity of a geotextile is equal to its hydraulic conductivity divided by its thickness.

Mechanisms by which fluids pass through thin geotextiles and through geotextiles with large openings are not similar to mechanisms by which fluids seep through soils. In addition, the determination of a hydraulic gradient (hydraulic head divided by geotextile thickness) would be extremely inaccurate in such cases. Therefore, the only suitable means of evaluating the ability of thin geotextiles or geotextiles with large openings to pass fluids is to use the permittivity, measured directly in a flow test, as already indicated.

Typical values of hydraulic conductivity of thick nonwoven geotextiles are 10^{-5} to 10^{-2} m/s (10^{-3} to 1 cm/s). Since their thicknesses range between 1 and 5 mm, their permittivities range approximately between 10^{-3} and 10 s^{-1}. Typical values of permittivity for woven geotextiles are also 10^{-3} to 10 s^{-1}.

Permeability Within the Plane. Some thick geotextiles and geotextile-related products can also convey fluids within their plane. The amount of water conveyed under a given hydraulic head is proportional to the hydraulic transmissivity of the geotextile. The transmissivity of thin geotextiles such as wovens or heatbonded nonwovens is negligible, while the transmissivity of needlepunched nonwovens, geomats, geonets, and geocomposites with thick open cores is high, thus enabling these geotextiles and related products to be used as drains. It should be pointed out that the transmissivity of some of these products is significantly affected by their compressibility.

The hydraulic transmissivity is equal to the in-plane hydraulic conductivity of the geotextile multiplied by its thickness, if the flow is laminar. This is the case of needlepunched nonwoven geotextiles. With most thick geotextile-related products the flow is turbulent except under very low hydraulic gradients, and the hydraulic conductivity obtained by dividing the transmissivity by the thickness is an equivalent hydraulic conductivity.

Typical values of hydraulic transmissivity are: 10^{-7} to 10^{-5} m²/s for needlepunched nonwoven geotextiles; 10^{-5} to 10^{-3} m²/s for geomats, 10^{-4} to 10^{-3} m²/s for geonets; and up to 10^{-2} m²/s for some geocomposites. In these ranges of values, the lower values are related to the geosynthetic subjected to a compressive stress of the order of 1 MPa (20,000 psf) which is the compressive stress at the base of a 60 m (200-ft) high embankment. Typical curves showing the influence of compressive stress on transmissivity are given in Figure 8-3.

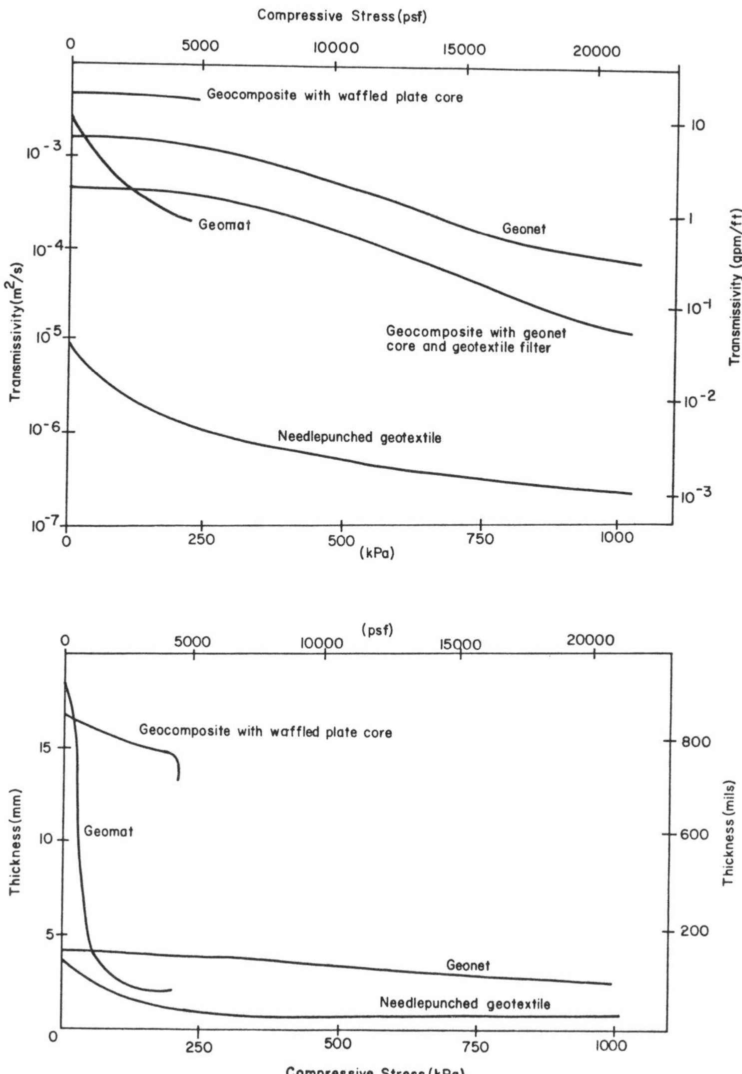

Figure 8-3. Influence of compressive stress on transmissivity and thickness of various geotextiles and related products typically used to convey water within their plane (combining data from Giroud [1984b] and Koerner [1986].)

Mechanical Properties

Mechanical properties of geotextiles and geotextile-related products include: tensile behavior, compressive behavior, resistance to concentrated stresses, and mechanical interaction with soil.

Tensile Behavior. This is certainly the most important aspect of the mechanical behavior of a geotextile or a geotextile-related product.

The Tension-Elongation Curve. The tensile behavior of a geotextile can be characterized by the plot of the force per unit width, also called tension (expressed in kN/m or lb/in.), versus the strain, also called elongation (dimensionless, usually expressed in %) (Figure 8-4). This plot is obtained by performing a tensile test, which consists of subjecting a rectangular specimen of geotextile to an increasing elongation in one direction and recording the resulting tensile force, until failure occurs. Typical features of this plot are the tensile strength (maximum tension, α_{max}), the strain at failure (ϵ_f), and the secant modulus (expressed in kN/m or lb/in.) for a considered strain (defined as the tension/strain ratio at the considered strain).

Note that the geotextile modulus is expressed with the dimension of a force per unit width (kN/m or lb/in.) and not the dimension of a stress, i.e., a force per unit area (kPa or psi), for the sake of practicality and to prevent errors caused by the use of the thickness, which is difficult to measure.

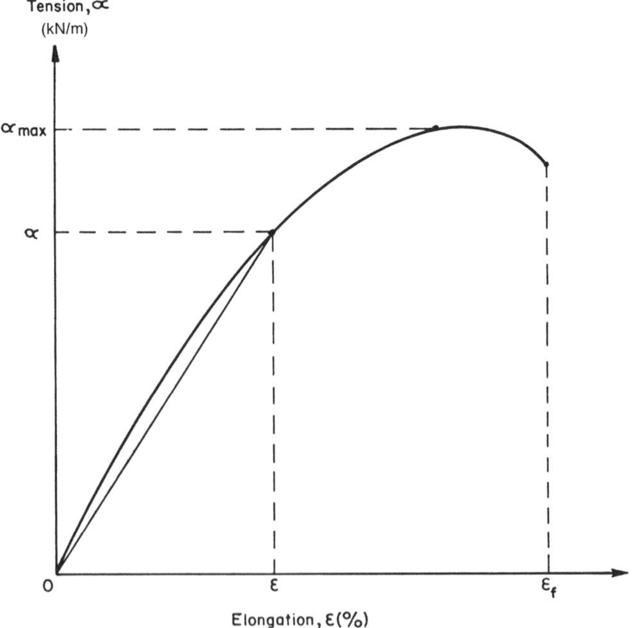

Figure 8-4. Tension-elongation curve. The modulus at elongation ϵ is $J = \alpha/\epsilon$. Note: To calculate the modulus, ϵ should be dimensionless, e.g., 0.05, not 5%.

Types of Tensile Tests. Tensile test results are representative of the in-field behavior of the geotextile or related product only if the force applied during the test is uniformly distributed throughout the width of the specimen. For nonwoven geotextiles this is possible only if the width of the specimen is larger than its height (which is the gauge length between the clamps of the tensile testing machine), and if the specimen is clamped over its entire width (Figure 8-5). This requirement is fulfilled by wide-strip tensile tests with width/height ratios ranging between 2 and 5. This requirement is not fulfilled by traditional tests of the textile industry such as tests on narrow strips (which should never be performed on geotextiles) and the grab test, performed with clamps narrower than the specimen (Figure 8-5). However, the grab test (with its result expressed as a force, not a tension) should not be eliminated because it is useful to evaluate geotextile resistance to concentrated stresses and, as a quality control test, to evaluate manufacturing consistency.

Typical Values of Tensile Characteristics. Geotextiles and geotextile-related products can be designed to have high tensile strength and modulus. Parameters governing tensile strength and modulus include: the type

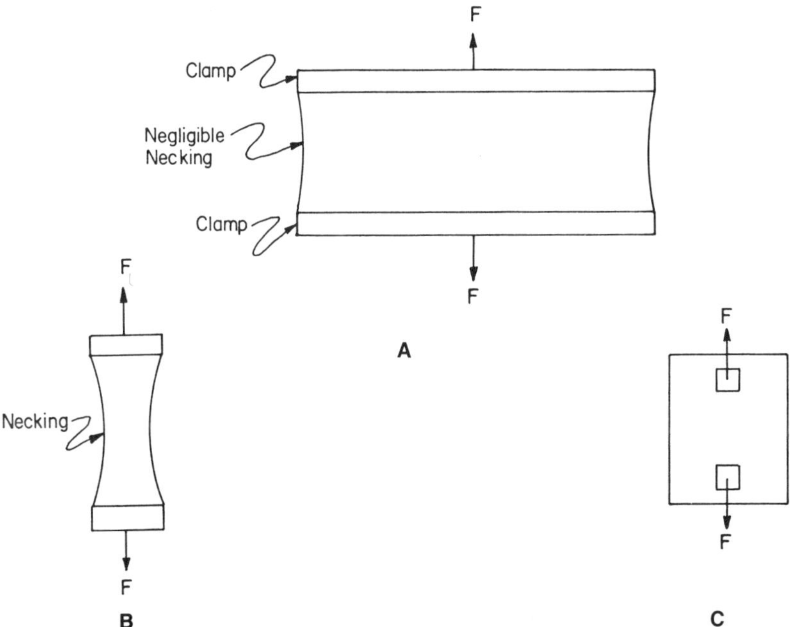

Figure 8-5. Three types of tensile tests: (A) wide-strip tensile test; (B) strip tensile test; (C) grab test.

of geosynthetic (i.e., woven or nonwoven geotextile, geogrid, etc.), the amount of polymer characterized by the mass per unit area, the type of polymer, and the degree of orientation of the polymer. Typical values of tensile characteristics are given in Table 8-1.

Table 8-1
Typical Values of Tensile Characteristics of Geotextiles and Geotextile-Related Products*

Type of Geotextile or Geotextile-Related Product	Elongation at Failure ϵ_f %	Level of Tensile Characteristics	Strength α_{max} kN/m (lb/in.)		Secant Modulus at $\epsilon = 5\%$[†] J_5 kN/m (lb/in.)	
Heatbonded nonwoven geotextile	50–100	Typical	20	(120)	50	(300)
Needlepunched nonwoven geotextile	50–100	Typical	20	(120)	20	(120)[‡]
		High	100	(600)	100	(600)[‡]
Woven geotextile	10–25	Typical	25	(150)	300	(1,800)
		High	80	(500)	1,000	(6,000)
		Very High	500	(3,000)	5,000	(30,000)
		Low	20	(120)	150	(900)
Geogrid	10–15					
		High	100	(600)	1,000	(6,000)
		Very high	200	(1,200)	2,000	(12,000)
Webbing	10	Typical	100	(600)	1,000	(6,000)
Geonets	>100	Typical	5	(30)	§	
Geomats	>100	Typical	1	(5)	§	

* Note that the elongation at failure is governed by the type of product, while strength and modulus are governed also by the amount of solids in (i.e., the mass per unit area of) the geotextile. For example, if two identical geotextiles are bonded together, back to back, the strength and the modulus would be multiplied by two, while the elongation at failure remains the same.

[†] The 5% elongation has been selected here only for illustrative purposes.

[‡] The modulus of needlepunched nonwovens, which are very compressible, increases significantly if the compressive stress increases.

§ Geonets and geomats are used for applications where tensile properties are not important. It would be irrelevant to give their modulus.

Factors Affecting Tensile Characteristics. Tensile characteristics of geotextiles and geotextile-related products are significantly affected by several factors such as temperature, velocity, and/or duration of load application, and confinement in soils. These factors are briefly discussed in the following.

The mechanical properties of all polymers are very sensitive to temperature: when the temperature increases, strength and modulus decrease while the elongation at failure usually increases. It is therefore important that designers identify situations where geotextiles and related products are exposed to high temperatures. Fortunately, in most applications, geotextiles and geotextile-related products are buried in the ground where temperature variations are small.

The mechanical properties of all polymers are also very sensitive to velocity of load application. As a result, strength and modulus of geotextiles and related products will be higher if the load is applied rapidly. (This effect is significant for geogrids whose properties are mostly governed by the polymer, while it is negligible for nonwovens whose properties are mostly governed by fabric construction.) It is therefore important that velocity of tensile testing be clearly specified. A related effect is creep: geotextiles and geotextile-related products exhibit an elongation increasing with time when they are subjected to a constant tension. This effect must be taken into account when geotextiles and geotextile-related products are used for soil reinforcement, especially if creep takes place during a long period of time after loading (which may occur with polypropylene). Polyester exhibits little creep, and, with polyethylene, it appears in many practical cases that most of the deformation due to creep takes place within a few hours after loading, i.e., often when construction is still in progress. Therefore, in those cases, postconstruction deformation can be expected to be small if the design has been made with a proper allowable tension in the geotextile or related product to preclude creep rupture or excessive deformation during construction.

The tensile behavior (including creep) of some geotextiles is affected by their confinement in the soil, which involves two factors: confinement stress and interlocking. The strength and the modulus of some geotextiles and geotextile-related products increase when the confinement stress increases, which is analogous to the effect of confinement on soil strength. This effect is mostly significant for compresssible geotextiles such as needlepunched nonwovens where stresses normal to the plane of the fabric increase friction and interlocking between fibers or filaments, thereby reducing the deformability and increasing the modulus of the geotextile. The tensile behavior of some geotextiles and geogrids may even be affected by the interlocking of soil particles with the fabric or grid structure. Interlocking restricts movements of the linear elements (e.g., fibers or strands) and thereby increases the modulus of the geotextile or the geogrid.

From this discussion it may be concluded that tensile characteristics of geotextiles and geotextile-related products used for design should include the effects of temperature, time (i.e., loading velocity and creep), and confinement by means of tests or corrective factors.

Compressive Behavior. The compressive behavior of geosynthetics is not usually a concern, except for geosynthetics used to convey fluids within their plane since transmissivity can be significantly affected by compressive stress if the geosynthetic is compressible. The compressibility of geosynthetics typically used for fluid transmission is illustrated by Figure 8-3.

Resistance to Concentrated Stresses. Properties related to the ability of geotextiles to withstand damage caused by concentrated stresses are typically the grab strength, the tear strength, the puncture strength (all three expressed in kN or lb), the impact energy (expressed in J or ft-lb), and the bursting resistance (expressed in kN/m^2 or psi). There are standard tests to evaluate grab, tear, puncture, impact, and bursting resistance, and it is also possible to develop a test specific to a given project, e.g., puncture by a given type of stone.

Soil-Geosynthetic Mechanical Interaction. Mechanical interaction is characterized by the interface shear strength between a geosynthetic and a material in contact, such as soil, concrete, asphaltic concrete, or another geosynthetic. Interface shear strength is expressed in kN/m^2 or psi. It can also be expressed using parameters of Coulomb's law: cohesion (expressed in kN/m^2 or psi), and friction angle (expressed in degrees). A high interface shear strength is required when a geosynthetic (such as geotextile or geogrid) is used to reinforce a soil. Of course, a low interface shear strength should be prohibited when a geosynthetic is in a location where it could act as an undesirable slip surface.

Mechanisms governing interface shear strength are:

1. Friction between the geosynthetic and a solid material with a flat surface such as concrete, asphaltic concrete, and other geosynthetics.
2. Interlocking, friction and/or cohesion, between a geosynthetic and a soil exhibiting friction and/or cohesion. Interlocking is possible if soil particles are smaller than the geosynthetic openings.

For example, small clods of soil interlock with geomats (openings on the order of 10 mm ($\frac{1}{2}$ in.)), and ballast interlocks with geogrids (openings on the order of 10 to 50 mm ($\frac{1}{2}$ to 2 in.)). Interlocking of a geotextile with coarse particles (e.g., gravel) is also possible, regardless of its opening size, if the geotextile is capable of deforming locally to follow the shape of the particles while retaining its overall tensile characteristics.

The interface shear strength between a geosynthetic and soil is smaller than, or equal to, soil shear strength, depending on geotextile surface roughness (which governs friction and cohesion) and opening size or deformation capability (which governs interlocking).

Endurance Properties (Durability)

The term durability indicates the resistance to progressive deterioration. Durability of geotextiles or geotextile-related products depends on the action exerted by external forces and on the raw material (usually polymer with additives) used to make the geotextile or the geotextile-related product. The type of action exerted by external forces can be mechanical (abrasion, fatigue) or physico-chemical-biological (degradation).

Abrasion and Fatigue. Abrasion and fatigue resistance of a geotextile or a geotextile-related product in a given application can be evaluated by tests simulating actual environmental and loading conditions. Presently there are no widely accepted procedures for such tests. Geotextiles with high abrasion resistance are available for use in contact with ballast in railroad tracks or rocks in bank protection.

Physico-Chemical-Biological Degradation. Physico-chemical-biological degradation of geotextiles and geotextile-related products can be caused by external agents, such as: contact with soil, atmospheric conditions (including exposure to the sun), and contact with chemicals or biological agents. Most polymers used to make geosynthetics have a high resistance to physico-chemical-biological actions from soils that are free from industrial chemicals.

Geotextiles are exposed to the sun more or less briefly during installation, temporarily in some applications (such as erosion control applications), and permanently in a few applications (such as silt fences). Woven and nonwoven geotextiles are usually very sensitive to ultra-violet (UV) light and should not be exposed to sunlight for a period of time exceeding a few weeks to a few months depending on the type of polymer and UV resistant additives used to manufacture the geotextile. For longer duration of exposure the geotextile should be coated or covered with a protective material (asphalt or resin spray, concrete, soil, etc.) Webbings, geonets, and geogrids have a better outdoor durability than wovens and nonwovens because they usually contain carbon black (which impedes penetration of UV light) and are usually made of coarse elements.

Resistance of geotextiles and geotextile-related products to chemicals is a concern only in special applications such as waste-management projects. The main parameter is the type of polymer used to manufacture the geotextile or the geotextile-related product.

Conclusions on Geotextile Properties

Geotextiles have many properties similar to those of soils. For example: all geotextiles are permeable; some, like sand, can retain fine soil particles; most have high friction coefficients. Due to these similarities, properties of geotextiles can be easily understood by geotechnical engineers. The consistency of symbols and units used for geotextiles [International Geotextile Society, 1985] with those adopted by the International Society for Soil Mechanics and Foundation Engineering further reveals the role of geotextiles in geotechnical engineering.

In addition, geotextiles have properties that soils do not have, which explains their success: they have a high tensile strength, they are continuous, and they are structurally stable, thus they cannot be dispersed by forces of low intensity such as rain or wind. The combination of tensile strength and continuity is important because it enables geotextiles to prevent local failures by homogenizing soil deformation, thereby improving the reliability of soil structures.

Note: The preceding section is an updated version of previously published work [Giroud, 1983; Giroud et al., 1985].

FUNCTIONS AND APPLICATIONS OF GEOTEXTILES

Geotextile and geotextile-related products are used in a wide variety of applications in which they perform certain functions. As will be shown later, the first step of the design of a structure incorporating geotextiles and/or geotextile-related products is to identify the functions of the geotextiles and/or the geotextile-related products governing the performance of the considered structure. It is therefore necessary for the designer to have a good knowledge of the functions that geotextiles and geotextile-related products can perform.

The Concept of Geotextile Function

Definition of Geotextile Function. The following definition has been proposed: "A geotextile function is a specialized action of the geotextile which is required to achieve a design purpose and results from a unique combination of geotextile properties" [Giroud, 1986]. This definition applies to geotextiles as well as geotextile-related products.

Relationships Between Functions and Properties. Relationships between functions and properties are given in Table 8-2. Also indicated in Table 8-2 are the design purposes achieved by geotextiles and geotextile-related products.

Table 8-2
Relationships* Between Properties and Functions of Geotextiles[†] and Their Purpose and Location

Properties[‡]	Functions	Purposes
Thickness	Fluid transmission[§]	Removes excess water
Permeability	Filtration	Prevents piping
Continuity	Separation	Prevents mixing
	Protection	Prevents damage
	Tensioned membrane[‖]	Provides reinforcement
Tensile Strength	Tensile member[‖]	Provides reinforcement
Friction		

From [Giroud, 1986b].

* Only the principal relationships are shown (e.g., the thickness also has some influence on the filtration and the protection functions).

[†] This table is applicable not only to geotextiles, but to all geosynthetics except geomembranes.

[‡] The three key properties are underlined.

[§] The fluid transmission function is often called drainage function, which can be misleading (see Table 3).

[‖] Tensioned membrane and tensile member are the two reinforcement functions.

Most properties indicated in the left column of Table 8-2 are self-explanatory. However, the following comments are necessary:

- The permeability involved in the "fluid transmission" function is the permeability within the plane of the geotextile, while the permeability involved in the filtration function is the permeability normal to the geotextile.
- Continuity is an essential property which results from the structure of geotextiles and geotextile-related products. It means that geotextiles and geotextile-related products cannot be dispersed, while granular materials can be dispersed since they are made of discrete particles. In order to retain their continuity, geotextiles and geotextile-related products must have appropriate resistance to tear, puncture, burst, localized tensile stresses (grab), etc.
- The term tensile strength has been chosen for the sake of simplicity. It refers to all pertinent aspects of the tensile behavior such as strength, modulus, limited creep, etc.
- The term friction has also been chosen for the sake of simplicity. It encompasses all mechanisms by which shear stresses are transferred between soil and geotextile or geotextile-related product, such as friction, cohesion/adhesion, and interlocking.

Presentation of the Functions

The six functions presented in Table 8-2 and illustrated in Figure 8-6 can be defined as follows.

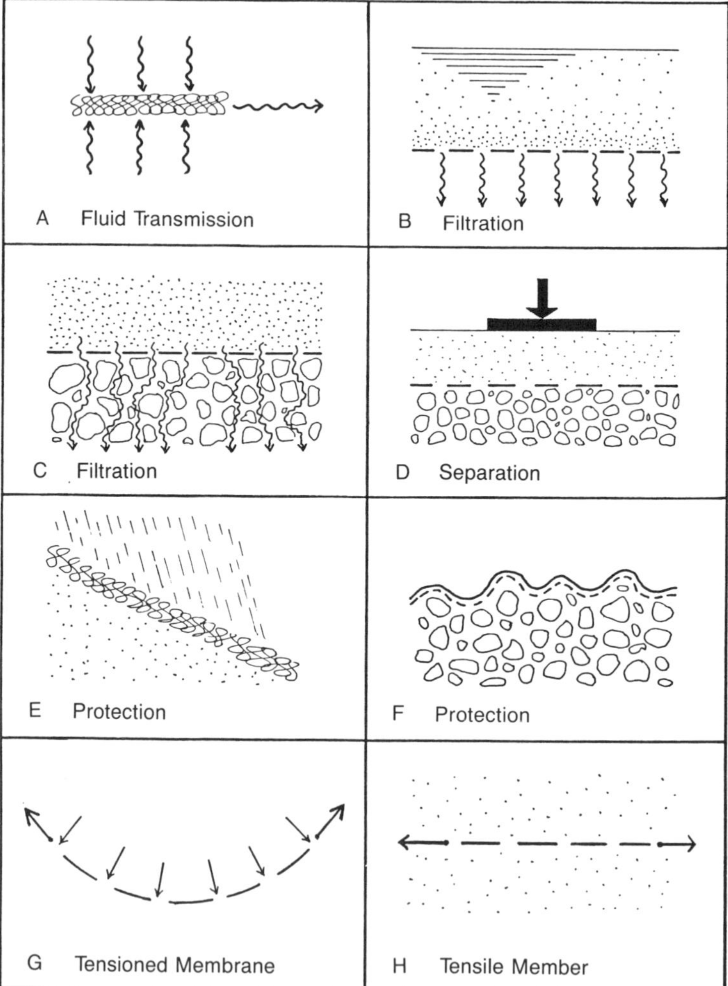

Figure 8-6. Functions of geotextiles and geotextile-related products.

Fluid Transmission. A geotextile, a geonet, or a geocomposite provides fluid transmission when it collects a liquid or a gas and conveys it, within its own plane, toward an outlet (Figure 8-6a). (Note: this function is usually called "drainage" which creates a lot of confusion with the "drainage applications" discussed in Table 8-3; therefore the terms "fluid transmission," or simply "transmission," are recommended.)

Filtration. A geotextile acts as a filter when it allows liquid to pass normal to its own plane, while preventing most soil particles from being carried away by the liquid current. Two cases can be considered:

- A geotextile placed across a flow of liquid carrying fine particles stops most of the particles (which therefore accumulate on the filter) while allowing water to pass through (Figure 8-6b).
- A geotextile, placed in contact with a soil, allows water seeping from the soil to pass through, while preventing any movement of soil particles (with the exception of a very small amount of the finest particles located near the filter) (Figure 8-6c).

Separation. A geotextile placed between a fine soil and a coarse material (gravel, stones, blocks, slabs, boards, etc.) acts as a separator when it prevents the fine soil and the coarse material from mixing under the action of applied loads (Figure 8-6d).

Protection. A geosynthetic protects a material when it alleviates or distributes stresses and strains transmitted to the protected material. Two cases can be considered:

- *Surface protection*—A geotextile or a geomat placed on the soil prevents its surface from being damaged by such actions as weather, light traffic, etc. (Figure 8-6e).
- *Interface protection*—A geotextile, placed between two materials (such as asphalt overlay/cracked pavement, or geomembrane/stony subgrade), prevents one of the materials (e.g., the overlay or the geomembrane) from being damaged by concentrated stresses applied by (or strains imposed by) the other material. The term "cushion" is typically used for a geotextile placed between a geomembrane and a subgrade (Figure 8-6f).

Tensioned Membrane. A geotextile or a geogrid functions as a tensioned membrane when it is placed between two materials having different pressures, and its tension balances the pressure difference between the two materials, thus strengthening the structure (Figure 8-6g).

Tensile Member. A geotextile or a geogrid acts as a tensile member when it provides tensile modulus and strength to a soil with which it is interacting through interface shear strength (i.e., friction, cohesion-adhesion, and/or interlocking between geotextile and soil) (Figure 8-6h).

Table 8-3
Relationships* Between Applications and Functions of Geotextiles

Application Category	Application Area	Application Type	Fluid Transmission	Filtration	Protection	Separation	Tensioned Membrane	Tensile Member
					Functions			
Hydraulic Applications	Drainage and Ground Water	a. Geosynthetic drains without filter (thick nonwoven)	X					
		b. Geosynthetic drains with filter (geocomposites)	X	X				
		c. Gravel drains		X				
		d. Pipes		X				
		e. Ground water recharge		X				
	Erosion Control	f. Bank revetments		X				
		g. Erosion mats			X			
		h. Silt fences		X			X	
		i. Retaining structures		X			X	
Geosynthetic Construction	Containers	j. Concrete forming		X			X	
		k. Sand tubes (hydraulic fill)		X			X	
		l. Gabions					X	
		m. Sand bags					X	
	Geomembrane Support	n. Bridging					X	
		o. Cushion			X			
		p. Slip surface			X			
Geotechnical Structures	Roadways†	q. Landing mats			X			
		r. Asphalt overlay			X			
		s. Unpaved roads (large deflection)				X	X	
		t. Base courses (small deflection)				X		X
		u. Ballast				X		X
		v. Asphalt pavement reinforcement				X		X
	Soil Reinforcement	w. Faced reinforced walls						X
		x. Unfaced reinforced walls						X
		y. Slopes						X
		z. Embankments						X

From [Giroud, 1981].
* For each application type, only the most important functions are indicated (e.g., the table does not show that the fluid transmission function may be involved when a geotextile is placed under flat concrete slabs in a bank revetment, or when a geotextile is used in a railroad track on a saturated subgrade, etc.)
† Here, roadways are considered to include all types of traffic-supporting structures such as roads, parking lots, staging areas, railroad tracks, etc.

The Reinforcement Functions. As indicated in Table 8-2, geotextiles provide reinforcement to geotechnical structures through both the "tensioned membrane" and "tensile member" functions. In fact, these are often combined into one function, called "reinforcement." However, to be consistent with the definition of geotextile function given earlier, it is more appropriate to consider them as two different functions ("the reinforcement functions") since they involve different combinations of geotextile properties.

Applications of Geotextiles and Related Products

Functions and Applications. It would be simple for both professors and designers if only one function were performed in each type of application. In reality, a geotextile often performs several functions, as indicated in Table 8-3. From a practical viewpoint, it is interesting to group the application types into six areas: drainage, erosion control, containers, geomembrane support, soil reinforcement, and roadways (defined here to include all types of traffic supporting structures). From Table 8-2, the dominant functions appear to be:

- In hydraulic applications (drainage and erosion control): the fluid transmission and filtration functions.
- In geosynthetic construction (containers, geomembrane support): the tensioned membrane and protection functions.
- In geotechnical structures (roadways and soil reinforcement): the tensile member and separation function, and, to a lesser extent, the tensioned membrane and protection functions.

Application types mentioned in Table 8-3 are illustrated in Figure 8-7.

Functions of Geosynthetics and Other Materials. Relationships between functions and materials likely to perform them are presented in Table 8-4. In this table, not only geotextiles, but all types of geosynthetics are considered and one function has been added, "fluid barrier," which can be performed by geomembranes or by low-permeability soils such as clay. Also, the protection function has been separated into two parts: surface protection (such as the use of geosynthetics for erosion control mats, temporary landing strips, etc.) and interface protection (such as use of geotextile between asphalt overlay and cracked pavement, or as a cushion between geomembrane and stony subgrade, etc.).

The following discussion refers to the materials presented in Table 8-4, as well as the three key properties presented in Table 8-2, namely: permeability, continuity, and tensile strength.

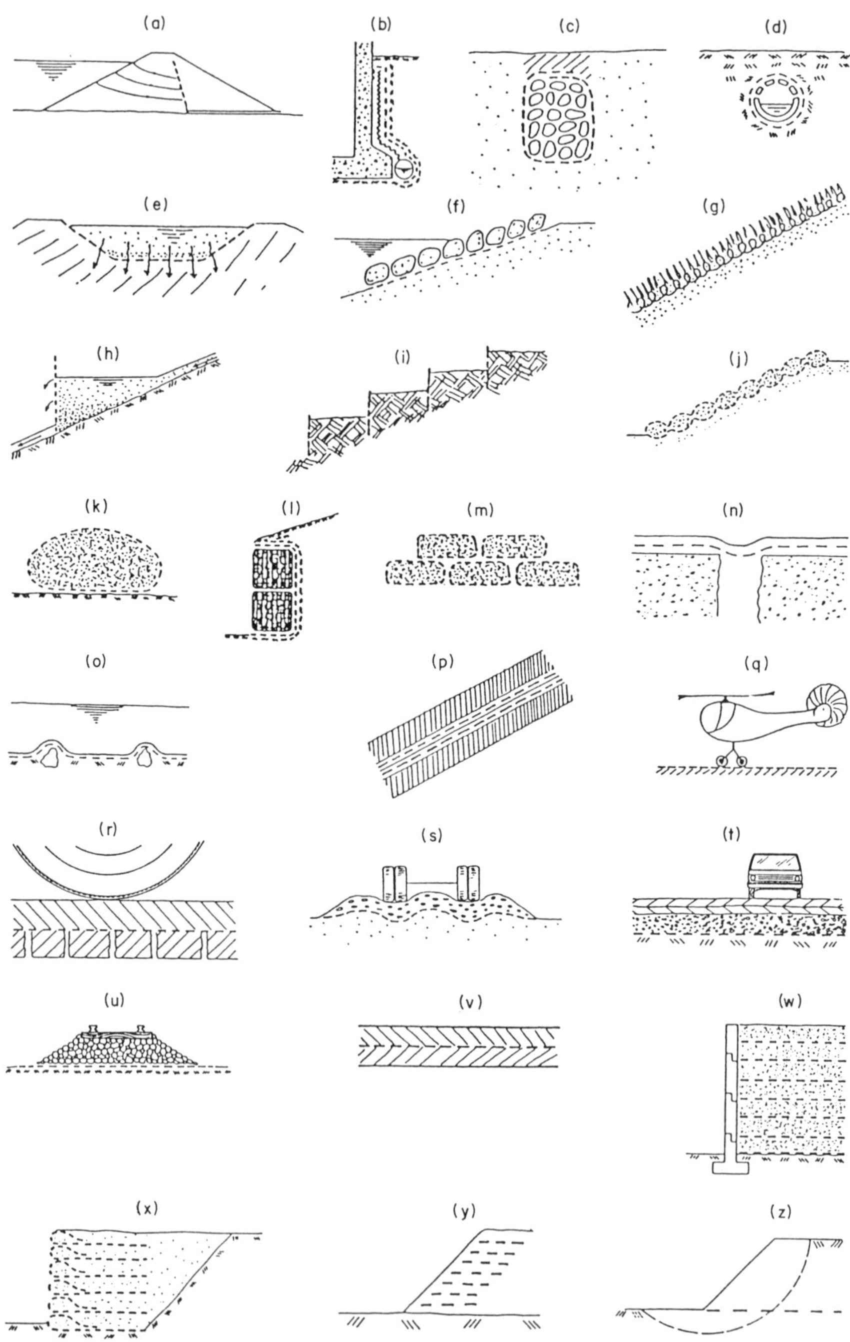

Figure 8-7. Examples of applications of geotextiles and geotextile-related products. Letters a through z refer to Table 8-3.

Table 8-4
Relationships Between Functions and Materials

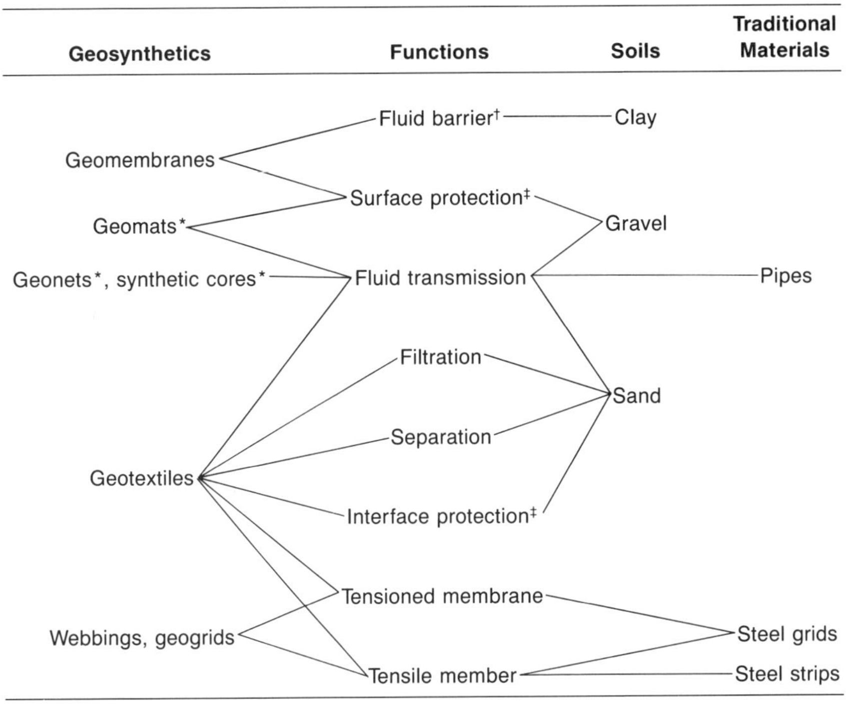

Geosynthetics	Functions	Soils	Traditional Materials
Geomembranes	Fluid barrier[†] ——— Clay		
Geomats[*]	Surface protection[‡]	Gravel	
Geonets[*], synthetic cores[*]	Fluid transmission	Gravel	Pipes
Geotextiles	Filtration	Sand	
Geotextiles	Separation	Sand	
Geotextiles	Interface protection[‡]		
Webbings, geogrids	Tensioned membrane		Steel grids
Webbings, geogrids	Tensile member		Steel strips

From [Giroud, 1986].

[*] Geomats, geonets, and synthetic cores (such as corrugated or waffled plates, and molded structures), performing the fluid transmission function, are typically associated with geotextiles, performing the filtration function, to form drainage geocomposites.

[†] The fluid barrier function (e.g., liners for dams and reservoirs) was not included in Table 8-2, since it is not perfomed by geotextiles.

[‡] The protection function has been separated into two parts: (1) surface protection (e.g., erosion control mats, landing strips); and (2) interface protection (e.g., geotextile between asphalt overlay and cracked pavement, geotextile as a cushion between geomembrane and stony subgrade).

Permeability. Regarding permeability, soils offer a wide range of possibilities from very low (clay) to very high (gravel), and they are well-suited to perform the hydraulic functions, i.e., fluid barrier, fluid transmission and filtration. Consequently, to perform these functions, the designer may choose between geosynthetics or soils. In addition, to perform the fluid transmission function, choices may be made between pipes and synthetic drainage products such as geonets, geomats, and synthetic cores. Drainage geocomposites, which are composed of a synthetic core (such as geomat, geonet, waffled or corrugated plate, or molded plastic structure) between two geotextiles, perform two functions: filtration which is performed by the geotextiles, and fluid transmission which is performed by the core.

Continuity. Soil layers have some measure of continuity when they are carefully placed and when they are not dispersed by erosion, traffic, etc. They can therefore perform functions where continuity is required such as surface protection (gravel), and filtration, separation and interface protection (sand). However, geotextiles have the advantage of better continuity, which facilitates their installation and enhances their performance. For example, geotextiles are used for:

- Filtration, especially when the installation of a sand filter is difficult, e.g., in French drains.

- Separation, especially when the continuity of a sand layer could be disrupted by large deformations, e.g., in unpaved roads.

- Interface protection, especially when a thin layer of sand would lack the required continuity, e.g., in asphalt overlays (where the continuity of a sand layer would be progressively disrupted by repeated traffic loads) or in protecting a geomembrane on a slope (where the continuity of a sand layer would be disrupted by sloughing, erosion, etc.).

Tensile Strength. Soils, obviously, cannot provide the two reinforcement functions, tensioned membrane and tensile member, since tensile strength is required for these two functions. Regarding the tensile member function, the only alternative at the present time to geosynthetics are steel strips and rods that have been used as tensile members in thousands of structures in the past twenty years. The advantage of steel is that it is very strong. The disadvantages are that steel is heavy, inflexible, and corrodible.

The tensioned membrane function is the only function for which there is virtually no alternative to geotextiles and geotextile-related products such as geogrids and webbings: soils lack tensile strength and steel strips lack bi-dimensional continuity. Steel grids or steel fabrics could theoretically be used but, to be economical, they would have to be thin and would corrode. The tensioned membrane function appears therefore as the most original function of geotextiles, resulting from a unique combination of the two key properties of geotextiles, continuity and tensile modulus/strength.

No wonder that geosynthetics are extensively used in unpaved roads where the tensioned membrane function plays an essential role.

Note: The preceding section is an updated version of previously published work [Giroud, 1986].

DESIGN

A Rational Approach

Some official organizations in the United States and abroad have prepared standard geotextile specifications for typical applications. While standard specifications may be useful as a first approximation, geosynthetic applications of geotextiles and other geosynthetics must be designed, like any other engineering structure or system, using a rational approach based on engineering principles.

A geosynthetic used in a given application should perform certain functions, survive the construction process, and be sufficiently durable to perform the intended functions over the design life. Therefore, the designer must determine the geosynthetic properties governing the performance of the structure, permitting the construction, and ensuring the required durability.

Regarding performance, the designer must:

1. Identify the functions of the geosynthetic governing the performance of the considered structure (which requires a good knowledge of the principles of soil mechanics as well as the functions that a geosynthetic can perform).
2. Identify the relevant geosynthetic properties, using the relationships between functions and properties presented in Table 8-2.
3. Determine the required values of the geosynthetic properties by using existing methods of design, when available, or by developing methods specific to the structure being designed.

Regarding construction, the designer must:

4. Select a method of construction and identify the various types of stresses likely to be applied to the geosynthetic during construction.
5. Identify the geosynthetic properties governing geosynthetic survivability through the selected method of construction.
6. Determine the minimum required values of the geosynthetic properties using available information on survivability or conducting specially developed tests.

Regarding effects of the environment, the designer must:

7. Identify the environmental conditions to which the geosynthetic will be subjected, and establish durability requirements.
8. Identify the geosynthetic properties governing its durability in the identified conditions.

9. Determine the minimum required values of the geosynthetic properties by using available information on durability or by conducting tests.

As a final step, the designer must combine the required values of the geosynthetic properties related to performance, construction, and durability, and write the project-specific specifications.

In Steps 3, 6, and 9, the required values of the geosynthetic properties are determined by analytical, analogical, or empirical design methods. A discussion of these three types of design methods follows.

Analytical Design Methods

Definition of an Analytical Design Method. An analytical design method consists of performing a theoretical analysis of the considered structure. The calculations performed may be entirely original or may extensively use published relevant theoretical results presented as equations, charts, and tables.

To obtain the values of materials' properties relevant to an analytical design method, it is necessary to conduct tests giving the constitutive laws of all the materials, such as soils and geotextiles, incorporated into the structure. Such tests are ideally conducted with uniform boundary conditions in order to ensure uniform stress and strain fields in the considered materials (soil, geotextile, etc.), which allows a direct measurement of stresses and strains (since stresses and strains at the boundary—where they are easy to measure—are equal to stresses and strains inside the material—where they would be very difficult to measure). As a result, the constitutive law of the tested material is given directly by the test, without making any assumption for the interpretation of the test. A constitutive law is typically represented as one or more stress-strain (or tension-strain) curves, corresponding to one or more strain-time paths, and their envelopes (if limit equilibrium is considered).

The foregoing considerations are related to mechanical design problems, which are the most likely to be solved using analytical design methods. However, similar considerations could be made for hydraulic design problems, such as flow rate evaluation, and physical design problems, such as filtration of granular materials.

Example of Analytical Method in Soil Mechanics. There are many analytical methods used routinely in soil mechanics. For example, the stability of an embankment constructed on soft soil is typically studied using a limit equilibrium analysis with a circular slip surface. The constitutive law of the soil used in such study is the Coulomb's law; and the values of the two soil properties involved in Coulomb's law, the cohesion and the friction angle, are typically determined using a triaxial test, where stresses and strains are uniform.

Examples of Analytical Methods in Geotextile Engineering. An embankment constructed on soft soil can be reinforced by one or more layers of geotextile or geogrid located at, or near, the soft soil-embankment interface. As indicated in Chapter 9, the stability of a geotextile-reinforced embankment on soft soil can be studied using an analytical method similar to the limit equilibrium method used in soil mechanics and discussed earlier. The relevant geotextile characteristic, its tension-elongation curve, is obtained in performing a wide-strip tensile test, i.e., a tensile test with uniform boundary conditions. (The grab test, often used because of its simplicity, is not appropriate in this case because the test's boundary conditions are not uniform.)

In the past five years many analytical design methods have been developed for geotextile applications, and it may be inferred that geotextile engineering has become a scientific discipline [Giroud, 1986]. Presently available analytical design methods are related to the following geotextile applications: filters using woven or nonwoven geotextiles; drains made of thick needlepunched nonwoven geotextiles, geonets, geomats, or geocomposites; use of geotextiles or geogrids, placed on top of a soil cavity or depression, to support a geomembrane, a soil layer, a road, etc.; unpaved roads reinforced with geotextiles or geogrids; embankments on soft soils with geotextile or geogrid reinforcement; vertical walls or steep slopes reinforced with multiple layers of geotextiles or geogrids; etc. A detailed discussion of analytical design methods can be found in Giroud et al. [1985] with emphasis on similarities between design methods used in soil mechanics and design methods used for geotextile applications. Design methods for geotextile applications can be found in Bonaparte and Christopher [1987], Bonaparte et al. [1985], Christopher and Holtz [1985], Giroud [1984], Koerner [1986], and in the Proceedings of International Conferences on Geotextiles [1977, 1982, 1986], and in *Geotextiles and Geomembranes,* an international journal.

Analogical Design Methods

Definition of an Analogical Design Method. An analogical design method consists of comparing real situations with model tests. Model tests differ more or less from the considered real situations by the scale and the boundary conditions. Model tests can be done in a laboratory (where the scale of the model is usually, but not necessarily, small) or in the field (where the scale of the model is often large, sometimes full-scale). Unless the scale of the model is one to one, it is necessary to make assumptions regarding the behavior of the materials involved in order to interpret the tests.

Example of Analogical Method in Soil Mechanics. Model tests are typically used in soil mechanics for deep foundations. If possible, the first of a series of identical piles or drilled shafts is subjected to a loading test. In this case, the scale is one to one, and interpretation is straightforward. Often, unfortunately, a full-scale test is impractical because it would involve excessive loads, and hence a reduced scale (such as $1/2$ or $1/3$) is used. In this case, a number of assumptions are required on deep foundation-soil interaction in order to extrapolate the observations to the scale of the actual foundations in order to predict their performance. Guidance in making the assumptions necessary for interpretation can be provided by measurements of stresses and displacements at various locations in the soil and the deep foundation, in addition to the basic measurement of load and settlement at the top of the foundation.

Examples of Analogical Methods in Geotextile Engineering. Analogical methods are particularly useful in geotextile engineering because there are many situations where localized concentrated stresses govern the performance of the geotextile and where full-scale model tests can be made. Examples are:

- The performance of a geotextile acting as a cushion to prevent puncturing of a geomembrane by a stony subgrade can be evaluated using a pressure vessel (Figure 8-8) where a full-scale test can be made in the laboratory, using the actual geomembrane and geotextile, the actual stones, and the actual liquid pressure.
- The performance of a geotextile acting as a separator can be evaluated using a full-scale laboratory model test simulating actual conditions. For example, the test can evaluate the ability of a geotextile to retain fine soil particles under the action of repeated loads due to traffic.

In both of these cases, the use of an analogical method is the only choice. Analytical methods have not been developed in those cases because of the extreme complexity of stress distribution in the first case, and the lack of knowledge regarding the effect of repeated actions on granular materials and geotextiles in the second case. Analogical methods can also be used as a complement in cases for which analytical methods are available. For example, filtration tests can be conducted to simulate a real situation where it appears necessary to verify a design conducted with filter criteria (i.e., with an analytical method). In these

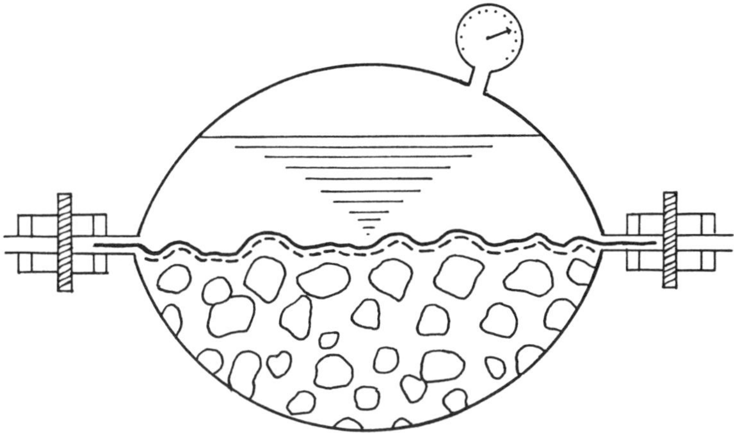

Figure 8-8. Pressure vessel to test the ability of a geotextile to protect a geomembrane from being punctured by stones when subjected to liquid pressure.

laboratory tests, the actual liquid flows through the actual soil and the geotextile filter to be evaluated. These are full-scale tests as far as the soil/geotextile geometry is concerned, but the hydraulic gradient in the test is usually much larger than in reality, as an attempt to accelerate any potential clogging mechanism. As a result, the interpretation of these tests is questionable because mechanisms pertaining to particulate mechanics (such as clogging by fine soil particles) are more difficult to extrapolate than those pertaining to mechanics of continua (such as stress distribution).

Another example of analogical method using model tests is the laboratory tests which consist of immersing geotextile samples in liquid chemicals to evaluate the compatibility between geotextiles and chemicals in a waste storage facility. Such tests are usually conducted at high temperature in an attempt to accelerate chemical reactions. Again, interpretation is difficult because there is no well established basis from which to infer the actual velocity of geotextile deterioration from the accelerated observations made in the laboratory.

Empirical Design Methods

Definition of an Empirical Design Method. Empirical design methods are established on the basis of experience gained in observing the behavior of actual structures. Part of the experience may have been gained through systematic full-scale testing done in the past. From a practical standpoint, the use of empirical methods involves formulas ("empirical formulas"), charts, or simply a classification of situations likely to occur.

In many cases, no tests are required to use an empirical method of design: a description of the actual situation is sufficient to use the formula or the classification, thereby obtaining the solution to the problem. In other cases, simple measurements, such as index tests, are necessary.

Semi-empirical methods are a blend of empirical and other types of design methods. Examples include:

- The use of empirical coefficients in equations established by a theoretical analysis.
- The use of an empirical formula to extrapolate the results of a model test.
- A combination of both.

Examples of Empirical Methods in Soil Mechanics. Soil mechanics is now an advanced scientific discipline and empirical methods are used less and less. However, there are still a few examples, such as:

- Formulas or charts relating the number of blows in a Standard Penetration Test (SPT) to the bearing capacity of a foundation.
- Charts relating the required thickness of a road pavement to the California Bearing Ratio (CBR) of the subgrade soil (which is obtained from a simple test).
- The group index method for road pavement design which is based on a classification of traffic volume and subgrade soils, and on the result of simple index tests (particle size distribution, plastic limit, and liquid limit).

The classical filter criteria, used in soil mechanics, are probably the best example of a design method which results from a blend of analytical, analogical, and empirical considerations. The original basis is analytical, which many geotechnical engineers were never told (see [Giroud, 1982]), but modifications were made as a result of extensive laboratory model testing and field experience.

Examples of Empirical Methods in Geotextile Engineering. Empirical and semi-empirical design methods are useful for geotextile applications where there are conditions which elude analytical design methods, such as concentrated stresses, repeated loads, time-dependent effects, etc. Examples are as follows:

- In the USA, a task force composed of members of the Federal Highway Administration (FHWA) and representatives of other organizations has established geotextile "survivability" specifications (i.e., specifications regarding minimum properties that a geotextile should have to withstand stresses resulting from the construction process

[Christopher and Holtz, 1985]). These specifications, established on the basis of the experience of the task force members, are related to the most common geotextile applications (erosion control, drainage, and road construction), and consider various levels of stresses depending on the installation process and the use of the geotextile.

- The French Committee on Geotextiles and Geomembranes [1981] has prepared an empirical design method for unpaved roads where eighty-eight different cases are considered, depending on traffic, expected performance, and soil properties, with minimum recommended geotextile properties in each case.

In the latter case, analytical methods of design are also available, as already indicated. However, all these apparently analytical methods can be considered to be semi-empirical because they include assumptions based on field data to take into account the effect of repeated loads.

Note: The preceding section is an updated version of already published work [Giroud, 1983a; Giroud et al., 1985].

Acknowledgments

This chapter includes updated versions of already published work. The author is grateful to his former co-authors and co-workers, namely: A. Arman, J. R. Bell, R. G. Carroll, Jr., R. M. Koerner, and V. Milligan. In addition, the author is indebted to J. F. Beech, R. Bonaparte, J. E. Fluet, Jr., M. C. Langston, J. N. Paulson, and R. B. Wallace for valuable information provided during the preparation of this chapter, and J. F. Desnoyers, C. A. Pearce, and G. Saunders for assistance during the preparation of the manuscript.

REFERENCES

1. Bonaparte, R., and Christopher, B. R., "Design and Construction of Reinforced Embankments over Weak Foundations," *Symposium on Reinforced Layered Systems,* Transportation Research Board, Washington, D.C., USA, 1987.
2. Bonaparte, R., Holtz, R. D., and Giroud, J. P., "Soil Reinforcement Design Using Geotextiles and Geogrids," *ASTM Symposium, "Geotextile Testing and the Design Engineer,"* ASTM STP 952, Los Angeles, USA, June 1985.
3. Christopher, B. R., and Holtz, R. D., *Geotextile Engineering Manual,* by the Federal Highway Administration, Washington, D. C., USA, 1985.

4. Geotextiles and Geomembranes, an international journal, T. S. Ingold (Ed.), Elsevier Applied Science Publishers, London, UK.

5. Giroud, J. P., "Filter Criteria for Geotextiles," *Proceedings of the Second International Conference on Geotextiles,* Vol. 1, Las Vegas, Nevada, USA, 1982, pp. 103-108.

6. Giroud, J. P., "Introduction aux geotextiles," Annexe au *Cours pratique de mecanique des sols,* Volume 2, par J., Costet et G. Sanglerat, Dunod, Paris, 1983, pp. 416-442.

7. Giroud, J. P., and Carroll, R. G., Jr., "Geotextile Products," *Geotechnical Fabrics Report,* Vol. 1, No. 1, 1983, pp. 12-15.

8. Giroud, J. P., "Geotextiles and Geomembranes," *Geotextiles and Geomembranes,* Vol. 1, No.1, 1984a, pp. 5-40.

9. Giroud, J. P., *Geotextiles and Geomembranes. Definitions, Properties and Design,* IFAI, St. Paul, Minnesota, USA, 1984b, 325 p.

10. Giroud, J. P., Arman, A., and Bell, J. R., "Geotextiles in Geotechnical Engineering, Research and Practice," Report of the ISSMFE Technical Committee on Geotextiles, *Geotextiles and Geomembranes,* Vol. 2, No. 3, 1985, pp. 179-242.

11. Giroud, J. P., "From Geotextiles to Geosynthetics: A Revolution in Geotechnical Engineering," *Proceedings of the Third International Conference on Geotextiles,* Vol. 1, Vienna, Austria, 1986, pp. 1-18.

12. International Geotextile Society, Symbols for Geotechnical Engineering, Geotextiles and Geomembranes, List of Symbols adopted by the International Geotextile Society, *Geotextiles and Geomembranes,* Vol. 2, No. 3, 1985, pp. 243-262.

13. Koerner, R. M., *Designing with Geosynthetics,* Prentice Hall, Englewood Cliffs, New Jersey, 1986, 424 p.

14. *Proceedings of the International Conference on the Use of Fabrics in Geotechnics* (first international conference on geotextiles), Paris, France, ENPC, Paris, France, 1977, 528 p.

15. *Proceedings of the Second International Conference on Geotextiles,* Las Vegas, Nevada, USA, IFAI, St. Paul, Minnesota, USA, 1982, 1026 p.

16. *Proceedings of the Third International Conference on Geotextiles,* Vienna, Austria, 1986, 1331 p.

17. *Recommandations pour l'emploi des géotextiles dans les voies de circulation provisoire, les voies à faible traffic et les couches de forme,* Comité Français des Géotextiles et Geómembranes, Boulogne sur Seine, France, 1981, 39 p.

18. Scott, J. D., and Richards, E. A., *Geotextile and Geomembrane International Information Source,* BiTech, Vancouver, Canada, 1985, 506 p.

9

Soft Soil Stabilization Using High-Strength Geotextiles and Strip Drains

Robert M. Koerner
Bowman Professor of Civil Engineering
Drexel University
Philadelphia, Pennsylvania USA

Jack Fowler
U.S. Army Corps of Engineers
Waterways Experiment Station
Vicksburg, Mississippi USA

The use of high strength geotextiles coupled with polymeric vertical strip drains for rapid consolidation has ushered in a new era in construction on soft soils. By soft soils, it is meant soils of less than approximately 100 lb/ft^2 undrained shear strength, which heretofore were essentially impossible to build upon, at least in a timely and economic manner. These soils are generally saturated fine-grained silts, clays, and their mixtures (often with organic material) and are typical of river transported and/or dredged materials. With the Corps of Engineers' heavy involvement in the dredging of rivers, harbors, and port facilities and disposal of the dredged material, it is understandable that they should be interested in and be at the forefront in this emerging technology.

While the Corps has had notable successes in dealing with the disposal of soft dredged soils on equally soft subsoil conditions using geotextiles (e.g., Pinto Pass, Mobile, Alabama; Craney Island, Norfolk, Virginia), the field of geosynthetics is still rapidly developing. Design philosophy, construction methods, and development of new materials are constantly arising. Thus it was decided to critique the existing state-of-the-art in soft soil stabilization using geosynthetic materials.

The general topic is felt by the chapter authors to epitomize soil-structure interaction situations at the operational level. As will be seen much remains to be done insofar as research, design, and procedure is concerned, however, the chapter should set the tone for future efforts in this regard.

The emphasis of this chapter is on geosynthetic materials (primarily polymeric strip drains and reinforcement geotextiles) and their interaction and influence on construction on soft, compressible fine-grained soils. Upon reviewing the general problem, that is, essentially vertical and radial consolidation with strip drains, the necessary background information will be presented to set it in its proper context. A complete review of currently available strip drains is also included. Stability and reinforcement concerns using high-strength geotextiles are then presented. Included in this section are fabric-sewing considerations and the effects of holes (necessitated by installation of the strip drains) in the reinforcement fabric.

Installation and construction considerations are then described with major emphasis on areas of potential problems.

A section on recommendations concludes the chapter following the summary and conclusions. We believe one can currently design and build with confidence on these soft soils using geosynthetic materials. As with most new areas, however, the reliability can be improved and/or design factor of safety decreased with additional inquiry into several areas, which will be described.

OVERVIEW OF THE PROBLEM

Various Elements Involved

Saturated fine grained soils (typical of dredged river, port, and harbor materials) suffer from being both highly compressible and very weak. Thus they cannot be used, developed, or even worked upon in their in-situ state. The common expedients of (a) excavation and replacement or (b) driving deep foundations through them are generally impractical and uneconomical on a large scale and are thus prohibitive. As a result, surcharging is a common approach for large areas. While this will eventually solve the compressibility problem, the initial weak strength of the soil makes any construction on the site essentially impossible. The need for a ground covering geotextile, functioning as a reinforcement fabric, becomes an obvious necessity. Details (both design and construction) of such a geotextile are a major consideration in this chapter.

Assuming that a surcharge load can be placed without a slope stability failure when using such a geotextile, the problem becomes one of determining the time for consolidation to occur. In dealing with thick (over 10 ft–20 ft) deposits of saturated silts and/or clays, times for consolidation will usually be over 5 years, and often 10 to 50 years. This situation, being generally unacceptable, calls for radial consolidation methods, which form the second major consideration in this chapter. Focus is on polymeric strip drains, rather than the older method of sand drains.

Reinforcement Fabric

The fabric (note that the words "fabric" and "geotextile" will be used interchangeably) that will be mainly considered for this type of work can be classified as "high performance," versus the more common lightweight geotextiles that form the bulk of this industry. This is because only one layer of fabric can usually be placed at the site and the subsoils need as much immediate help as possible. Numerical examples illustrating this feature will be given. High-performance fabric versus conventional fabric generally calls for the fabric to be of high strength, high modulus, low elongation, and low creep. High survivability properties, i.e., puncture, impact, tear and burst resistances, will also be required.

Fabric seams and seaming methods play a pivotal role in the performance of the fabric. As will be seen, the seams unfortunately form a "weak link" in the reinforcement system. The effect of holes on the reinforcement performance of the seamed fabric will be an obvious concern. These holes are unfortunately required for the installation of the strip drains.

Strip Drains

To hasten consolidation of weak compressible soils, radial drainage is preferred over vertical drainage. Drains for this purpose were formerly made by using vertical columns of sand (called "sand drains" or "drain wells"), and now the current trend has swung heavily toward polymeric strip drains. Strip drains (also called wick drains) consist of tubular, mesh, or net drainage cores protected by geotextile sheaths surrounding them and acting as a filter. Some are completely unitized into a single system.

Design concepts for strip drains will be presented along with flow rate quantification, crush strength, and potential "kinking" and "smear" problems. The geotextile filter will also be assessed in light of typical requirements. Such strip drains will be compared to conventional sand drains by means of numeric examples.

Fabric Damage Assessment

The installation of the strip drains cannot be accomplished on the soft soils being considered without the reinforcement fabric and a drainage soil layer being placed first. The thickness of this soil layer is site specific ($\simeq 2$ ft to 4 ft) and is made using granular soil since it will eventually serve as a horizontal drainage blanket. The strip drain installation rig punches a steel lance (with the strip drain inside of it) through the drainage blanket layer and fabric and continues down to the design depth (see

Figure 9-1). At this depth the lance is removed leaving the strip drain behind. The damage to the fabric caused by this operation can be catastrophic. Considering that the installing lance is a minimum of 7 in. wide and the driving shoe on its end somewhat larger, the resulting hole can easily be 12 in. in length by approximately 6 in. in width. For a situation where the strength of the fabric (along with its seams) is already being challenged, the creation of holes is unfortunate. As this situation is nec-

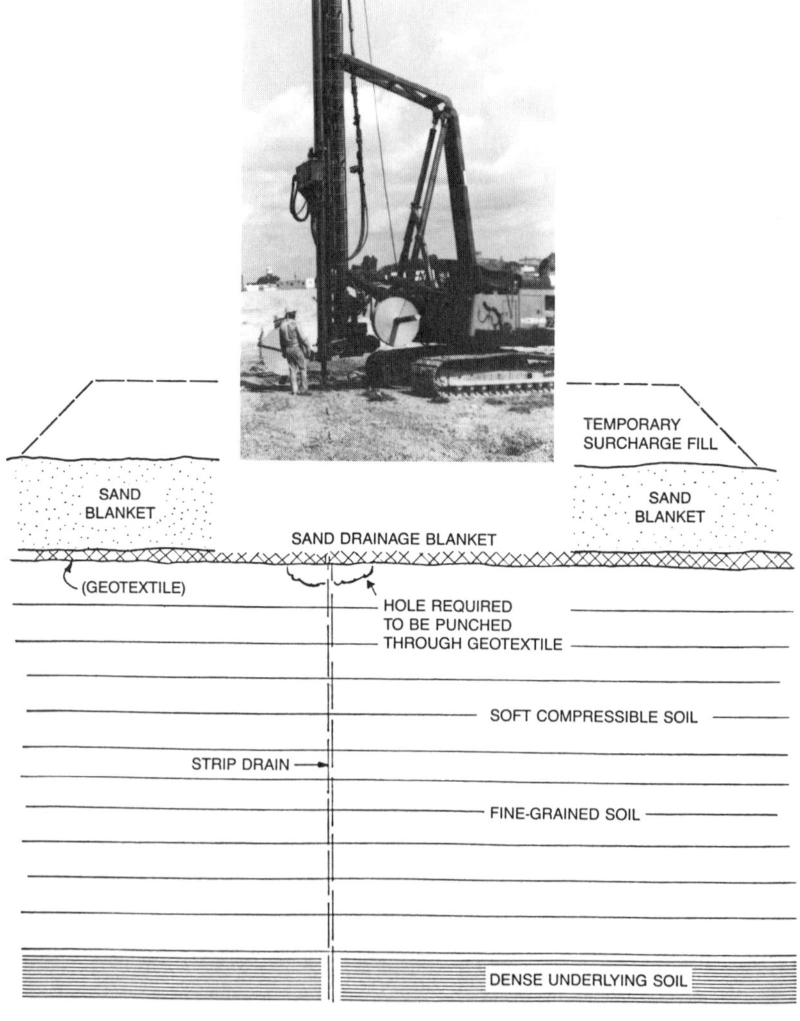

Figure 9-1. Typical cross section showing reinforcing fabric, strip drain, and method of installation.

essary, it must and will be carefully addressed. Numeric examples illustrating how to include such holes in the stability analysis will also be provided.

Implications to Project Design

Both of the forementioned considerations (reinforcement fabric and strip drains) will be presented by means of an analytic approach followed by numeric examples. In this way one could use the information via a sensitivity analysis (for a specific site) to note the optimal fabric and strip drain designs. Such work would be done in the design stage so that the resulting behavior can be reflected in the bid plans and specifications and eventually into project inspection and subsequent maintenance.

Implications to Project Performance

Since building on these very soft soils is brinkmanship-at-its-best, the utmost concern must be given to design, plans, specification, monitoring, and observed field performance. This is indeed a new "art," where we learn from project to project. This somewhat volatile state will no doubt continue for the near future. However, the work should proceed since the options to it (e.g., excavation and replacement or deep foundations) are not acceptable either environmentally or economically. Furthermore, numerous successes have indeed resulted. It is these successes that are the stepping-stones on which this chapter is based.

BACKGROUND

Construction on Extremely Soft Soils

Extremely soft soils are characterized by high water content and fine-grained soil, thus both high compressibility and low shear strength are to be expected. If any structural load is placed on such soils, the mass will either fail (perhaps in a rotational arc or by a translational mud wave) or settle considerably. For these reasons such soils have either been avoided entirely or worked with very slowly. In this latter category, the use of a gradually increasing surcharge fill is common. Here with lift thicknesses of as little as 3 in. per day, shear failures have been known to occur. The point being that the rate of surcharge fill placement must be linked to the soils' initial shear strength and the rate of strength increase as indicated by the dissipation of pore water pressure and corresponding increase in effective stress. The well known effective stress equation describes the process.

$$\sigma' = \sigma - u_w \tag{9-1}$$

where σ' = effective stress
σ = total stress
u_w = pore water pressure

By holding the total stress (σ) constant, any decrease in pore water pressure (u_w) will result in an increase in effective stress (σ'). Furthermore, as seen in the Mohr-Coulomb failure criterion this increase results in an immediate gain in shear strength.

$$\tau = c + \sigma' \tan \phi \tag{9-2}$$

where τ = soil shear strength
c = cohesion
ϕ = angle of shearing resistance

Thus as settlement proceeds, the soil gains in shear strength allowing yet greater surcharge load to be placed. The process continues in this manner until sufficient surcharge load is in place for the equivalent permanent load. At this time the surcharge load (or a portion of it) is removed and the permanent system is built. Usually, rebound during the time between surcharge load removal and the construction of the permanent system is not a major problem.

Vertical Consolidation Mechanisms

As a load is applied to a saturated soil mass it is initially held by excess pore water pressure (considered to be 0% consolidation) and gradually shifts completely to effective stress (considered to be 100% consolidation) according to Equation 9-1. Well established is that the amount and time for this consolidation process to occur are as follows [Koerner, 1984].

$$\Delta H = H_1 \frac{C_c}{1 + e_1} \log\left(\frac{p_1 + \Delta p}{p_1}\right) \tag{9-3}$$

$$t = \frac{H_2^2 T_v}{c_v} \tag{9-4}$$

where ΔH = amount of settlement
H_1 = thickness of consolidating layer

C_c = compression index
e_1 = initial void ratio
$p_1 = \sigma'_v$ = effective vertical stress
Δp = increment of load being added
t = time for settlement to occur
H_2 = maximum drainage path length
T_v = vertical time factor
c_v = coefficient of vertical consolidation

In the preceding equations H_1, e_1, H_2, and p_1 are obtained from the site's geometry and initial conditions, while C_c and c_v are obtained from laboratory testing. T_v is a constant (and a function of the percent consolidation) leaving Δp as the "forcing function" which mobilizes the entire process, i.e., ΔH and t.

Theoretically, the time vs. settlement curves will be parabolically shaped. However, the amount of settlement and the time for settlement to occur varies widely from soil to soil.

Time Rate of Settlement (Example Problem)

As an example of the just-described process, consider the vertical consolidation of 66 ft of organic silty clay (OH) under a surcharge load of 20 ft of soil weighing 120 lb/ft³. The laboratory-determined properties of the soil are:

$\gamma = 70$ lb/ft³
$e_1 = 0.45$
$C_c = 0.32$
$c_v = 0.005$ in.²/min

Substituting these variables into the foregoing equations with:

$p_1 = 33(70) = 2210$ lb/ft²

$\Delta p = 20(120) = 2400$ lb/ft²

1. The total amount of consolidation settlement will be:

$$\Delta H = H \frac{C_c}{1+e} \log \frac{P_1 + \Delta p}{p_1}$$

$$= 66(12)\left(\frac{0.32}{1+0.45}\right)\log\left(\frac{2{,}210 + 2{,}400}{2{,}210}\right)$$

$$= 56 \text{ in.}$$

2. The time for 90% of this consolidation ($\therefore T_v = 0.848$) will be calculated on the basis of single (top only) and double (top and bottom) drainage.

- *Single drainage*

$$t_{90} = \frac{H_2^2 T_v}{c_v}$$

$$= \frac{(66 \times 12)^2(.848)}{0.005} \frac{1}{(60)(24)(365)}$$

$$t_{90} = 202 \text{ years}$$

- *Double drainage*

$$t_{90} = \frac{H_2^2 T_v}{c_v}$$

$$= \frac{(33 \times 12)^2(.848)}{0.005} \frac{1}{(60)(24)(365)}$$

$$t_{90} = 50.6 \text{ years}$$

Easily seen by this example (which is taken from an actual project) is that settlements can be enormous and take extremely long times to occur.

Genesis of Vertical Consolidation Techniques

A number of different techniques have been used to mobilize the required pore water pressures illustrated in the previous problem. They are shown schematically in Figure 9-2 and explained in the following.

Surcharging with Soil. The oldest, simplest, cheapest, and most common method of mobilizing pore water pressures in saturated soils for the purpose of consolidating them is by using soil piled as high as possible, i.e., a soil surcharge. For elevations above the drainage blanket (which must be granular), almost any soil will do. Since such a surcharge must be spread in relatively thin layers and be trafficable by the equipment bringing the fill to the placement site, some type of well graded sand, silt, and/or clay combination is necessary. A considerable range in soil type is possible. As mentioned previously, the surcharge fill is added at a rate commensurate with the ability of the consolidating subsoil to dissipate the mobilized pore water pressure. This is often in the range of 3 in. to 12 in. per day, i.e., 30 to 120 lb/ft^2 per day.

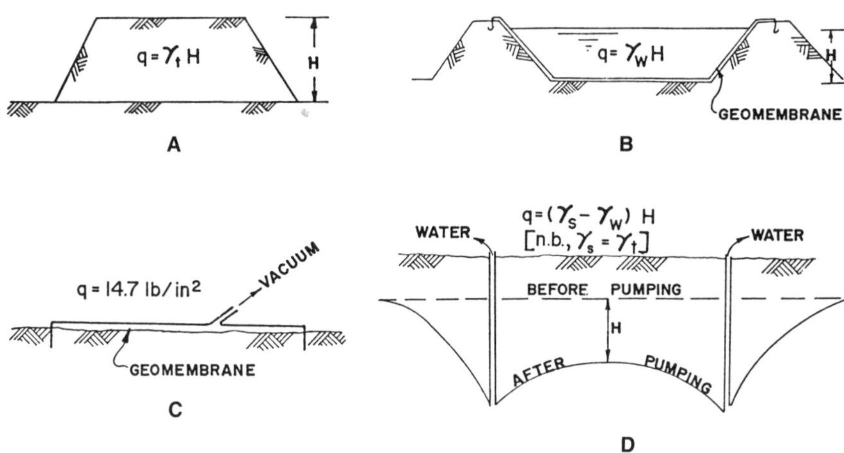

Figure 9-2. Various methods for applying surcharge loads to consolidate saturated fine-grained soils. (A) Surcharging with soil. (B) Surcharging with water. (C) Surcharging by vacuum. (D) Dewatering.

Upon adequate consolidation of one zone of the site, the surcharge fill can be leap-frogged ahead to an adjacent zone depending on the project size, surcharge fill availability, and time for consolidation to occur. When the project is completed, the surcharge fill soil serves no further useful function and can be taken off site.

Surcharging with Water. Water placed directly on the saturated soil which is to be consolidated serves no useful function. Its only effect would be to increase the pore water pressure in direct relationship to the height of water impounded. However, if a geomembrane (or pond liner) is first placed on the ground surface and has its ends contained within an enclosing impoundment dike, water can be very effectively used to mobilize pore water pressures. This technique has been used on a number of large sites which have an ample supply of water nearby, e.g., Elizabeth, New Jersey port facilities stabilization. On that project, containment dikes were constructed in a box-like fashion, the geomembrane was placed and anchored at the top of the dikes, and then river water was used for the surcharge load. When one zone was completed, the system was moved to an adjacent one. It was very effective both technically and economically.

Surcharging by Vacuum. While this method is not used very often, it is possible to place a geomembrane on the ground surface, toe it in around the periphery of the site and apply a vacuum to its underside. The maximum amount of surcharge that can be theoretically mobilized is $14.7 \times 144 = 2{,}120$ lb/ft^2 which is approximately equal to a 20-ft high soil surcharge. Lateral escape routes of the vacuum as well as a relatively high cost of implementation limit the technique.

Dewatering. By lowering the watertable using vacuum wellpoints surrounding the site, the effective weight of the soil is increased by an amount equal to $(\gamma_t - \gamma')$. This amounts to about 40 to 60 lb/ft^3 which is one-half the weight of the soil. Thus a 10-ft dewatering system is equivalent to about 5-ft of soil surcharge. The method has been used where surcharge soil is unavailable or where the surcharge fill height is objectionable, e.g., at airports. It is, however, quite expensive since the dewatering pumps must be kept operating on an around-the-clock basis.

Benefits of Radial Consolidation

The results of the example problem clearly show that the time for consolidation is typically too long for most situations, and that the culprit is the drainage path length "H_2" in Equation 9-4. If this value could be decreased (note that it is squared in the relationship), then marked decreases in the time for consolidation to occur can be realized. Since it is essentially impossible to intersperse horizontal drains in deep strata, the concept of vertical columns of high drainage material was attempted. Kjellman reported of using "cardboard wicks" threaded into the soil by a "stitching machine" in a 1938 paper. Several attempts were made in Europe using this technique and the machine was brought to Canada where several more attempts were made. Severe problems were encountered, however, when the cardboard became wet. The wicks were hydraulically clamped to a lancing device and often failed by pulling apart in tension. Driving through any type of obstacle was also a problem which tore the cardboard wicks.

Emphasis shifted to vertical columns of sand installed by a closed mandrel or through a hollow stem auger. These sand drains were readily accepted by the geotechnical engineering profession. They were particularly helped via Barron's classic paper [1948] on the theoretical concepts of radial drainage published in ASCE Transactions. This study continues today to be the basis of radial consolidation theory and design.

Radial Consolidation Mechanisms

It is of interest to compare the governing differential equations for vertical flow (Terzaghi-type) and for combined vertical and radial flow

(Barron-type) and radial flow only (Barron-type). These three equations are listed in order following where c_v and c_h are the vertical and horizontal coefficients of consolidation.

- Vertical drainage only (cartesian coordinates)

$$\frac{\partial u}{\partial t} = c_v \left[\frac{\partial^2 u}{\partial x^2} + \frac{\partial^2 u}{\partial y^2} + \frac{\partial^2 u}{\partial z^2} \right] \tag{9-5}$$

- Vertical and radial drainage (polar coordinates)

$$\frac{\partial u}{\partial t} = c_v \frac{\partial^2 u}{\partial z^2} + c_h \left[\frac{1}{r}\frac{\partial u}{\partial r} + \frac{\partial^2 u}{\partial r^2} \right] \tag{9-6}$$

- Radial drainage only (polar coordinates)

$$\frac{\partial u}{\partial t} = c_h \left[\frac{1}{r}\frac{\partial u}{\partial r} + \frac{\partial^2 u}{\partial r^2} \right] \tag{9-7}$$

As noted earlier, Equation 9-5 for vertical drainage only, results in a time for consolidation of,

$$t = \frac{H_2{}^2 T_v}{c_v} \tag{9-4}$$

while Equation 9-7 for radial drainage only, results in Equation 9-8

$$t = \frac{d_e{}^2 T_h}{c_h} \tag{9-8}$$

where d_e is the drain well spacing. Equation 9-6 for combined vertical and radial flow is used only for relatively thin compressible layers.

The differences in the predicted times for consolidation between radial and vertical drainage (Equation 9-4 vs. Equation 9-8) can be enormous. Not only is the drainage path length (H_2 or d_e) decreased significantly (again note that it is a squared value), but also "c_h" is often much greater than "c_v". This occurs whenever stratification of the in-situ deposit is present, as it generally is with water transported and deposited soil. The differences between the two time factors (T_v and T_h) are not very significant as Figure 9-3 indicates. Also note on the sketches of Figure 9-3 the definitions of n, d_e, and d_w when using sand drains.

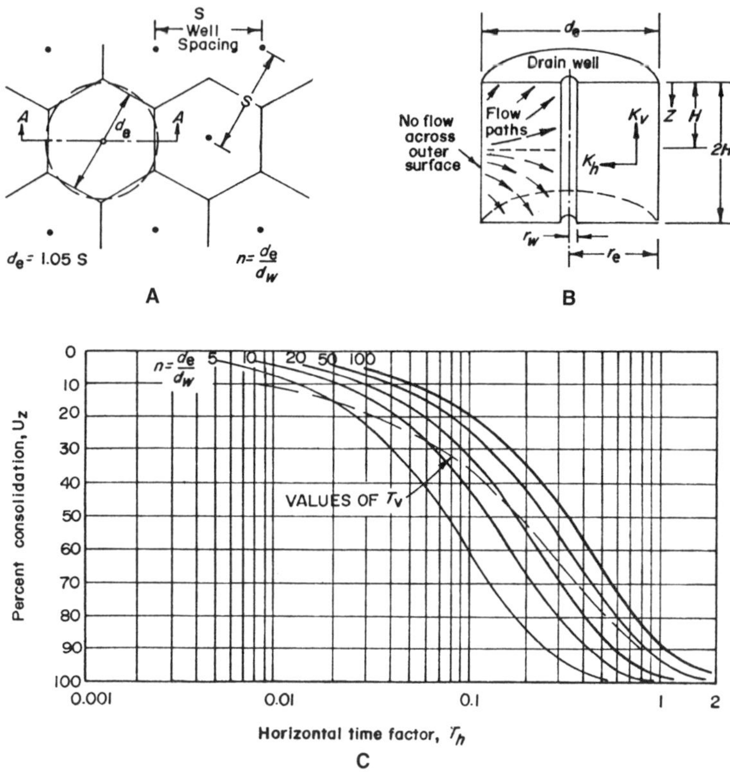

Figure 9-3. Theoretical results for radial consolidation to vertical drain wells. (A) Plan of drain well pattern. (B) Section A-A. (C) Values of T_h as a function of U for various ratios of sand spacing to sand drain size. [After Barron, 1948.]

Time Rate of Settlement (Example Problem Continued)

To continue the example problem given earlier, but now for radial drainage using *sand drains*, the horizontal coefficient of consolidation (c_h) must be obtained. This can be done in a number of ways. The simplest one is to take an undisturbed sample, rotate it 90°, and trim it to fit the laboratory consolidometer. The test is then performed in a standard manner. For the example problem to follow (which continues the actual project mentioned earlier) $c_h = 0.010$ in.^{2}/min will be used. This is the minimum that can be anticipated in such a soil type, i.e., $c_h = 2\,c_v$. Thus it represents a worst case scenario which results in the maximum predicted consolidation times.

For the calculations in this type of problem, it is customary to assume a given sand drain diameter and then calculate, for a series of spacings, the resulting consolidation times. Thus for $d_w = 12$ in., and $d_e = 25$ ft, $n = d_e/d_w = 25$ and from Figure 9-3, $T_h = 0.72$, therefore

$$t_{90} = \frac{t_h d_e^2}{c_h}$$
$$= \frac{(0.72)(25 \times 12)^2}{0.010} \; \frac{1}{(60)(24)(365)}$$
$$= 12.4 \text{ years}$$

For a range of spacings (and, for that matter, of different diameters if they are desired), Table 9-1 is provided. The time for consolidation can be seen to decrease drastically, e.g., for $d_e = 5'$; the $t_{90} = 66$ days!

Table 9-1
Spacing Ranges

d_w (ft)	d_e (ft)	n	T_h	r_e (ft)	t_{90} Years	Months	Days
1	25	25	0.72	12.5	12.4	149	4460
1	15	15	0.66	7.5	4.0	48	1400
1	10	10	0.46	5.0	1.2	14	420
1	5	5	0.27	2.5	0.18	2.2	66

Genesis of Radial Consolidation Techniques

There have been many attempts at providing for radial consolidation. These are reviewed in this section.

Cardboard "Wicks." Kjellman unfortunately called them "wicks." Perhaps cardboard does have some wicking ability, but clearly the drainage action is the induced pore water pressure which "pushes" the water up and/or down the vertical drainage system. As mentioned previously the cardboard wicks were not practical due to the low wet strength and inadequate installation equipment.

Sand Drains. Vertical columns of free draining sand (6 in. to 30 in. diameter) at 10 ft to 30 ft centers have been the "workhorse" of radial consolidation methods. Many millions of linear feet of these sand drains have been driven and the literature is replete with references. However, there are major problems known to exist:

1. Discontinuous sand drains are sometimes created by removing the installation mandrel or auger too fast.
2. Discontinuous sand drains can be created by running out of soil in the skip supplying sand to the mandrel or auger.
3. Sand drains provide no reinforcement against a shear failure which can easily be mobilized in the soft foundation soil.
4. Some nominal amount of resistance is offered by the sand column to the escaping water.
5. The effect of side wall smear due to installation and withdrawal of the mandrel or auger is unknown and very difficult to quantify.
6. Unavailability of sand (which must be properly graded since it has to serve as its own filter) is sometimes a problem.
7. Cost (particularly of the transportation) of the sand in some locations is high.

Geotextile Wrapped Sand Drains. The Chioyda Drain Pack® method was the first to use synthetic, polymer-based, materials with the concept of radial drainage. Here geotextile "stockings" were filled with sand and driven (four at a time) within a mandrel to the desired depth. The outer steel mandrel was withdrawn, leaving the sand-filled fabric stocking behind. The diameters were typically 4 in. to 6 in. This design recognized the need for a separate filter and it also gave shear strength to the entire system in the form of tensile reinforcement by the geotextile filter. However, it never "caught-on" due to the introduction of newer systems not using sand at all.

High Transmissivity Geotextiles. This concept (originally called the Colbond® system) uses a thick needlepunched nonwoven geotextile possessing good in-plane transmissivity characteristics. The 4-in. wide strips are sometimes used inside a separate geotextile stocking (as a filter) and lanced into the soil to the appropriate depth. They have not been particularly successful due to their sharply decreasing flow rate at high normal pressures. Other than this limitation, however, most of the objections to sand drains listed earlier have been eliminated.

Polymeric Strip Drains. This new generation of materials entered the geotechnical community in about 1980 and are the modernized version of Kjellman's cardboard wicks. They consist of a core material (of various configurations) protected by a geotextile filter. They have essentially revolutionized radial consolidation by eliminating virtually all of the objections to sand drains (perhaps not the smear) in an economic and efficient manner. They (and their many variants) are the focus of the next section.

CURRENT POLYMERIC STRIP DRAINS

Called "wick" drains by many people, these high drainage polymeric systems do not wick anything! They drain, and only drain, by (a) gravity, or (b) pressure. Using the technique of radial consolidation mobilized by a surcharge load the phenomenon is clearly pressure flow and when this pressure is no longer available, drainage will stop. *Therefore, it is recommended to stop calling them "wick" drains and start using a more accurate term like "strip" drains.* This is the term that will be used in this chapter.

Types, Characteristics and Manufacturers

Probably using Kjellman's cardboard drains as the model, almost all current strip drains are about 100 mm ($\simeq 4$ in.) in width (see Figure 9-4). Thereafter nothing is similar. As seen in Table 9-2, thickness varies from 1.5 to 10 mm, and the cores can be made from ribs, tubes, nubs, or slots. The geotextile filter covering is usually a heat set or needled nonwoven geotextile. This geotextile acts as a true filter allowing water to pass into the central drainage core, but retaining the outside soil from piping. Such piping would, if allowed to occur, either clog the geotextile or allow soil into the core reducing or completely blocking its drainage ability. For the unitized systems (like Desol) with no geotextile or paper

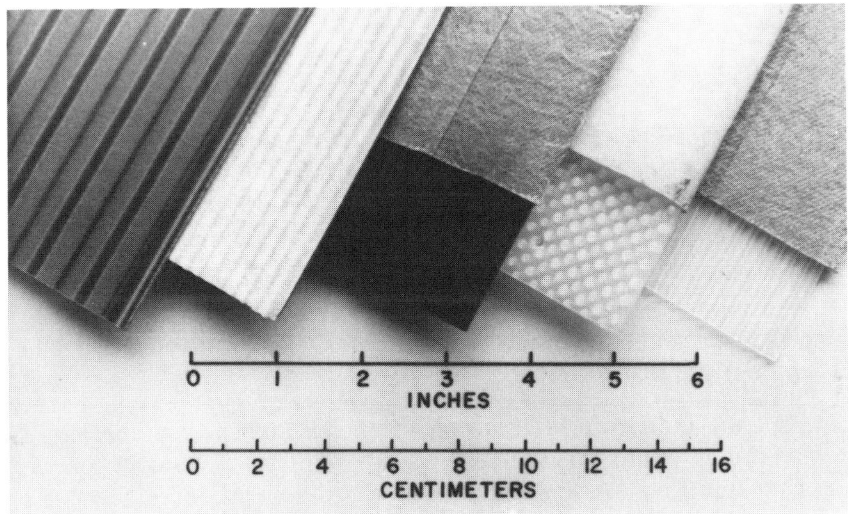

Figure 9-4. Various types of polymeric strip (or wick) drains.

Table 9-2
Details of Various Commercially Available Strip (Wick) Drains

Type	Manufacturer or Sales Agent	Material	Core Characteristics Shape	No. Channels	Covering Filter	Size (in mm)
Kjellman	SGI[1]	paper		10	paper	100 x 3.5
Geodrain	SGI[1] (Wager)	polyethylene		54	paper	92 x 4.0
Alidrain	Vibroflotation[2]	plastic (?)		studs	geotextile	100 x 6.0
Castle	Japanese (Harguim[3])	plastic (?)		36	paper	94 x 2.6
Bando	Japanese (Fukuzawa[4])	PVC		48	paper	96 x 2.9
Mebra Drain	Dutch[5,6,7]	polyethylene		38		
- ester					geotextile	95 x 3.3
- pap					paper	95 x 3.2
- prop					geotextile	95 x 3.4
- sol					geotextile	95 x 2.0
Colbond[8]		polyester	geotextile	∞	geotextile	100 x 6.0
PVC		PVC		11	unitized	100 x 1.5
Amerdrain	ICE[6]	polypropylene		8	geotextile	92 x 10
Hitek	Burcan[9]	polyethylene		studs	geotextile	100 x 6.0
Desol	Recosol[10]	polyolefine		24	unitized	92 x 2.0

1. Swedish Geotechnical Institute; Terrafigo AB, Kungsgatan 32, S-111 35, Stockholm, Sweden.
2. Vibroflotation Foundation Co., U.S. Steel Building, 39th Floor, 600 Grant Str., Pittsburgh, PA 15219.
3. Harguim Intl. Corp., 3112 Los Feliz Blvd., Los Angeles, CA 90039.
4. H. Fukuzawa Assoc., 6129 Queenridge Dr., Rancho Palos Verdes, CA 90274.
5. Geotechnics Holland BV, P.O. Box 270, 6950 AF, Dieren, Holland.
6. International Construction Equipment Co., 301 Warehouse Dr., Mathews, NC 28105.
7. L. B. Foster Co., 415 Holiday Drive, Pittsburgh, PA 15220.
8. Koerner and Welsh Book.
9. Burcan Mfgr. Inc. - 111 Industrial Drive, Whitby, Ontario, Canada L1N5Z9
10. Recosol, Reinforced Earth Co., Rosslyn Center, 1700 North Moore Str., Arlington, VA 22209-1960

filter, the exterior of the strip drain must be perforated to allow for water inflow. However, as with the geotextile voids, these perforations are critical for they must serve two cross purposes; adequate permittivity and soil retention.

Current manufacturers' addresses of the strip drains listed in Table 9-2 are also given. There is considerable mobility in this market, however, and changes can be expected.

Strip Drain Spacing Design

Two different design methods will be presented in this section to determine the spacing of polymeric strip drains. The equivalent method (Koerner's) and that of Hansbo.

Equivalent Sand Drain Method. Since much is known about the design of sand drains, one method to design strip drains is to obtain an equivalent sand drain diameter and then design accordingly [Koerner, 1984]. As an example problem, determine the 90% consolidation of a soil stratum with $c_h = 0.010$ in.^{2}/min using Bando strip drains at varying spacings.

For the solution one must measure the cross-sectional area properties of a Bando strip drain,

which are   length = 96 mm
                         width = 2.9 mm
                    void area = 92%

and calculate the equivalent void diameter of a sand drain (as an estimate use a porosity of 0.3).

- Strip drain void circle diameter

$$d_v = \sqrt{(4)(96)(2.9)(0.92)/\Pi}$$
$$d_v = 18.0 \text{ mm}$$
$$d_v = 0.71 \text{ in.}$$

- Equivalent sand drain diameter

$$d_{s.d.} = d_w = d_v/0.3$$
$$= 60 \text{ mm}$$
$$= 2.4 \text{ in.}$$

Now design proceeds as usual:

$$t_{90} = \frac{d_e{}^2 T_{h90}}{c_h}$$

Using, for example, $d_e = 24$ in., thus $n = d_e/d_w = 10$ and $T_h = 0.46$, so:

$$t_{90} = \frac{(24)^2(0.46)}{(0.010)(60)(24)}$$
$$= 18.4 \text{ days}$$

Continuing this process for spacings up to 90 in., the upper curve of Figure 9-5 results. This is the necessary design curve which could easily be extended for other values of percent consolidation and to other types of strip drains.

Hansbo's Method. The second approach toward strip drain design is more straightforward than the proceeding approach and is the preferable one. As developed by Hansbo [1979], the time for consolidation is given

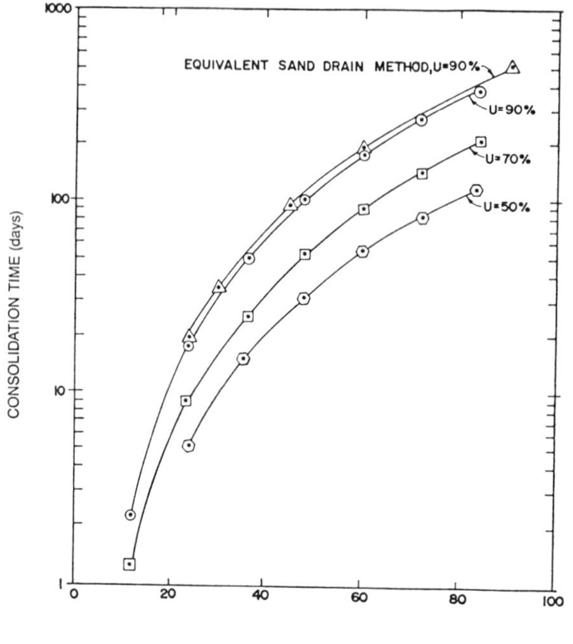

Figure 9-5. Results of example problem.

by the following equations which are illustrated by means of continuing the example problem:

$$t = \frac{D^2}{8c_h} \left[\frac{\ln(D/d)}{1-(d/D)^2} - \frac{3-(d/D)^2}{4} \right] \ln\left[\frac{1}{1-U} \right]$$

This can be simplified, since d/D is small, to:

$$t = \frac{D^2}{8c_h} \left[\ln(D/d) - 0.75 \right] \ln\left[\frac{1}{1-U} \right]$$

where   t = time for consolidation
   c_h = coefficient of consolidation for horizontal flow
   d = equivalent diameter of strip drain ($\simeq C/\Pi$)
   C = circumference of strip drain
   D = sphere influence of the strip drain (for a triangular pattern use 1.05 spacing, for a square pattern use 1.13 spacing)
   U = average degree of consolidation

To illustrate the procedure, we continue with the previous example. Calculate the times required for 50%, 70%, and 90% consolidation of a saturated clayey silt soil using strip drains at various triangular spacings. The strip drains measure 100×4 mm and the soil has $c_h = 0.010$ in.^{2}/min.

In the preceding formula;

$$d = C/\Pi$$

$$d = \frac{100 + 100 + 4 + 4}{\Pi}$$
$$= 66.2 \text{ mm}$$
$$= 2.61 \text{ in.}$$

so

$$t = \frac{D^2}{8(.010)} \left[\ln (D/2.61) - 0.75\right] \left[\ln \frac{1}{1-U}\right]$$

which results in Table 9-3 for consolidation times in minutes (the equivalent number of days in parentheses). These values are now plotted on Figure 9-5 resulting in the required design curves. Note that the D spacings must be decreased by 1.05 using a triangular strip drain pattern. When compared to the results using the equivalent sand drain method (for 90% consolidation) these values are seen to be approximately the same. The only difference is in the thickness of the strip drains used in the two examples (2.9 mm vs. 4.0 mm) which hardly matters in the foregoing calculations.

Table 9-3
Consolidation Times

D \ U	50%	70%	90%
84"	166,000 (116)	289,000 (201)	553,000 (384)
72"	115,000 (80)	200,000 (139)	383,000 (266)
60"	74,000 (52)	129,000 (90)	247,000 (172)
48"	43,000 (30)	75,000 (52)	143,000 (100)
36"	21,000 (15)	35,000 (25)	70,000 (49)
24"	7,000 (5.1)	13,000 (8.8)	24,000 (27)
12"	970 (0.7)	1,000 (1.2)	3,000 (2.2)

Relevant Properties

While obviously the entire strip drain system is important, one can focus on the core separately from the filter. Concerning the core, the following aspects need careful consideration and quantification.

- Flow rate capability of the core at the applied normal pressure it will be functioning at.
- Core breakdown pressure.
- Sustained load (creep) characteristics of the core.
- Flow channel crossover possibility if selected channels become blocked.

Concerning the geotextile filter covering of the core or the holes in the unitized body type of strip drains, the following aspects are important.

- Sufficient void or open space to handle at least the rate of the water being expelled from the consolidating soil. This, of course, is the property called permittivity.
- Sufficiently tight voids so that the adjacent soil is retained and will not pipe into the core. This, of course, is directly opposed to the permittivity feature and is very complex since we are dealing with cohesive soils.
- Sufficient tensile strength (vis-a-vis the span between contact points of the core), so that geotextile failure into the core will not occur.
- Sufficiently low elongation and high modulus so that deformation does not markedly reduce flow in the core.
- Adequate creep resistance so that the core flow is not reduced during the strip drains' service life.

Table 9-4 gives the authors' assessment of the foregoing listed concerns regarding strip drains that are required for a confident design. As seen, there is much room for improvement in the state-of-the-art.

Flow Rate Quantification

The flow rate capacity of strip drain cores can be measured in the laboratory. When obtained, this flow rate is compared to the water outflow of the consolidating soil to calculate a flow rate factor of safety.

The device used at Drexel University is one whereby the water flow through the core is under a constant hydraulic head while the sample is subjected to a desired applied normal pressure (see Figure 9-6). The re-

Table 9-4
Strip Drain Assessment Regarding Various Design Items

Property Under Consideration	Design Status	Experimental Status	Current Data Base
Core Properties			
Flow Rate	possible	possible	poor
Breakdown (Crush) Pressure	good	good	poor
Creep Characteristics	weak	weak	none
Flow Crossover Properties	observation	good	adequate
Filter Properties			
Flow Quantification	good	good	poor
Retention Characteristics	weak	weak	none
Tensile Strength	good	good	good
Elongation and Modulus	good	good	good
Creep Characteristics	good	good	good

Figure 9-6. Drexel University's device for obtaining flow rates of strip drains.

sulting hydraulic gradient can be varied from 0.06 to 2.0, but it should be recognized that strip drains function under a pressure head of varying magnitude. Applied normal pressures up to 30 lb/in.^{2} can be mobilized with this system and sustained indefinitely. This pressure is approximately that which would be developed at the base of a 100-ft-long strip drain.

The flow rate response for Amerdrain 407 (this is the strip drain being used at a recent project in Baltimore, Maryland) is given in Figure 9-7. Since these particular strip drains are about 35 ft long, the applied normal effective pressure is approximately;

$$\sigma'_h = \gamma' z K_o$$
$$= (70)(35)(0.5)$$
$$= 1225 \ lb/ft^2 = 8.5 \ lb/in.^2$$

and the available flow rate at a hydraulic gradient of 1.0 is about 1.0 gal/min. Under a pressure gradient this is probably somewhat higher.

This value is now compared to the amount of water (per strip drain spacing) coming from the site during its consolidation process. For exam-

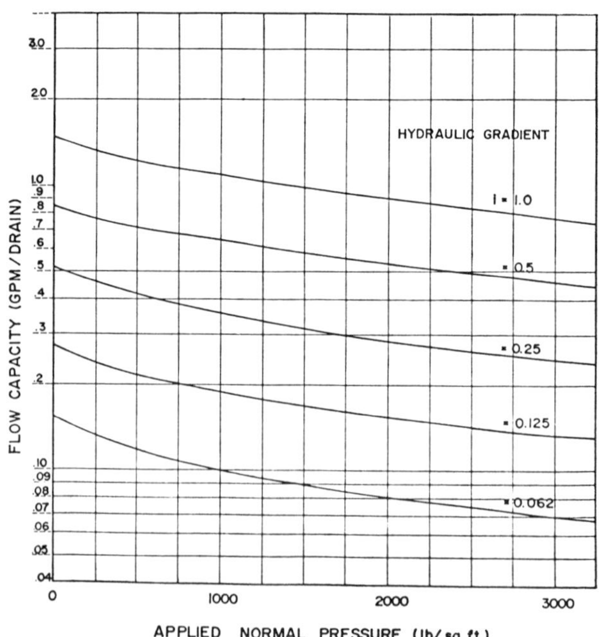

Figure 9-7. Flow rate behavior of Amerdrain Strip Drain under different gradients and pressures.

ple, if the strip drains were 35 ft deep on 5 ft centers and were due to consolidate 8 ft within 30 days, the amount of expelled water would be equal to the following:

flow rate = drainage area $\times$ settlement $\div$ time

$$\begin{aligned}
q_{reqd} &= \frac{\Pi(5)^2}{4}(8)\left(\frac{1}{30}\right) \\
&= 5.23 \text{ ft}^3/\text{day} \\
&= \frac{(5.23)(7.48)}{(24)(60)} \\
&= 0.027 \text{ gal/min}
\end{aligned}$$

Thus, the factor of safety for flow rate is as follows:

$$FS = \frac{q_{available}}{q_{required}}$$

$$FS = \frac{1}{0.027}$$

$$FS = 37$$

Miscellaneous Concerns

There are two items using strip drains which are not easily quantified as was the case of the flow rate calculations in the previous section. These are "kinking" and "smear." Kinking refers to the buckling and crimping of the strip drain while its axial dimension is drastically decreased during soil consolidation. In the previous problem the axial deformation would be $(8/35)(100) = 23\%$. If this occurred in a localized section, it could cause kinking and a potential blockage of flow. This phenomenon has been modeled by Geotechnics Holland B.V. and reported by Cortlever [1983] and Van deGriend [1984]. Figure 9-8 shows the laboratory test device and a few of the deformed strip drain shapes. Even more alarming is the reduction of flow as shown in Table 9-5. Easily seen is that "kinking" is a real problem.

The second item considered in this section is "smear." Smear is the physical remolding of the in-situ soil as the strip drain assembly is lanced into the ground and then the empty lance removed. Since "c_h" is being used in the design, the effect this installation procedure has on the performance of the system is not known. (This aspect of radial consolidation has a direct parallel with sand drains where the problem has been investi-

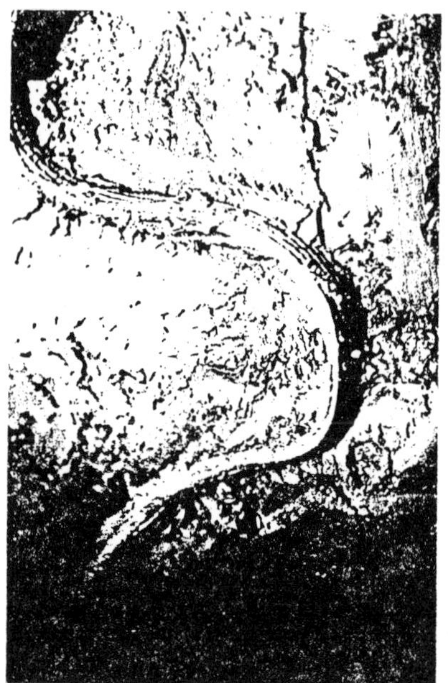 

Figure 9-8. Deformed strip drains after compression of 44% of original heights. [After Cortlever, 1983.]

Table 9-5
Strip Drain Flow Rates Before and After Compression*

Strip Drain	Before Compression		After Compression		Cause of Reduction
	cc/min	gal/min	cc/min	gal/min	
Bidim	1130	0.30	2	0.01⁻	compression of geotextile
Desol	1908	0.50	76	0.02	buckling (kinking)
Mebra	2640	0.70	2080	0.55	negligible
Colbond	2355	0.62	1620	0.43	compression of total system
Geodrain	2160	0.57	1085	0.29	geotextile pressed into flow channels

* After Cortlever, 1983.

gated for years. Unfortunately, no well-defined quantification procedure has been developed.)

It is a difficult phenomenon to model in the laboratory (scale effects are horrendous), and equally difficult to assess in the field. The philosophy generally used is to keep the installation assembly as small and as "streamlined" as possible. In this way the effect should be minimized. Additionally, the spacing of the strip drains should be made somewhat closer than that required by the design formulas.

Comparison of Strip Drains to Conventional Sand Drains

Contrasting sand drains to the alternate of polymeric strip drains, a number of interesting features are revealed. The strip drains, because they consist of plastic fluted or nubbed cores which are surrounded by geotextile features, have considerable tensile strength. Typically, the breaking strength of a 4-in. wide strip drain is 1,000 to 3,000 lb. When threaded throughout a site on centers of 3 ft to 6 ft, they offer a sizeable reinforcing effect. Furthermore, they do not require any sand to transmit flow, nor large construction equipment for installation. A rig called a "sticker" is used for installation and is relatively lightweight in comparison to sand drain installation cranes. Thus the likelihood of a shear failure during construction is somewhat reduced.

Regarding a potential disadvantage of strip drains over sand drains, the phenomenon of "kinking" was discussed. It can be very serious and further evaluation on a site specific basis is warranted. A second problem (also occurring with sand drains) is the effect of soil smear. This includes the distortion of the soil due to installation, withdrawal, and collapse of the in-situ soil on the strip drain. Its effect is mainly on the horizontal coefficient of consolidation (c_h) and it is yet to be understood.

Current Status of Strip Drains

In summary, it is felt that strip drains offer so many advantages over sand drains that they (strip drains) will be used almost exclusively in the future. Strongly in their favor are the following items:

- Tensile strength is definitely afforded to the soft soil by installation of the strip drains. It is, however, a difficult, three-dimensional, problem to quantitatively assess.
- Properly graded sand is getting to be an expensive commodity in many areas, particularly where long transportation distances are involved.
- Unlike sand drains, there is no resistance to the flow of water once it enters the strip drain.
- Construction equipment is generally small imparting low ground contact pressures on the soft soils when installing strip drains.
- Installation is simple, straightforward, and clean.

STABILITY AND REINFORCEMENT CONCERNS

This particular section of the chapter focuses on the inherent instability of the soft in-situ soils being consolidated and on the basic method to be used for their stabilization. This, of course, involves the use of a high-performance fabric (geotextile) placed directly on the surface of the soft soil.

Overview

The soft soils under consideration have extremely low bearing capacities. For example for a soil having 100 lb/ft^2 undrained strength, the bearing capacity is:

$$q_O = 5.7\ c$$
$$= 5.7\ (50)$$
$$= 285\ \text{lb/ft}^2$$
$$= 2.0\ \text{lb/in.}^2$$

This is inadequate for anything but the lightest of low contact pressure construction equipment and just barely supports the weight of an individual. Clearly, these soft soils need strengthening and this is precisely what the geotextile is intended to provide.

Geotextile (Fabric) Reinforcement Requirements

Placed directly upon the in-situ soil is the reinforcement geotextile. Due to the soft nature of the soil it is mandatory to place the fabric with the least disturbance possible. If a surface crust is at the site it should be kept intact as it can sometimes support laborers and sewing equipment and thereby expedite fabric placement. Thus a single fabric layer of maximum achievable strength is the targeted geotextile. Furthermore it is necessary that seams be sewn since overlaps would be enormously large and difficult to accurately control.

Sand Drainage Blanket Considerations

A sand drainage blanket placed directly above the reinforcement fabric is necessary for a number of reasons. Its purposes are to

1. Provide a working platform for the strip drain installation rig.
2. Laterally drain the water which will come up from the strip drains after surcharging begins.
3. Initially tension the fabric thereby mobilizing a portion of its strength.
4. Initiate the surcharge fill itself.

A high permeability sand and/or gravel is necessary because of both the drainage requirement and the fact that after settlement it will be beneath the completed structure. Usual thicknesses of the sand blanket are from 2 ft to 4 ft, but this is actually a designed value which will be discussed later in the slope stability section.

Surcharge Fill Considerations

After completion of the strip drain installation (which penetrates the drainage blanket, the reinforcing fabric, and the soil to be consolidated) an additional surcharge fill is applied above the sand drainage blanket. This usually consists of on-site soil, but can be different as described earlier. When soil is being used it is placed in horizontal layers until the final height is reached. The final height depends upon the intended use of the site. For example, if 600 lb/ft^2 loads are anticipated after surcharge removal, approximately 8 ft of soil at 115 lb/ft^3 should be used (some amount of rebound usually occurs). After completion of about 90% consolidation, the surcharge load is removed and moved to another portion of the site or off the site completely.

Slope Stability Considerations

Each of the foregoing stages is carefully controlled so as not to create a stability failure. Since instability usually starts as the fill is progressing it is analyzed as a slope stability failure. There are three stages of particular concern; during placement of the drainage blanket, during strip drain installation, and immediately after each lift of surcharge fill placement. Each case will be illustrated by means of example problems. The following assumptions will be used throughout.

1. Circular arc failures will occur.
2. Soil strength is based on its undrained shear strength, i.e., there are no separate "c" and "ϕ" components.
3. Fabric strength will be used as a working stress, i.e., ultimate strength divided by a suitable factor of safety.
4. Fabric strength is horizontal with a moment arm vertical to the center of the slip circle (this is a conservative approach versus the use of the slip circle's radius).

Using these assumptions, the configuration shown in Figure 9-9 results in the following equation for factor of safety [Koerner, 1986].

$$FS = \frac{\tau_f R + \tau_e R + T_a y}{W_f x_f + W_e x_e}$$

where FS = factor of safety
τ_f = shear strength of the foundation soil (times arc length)
τ_e = shear strength of the embankment soil (times arc length)
R = radius of slip circle
T_a = allowable tensile strength of the fabric
y = moment arm of the fabric
W_f = weight of soil within the foundation failure arc
W_e = weight of soil within the embankment failure arc
x_f = moment arm of W_f
x_e = moment arm of W_e

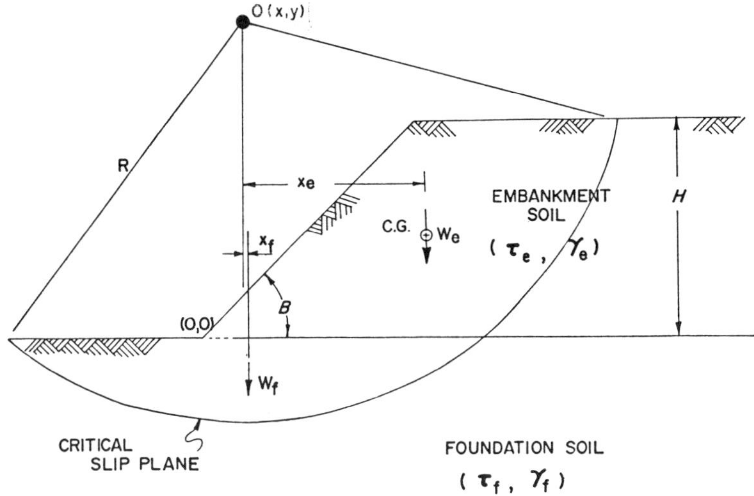

Figure 9-9. Cross section and nomenclature used in slope stability analyses.

Slope Stability (Example Problems)

As mentioned previously, three situations are included in this section. They are shown with the required data in Figures 9-10 and 9-11. Each is self descriptive and uses soil parameters which are common to types of situations in stabilizing dredged soils. These problems were solved on a micro-computer and are available for use and extension to other related situations. The three situations which will be analyzed follow. They each have variations without, then with, a fabric reinforcement layer.

Sand Drainage Blanket.

(a) Sand blanket overlying soft soil without geotextile.
(b) Sand blanket overlying soft soil with geotextile.
(c) Sand blanket overlying soft soil without geotextile with a bulldozer at the edge.
(d) Sand blanket overlying soft soil with geotextile with a bulldozer at the edge.
(e) Sand blanket overlying soft soil without geotextile with a loaded dump truck at the edge.
(f) Sand blanket overlying soft soil with geotextile with a loaded dump truck at the edge.

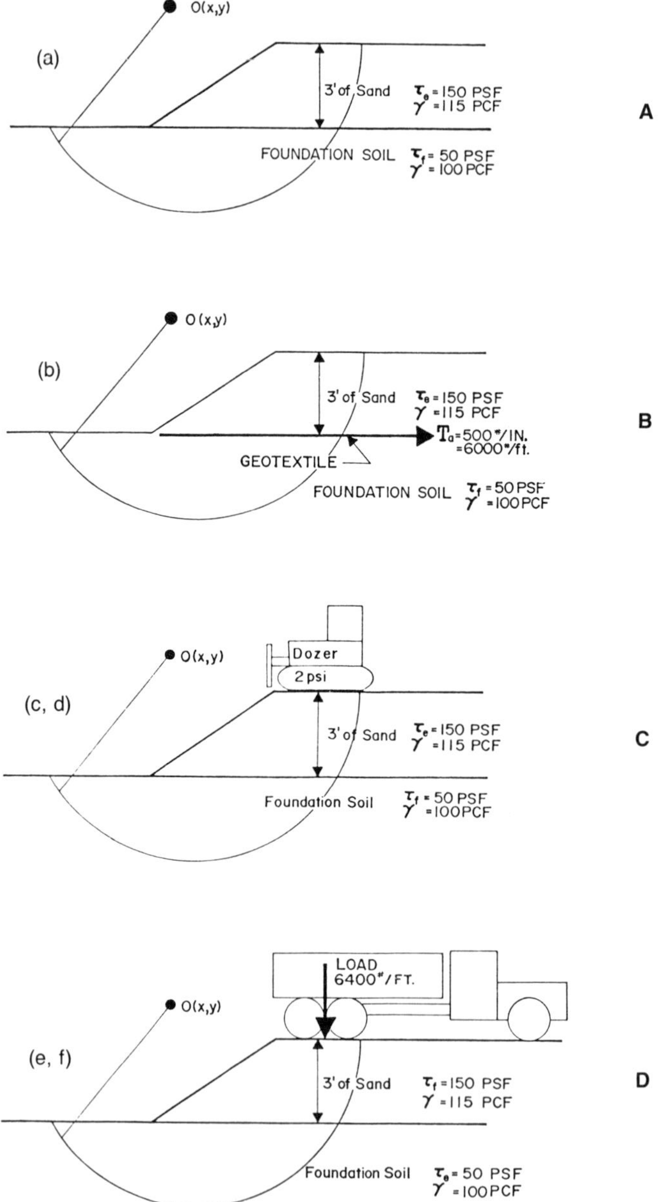

Figure 9-10. Stability analysis of sand blanket: (A) Without geotextile (a). (B) With geotextile (b). (C) With a light bulldozer (c, d). (D) With a fully loaded dump truck (e, f).

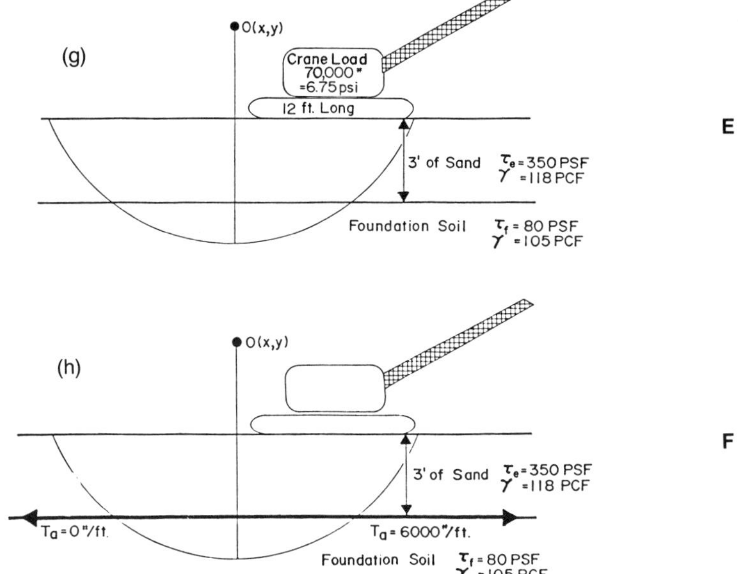

(E) Without geotextile (g). (F) With geotextile (h).

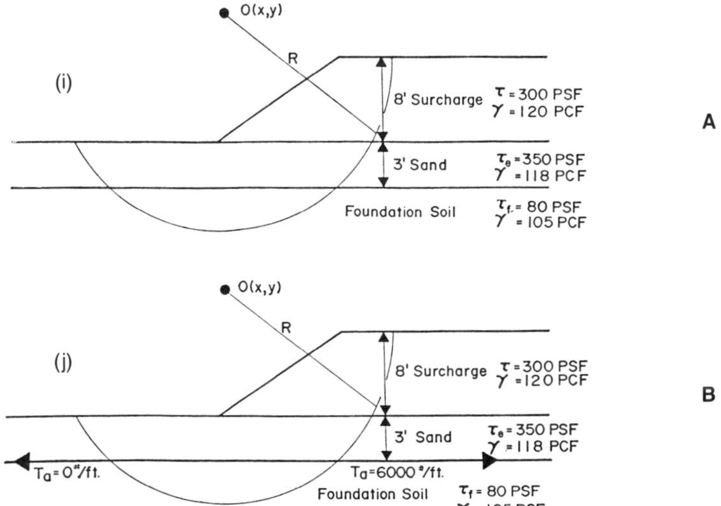

Figure 9-11. Stability analysis of complete surcharge. (A) Without geotextile (i). (B) With geotextile (j).

Sand Blanket with Strip Drain Installation Rig.

(g) Crane on sand blanket overlying soft soil without geotextile.
(h) Crane on sand blanket overlying soft soil with geotextile.

Sand Drainage Blanket and Surcharge Load.

(i) Surcharge on sand blanket overlying soft soil without geotextile.
(j) Surcharge on sand blanket overlying soft soil with geotextile.

The sketches illustrating these situations are given as Figures 9-10 and 9-11 with the resulting factors of safety given in Table 9-6. While the data speak for themselves, the necessity of the fabric is immediately apparent, as well as the general low factors of safety that the various situations bring about.

Table 9-6
Factor of Safety Stability Analysis Results of Problems Shown in Figures 9-10 and 9-11

Problem (Example Number)	No Reinforcement Geotextile	Geotextile Above Soft Soil $T_{allow} = 500$ lb/in.
sand blanket alone (a) and (b)	0.8	3.4
sand blanket and dozer (c) and (d)	0.6	2.0
sand blanket and truck (e) and (f)	0.2	0.8
sand blanket and crane (g) and (h)	0.8	1.1
sand blanket and surcharge (i) and (j)	0.9	1.4

Anchorage Requirements

In order to mobilize the strength of the fabric, sufficient embedment length in the soil beyond the failure plane is required. This is called the anchorage length. While some information is available, determination of anchorage length is a difficult problem and much research remains to be done. Recommendations at this time are thought to be very conservative, but this is not known for sure.

Example. What is the required anchorage length of a geotextile stressed to 250 lb/in. in a soil of maximum shear resistance of 125 lb/sq ft using an efficiency of 0.75. Use a factor of safety of 1.5.

Solution. Considering forces in the x-direction:

$$\Sigma F_x = 0; \; 2\tau E \; L_e = T(FS)$$
$$2(125)(0.75) \; L_e = 250(12)(1.5)$$
$$L_e = 24\,'$$

This analysis assumes that the shear strength is mobilized uniformly over the fabric in the anchorage zone. This is almost surely *not* the case. It is probably high near the slip plane and then falls off rapidly within the anchorage zone. Exactly how, remains for future research. Until then, the long lengths as just computed must be used.

Fabric Reinforcement Details

This section focuses on the specific (or certainly desirable) properties of the reinforcement fabric used between the in-situ soil and the sand drainage blanket.

Stress vs. Strain Characteristics. From the results of the example problems given earlier, it is seen that the in-situ soil needs strengthening. Thus on a conceptual basis the strongest fabric available should be used. This implies high strength, low elongation, high modulus, and high toughness. Note that we are not considering multilayers of lower-strength fabric due to logistics and deployment problems. One layer of high performance fabric is the goal. Such fabrics are commercially available from a number of manufacturers in the 1,000 lb/in. tensile strength range. While strengths greater than this can certainly be made, the load transfer across the seams often becomes the weak link in the system. Sewn seam strengths of approximately 1,000 lb/in. are the maximum currently attainable. Thus higher strength fabric becomes excessive (and costly) since it cannot be effectively mobilized, at least for areal-type stabilization projects. For linear-type projects sewn seams can be avoided in the high-stress direction.

Strain vs. Time (Creep) Characteristics. Strain with time under constant load can pose a problem, particularly when accompanied by stress relaxation while the system is under surcharge load. A number of analytic methods are potentially available to evaluate this situation, e.g.,

- Rate process theory (using thermodynamic modeling)
- Rheology (using spring and dashpot combinations)
- Three-element modeling (using experimentally obtained constants m, α, and A)

The last method is recommended for analysis purposes. It is illustrated by means of an example problem.

Example: Given a soft clayey silt soil having experimentally determined creep properties of $m = 0.80$; $\alpha = 5$; $A = 0.0030$ %/min and a fabric having similarly obtained creep properties of $m = 0.88$; $\alpha = 3.5$; $A = 0.0015$ %/min. Determine the creep strain up to 2 years if the system is acting at 50% of its ultimate strength.

Solution: The relevant equations are as follows, where D is 0.50, ϵ_1 is the initial (or elastic) strain and t is the time. Substituting the given values in the appropriate equations results in Figure 9-12.

$$\epsilon = \epsilon_1 + \frac{A}{1-m} \, e^{\alpha D} \, (t^{1-m} - 1)$$

- For the soil

$$\epsilon = \epsilon_1 + \frac{0.0030}{0.20} \, e^{2.5} \, (t^{0.2} - 1)$$

$$= \epsilon_1 + 0.183 \, (t^{0.2} - 1)$$

- For the fabric

$$\epsilon = \epsilon_1 + \frac{0.0015}{0.12} \, e^{1.75} \, (t^{0.12} - 1)$$

$$= \epsilon_1 + 0.0719 \, (t^{0.12} - 1)$$

The key features to realize from this example are that creep strains are (a) predictable upon having the required experimental data base, and (b) they are within reason using stress levels of 50% or less of the ultimate strength of the fabric. This, in turn, should set the design mode for the fabric in that a working stress based on a FS ≥ 2.0 is required. Thus, for a 1,000 lb/in. fabric, a maximum design value of 500 lb/in. should be used.

Survivability Considerations. A number of scenarios can be envisioned where the fabric might be damaged before or during installation. Those conditions sometimes outweigh the functional design considerations already presented and are called survivability criteria. They are essentially

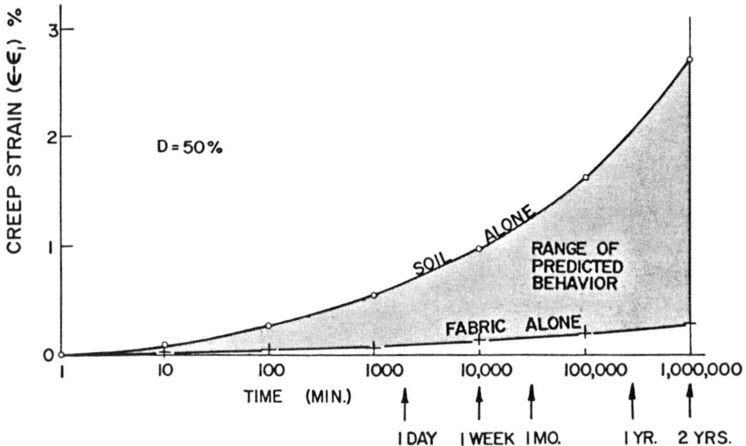

Figure 9-12. Results of example problem predicting creep strain according to the three-element model.

minimum values to be used under any consideration. A number of short numeric examples are illustrated giving order-of-magnitude results for typical situations. They are adopted from Koerner [1986].

Puncture Analysis. Consider a 1.0-in. diameter bush pushing up through the fabric as the sand blanket and surcharge (11 ft at 100 lb/ft^3) are placed upon it. For a fabric of opening size 0.0024 in., the actual puncture force is:

$$T_{act} = \Pi \, d_i d_a \, p' S'$$

where d_i = diameter of fabric opening
           d_a = diameter of soil particles above fabric
           p' = pressure
           S' = shape factor $= \Pi \, (0.0024)(1.0)(11 \times 100)(1.0)$

therefore

$$T_{act} = \Pi \, (0.0024)(1.0)(11 \times 100)(1.0) = 8.3 \text{ lb}$$

If a FS of 3.0 is used, then;

$$T_{reqd} = T_{act} \times FS$$
$$= 25 \text{ lb}$$

which is usually satisfied by high performance fabrics.

Impact Resistance. Consider a 200-lb sewing machine accidentally falling directly on the fabric from a 3 ft height. This is an energy of 600 ft-lb which is reduced due to the yielding of the soft soil beneath the fabric by a reduction factor (R.F.) of 21 (see Koerner for details [1986]).

$$E_{reqd} = E_{max}/R.F.$$
$$= 600/21$$
$$= 28.5 \text{ ft-lb}$$

This is also within reason of most high performance fabrics.

Tear Resistance. Holes often occur in the field deployed fabric and if a vehicle subsequently runs over the area, there is a tendency to propagate a tear. The tear forces can be quite high particularly on soft subsoils beneath the fabric. While difficult to quantify, tear forces of 100 to 500 lb should be capable of being resisted.

Burst Resistance. Here two situations can be envisioned:

1. The fabric being pushed up into the overlying sand voids
2. The mud wave ballooning the fabric as sand is placed over it.

The first instance should be quite low while the second could mobilize burst stresses in the 100 to 400 lb/in.^{2} range. High strength fabric should be able to handle these values.

Fabric Manufacture Considerations. Included in this section are general comments on fabric manufacture which will probably best suit the conditions described up to this point.

Polymer Type. While polyester holds advantages over polypropylene in lower creep susceptibility, better UV resistance and higher temperature stability, polypropylene is less expensive than polyester. However, none of the above arguments are overwhelming enough to specify one over the other. Far better, is to design the fabric appropriately and then let the free market decide on the type of polymer.

Fabric Style. In comparison to woven fabrics both the nonwovens and knits suffer in their low initial modulus and limiting strength to the levels envisioned. Certainly, the multifilament woven fabrics offer distinct advantages and are generally recommended for this application.

Stiffness. While stiff fabrics are generally desirable on soft soils, flexibility is important in sewing as far as constructability and handleability is concerned. Treated in ASTM D-1388 stiffness is defined as resistance of the fabric to bending. The value must be tuned to site specific conditions.

Tear Stops. As will be seen later relatively large holes are required due to the strip drain installation. These holes, at spacings of 5 ft to 20 ft might cause the fabric to fail between them when it is placed under high stress. To gain a degree of safety for such situations, tear stops can be manufactured into the fabric at random orientations to the warp and weft directions. As was seen, however, the seam is the critical link and here tear stops are not possible.

Sewing Considerations

A topic of major importance is the sewing of the longitudinal and transverse ends of the fabric from roll to roll. It is a critical part of the entire process and undoubtedly the weak link in the system.

Variables Involved. In considering the field sewing of geotextiles à number of details must be addressed, including:

- Thread type, where the choices are Kevlar®, nylon, polyester, and polypropylene (in order of decreasing strength and decreasing cost).
- Thread tension which is usually adjusted in the field so as to be sufficiently tight but not cutting the fabric.
- Stitch density, where 2 to 4 stitches per inch are customary.
- Stitch type, using single or double thread.
- Seam type, where a number of possibilities are available (see Figure 9-13), the strongest being the butterfly type.
- Number of rows as shown in Figure 9-13.

The sewing of geotextiles has rapidly advanced to the point where all fabric construction on soft ground should consider its use. Tensile seam strengths of 1,000 lb/in. can be attained and productivity has reached a point where sewing is no longer an obstacle for rapid progress of the work.

Longitudinal Seams. These are the seams made along the long axis of two adjoining fabric sheets. They are typically made by unrolling two layers on top of one another and then seaming one side of them. Three, four or more layers can be seamed by sewing alternate sides. The connected fabric panel is then deployed transversely to cover the site. This results in the seam being on the bottom side adjacent to the in-situ subsoil. Seams can generally be made quite well using portable sewing machines (electric or air-driven) and a team of skilled laborers. Since the

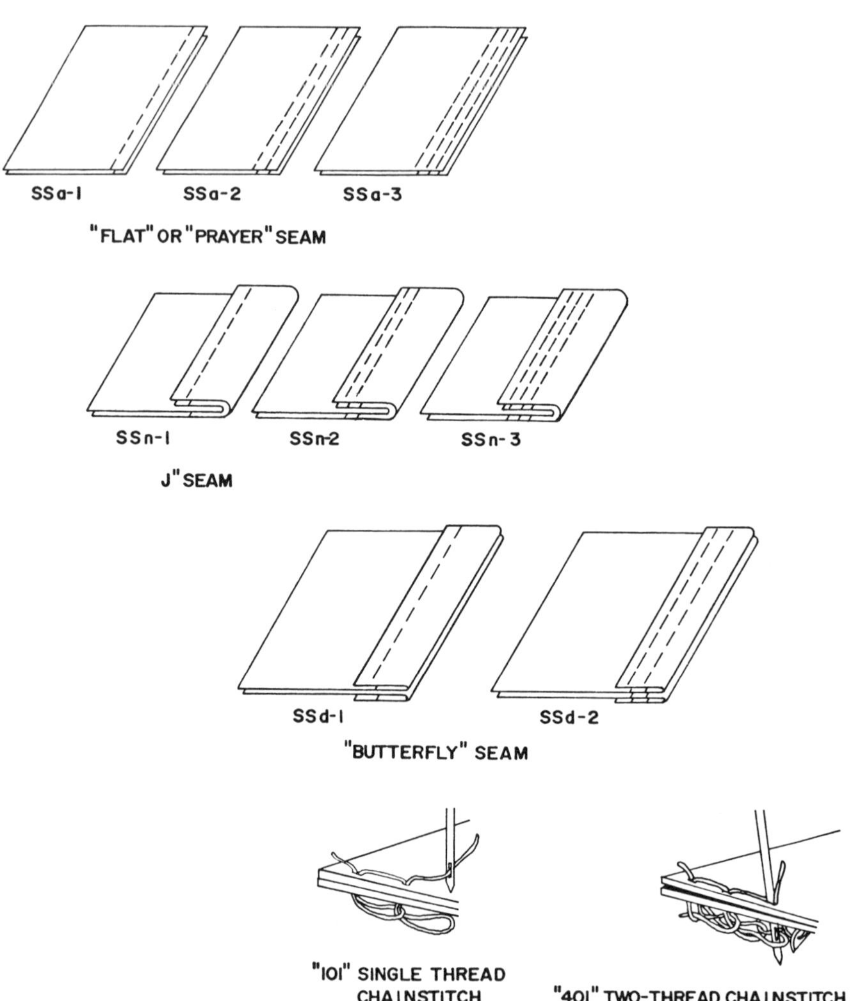

Figure 9-13. Various types of sewn seams for joining geotextiles.

sewing machine weighs about 100 pounds it is very helpful if a lightweight vehicle can support it and travel along with the seaming crew. This, of course, assumes that the in-situ soil can support such a vehicle. When done in this manner, the quality assurance of such seams can be quite high. Without it, things tend to get very sloppy since the sewing crew must work from previously placed fabric which is often very "spongy."

Transverse Seams. At the long ends of the fabric sheets the transverse edge must also be seamed. These seams are difficult seams to make.

There is too little fabric to properly grab onto, it is held in place by the longitudinal seams and the "runs" are short, i.e., they are limited to the width of the fabric which is 10 ft to 15 ft. Furthermore, it appears almost impossible to run the transverse seam completely into the longitudinal seam at each end. Thus a gap of 6 in. to 12 in. invariably arises. Furthermore, the seam looks up, and generally looks uneven and ragged. Rigid inspection is required on these particular seams.

Influence of Holes in Fabric

Of necessity, the installation of the strip drains will require punching through the fabric at whatever strip drain interval is specified. While these holes could ideally be as small as the strip drain itself (approximately 4.0 in. by 0.5 in.), the installation lance produces a much larger size. Furthermore, the driving shoe on the end of the lance to which the strip drain is attached meets the fabric first and must be pushed through it. Different types of driving shoes will cause different sizes and shapes of holes. Typically, the holes will be 6 in. to 18 in. in dimensions and circular, elliptical, rectangular or diagonal in shape.

This section concerns itself with quantifying the effect of these holes on the fabric strength and the resulting influence on the stability of the site.

Fabric Strength. Recent research [Nowatzki and Pageau, 1984] has shown that holes in fabric (exposed as a percent reduction in cross section) significantly decrease fabric strength. For woven fabrics, stressed in the warp, fill, or bias directions, the reduction in tensile strength is approximately linear with reduction in cross section. It amounts to 0.75% strength reduction per 1.0% reduction in cross section in the warp or fill directions, and 1.8% strength reduction per 1% reduction in cross section in the bias direction. Both of these reductions are quite severe and definitely of concern. While statistically it can be challenged (due to insufficient data), the stability analysis to follow will use a 1% strength reduction for a 1% reduction in cross section. This leads to the data of Table 9-7 where significant fabric strength reductions can be seen for large holes at closely spaced centers.

Nowhere in the literature is mentioned the strength reduction from holes punched along the fabric's sewn seams (vs. within the fabric itself). The thread's unraveling which might be created is even more critical. Hopefully, failure would not be sudden, but rather progressive as the thread yields and unravels. Research in this area is definitely needed and should be initiated as soon as possible.

Stability. Since the strength of the fabric plays a key role in the overall site stability, its reduction due to holes must be included in the analysis. Table 9-7 gave a range of reductions on the basis of various spacings and hole sizes. For the example problems to follow, 12 in. holes at 5 ft strip drain spacings will be used. This amounts to a 20% reduction in fabric strength, i.e., 80% of the strength is remaining in its in-plane condition.

Example. Recalculate the stability problems worked on page 419, on the basis that the fabric's allowable strength is reduced from 500 lb/in. to 400 lb/in., i.e., use 80% of the intact fabric strength.

Solution. The resulting factors of safety are shown in Table 9-8.

Table 9-7
Fabric Strength Reduction Due to Holes

Strip Drain Spacing (ft.)	Hole Size (in.)	Reduction in Fabric Cross Section (%)	Fabric Strength Remaining (%)
20'	6	2.5	97.5
	12	5.0	95.0
	18	7.5	92.5
15	6	3.3	96.7
	12	6.7	93.3
	18	10.0	90.0
10	6	5.0	95.0
	12	10.0	90.0
	18	15.0	85.0
5	6	10.0	90.0
	12	20.0	80.0
	18	30.0	70.0

Table 9-8
Answer to Example Problem

Problem	No Reinforcement Geotextile	Geotextile at Full Strength $T_a = 500$ lb/in	Geotextile at 80% Strength $T_a = 400$ lb/in
sand blanket alone	0.8	3.4	2.8
sand blanket and dozer	0.6	2.0	1.7
sand blanket and truck	0.2	0.8	0.7
sand blanket and crane	0.8	1.1	1.0
sand blanket and surcharge	0.9	1.4	1.3

Drainage. With large holes in the fabric, the soft subsoils easily extrude up into the drainage blanket. The situation is analogous to a pressurized tube of toothpaste with the cap suddenly removed. The rapidly withdrawing lance even aids the upwardly flowing material and leaves it deposited in a circular pile around the in-place strip drain. The influence on the drainage capability of the sand blanket is difficult to assess, but its impact is definitely a negative one. It should be evaluated in greater detail.

STRIP DRAIN INSTALLATION CONSIDERATIONS

This section is presented to describe the variations used to install strip drains. It is intended to preview the next section which is specifically oriented toward a single project.

Installation Lance

The strip drain is threaded into a metal lance for its insertion into the soil to be consolidated. The lance is generally made from steel, but it can also be of aluminum to lighten the load. This load is a major overturning force and a definite stability consideration. Its internal size must be larger than the strip drain itself with at least 1 in. clearance on all sides. Thus a 6 in. by 2.5 in. rectangular section would be adequate. For greater structural stiffness, however, the cross sections of the lances used are usually diamond-shaped with the long dimensions 6 in. to 8 in. and the short dimension 3 in. to 5 in.

Some earlier model lances were solid bars with the strip drain hydraulically clamped to the bottom of them. After insertion, the clamp was disengaged and the lance withdrawn. In hard driving, however, the strip drains had a tendency to tear and this type of lance was discontinued in favor of the hollow stem lance.

Type of Shoe

The strip drain only fills part of the hollow lance and the soft soil may easily get pushed up into the remaining open space. The danger here is that the strip drain does not release from the lance at its intended depth due to wedged soil within the lance. To avoid this, the strip drain must be connected to an expendable driving shoe which covers the open area of the lance. Figure 9-14 shows sketches of a number of driving shoes. In particular note the reinforcing bar which is the simplest and least expensive method. It, unfortunately, does not cover the entire bottom of the lance and allows the soft soil to enter it. This type of connection is *not*

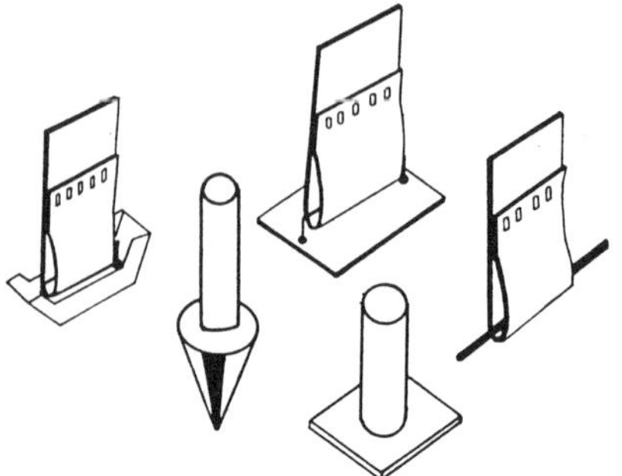

Figure 9-14. Various types of driving shoes used during strip drain installation.

recommended for use in soils of unconfined compression strength of less than 200 lb/ft^2. One of the other types shown which cover the entire bottom of the lance is recommended.

For penetrating stiff layers or puncturing high strength fabrics, it is necessary to have a pointed end on the bottom of the shoe. Advantages are the hole size in the fabric will be small thereby minimizing the strength decrease in the fabric and the installation force for penetration will be decreased thus a lighter installation rig can be used. Both of these features are very important and should easily offset the increased cost of a pointed driving shoe.

Effect of Smear

Whatever the configuration of the installation lance and the type of driving shoe, the effect on the in-situ soil is to smear it downward during lance insertion and then upward during withdrawal. The net effect is illustrated in Figure 9-15. Here it is seen that the influenced zone might be 4 in. to 6 in. beyond the strip drain itself. How the escaping pore water "fights its way" into the strip drain through this zone is of major (and completely unanswered) concern. The design parameter most seriously affected is the horizontal coefficient of consolidation (c_h) as seen earlier. No method is known as to how to handle this situation except to grossly estimate a reduction factor for "c_h" which would result in an increased time for consolidation. Whether this reduction factor is 2 or 5 or 10 is simply not known and awaits further research.

Type of "Driving Hammer"

The lance can usually be inserted into the soft soil by its static weight and/or hydraulic pressure. When inserting it through stiff upper soils or through high strength geotextiles, however, pile driving methods must be used. Both single- and double-acting impact hammers and vibratory

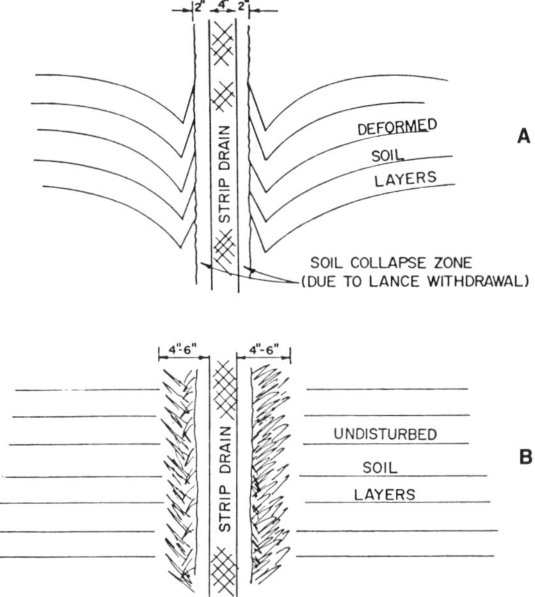

Figure 9-15. Smear effects on soil due to strip drain installation. (A) Idealized smear configuration. (B) Probable smear zone.

methods have been used. It should be noted that the forces required are generally far less than with conventional pile driving projects. Energies of 2,000 to 10,000 ft-lb are generally sufficient vs. greater than 15,000 ft-lb for driving conventional piles.

Type of Rig

Most rigs used for strip drain installation are crawler mounted cranes with wide tracks exerting relatively low ground pressures. This usually amounts to 5 to 10 lb/in^2. While such rigs are semi-stable when walking on a 2 ft to 4 ft thick sand blanket supported by a reinforcement fabric, they are very unstable during the actual driving of the strip drain. It is certainly desirable to minimize ground contact pressure and the overhanging lance/hammer weight.

Special Equipment

The strip drain threading system can be via a roll on top of the lance or via a set of sheaves from ground level up to the top of the lance. Using either method, it is then threaded down through the lance and connected to the driving shoe. Each system is slightly different and usually trouble-free. Problems can arise, however, with weak tensile strength strip drains which tear apart requiring splicing and rethreading. Equipment choices are best left to the contractor.

The method used to splice strip drains is dependent on the type of drain used (see Table 9-2). For cores separate from the geotextile filter, they can be stapled together and then the geotextile covered back over the core. For unitized strip drains, the entire system is stapled together in a 6 in. to 12 in. overlap joint fashion. The influence of this connection on the flow rate has never been quantified. It is of great concern for unitized strip drain systems.

Sequence of Operations

The installation of an individual strip drain is quite straightforward and cycle times of 1 to 3 minutes should be capable of being realized for lengths of 50 ft or less. Of greater concern, performance-wise, is the orientation and sequencing of the work. There seems to be no procedure generally specified other than depth and spacing. The orientation of the strip drain seems to be at the contractors convenience, as is the sequence of operation. While orientation might not prove to be a problem, it seems as though a systematic procedure of advancing one or two rows at a time across the entire site should be followed. Furthermore, it should be staged to follow the fabric installation and sand blanket placement, and precede the surcharge filling operations. Certainly, a random zone-by-zone installation procedure should be avoided.

MARYLAND PORT AUTHORITY'S "SEAGIRT" PROJECT

To more specifically illustrate the stabilization of soft dredged soil using a reinforcing fabric and strip drains, the Maryland Port Authority's (MPA) "Seagirt" project will be described.

Overview

The Seagirt stabilization project is a 113-acre disposal area for dredged spoil from the recently opened I-95 tunnel under Baltimore Harbor. The site is shown in Figure 9-16 and is being prepared for an extension to an existing container facility. The MPA contractor is C. J. Langenfelder and Son, Inc. of Baltimore for $10.9 million on a 600 calendar day fast-track schedule. The existing site contains 20 ft to 35 ft of dredged soil (slurry) consisting predominantly of "very soft silts mixed with variable amounts of clay and fine sand." Its SPT resistance is generally a "push" or "weight of rod," i.e., its strength is too low to be measured by conven-

Figure 9-16. Aerial view of the Maryland Port Authority's "Seagirt" stabilization project.

tional methods. The existing water content (50% to 130%) is typically 50% to 150% *above* the liquid limit. The site is contained by a roadway on the inland side and a cellular sheet pile bulkhead on the water sides. An "alligator cracked crust" 3 in.–12 in. deep exists on the ground surface, allowing one to walk over most of the areas to be stabilized.

The design consultants are STV/Lyon Assoc., Inc. of Baltimore, Maryland. The overall goal is to ready the site for an extension of the existing containerized staging area (loading, parking, and unloading of containerized truck trailors) with anticipated ground surface loads of approximately 600 lb/ft².

Fabric Type, Seams, and Design

The project specifications call for the fabric to have a 1,010 lb/in. minimum strength in both the warp and the fill directions, a 10% modulus in the warp direction of 2,780 lb/in., an elongation at failure between 15% and 35%, a soil-to-fabric friction angle of 30°, a stiffness of 30,000 mg/cm and an EOS of 0.0165 in. max (i.e., 0.0042 mm or < #400 sieve). It can be made from polypropylene or polyester, with polypropylene being "properly" UV stabilized. Furthermore, the minimum unseamed width is 12 ft and the seams must be capable of 600 lb/in. Samples for testing must be taken at each 25,000 sq yd, or less, and tested by an independent testing laboratory.

The high performance geotextile actually being used (the project is just being concluded as of this writing) is Nicolon's 62809 woven fabric consisting of multifilaments of UV stabilized polypropylene in the warp direction and polyester in the fill direction. Its mass per unit area is 30 oz/yd^2 and has a wide width tensile strength of 1,100 lb/in. and 1,300 lb/in. in the warp and fill directions, respectively. The fabric is shipped in 1,200-lb rolls which are 16.5 ft wide and 270 ft long. It is deployed at the site by wide track dozers and shifted into position by a team of 8 to 12 laborers. Fortunately, the soil surface crust supports this type of activity.

Seaming is done using a heavy-duty electric sewing machine hung on a small Steiner farm tractor. The sewing machine is powered by a Homelite HG 1,400 portable generator mounted on the rear of the tractor. Again the crust of the upper surface of the dredged spoil is an advantage for the tractor and seaming crew can work directly on it. This avoids working on the previously placed fabric which sometimes can feel like a "waterbed."

The longitudinal joints are double 'J-stitched' wherein two fabric layers are lapped, folded over and the resulting four thicknesses of material (about 1/3 in.) are stitched together. This seam is designated SSn-2 in Figure 9-13. Originally polyester thread was used, but the job was finished using Kevlar thread. The transverse joints are double "prayer-stitched" between ends of abutting fabric rolls. This is designated SSa-2 in Figure 9-13. The field crew consists of 6 to 8 laborers who shape and fold the fabric seams for the sewing machine operator.

Drainage Blanket Details

Approximately 250,000 cu yd of drainage sand is placed over the fabric in a single lift which is 2.5 ft thick. The specifications call for gradation as found in Table 9-9. The sand is truck-hauled in from off-site and spread by two Caterpillar D3B's and one Komatsu D31P wide track dozers. Specifications call for contact pressures of these small dozers not to exceed 3.0 lb/in^2.

This sand layer is capped with 6 in. of crushed slag to provide a stabilized working surface over the sand blanket. Specifications call for either crushed rock or slag meeting AASHTO M43, size number 57 as shown in Table 903 of the MSHA Specifications. It is placed similarly to the sand and is leveled by the small dozers dragging their blades.

Strip Drain Type and Design

Strip drains are being installed by a subcontractor, Geotechnics, Inc. of Bay St. Louis, Mississippi, who works off of the slag surface with a modified Koehring 266 hydraulic excavator. It has 42-in. wide crawler tracks and exerts 6 lb/in.^{2} ground pressure. The bucket and boom have been replaced with a 56 ft high set of leads and a lance. Approximately 3,000,000 lin ft of strip drains are being installed, *making this project the largest such installation in this country to date.* To punch through the geotextile, the rig's 15 ton static force must be augmented by a short burst of vibratory power. Once through the geotextile, the lance is pushed easily through the soft soil and withdrawn leaving the strip drain and driving shoe behind.

Originally, a reinforcing bar was used to attach the strip drain to the end of the lance, but subsequently a flat plate driving shoe was used (see Figure 9-14). The resulting holes in the geotextile are shown in Figure 9-17. Also shown is the hole resulting from arrow-pointed shoe which was specially made for this project (see Figure 9-17). Typical hole sizes in the geotextile are shown in Table 9-10. After the lance is withdrawn, a laborer cuts the strip drain with a knife, doubles the loose end through the metal strap on the back of the driving shoe and staples it (often he just folds it) to the new section of strip drain protruding through the lance. The laborer then hand rotates the strip drain roll to snug the driving shoe up against the bottom of the lance for the next insertion. Photographs of various phases of the installation are shown in Figure 9-18. Typical hole sizes are shown in Table 9-10.

With no interruptions, one strip drain can be installed every 30 sec as the rig rotates from side to side inserting two or three rows from a single

Table 9-9
Gradation Specifications

Sieve Size (U.S. Std.)	Percent Passing (by Weight)
3/8 "	100
#4	95–100
#10	70–100
#40	15–65
#100	0–32
#200	0–15

Table 9-10
Hole Sizes in Reinforcing Fabric from Strip Drain Installation

Type of Shoe	Shape of Hole	Dimensions of Hole (Length by Width)
Reinforcing Bar	Elliptic	9" x 5"
Flat Plate Shoe	Rectangular	6" x 6"
Arrow-Pointed Shoe	Rectangular	6" x 4"

Figure 9-17. Various hole sizes and shapes reinforcing fabric from strip drain installation. (A) Using reinforcing bar. (B) Using arrow-pointed shoe. (C) Comparison of hole sizes in geotextiles (reinforcing bar at left; arrow-pointed shoe at right).

setting. The strip drain spacing is on 5 ft centers in a triangular pattern. Daily production (including downtime to change reels of strip drain or from major movement of the rig) ranges from 10,000 to 18,000 lin ft per 10-hour day.

The strip drain being used is Amerdrain Type 407 manufactured by the American Wick Drain Company, a division of ICE. See Table 9-2 for its properties and Figure 9-7 for its flow rate under load. Each reel consists of 1,000 lin ft and weighs about 30 lb.

Figure 9-18. Strip drain installation. (A) Installation rig. (B) Strip drain and shoe at bottom of driving lance. (C) Point behind lance used to penetrate geotextile beneath sand blanket.

The specification, regarding strip drains, is detailed in a number of areas, e.g.,

1. Strip drain must be Amerdrain Type 407, Mebradrain, Desol band drain, or as approved by the Engineer.
2. A trial or test section of the proposed installation was required to be done with the engineer present and with his approval before job commencement.

3. The contractor must demonstrate that strip drain splicing will not decrease the strength nor reduce the flow capacity of the completed strip drain.

Surcharge Fill Details

Once the strip drains are installed a 7-ft to 9-ft earth surcharge fill is placed above the sand drainage blanket in order to mobilize pore water pressure in the subsoil. According to the specification, this surcharge fill material is to be brought from off-site and must meet the requirements of Section 205 of the MSHA Specification for Type II Borrow Excavation. It is to be placed in three lifts of 2 ft to 3 ft each with approximately 30 days between successive lifts. A 30-day additional time period is mentioned as being a possibility if piezometer and settlement indicators indicate a slow dissipation of excess pore water pressure. Monitoring will be described in the following sections.

Field Monitoring

Field monitoring is required according to the plans and specifications, and consists of the following devices:

1. Settlement indicators, Borros type or equal (150 lin ft required), to monitor settlements at various depths of the compressible soil.
2. Settlement plates (27 required), which are placed directly on the reinforcement fabric after strip drain installation is completed (this requires excavation placement and then backfilling of the 3 ft sand drainage blanket).
3. Piezometers (1,290 lin ft required), to be of the pneumatic type with the tip located in the compressible soil.
4. Slope inclinometers (400 lin ft required), to be located within the compressible soil.
5. Pressure load cells (19 required), to be installed within the compressible soil.

Field Performance

Without question, the surcharge is inducing *excess pore water pressure* in the subsoil, which is finding its way to the sump pump areas (see Figure 9-19). Water is simply pouring into these areas and is directly related to the nearness of the surcharge fill. Presumably, the route that the water is taking is into the strip drains, up into the sand drainage blanket and then laterally to the perforated 55 gal drums. Here it is pumped through

Figure 9-19. Water inflow from surcharged strip drains through sand drainage blanket.

hoses on the ground surface into the adjacent waterways (see Figure 9-19).

The object of the field monitoring devices noted previously is to quantify this performance. Questions such as What part of the subsoil is consolidating most rapidly? What is its variation in location and depth? Are settlements and pore pressure dissipation correlated with one another? Are total stresses eventually related to surcharge loads? Are lateral deformations contributing to the vertical settlement? are all capable of being answered by the job-required instruments. They are, indeed, the proper types of instruments for answering these questions.

At the time of this writing, however, results are not available for a critique of how successful they functioned, nor how effective they were in providing answers to the questions posed. Currently, strip drains are still being installed and surcharge is still being placed. Thus only miscellaneous comments by the field personnel can be offered. They are as follows:

1. The settlement plates are working as anticipated.

2. The piezometers are working as anticipated.

3. Settlements have been recorded up to 9 in. (as of Dec. 15, 1985), which approximately correlate with decrease in excess pore water pressure.

4. Other instrumentation was also installed and is currently being monitored.

SUMMARY AND CONCLUSIONS

Construction on Soft Soils

A new era of soft soil construction is upon us. Proper use of geosynthetics is allowing us to work on soils heretofore impossible to handle. Previous options of "excavate and replace" or "avoid by using deep foundations" are no longer necessary. Indeed the use of high-performance geosynthetics has probably saved lives [Robison, 1985]. Today, soils with shear strengths lower than 200 lb/ft^2 (and down to 25 lb/ft^2) are capable of being stabilized. Such soils as these have water contents significantly higher than their liquid limits, the value at which (by definition) the soil has negligible shear strength. The fact that these techniques are cost effective demands that this technology be used to its fullest extent.

Reinforcing Fabric

The key toward construction on very soft soils is high performance fabric (geotextiles). Tensile strengths of higher than 1,000 lb/in. are being used. This allows for a single layer of fabric to be deployed and worked upon immediately. With a minimum amount of sand placed directly on the fabric, one is out of the water and into the consolidation phase. From this point on, the situation becomes more stable rapidly.

Strip Drains

To hasten consolidation, the concept of radial flow is utilized. Rather than using the older type of sand drains, however, polymeric strip (or wick) drains are now used. Lightweight, high tensile strength, light installation equipment, fast and economical; all these are appropriate to describe these new materials. Indeed, strip drains have essentially caused the demise of sand drains in the space of a few short years.

Constructability

The hallmark of these materials is that they can be constructed. The Corps of Engineers, the California Dept. of Transportation, and now the Maryland Port Authority have clearly shown that the work is getting better, faster, and less costly. Conversations with the general contractor in Baltimore clearly show that they will follow this type of construction since it is a "winner."

Conclusions

Unfortunately, the last aspect of this type of soft soil construction to catch-up is design. This is not unlike all of geosynthetics, however, where manufacturing and successful use has preceded design. Perhaps with projects like the one described here, and with analyses as presented in this chapter some of the gap has been filled.

RECOMMENDATIONS

Additional research and development in the use of geosynthetic materials for soft soil construction and stabilization still remains to be done. This last section outlines some directions where further inquiry should be directed.

Regarding Design Method

Here is where the majority of the work in this area remains to be done. While the designs presented here are felt to be reasonable, they must be further tuned in order to "build-with-confidence." In the fabric area better quantification of the effects of holes is needed along with an assessment of the sewn seam strength with and without holes. If these are indeed the limiting factors then the use of extremely high strength fabrics may be uneconomical. In the area of strip drains a number of features need investigation, recall Table 9-4. A separate thrust in addition to FHWA's current efforts [see Reference 3] should be initiated.

Regarding Test Methods

Much work is required in order to establish valid index and performance tests for both high strength fabrics and strip drains. It is foolish to "design by function" [Koerner, 1986] and then compare the required values to actual material properties which are only loosely defined or known. At the minimum we need work on the following:

- Wide width tensile tests on high performance fabrics.
- Proper seam testing procedures.
- Strip drain flow information in the nondeformed and deformed (kinked) condition.
- Potential reinforcing benefits of strip drains after they are installed.

Regarding Constructability

Perhaps the greatest need in this category is ways and methods to increase seam strength. A number of alternatives in addition to sewing come to mind, e.g., mechanical, heat, ultrasonic, adhesives, etc. A major thrust is needed so as to better utilize the high-performance-reinforcing fabric's strength.

Regarding Plans and Specifications

Focus should be on performance characteristics and not on specific materials and "or equal" concepts. For design projects such as these we must have confidence in our methods, state our requirements, and let the manufacturers and contractors propose their best solutions. In this manner we will achieve the best of new and innovative solutions, at the most economical price.

REFERENCES

1. Barron, R. A., "Consolidation of Fine Grained Soils by Drain Wells," *Trans. ASCE,* Vol. 113, 1948, pp. 718–754.
2. Cortlever, N. G., "Investigation Into the Behavior of Plastic Drains in Rotterdam Harbour Sludge," *Geotechnics Holland,* B.V., April 1983, 16 pp.
3. Federal Highway Administration, Dept. of Transportation, Grant No. DTFH61-83-R-00059 to Haley and Aldrich, Inc., Cambridge, Mass.
4. Hansbo, S., "Consolidation of Clay by Band-Shaped Prefabricated Drains," Ground Engineering, July 1979, pp. 16–25.
5. Koerner, R. M., *Construction and Foundation Methods in Geotechnical Engineering,* McGraw-Hill Book Co., NY, 1984.
6. Koerner, R. M., *Designing with Geosynthetics,* Prentice Hall, Inc., Englewood Cliffs, NJ, 1986.
7. Nowatzki, E. A., and Pageau, S. R., "The Effect of Holes on the Tensile Performance of Geotextiles," *Geotechnical Fabrics Report,* Spring 1984, pp. 12–15.
8. Robison, R., "Engineering with Fabric," *Civil Engineering/ASCE,* Dec. 1985, pp. 52–55.
9. Van deGriend, A. A., "Deformation of Plastic Drains," *Weg and Water,* No. 1, Oct. 1984, Holland, 7 pp.

Conversion Factors
(U.S. Customary Units)

British Absolute	To Convert from/to, Multiply by	British Gravitational	To Convert from/to, Multiply by	SI Units
		Mass		
pounds (lb)	× 0.031081	= slugs	× 14.59390	= kilogram (kg)
		Force		
poundal	× 0.031081	= pound (lbf)	× 4.448222	= newton (N)
		kip (1,000 lbf)	× 4.448222	kilonewtons (kN)
		ton (2,000 lbf)	× 8.896444	kilonewtons (kN)
		Pressure or Stress		
poundals per sq ft	× 0.031081	= pounds per sq ft (lbf/sq ft)	× 47.88026	= newtons per sq meter or pascal (N/m^2 or Pa)
		kips per square foot	× 47.88026	= kilonewtons per square meter (kN/m^2 or kPa)
		tons per square foot	× 95.76052	= kilonewtons per square meter (kN/m^2)
		Mass Density		
pounds per cu ft (pcf)	× 0.031081	= slugs per cubic feet	× 515.3788	= kilogram per cubic meter (kg/m^3)
		slugs per cubic feet	× 0.5153788	= megagrams per cubic meter (Mg/m^3)
		Unit Weight		
poundals per cubic ft	× 0.031081	= pounds per cu ft (lbf/cu ft)	× 0.1570875	= kilonewtons per cubic meter (kN/m^3)

Subject Index

A

Acceleration, 78, 93
Algorithm, 44
  explicit (one step), 48, 50
  integration, 42
    for elastic-plastic materials, 43, 44
  return mapping, 46, 48
Allowable
  factor of safety, 287, 289, 294, 296
  settlements (deformation), 310, 312
Anchors, in sand, 169, 173
Anisotropic soil, 2, 31
Anisotropy
  cross-anisotropy, 14, 37, 41
Angle
  shearing resistance, 182, 396
  internal friction, 182
  mineral friction, 182
Apparatus, direct shear, 161
Apparent opening size (AOS), 364, 365
Applications, geotechnical engineering, 1,
    35, 42, 60, 66, 109, 141, 163, 180,
    242, 246, 253, 255, 261, 265, 267,
    276, 279, 282, 285, 286, 290, 294,
    333, 349, 374, 382, 397, 398, 403,
    406, 409, 416, 444
Associative flow rule, 24, 26
ASTM, 355, 356

B

Batter piles, 237, 238
Bearable deformations, 310, 312
Bearing capacity, of shallow foundations
  central and inclined loading, 227, 231
  centrally vertical-loaded strip footing,
    219, 225
  examples, 242, 246, 253, 225, 261,
    264, 267, 276, 290, 294
  inclined footing, 249, 251
  modes of failure, 213, 215
  rectangular footing, 231, 242
  shear resistance of the overburden, 232,
    233
  strength parameters, 255, 261
    influence of compressibility, 259, 261
    scale effect, 256, 258
  strip footing near a slope, 251, 252
  ultimate bearing capacity, 218
  ultimate load criterion, 218
Behavior, of interface, 149, 161
Behavior, soil, 3
BFGS iterations, 105, 109
Body forces, 83, 89
Broyden iterations, 109, 110
Boundaries, 90, 92
Boundary conditions, 90, 92, 328
Boundary element formulation, 322
Boundary value problem, 43
Boussinesq's solution, 298
Bulk modulus, 31, 44, 130

C

Capped yield models, 54
Case histories, 122, 142, 434, 444
Categories, of geotechnical models, 3
CBR, 388
Clay, soft, construction on, 395
Closed-form solutions, 43, 89
Coefficient, subgrade reaction, 312, 313
Cohesion, 55, 396
Cohesionless soils, 181, 210
Compatibility conditions, 328
Compression, isotropic, 186, 199
Compressible fluid, 91
Compression triaxial tests, 181, 203
Cone penetration test, 266, 267
Confining stress, 182
Consolidation, 109, 114, 396, 404

Consolidation, radial, 400, 402
Consolidation, vertical, 396, 398, 400
Constants, of integration, 349, 351
Constitutive behavior, 8, 10
Constitutive laws, 8, 10, 34, 54
Construction, on soft soils, 395, 442
Contact pressure, 312, 313
Convergence, 47, 50, 100, 109
Coulomb
    Mohr-Coulomb yield criterion, 20, 396
Coupled behavior, 86, 160
Cyclic loading, 153, 154
Cyclic response of interfaces, 147, 176
Cross-anisotropy, 14, 37, 41

D
Damping
    geometric, 329
    material, 329
Damping matrix, 94
Damping ratio, 325
Dams, dynamic analysis of, 125, 141
Darcy, permeability tensor, 91
Deep foundations, 322
Deformation, problems of, 296
    in clays, 306, 310
    in sands, 302, 306
Degree of saturation, effect of, 191, 194
Degradation, of granular materials, 181,
    182, 209
Density, effect of, 185, 189
Density, relative, 185, 193
Depth factor, 233
Derivatives, of invariants, 69
Design, 383, 389
Design, of shallow foundations
    allowable factor of safety, 287, 297
    allowable settlements, 312
    field tests, 265, 266
    laboratory tests, 161, 185
Deviatoric plane, 15
Deviatoric stress, 16
Deviatoric tensor invariants, 67
Differential settlements, 312
Dilatancy, 57, 182, 206
Dilatant materials, 26, 57
Direct shear test, 161
Direction cosines, 328
Displacement, 90, 92, 171
Drains
    fabric utilization, 377
    sand, 415, 416
    strip, 393, 405, 408

Drainage conditions, 88
Drucker-Prager yield criterion, 22, 23
Dutch cone penetrometer, 266
Dynamic analysis of piles, 322, 333, 338,
    348
Dynamic analysis
    of nuclear power plants, 123, 176
Dynamic behavior of a dam, 125
Dynamic behavior of pile groups, 322,
    350
Dynamic loading, 286, 287
Dynamics of porous media, 77
    applications, 86
        drained conditions, 88
        finite element equations, 92
        pseudo-static loading conditions, 88
        undrained conditions, 87
        time integration, 93
        without inertia and convective effects,
            86
    average quantities, 79
    balance laws, 81, 83
        balance of mass, 81
        balance of linear momentum, 82
        entropy inequality, 83
    basic theory, 77
    constitutive assumptions, 84
        equations of state, 84
        momentum supplies, 85
        solid porous skeleton, 85
    coupled field equations, 86
    kinematics, 78
Dynamic pile formula, 4

E

Earth dam, 125, 141
Earth pressure, 167, 168
Earthquake, San Fernando, 125
Effective stress, 55, 76, 116, 120, 396
Elasticity, 4, 10
Elastic behavior, 4, 10
Elastic-plastic rate equations, 31
Elements, finite, 111, 115, 164, 165, 167,
    170
Elliptic equations, 89
Embankments, soft soils, fabric use in,
    395, 442
Empirical rules, 3, 387, 388
Endurance, fabrics, 373
Engineering judgment, 2, 6
End restraint, effect of, 195, 196
Entropy, inequality, 83

Erosion control, fabric use in, 378
Examples, dynamics of porous media,
    109, 141
Excess pore pressures, 76, 86, 87, 396

F

Fabrics
    endurance tests, 373
    hydraulic properties, 365, 367
    mechanical properties, 367, 373
    physical properties, 363, 365
Fabrics, applications of, 379, 382
Fabrics, functions of, 374, 379
Fabric, reinforcement, 393, 423, 426
Fabric thickness, 363, 364
Factors, bearing capacity, 219, 223
Factor of safety, 287, 289
Failure criterion
    Drucker-Prager, 22
    Huber-Von Mises, 20
    Mohr-Coulomb, 20
    Tresca, 19
Field tests
    cone penetration, 266, 267
    plate, 265, 266
    pressuremeter, 272, 273
    standard penetration test, 388
Fill, 417, 422
Filter, 376, 378
Finite element formulation
    for dynamic analysis, 89, 109, 153, 160
Finite element method
    interface, representation of, 149, 176
Finite difference, 93
Flow rate, 410
Flow rules, 23
    associated-, 24
    nonassociated-, 24
Fluid
    compressible, 91
    incompressible, 91
Foundations
    deep, 322, 351
    shallow, 212, 315
Fourier transform, 132, 138
Frequency, 133, 141, 161, 164

G

Geomaterials, behavior of, 9, 10
Geotechnical construction,
    fabrics used in, 379, 389, 391, 443
Geotechnical models, categories of, 3, 5
Grab tensile test, 369, 370

Geotextiles and related products
    applications, 379, 382
    conventional, 358, 359
    definitions, 355, 356
    description, 361, 362
    design, 384, 389
        analogical methods, 385, 387
        analytical methods, 384, 385
        empirical methods, 387, 389
    functions, 374, 379
    geocomposites, 361
    properties, 363
        endurance, 373
        hydraulic, 365, 367
        mechanical, 367, 373
        physical, 363, 364
    types, 357, 358
Grain size, effect of, 197, 198
Green's function, 326, 327
Ground motion acceleration spectrum, 132

H

Hardening
    kinematic, 30, 31
    isotropic, 27, 30
Hardening rule, 57
High pressures, 181, 206
Huber-Von Mises criterion, 20, 21
Hydraulic properties, of fabrics, 365, 369
Hyperbolic equations, 89
Hysteresis, 65
Hysteretic damping, 325

I

Idealization, material model, 9
Incompressible fluid, 91
Incremental stiffness, 160
Index, rigidity, 216, 218
Inelastic behavior, 8, 12
Initial conditions, 90, 92
Initial normal stiffness modulus, 156
Initial shear stiffness, 154
Initial stresses, 98
Integration, constants of, 349, 351
Integration, time, 93, 100
Interfaces
    combined normal and shear behavior,
        160
    dynamic analysis, 151
    models, 151, 152
    normal behavior, 155
    reloading, 157
    unloading, 156

virgin loading, 156
practical problems
  anchors in sand, 169, 173
  buried pipe, 165, 167
  nuclear power plant, 173, 176
  tied-back wall, 167, 168
shear behavior, 153, 155
static and cyclic response of, 147, 148
static analysis, 149, 150
Interface behavior, representation of, 147, 177
International Geotextile Society, 374
Invariants of stress, 12, 13, 67, 70
Isotropy, 19, 31
Isotropic hardening, 27, 30
Iteration
  modified Newton-Raphson, 102, 104
  Newton-Raphson, 101, 102
  quasi-Newton, 104, 109
    BFGS, 105, 109
    Broyden, 109, 110

J

Jacobian, determinant, 109
Joint elements
  normal, 155, 160
  shear, 152, 154
  stiffness, 151
Judgment, engineering, 2, 6

K

Kinematics, 78, 79
Kinematic hardening, 30, 31

L

Lamé's constants, 85
Lateral earth pressure, 167, 168
Laterally loaded piles, 322
  differential equations for, 325, 326
  cyclic loads on, 322, 349
Linear behavior, 10
Loads, dynamic, 70, 147, 286, 287
Local effects, hardening, 27, 28
Local overstress, 27
Load reversal, 10, 11, 65, 66

M

Maryland port authority, 434
Mass density, 80, 86, 324, 325
Mass matrix, 93

Material behavior
  ideal, 8
  modeling of, 9
  nonlinear, 1, 2, 8, 10
  real, 8
Material constitutive tensor, 85
Material symmetries, 13
Matrix, stiffness, 92
Mechanical properties
  of fabrics, 367, 373
  of soil, 8, 42
Methods
  analytical, 34, 384
  boundary element, 322
  empirical, 34, 384
  finite element, 3, 45, 76, 147
Microstructure, 11
Minerals, 194
Mineral composition, 189, 201
Modeling, in geotechnical engineering, 3, 89, 152
Models, interface, 151
Modified Newton-Raphson iteration, 102, 104
Modulus
  bulk, 31, 44, 130
  shear, 31, 44, 130
  subgrade, 312
  Young, 324
Mohr-Coulomb criterion, 20
Mohs hardness, 185
Motion, input ground, 132

N

Newmark's method, 95
Newton-Raphson iteration, 101, 102
Non-associative flue rule, 24
Nonhomogeneities, 276, 286
Nonlinear analysis, 125
Nonlinear behavior, 10
Nonlinear materials, 31
Nuclear power plant structures, dynamic analysis of, 173, 176
Numerical methods, for shallow foundations, 314, 317
Numerical stability, 45

O

Objectivity requirements, 12, 13
Observational method, 2
Oedometer, 300
  tests, 300, 308

One-dimensional wave propagation, 114, 120
Overconsolidated clay, 308

P

Parabolic equations, 322, 335, 349
Partial safety factors, 295, 297
Penetration tests
  Dutch cone, 266, 267
  standard, 388
Phases of soil, 1, 8
Piles, 322
  axial loading, 332, 333
  batter, 338
  lateral loading, 330, 332
Pile group, 322, 335, 349
Pipe, buried, 165, 166
Plasticity, 10, 31
Plastic flow rules, 23, 26
Plastic hardening rules, 26, 27
  isotropic hardening, 27, 30
  kinematic hardening, 30, 31
Plasticity models, a simple model for
    soils, 54
  flow rule, 57
  hardening rule, 57, 58
Plastic potential, 24
Plastic strains, 23, 27
Plastic stress relaxation, 34
Plate tests, 265, 266
Poisson's ratio, 217
Polymers, 362
Pore pressures, 76, 86, 87
Pore pressure coefficients, 308
Porous media, dynamics of, 76
Port Allen lock, 4, 5
Predictive capability, in geotechnical
    engineering, 1, 7
Pressure insensitive materials, 19, 20
Pressure sensitive materials, 21
Pressuremeter tests, 272, 273
Principal stress, 15, 17
Principal stress space, 15
Propagation, wave, 114, 120
Punching shear, 214, 215

Q

Quantity of flow, 410
Quasi-Newton iterations, 104, 110
  BFGS, 105, 109
  Broyden, 109, 110

R

Radial consolidation techniques, genesis of
    cardboard "wicks," 403
  geotextile wrapped sand drains, 404
  high transmissivity geotextiles, 404
  polymeric strip drains, 404
  sand drains, 403
Ramberg-Osgood parameters, 153, 159
Rate of consolidation, 397, 402, 403, 407,
    409
Reactions, distribution of, 312, 313
Reinforcement, fabrics and, 379
Relative density, 185, 193
Reloading curves, 153, 154, 157
Requirements, objectivity, 12, 13
Response spectra, 135, 141
Retaining structures, 147, 148, 167, 173
Rigidity index, 216, 218
Road reinforcement, fabrics used in, 379
Rocks, 182
Rupture, of grain contacts, 181, 209

S

Safety factor, 287, 289
Sand, 63, 66, 181, 212
Sand drains, 415, 416
Santa Felicia dam, 125, 141
Saturation, degree of, 190, 191
Saturated soils, 76
Scale effect, 256, 258
Seismic
  accelerographs, 124, 126, 128, 141
  analysis, 120
Settlement, 296, 312, 334, 336
Shallow foundations, 212, 314
  examples of, 242, 246, 253, 255, 261,
      264, 267, 276, 290, 294
Shape factors, 231, 232
Shear modulus, 31, 39, 130
Shear strength, of granular material, 181,
    209
Shear tests
  combined normal and shear, 162
  cyclic, 162
  static, 161
Sign conventions, for stresses, 21, 23
Simple shear tests, 65
Slip element. *See* Interfaces.
Soft soils, embankments over, fabric use
    in, 391, 444

Soils, 1, 3, 76
  cohesionless, 54, 63, 66
  cohesive, 55
  theory of mixtures, 77
Soil behavior, 2, 3, 8, 10, 76, 77
  modeling for, 3, 4, 77, 151, 160
Soil-structure interaction
  in piles, 322
  in retaining structures, 167, 172
Stability
  time integration, 93
  numerical, 45, 95
Stabilization, of soft soils, 391, 444
Stability, slope, 419, 423
Standard penetration test, 388
Strains
  recoverable, 23
  volumetric, 23, 25
Stress
  deviator, 16
  effective, 55, 76
  invariants, 12, 13, 67, 70
  levels, 183, 184
  principal, 17
  sign conventions, 21, 23
  tensor notation for, 15
  total, 396
Stress-strain behavior, 38
Stress-strain curves, 11, 39, 41
Stress path, 17
Stress tensor, 15
Stress vector, 15
Strip drains, 393, 405, 442
  installation considerations, 431, 434
  spacing design, 406, 408
  types, 405
Subgrade reaction, 312, 313
Surcharge loading, 398
Synthetic fabrics, 355, 356, 391, 392

T

Tensor, invariants, 67
Tests, high stress level, 201, 203
Tied-back walls, 167, 168
Time integration, 93
Transformation matrix, 12, 13
Tresca criterion, 19, 20

Triaxial compression tests, 181, 207
Triaxial stress state, 60, 63, 64
Two-phase porous media, 77

U

Ultimate bearing capacity, 213
  shallow foundations, 212
Ultimate load criterion, 218
Undrained behavior, 87
Uniformity coefficient, 189, 206
Unloading, 65, 153, 156

V

Velocities, 78, 81, 119
Vertical and radial consolidation, 396, 402
Vertical consolidation techniques, genesis
  of
  dewatering, 400
  surcharging by vacuum, 394, 400
  surcharging with water, 399
  surcharging with soil, 398, 399
Vibration of piles
  lateral, 330, 332
  axial, 332, 333
Volume changes, 27, 187
Volumetric strains, 27, 187
Von Mises criterion, 20

W

Wave propagation, 114, 120
Walls, retaining, 148, 167, 173
Wave speed, 100
Winkler foundation, 312

Y

Yield criterion, 11, 23
  Drucker-Prager, 22
  Huber-Von Mises, 20
  Mohr-Coulomb, 20
  Tresca, 19, 20
Yield function, 55
Yield surface
  Drucker-Prager, 23
  Huber-Von Mises, 20
  Mohr-Coulomb, 22
  Tresca, 19
Young's modulus, 217

Author Index

Abdel-Ghaffar, A. M., 128, 130, 142, 145
Abel, P. G., 318
Aboul-Ella, F., 323, 353
Adams, 183
Agbabian, M. S., 151, 179
Ahrens, 183
Akroyd, 183
Appel, G. C., 150, 178
Apsel, R., 323, 351
Argyris, J. H., 71
Arman, A., 389–390
Atkin, R. J., 83, 86, 142
Atkinson, J. H., 3, 6
Baguelin, F., 273, 316
Baker, R., 25, 71, 318
Baladi, G. Y., 54, 71
Balsley, 183
Banerjee, P. K., 322–324, 326–328, 330, 333, 338, 344, 351–352
Banks, D. C., 183, 217, 318
Barksdale, 183
Barron, R. A., 400–402, 445
Bathe, K. J., 77, 86, 89, 92, 106, 113, 142–143
Bazant, Z. P., 82, 142
Beech, J. F., 389
Bell, J. R., 389–390
Biot, M. A., 76–77, 80, 114, 116, 142
Bishop, A. W., 183, 255, 316
Bjerrum, L., 255, 303, 308, 310–312, 316–317
Blinde, A., 318
Boehler, J. P., 12, 71
Bogard, D., 152, 179
Bonaparte, R., 385, 389
Borg, 183
Boussinesq, 298, 302, 307, 312–313
Bowden, F. P., 149, 177

Bowen, R. M., 77–78, 82–83, 85, 114, 116, 142–143
Bransby, P. L., 3, 6
Brekke, T. L., 149, 179
Bridgman, 183
Brodlie, K. W., 77, 101, 106, 143
Bromwell, L. G., 182, 192, 210
Broyden, 104–105, 109–110, 113–114
Buisman, A. S. K., 220, 224, 278–280, 282, 302–303, 317
Burland, J. B., 54, 74, 311, 317
Burns, J. W., 166, 177
Butterfield, R., 328, 338, 352
Byerlee, 183
Caquot, 220, 224, 238, 248, 317
Carroll, R. G., Jr., 363, 389–390
Casagrande, A., 3, 6
Cauchy, 11
Cayley, 67
Chan, 183
Chang, C. Y., 54, 71
Cheang, L., 152, 179
Chen, P. J., 77, 114, 116, 143
Chen, W. F., 43, 71
Christensen, N. H., 225, 318
Christian, J. T., 3, 6, 314, 317
Christopher, B. R., 385, 389
Cimento, A. P., 86, 106, 142
Clausen, C. J. F., 167–168, 177
Clough, G. W., 4–6, 150, 152, 177, 183, 184, 207
Coon, M. D., 54, 71
Cortlever, N. G., 413–415, 444
Cosserat, E., 11, 71
Cosserat, F., 11, 71
Coulomb, C. A., 18, 20–21, 23, 56, 71, 147–152, 372, 396
Coyette, I., 315
Craine, R. E., 83, 86, 142

Crandall, S. H., 149, 177
Crochet, M. J., 143
Cryer, C. W., 78, 82, 110, 143
Dafalias, Y. F., 35, 71
Darcy, H. J., 81, 86, 91, 143, 365
Davies, T. G., 322–324, 326–327, 333, 344, 351, 354
Davis, E. H., 330
De Beer, E. E., 183, 212, 222–223, 225, 235, 256, 258, 265, 303, 317
De Poorter, 225, 317
de Souza, 183
Deere, D. U., 182, 192, 210
Dennis, J. E., Jr., 105, 109, 143
Desai, C. S., 3, 6, 25, 71, 147, 149–153, 155–156, 159–162, 165, 167–169, 177–180, 314, 317
Desnoyers, J. F., 389
DiMaggio, F. L., 54, 71
Dikmen, S. U., 77, 82, 121, 143–144
Dobry, R., 179
Driscoll, R. M., 330
Drucker, D. C., 22, 26, 31, 42–43, 54, 71
Drumm, E. C., 153, 155, 159, 161–162, 177, 179
Duncan, J. M., 4–6, 54, 71–72, 150, 152, 177–179
Engelman, M. S., 77, 109, 143
Eringen, A. C., 77, 82, 143
Euler, 48
Evans, R. J., 54
Fardshisheh, F., 72
Farhoomand, 183
Feagin, 337–338
Finn, N. D., 121, 144
Fishman, K. L., 160, 179
Fletcher, 105
Fluet, J. E., Jr., 389
Focht, J. A., 4, 6
Foo, S. H. C., 322, 353
Fowler, J., 391
Frantziskonis, G., 178
Frohlich, 308
Fumagalli, 183
Galagoda, H. M., 178
Galerkin, 93
Garber, M., 314, 316, 318
Gazetas, G., 352
Ghaboussi, J., 54, 71, 77, 82, 121, 143–144, 150–151, 179
Gibson, R. E., 71, 323, 352

Giroud, J. P., 355, 357, 363, 367, 374–375, 378, 381–382, 385, 389–390
Golder, 183
Goldfarb, 105
Goodman, R. E., 149, 151, 179
Gordon, 183
Gourlay, A. R., 77, 143
Gray, W. G., 77, 80, 82, 144
Green, A. E., 35, 71, 77, 144, 327, 348
Greenstadt, J., 77, 143
Griffith, D. U., 314, 318
Griggs, 183
Gurtin, M. E., 82, 144
Gussmann, P., 316, 318
Hall, J. R., 74
Hamilton, 67
Handin, 183
Hansbo, S., 406–407
Hansen, J. Brinch, 220, 225–226, 235–236, 248–249, 251–252, 295, 309–310, 318
Harget, C. M., 178
Hartman, H. G., 323, 354
Haskel, N. A., 326–327, 352
Hassanizadeh, M., 82, 144
Haley, 183
Heard, 183
Heller, L. W., 217, 318
Henkel, B. J., 308, 318
Henkel, D. J., 71
Herrmann, L. R., 151, 179
Hicher, P. Y., 54, 72
Hilber, H. M., 99, 144
Hill, R., 9, 27, 72
Hillis, 183
Hirschfield, 183
Holland, 183
Holloway, D. M., 150, 178
Holtz, R. D., 385, 389
Horn, H. M., 182, 192, 210
Huber, M. T., 20, 26–27, 29, 35, 72
Hughes, T. J. R., 94, 99, 113, 144–145
Idriss, I. M., 149, 153, 179
Ingram, J. D., 77, 82, 143
Insley, 183
Isenberg, J., 150–151, 173–174, 179–180
Jamiolkowski, M., 308, 318
Janbu, N., 299, 318
Jaeger, 183
Jézéquel, J. F., 316
Johnson, L. D., 178

Kahl, H., 318
Katona, M. G., 151, 165-166, 179
Kausel, E., 151, 179, 326-328, 341, 343-350, 352
Kaynia, A. M., 323-324, 330, 341, 343-350, 352
Kerisel, 220, 300-302, 317-318, 323
Kishida, 265
Kjellman, 400, 403-405
Koerner, R. M., 184, 194, 197, 205, 208, 210, 367, 385, 389-391, 396, 406, 418, 426, 444-445
Kramer, J. M., 74
Krieg, D. B., 43, 72
Krieg, R. D., 43, 72
Krishnan, R., 323, 352
Krizek, R. J., 82, 142
Kronecker, 85, 93
Krsmanovic, 183
Kuhlemeyer, R. L., 323, 352
Kulak, R. F., 74
Kummeneje, 317
Lacroix, Y., 3, 7
Lacy, S. J., 109-110, 125, 129-130, 144
Ladanyi, B., 238, 256, 318
Lade, P. V., 54, 72
Lambe, T. W., 298, 302, 318
Lamé, 85
Langston, M. C., 389
Lee, K. L., 183, 184, 210
Lee, L., 151, 179
Lee, M. K. W., 121, 144
Lee, P. C. Y., 77, 80, 144
Lee, S. S., 149, 177
Leps, T. M., 3, 7
Leslie, 183
Leussink, H., 297, 318
Liam, 121, 145
Lightner, J. G., 177, 179
Little, R. W., 146
Liu, I. S., 12, 72, 144
Lockett, R. R., 114, 116, 143
Loret, B., 35, 43, 72
Lundborg, 183
Lundgren, H., 225, 295, 318
Luong, M. P., 54, 72
Lysmer, J., 146, 327, 353
Malkus, D. S., 113, 144
Malvern, L. E., 144
Mana, A. I., 167-168, 179
Mandel, J., 11, 72-73, 110, 144, 278-282, 318

Mann, 183
Mansour, C. I., 4, 6
Marachi, 183
Marsal, 183
Marsland, A., 266, 318
Martens, A., 303, 317
Martin, P. O., 121, 144
Matlock, H., 152, 179, 322, 353
Matthies, H., 102, 106, 144
Maurer, 183
Maxwell, 183
Mayer, A., 217, 318
Mazanti, 183
McIver, 183
Melan, E., 73
Menard, L., 272-274, 302, 318
Meyerhof, G. G., 220, 222-223, 225-226, 265, 282-283, 318
Milligan, V., 389
Mindlin, 326
Misa Tadashi, 318
Mitchell, J. K., 6
Miura, F., 180
Mohr, 20-21, 23, 56, 143, 147, 152, 182, 184, 203, 207, 210, 396
Mohs, 185, 203
Momen, H., 54, 72
More, J. J., 105, 109
Mortensen, K., 225, 318
Mroz, Z., 39, 54, 58, 73
Muhs, H., 222, 297, 318
Muqtadir, A., 161, 169, 177-179
Murphy, D. J., 181, 183-185, 195, 199, 210-211
Nagaraj, B. K., 156, 159-162, 177-178, 180
Naghdi, P. M., 35, 71, 77, 143-144
Nayak, G. C., 43, 73
Nemat-Nasser, S., 31, 54, 73
Newmark, N. M., 95, 98-99, 112, 134, 145, 149, 180
Newton, 95-96, 101-105, 113-114, 140-141
Ngo, D., 149, 180
Nguyen, Q. S., 48, 73
Nogami, T., 323, 353
Noll, W., 146
Novak, M., 323, 353-354
Nowatzki, E. A., 429, 445
Oliver, M. L., 82, 144
Olson, 183
Onat, E. J., 72

Ortiz, M., 43, 45–46, 48, 73–74
Osgood, 153–154, 159
Pageau, S. R., 429, 445
Parmalee, F. A., 353
Parola, 183
Patterson, 183
Paulson, J. N., 389
Pearce, C. A., 389
Peck, R. B., 2–4, 6, 183, 220, 303, 306, 310, 318
Penzien, J., 322, 353
Perloff, W., 302, 307, 318
Perumpral, J. V., 150, 178
Phan, H. V., 150, 178
Pietruszak, S. T., 54, 73
Pister, K. S., 144
Poisson, 259, 262, 300, 302, 325, 333
Popov, E. P., 43, 45, 73
Poulos, H. G., 183, 328, 330, 338, 341, 353–354
Prager, W., 22–23, 26–27, 30–31, 42–43, 71, 73
Prandtl, 218, 318
Prevost, J. H., 8, 43, 54, 65, 73, 76, 94–95, 98, 121, 125, 129–130, 134, 144–145
Ramberg, 153–154, 159
Ramelot, C., 222, 318
Randolph, M. F., 323, 354
Raphson, 95–96, 101–105, 113–114, 140–141
Read, H. J. E., 54, 74
Reinicke, K. M., 77, 114, 116, 143
Repnikov, 313, 318
Richard, R. E., 60, 74
Richards, E. A., 390
Richards, R. M., 166, 177, 355
Richart, F. E., 146
Roberts, 183
Robison, R., 442, 445
Roesset, J. M., 121, 144, 327, 352
Rohani, B, 54, 71
Roscoe, K. H., 54, 74
Rowe, P. E., 54, 74
Rowe, P. W., 188, 197, 211
Salencon, J., 278–280, 318
Sandler, I. S., 54, 71
Sato, T., 180
Saunders, G., 389
Sayed, S. M., 1, 4, 6
Scheele, F., 161, 169, 178, 180
Scheffley, C. F., 353

Schields, R. H., 316
Schiffman, R. L., 80, 146
Schnabel, P. B., 121, 146
Schofield, A. N., 3, 6, 54, 74
Schreyer, H. L., 43, 74
Schultze, E., 313, 318
Scordelis, A. C., 149, 180
Scott, F. C., 122, 146
Scott, J. D., 355, 390
Scott, R. F., 3, 7, 19, 74, 128, 130, 142, 146
Seed, H. B., 121, 144, 146, 183, 184, 210
Sen, R., 322, 324, 327–328, 344, 354
Shanno, 105
Sheta, M., 354
Simo, J. C., 43, 46, 48, 73–74
Singth, R. D., 179
Siriwardane, H. J., 151, 153, 177–178, 317
Skempton, A. W., 308, 310, 318
Skermer, 183
Skinner, 183
Smith, G. F., 12, 74
Smith, I. M., 314, 316, 318
Sokolovski, V. V., 318
Somasundaram, S., 178
Sowers, 183
Spencer, A. J. M., 12–13, 74
Steel, T. R., 144, 146
Strang, G., 77, 102, 106, 143–144
Streeter, V. L., 121, 146
Tabaddor, F., 146
Tabor, D., 149, 177, 211
Tai, T., 183, 211
Taylor, R. L., 149, 179
Terzaghi, K., 2–4, 7, 76, 82, 146, 183, 218, 220, 224, 258, 267, 299, 303, 306, 308, 310, 318, 400
Thompson, W. T., 326–327, 354
Toki, T., 151, 180
Touati, A., 55, 72
Toupin, R., 77, 82, 146
Tresca, H., 19–20, 26, 74
Truesdell, C., 77–78, 82, 146
Tschebotarioff, G. P., 3, 7, 182, 192, 211
Turner, 183
Ukhov, 183
Valanis, K. C., 54, 74
Van de Griend, A. A., 413, 445
Van de Perre, S. J., 222, 318
Vaughn, D. K., 151, 173–174, 180

Velez, A., 352
Vermeer, P. A., 43, 74
Vesic, A. S., 183, 184, 207, 211,
    214–218, 222–224, 227, 247–248,
    250–252, 257, 259–261, 265, 284,
    286, 317–318
Von Arx, G. A., 323, 354
Von Karman, 183
Von Mises, R., 20, 26–27, 29, 35, 41,
    43, 49, 74, 167
Wallace, R. B., 183, 389
Wang, C. C., 12, 75
Wass, G., 323, 353–354
Wathugala, G. W., 178
Webb, 183
Weiss, 183
Welch, J. D., 182, 192, 211
Westergard, H. M., 15, 75
Whirley, B. G., 43, 75
Whitaker, S., 79, 146

Williams, J. H., 149, 177, 183
Williams, W. O., 82, 144, 146
Wilson, E. L., 86, 89, 92, 113, 142, 150,
    179
Winkler, 312–313, 322
Wolf, J. P., 149, 151, 180, 323
Woods, R. D., 74
Woodward, J. M., 217, 318
Wroth, C. P., 3, 6, 54, 74, 317, 323,
    354
Wu, T. H., 178
Wylie, E. B., 146
Yoder, P. J., 43, 75
Youash, 183
Young, 262, 313, 324
Zaman, M. M., 155, 160–161, 173, 176,
    178, 180
Zeevaert, L., 3, 7
Zienkiewicz, O. C., 43, 73, 89, 92, 113,
    122, 146, 150, 180